CANCER BIOMARKERS

THE WILEY BICENTENNIAL–KNOWLEDGE FOR GENERATIONS

Each generation has its unique needs and aspirations. When Charles Wiley first opened his small printing shop in lower Manhattan in 1807, it was a generation of boundless potential searching for an identity. And we were there, helping to define a new American literary tradition. Over half a century later, in the midst of the Second Industrial Revolution, it was a generation focused on building the future. Once again, we were there, supplying the critical scientific, technical, and engineering knowledge that helped frame the world. Throughout the 20th Century, and into the new millennium, nations began to reach out beyond their own borders and a new international community was born. Wiley was there, expanding its operations around the world to enable a global exchange of ideas, opinions, and know-how.

For 200 years, Wiley has been an integral part of each generation's journey, enabling the flow of information and understanding necessary to meet their needs and fulfill their aspirations. Today, bold new technologies are changing the way we live and learn. Wiley will be there, providing you the must-have knowledge you need to imagine new worlds, new possibilities, and new opportunities.

Generations come and go, but you can always count on Wiley to provide you the knowledge you need, when and where you need it!

William J. Pesce

WILLIAM J. PESCE
PRESIDENT AND CHIEF EXECUTIVE OFFICER

Peter Booth Wiley

PETER BOOTH WILEY
CHAIRMAN OF THE BOARD

CANCER BIOMARKERS

Analytical Techniques for Discovery

MAHMOUD H. HAMDAN
GlaxoSmithKline
Verona, Italy

BICENTENNIAL 1807 WILEY 2007 BICENTENNIAL

WILEY-INTERSCIENCE
A John Wiley & Sons, Inc., Publication

Published by John Wiley & Sons, Inc., Hoboken, New Jersey
Published simultaneously in Canada

For general information on our other products and services or for technical support, please contact our Customer Care Department within the United States at (800) 762-2974, outside the United States at (317) 572-3993 or fax (317) 572-4002.

Wiley also publishes its books in a variety of electronic formats. Some content that appears in print may not be available in electronic formats. For more information about Wiley products, visit our web site at www.wiley.com.

Library of Congress Cataloging-in-Publication Data:

Hamdan, Mahmoud, 1947-
Cancer biomarkers : analytical techniques for discovery / Mahmoud H. Hamdan.
p. ; cm.
Includes bibliographical references and index.
ISBN-13: 978-0-471-74516-7 (cloth)
1. Tumor markers. I. Title.
[DNLM: 1. Neoplasms–diagnosis. 2. Biological Markers. 3. Neoplasms–pathology. QZ 241 H211c 2007]
RC270.3.T84H36 2007
616.99'4075–dc22 2006025185

Printed in the United States of America

10 9 8 7 6 5 4 3 2 1

This book is dedicated to two groups of people:
To all patients who are suffering from any form of cancer.

To research scientists, doctors and nurses who are engaged in a long and hard battle to defeat these devastating malignancies.

Mahmoud Hamdan

CONTENTS

PREFACE

The idea of the early detection of various forms of cancer before they spread and become incurable has tantalized both physicians and research scientists for decades. Although spectacular advances in molecular medicine, genomics, and proteomics have been made, current efforts to combat cancer remain extremely disappointing. One main reason for the lack of such desired success is that in many cases, cancer is diagnosed and treated too late, when the cancer cells have already invaded adjacent tissues and established new colonies. Despite this rather pessimistic statement, it is encouraging to note that our increasing understanding of the biology of cancer, including genetic, molecular and cellular mechanisms is now providing clear objectives for the early detection of these malignancies. In a decade of unprecedented genomic and proteomic activities, much progress has been made towards the understanding of the various mechanisms responsible for the initiation and progression of various forms of cancer. Such progress had a direct impact on current activities dedicated to the search for sensitive and above all specific biomarkers for the early detection and diagnosis of various cancers. Technologies capable of performing parallel rather than serial analyses represent a central component in the search for such biomarkers. To perform such large-scale studies, contemporary biology has amassed a battery of methods to survey the global features of the cells, from DNA, RNA and proteins to small molecules. The correlation between what has been already achieved and what is remained to be done is still difficult to assess. However, we can anticipate that current research efforts to discover sensitive and specific cancer biomarkers will provide a valuable contribution to future efforts to combat these malignancies. I do not pretend that this text has captured all what I wanted it to capture. The incredible pace of technical advances and the amount of data generated by high throughput analyses renders such desire almost impossible. Having said this, I would like to think that I have managed to put together a number of arguments, which will interest a relatively wide audience. Apart from the introduction and the overview, the book deals with three main arguments: Chapter 2 deals extensively with various proteomic platforms relevant to biomarkers discovery. Other technologies are discussed within examples given in successive chapters. The second argument deals with some existing and some potential cancer biomarkers. This argument is discussed in chapters 3 and 4, while chapter 5 describes the emerging role of protein-protein networks and its impact on biomarkers discovery.

Over the last 20 years, both proteomic and genomic activities have begun to make extensive use of products of human origin. This new trend has raised many ethical and social issues particularly those involving the individual rights, including issues of consent. This means that researchers should carefully consider several aspects when designing studies in which samples of human origin are required. This aspect is described in chapter 6.

MAHMOUD HAMDAN

ACKNOWLEDGMENTS

The author is grateful to GlaxoSmithKline for the opportunity and support given to him during a period of over 17 years. I am sure this text has greatly benefited from an environment where extensive research efforts are dedicated to the discovery of innovative and efficacious therapies to a wide range of diseases, including various forms of cancer. The author is also grateful to Professor Keith Birkinshaw of the University of Wales, Aberyswyth for his continuous encouragement and the revision of some parts of the text.

MAHMOUD HAMDAN

ACKNOWLEDGMENTS

INTRODUCTION

Cancers have exacted and continue to exact a tremendous price on our society. Such price is difficult to quantify as it encompasses a wide spectrum of devastating effects, including death, suffering, stress, and economic losses. The fact that we are well aware of these effects represents the first step in the reaction of our modern society to combat these malignancies. For a long period of time, cancer has been considered mainly as a genetic disease, a concept which was partially sustained by the notion that one gene means one protein. This concept has gradually lost its meaning, thanks to a number of scientific achievements over the last two decades.

- Completed genome sequencing projects have revealed the precise number of genes in a number of species, including humans. One of the striking results of these projects is the relatively small number of genes, ranging from a few hundreds in bacteria to tens of thousands in mammalian species. The confirmation that there might be fewer than 30,000 protein-coding genes in the human genome is one of the key results of the monumental work presented by Venter et al. (2001) and Lander et al. (2001). This number represents only a one-third increase in genes number from a rather unsophisticated nematode (*Caenorhabditis elegans*) with about 20,000 genes.
- Genomic and proteomic knowledge accumulated before and after these sequencing projects have confirmed that such unexpectedly low number of human genes is capable of encoding a much higher number of products. In particular, the number of encoded proteins can be enormous, as the same gene can generate multiple protein products that differ as a result of combinatorial splicing, processing, and post- cotranslational modifications. Prior to the publication of the first draft of the human genome, Lander and Weinberg (2000) estimated the possible number of

proteins that can be encoded by an estimated 45,000 human genes. The complexity of the repertoire of the proteome was underlined by the following numbers:

- Proteome of a cell is ~5.000 polypeptides.
- Proteome of an individual as a snapshot is ~10^6 polypeptides.
- Proteome of an individual during the entire life span is ~10^7 polypeptides;
- Proteome of a species is ~10^8 polypeptides.

- Since its discovery in 1983 (Feinberg and Vogelstein, 1983; Gama-Sosa et al., 1983), the epigenetics of human cancer has been in the shadows of human cancer genetics. Recently, this area has become more visible with a growing understanding of specific epigenetic mechanisms, including hypomethylation, hypermethylation, loss of imprinting, and chromatin modification (Feinberg and Tycko, 2004). It is becoming more evident that epigenetic, as well as genetic, events might be central to the initiation and progression of cancer. Indeed, cancers often exhibit an aberrant methylation of gene-promoter regions that are associated with loss of gene functions. This DNA change constitutes a heritable state, not mediated by altered nucleotide sequence. Currently, the search for DNA methylation as biomarkers for various types of cancer is gaining momentum. Methylation biomarkers are particularly suited for situations where specific and sensitive detection is necessary, such as when tumor DNA is either scarce or diluted by excess normal DNA. This particular area of research has been strengthened by new generation of methylation detection methods, including various versions of methylation-specific PCR.
- The above-mentioned achievements would not have been possible without the employment of a battery of high-throughput technologies in both genomic and proteomic research. The use of DNA microarrays has provided one of the most powerful tools to investigate global gene expression in most aspects of human cancer. For the last 10 years, this technology has been used to obtain major insights into progression, prognosis, and response to therapy on the basis of gene-expression profiles. High-density oligonucleotide Genechips are currently produced by synthesizing several thousands of short oligonucleotides in situ on glass wafers, using a combination of photolithography and light-directed solid-phase DNA synthesis (Fodor et al., 1991; Lipshutz et al., 1999). Large-scale serial analysis of gene expression is another powerful technology capable of providing sensitive and comprehensive analysis of gene expression, including analyses in organisms with uncharacterized genomes (Velculescu et al., 1995). On the proteomic side there is a wide spectrum of methods capable of direct comparison between normal and tumor specimen. For over 20 years, two-dimensional gel electrophoresis 2-DE with and without mass spectrometry detection has been the workhorse for protein analysis in complex mixtures. However, difficulty of full automation and relatively large sample requirements has underlined the need for alternative/complementary approaches. Various MS-based approaches, including multidimensional liquid chromatography coupled with tandem mass spectrometry, various tagging and labeling strategies, the emerging role

of surface-enhanced laser desorption ionization (SELDI), and protein microarrays, are discussed in some details within this text.

NEW TREND IN BIOMARKERS DISCOVERY

The completion of a number of gene sequencing projects and recent advances in genomic and proteomic technologies, together with powerful bioinformatics tools, have a direct impact on the way the search for cancer biomarkers is conducted. More recent literature on the search for biomarkers for the prediction/prognosis of cancer indicates a new trend. Such trend can be loosely described as "combined or multiple biomarkers approach." This trend encompasses two distinct approaches:

(i) Combination of a number of single biomarkers for the same type of cancer to enhance the specificity and the prediction capability of the single components within such combination. The use of multiple biomarkers for the detection and prognosis of ovarian and prostate cancers are two representative examples of such approach. Since its development over 25 years ago, measurement of serum levels of cancer antigen 125 (CA-125) has become an integral part of the management of epithelial ovarian cancer. While the levels of this antigen at the time of diagnosis are of limited prognostic significance, the measurements of the same antigen are now performed almost routinely during the course of various treatments. The early attempts to evaluate CA-125 in combination with other potential biomarkers were conducted by Bast's group (Xu et al., 1991; 1993). Lewis X mucin determinant (OVX1) and a cytokine machrophage colony-stumulating factor (M-CSF) were evaluated for their ability to detect stage I ovarian cancer and to complement CA-125. To increase preoperative sensitivity for early stage ovarian cancer, Skates et al. (2004) used a panel of biomarkers composed of CA-125II, CA 15-3, CA 72-4, and M-CSF.

Since its approval by the FDA in 1986, prostate-specific antigen (PSA) has probably become the most useful tumor marker in clinical practice today and is now routinely used in the diagnosis and monitoring of prostate cancer. There is a general agreement among clinicians that the PSA test has the highest predictive value for prostate cancer, particularly when the malignancy is in its early stages. At the same time, there is a disagreement as to what level of PSA should prompt a prostate biopsy. In a study using reverse-phase protein microarrays (Paweletz et al., 2001), it was demonstrated that prostate cancer progression was associated with an increased phosphorylation of Akt together with a suppression of apoptotic pathways. In a second and more recent study (Kreisberg et al., 2004), an attempt was made to use the extent of phosphorylation of Akt(Ser^{473}) to predict poor clinical outcome in prostate cancer. This study was designed to establish whether increased phosphorylation of Akt and/or decreased phosphorylation of ERK (extracellular signal-regulated kinase) could be used to predict poor clinical outcome in prostate cancer.

(ii) The use of proteomic patterns and gene-expression profiling (gene signatures) can be considered as the second and more recent approach to biomarkers discovery. The first is commonly associated with the SELDI approach, which generates

MALDI (matrix-assisted laser desorption ionization) mass spectra containing a range of mass/charge ratios (*m/z*) from samples irradiated by a laser beam. These spectra can be generated from cancer and noncancer specimen, which can be analyzed and elaborated with software packages to establish differences between the two sets of mass spectra. Such differences are generally presented as a pattern containing a number of *m/z* values that happen to be different in the two samples. Despite a highly promising start, the use of these patterns as biomarkers is still hotly debated, particularly regarding their reproducibility and their clinical value, particularly in the absence of identification of the proteins they represent, the capability of the SELDI chips to capture low-abundance components without the need for prefractionation, and the danger of overfitting the generated data. These and other aspects of the SELDI approach are discussed in Chapter 2 of this book.

Till very recently, most published studies on marker genes have applied gene-expression profiling to single cancer types. We are already witnessing an increasing use of array technology in genomic research, together with an enhanced knowledge of the molecular mechanisms of cancer, which resulted in enhanced efforts toward developing multiclass classifiers capable of distinguishing between multiple common human malignancies. This approach holds much promise for the uniform, molecular, and database-driven classification of all human tumors. In a demonstration of such type of analysis, Ramaswamy et al. (2003) identified a gene-expression signature that was differentially expressed in metastatic tumors of diverse origins relative to primary cancers. This study demonstrated that the metastatic signature was also expressed in a subset of the primary tumors analyzed, leading the authors to hypothesize that such signature may represent a metastatic program that is encoded in primary tumors destined for metastasis.

THE EMERGING ROLE OF PROTEIN-NETWORKS

The central role that protein–protein interactions play in most biological processes and the realization that proteins exert virtually all their activities via interactions with other molecules, whether they are proteins, nucleic acids, lipids, carbohydrates, or small molecules, have inspired efforts to map protein–protein interactions on a proteome-wide scale. To date, most of the interactions that have been detected experimentally have relied on one of the two technologies: the yeast two hybrid system (Fields and Sternglanz, 1994) and tandem affinity purification/mass spectrometry (Rigaut et al., 1999). An alternative to experimental determination of protein–protein interactions is prediction by various computational genomic approaches (Marcotte, 2000; Xia et al., 2004; Droit et al., 2005). These approaches utilize information on individual protein interactions taken from publicly available databases and combine these data with sequence similarities within genomes and orthologies across genomes for in silico prediction of protein–protein interactions (Jansen et al., 2003; Yu et al., 2004; Lehner and Fraser, 2004).

It is reasonable to ask how the construction of these interaction maps is likely to impact future efforts for biomarkers discovery. The following general considerations may give a partial answer to this question: first, several technologies to study specific cellular functions or processes have been around for many years, such as enzymatic assays, complex purification, or subcellular localization. In most cases, the studies have focused on a small set of genes or proteins. The increasing amount of genomic data, however, has led to the availability of more new biological objects than could reasonably be studied using classical genetic or biochemical means. An initial approach was to screen for new essential genes, assuming that these genes would be more interesting than others. It soon became clear, however, that most genes could be deleted without obvious changes in phenotype, and that it was necessary to study combinations of proteins or subtle phenotypes to understand cellular functioning. Second, over the last decade, studies in cancer have validated the effectiveness of the microarray techniques in the identification of tumor subclasses and potential marker genes for diagnosis and treatment of some types of cancer (DeRisi et al., 1996; Sorlie et al., 2001). However, such powerful approach does not provide further information on possible association between the identified genes and other coregulated genes. On the contrary, topological features of the protein networks have been demonstrated to reflect the functionality of the interacting genes. For example, essential genes in yeast tend to be well connected and globally cantered in the protein network (Jeong et al., 2001; Wuchty and Almaas, 2005). In other words, protein-network analysis will place the genes identified in microarray experiments in a broader biological context. As protein networks reflect the functional grouping of these interacting or coordinately induced/suppressed genes, the roles of the subsets of coexpressed genes may be resolved using the combined data. This may be done by evaluating the topological features of the sets of genes identified by microarray experiments.

CLINICAL PROTEOMICS

Over the last decade there has been an impressive adaptation of mass spectrometry-based technologies in the realm of clinical proteomics. These technologies provide higher analytical capabilities, and employ automated liquid handling systems, fractionation techniques, and bioinformatics tools for greater sensitivity and resolving power, more robust and higher throughput sample processing, and greater confidence in analytical results. These technologies have been massively employed in two areas: first, mining of the proteome in various biological samples and second, identification of proteomic patterns and/or single proteins that can be developed as biomarkers for the prediction, prognosis, and response to therapy of various disease. Attempts to achieve this daunting task have to deal with a number of technical hurdles, including the prohibitive number of different protein entities and the range of protein concentration. Estimates of potential protein types can reach millions when considering posttranslational modification events, and the relative concentration range can span 10 orders of magnitude. Thus, researchers continue to develop and test higher resolution and higher sensitivity platforms. An additional level of complexity exists when one considers the heterogeneity of individuals, which is a significant confounding factor in the study design of successful biomarkers efforts. Solutions to these issues have come in two forms: "top down" proteomics in which intact proteins from sample and control are separated and compared to establish protein expression differences (e.g., 2-DE analysis with or without mass spectrometry detection). Alternatively, complex protein mixtures can be enzymatically digested prior to separation and differential expression determination using mass spectrometry in "bottom up" approaches. This latter approach has benefited from advances in tandem mass spectrometry and multidimensional liquid chromatography. Both approaches have also benefited from recent advances in "front-end" sample fractionation and separation strategies to reduce the complexity of native clinical samples (or cell lysates). Clinical proteomics also have to deal with two challenging problems: data analysis and testing/validation of potential biomarkers. Large-scale protein expression data sets, generated by MS-based technologies, pose a significant challenge in their biological interpretation. Such interpretation requires automated data analysis strategies to reveal biologically interesting features within large expression data sets that might otherwise go unnoticed. Several useful data mining strategies are already in place. At present, there are several publicly available software packages to map proteomic data sets, generated by searching collision-induced dissociation spectra of peptides against one or more major protein databases, including Swiss-Prot/TrEMBL (http://ca.expasy.org/sprot), International Protein Index (www.ebi.ac.uk/IPI/IPIhelp.html), and NCBI (www.ncbi.nlm.nih.gov). One of the highly useful online resources is the ExPASy Web site that provides an access to individual proteins. This resource can be searched for detailed information, including relevant published literature and referenced support material (http://ca.expacy.org). The Gene Oncology consortium is another global annotation initiative, which provides information on the biological roles, molecular functions, and biochemical properties of most proteins of model organisms and humans (www.geneontology.org).

Development of assays for testing potential biomarkers is one of the hurdles faced by clinical proteomics and other approaches in the area of biomarkers development. These assays have to be rapid, reliable, and, possibly, not expensive. At present, there are several methods for the development of such assays: (a) Immunohistochemistry relies on the staining of histological specimens for a particular biomarker. This approach is commonly used to predict outcomes among patients with different grades of disease by staining tissues from biopsy samples taken from cancer patients at the time of diagnosis. One of the main drawbacks of this approach is that it depends on biopsy samples that are collected invasively. In addition, the small percentage of tumor tissues obtained by biopsy can result in missing more aggressive tumor populations that reside nearby. (b) Enzyme-linked immunosorbent assay (ELISA) is a sensitive, high-throughput technique capable of providing quantitative measurements of analytes present in a physiological fluid such as serum and urine. In this approach, a primary antibody binds to the analyte, and an enzyme-linked secondary antibody binds to the previous complex. The enzyme activity is quantitatively measured by the addition of an appropriate substrate, and it is proportional to the amount of analyte present in the fluid. The advantage of ELISA is that it can be utilized by body fluids that are collected noninvasively. The increasing number of potential biomarkers and the trend of looking for panel rather than single biomarkers have underlined the need for new tools that can do for protein-expression profiling, what DNA chips have done for RNA expression analyses. This has led to a surge in the development of various formats of protein chips.

Standardization and stringency of various analytical approaches employed in clinical proteomics is another challenge, which has to be adequately addressed without loosing the advantages of comparing results generated by different techniques and different research groups. Extensive efforts have been made to standardize preanalytical, analytical, and postanalytical methods for biomarkers discovery. However, these efforts need to be further consolidated and refined to guarantee quality assurance in biomarkers research and, in particular, when potential biomarkers are tested in the clinic. A major problem associated with evaluating biomarkers in tissues, blood, or bodily fluids is that different procedures (sample collection, sample storage, sample processing) and different assay formats may yield different results. Therefore, assays and procedures have to be standardized, clear operating procedures should be developed for each type of sample and assay format, and the quality of the biomarker assay results should be monitored and assessed by different research groups. Although for some markers considerable progress has been made in the standardization of methods and assay protocols, efforts should be continued as only the stringent application of quality control systems enables a consistent assessment of the clinical value of biomarkers.

REFERENCES

DeRisi, J., Penland, L., Brown, P. O., et al. (1996) Nature Genet. **14**, 457.

Droit, A., Poirier, G. G., Hunter, J. M. (2005) J. Mol. Endocrinol. **34**, 263.

Feinberg, A. P., Tycko, B. (2004) Nat. Rev. Cancers **4**, 143.

Feinberg, A. P., Vogelstein, B. (1983) Nature **301**, 89.

Fields, S., Sternglanz, R. (1994) Trends Genet. **10**, 286.

Fodor, S. P., Read, J. L., Pirrung, M. C., et al. (1991) Science **251**, 767.

Gama-Sosa, M. A., S Lagel, V. A., Trewyn, R. W., et al. (1983) Mol. Biol. **11**, 6883.

Jansen, R., Yu, H., Greenbaum, D., et al. (2003) Science **302**, 449.

Jeong, H., Mason, S. P., Barabasi, A. L., et al. (2001) Nature **411**, 41.

Kreisberg, J. I., Malik, S. N., Prihoda, T. J., et al. (2004) Cancer Res. **64**, 5232.

Lander, E. S., Linton, L. M., Birren, B., et al. (2001) Nature **409**, 860.

Lander, E. S., Weinberg, R. A. (2000) Science **287**, 1777.

Lehner, B., Fraser, A. (2004) Genome Biol. **5**, R63.

Lipshutz, R. J., Fodor, S. P., Gingeras, T. R., et al. (1999) Nat. Genet. **21**, 2.

Marcotte, E. M. (2000) Curr. Opin. Struct. Biol. **10**, 359.

Paweletz, C. P., Charboneau, L., Bichsel, V. E., et al. (2001) Oncogene **20**, 1981.

Ramaswamy, S., Ross, K. N., Lander, E. S., et al. (2003) Nature Genet. **33**, 49.

Rigaut, G., Shevchenko, A., Rutz, B., et al. (1999) Nat. Biotechnol. **17**, 1030.

Skates, S., J., Horick, N., Yu, Yinhua, et al. (2004) J. Clin. Oncol. **22**, 4059.

Sorlie, T., Perou, C. M., Tibshirani, R., et al. (2001) PNAS USA. **98**, 10869.

Velculescu, V. E., Zhang, L., Vogelstein, B., et al. (1995) Science **270**, 484.

Venter, C., Adams, M. D., Myers, E. W., et al. (2001) Science **291**, 1304.

Wuchty, S., Almaas, E. (2005) Proteomics **5**, 444.

Xia, Y., Yu, H., Jansen, R., et al. (2004) Ann. Rev. Biochem. **73**, 1051.

Xu, F-J., Ramakrishnan, S., Daly, L., et al. (1991) Am. J. Obstet. Gynocol. **165**, 1356.

Xu, F-J., Yu, Y-A., Daly, L., et al. (1993) J. Clin. Oncol. **11**, 1506.

Yu, H., Luscombe, N. M., Lu, H. X., et al. (2004) Genome Res. **14**, 1107.

1

OVERVIEW

1.1. INTRODUCTION

Cancer exacts a tremendous price on society through devastating effects on patients and their families, tremendous economic costs in terms of direct medical care for its treatment, and the loss of capital because of early mortality. The idea of the early detection of various forms of cancer before they spread and become incurable has tantalized both physicians and research scientists for decades. Although such an objective is still too far, it is encouraging to note that our increasing understanding of the biology of cancer, including genetic, molecular, and cellular mechanisms, is now providing clear objectives for the early detection, prevention, and therapy of a number of cancer forms. Understandably, the question of finding specific and reliable biomarkers for the early detection of various forms of cancer is attracting both enthusiasm and scepticism. The enthusiasm is driven by the completion of genome sequencing for a number of species including humans and by the availability of a spectrum of high-throughput technical platforms for both proteomic and genomic analyses. The scepticism on the contrary is partially derived from some inflated expectations, which are frequently followed by disappointment when the original results of certain investigations could not be reproduced. This scepticism, however, is not directed toward the final objective of defeating these devastating diseases; instead, it can be looked upon as some form of cautious assessment of current achievements and an attempt to dampen likely overenthusiasm generated by recent successes in this area of research. In other words, there seems to be a general

Cancer Biomarkers: Analytical Techniques for Discovery, by Mahmoud H. Hamdan

agreement within the research community that impressive steps have been made in the direction of discovering new and more specific biomarkers, yet there is a continuous and legitimate debate within the same community on what is needed to be done to translate laboratory successes into concrete clinical applications. Regardless of which side one would take, it is encouraging to note that the search for cancer biomarkers is one of the areas that bring together the scientist's quest to understand the biology and the molecular basis of these devastating malignancies with the physician's dedication to relief suffering and improve the quality of life of his patients.

Although spectacular advances in molecular medicine, genomics, and proteomics have been made, current efforts to combat cancer remain extremely disappointing. One main reason for the lack of such desired success is that in many cases, cancer is diagnosed and treated too late, when the cancer cells have already invaded adjacent tissues and established new colonies. The capability for invasion and metastasis enables cancer cells to escape the primary tumor mass and colonize new terrain in the body where, at least initially, nutrients and space are not limiting (Hanahan and Weinberg, 2000). These distant settlements of tumor cells are the cause of 90% of human cancer deaths (Sporn,1996). Currently, there are a number of platforms leading to the search for new biomarkers in cancer research. On the proteomic side, we have a number of emerging technologies that are applied in the area of biomarkers discovery, including surface enhanced laser desorption ionization (SELDI) (Hutchens and Yip, 1993; Tang et al., 2004), mass spectrometry combined with two-dimensional liquid chromatography (Link et al., 1999; Washburn et al., 2001; Wang and Hanash, 2003) or two-dimensional gel electrophoresis (O'Farrel, 1975; Klose and Kobalz, 1995; Aebersold and Goodlett, 2001; Abersold and Mann, 2003; Hamdan and Righetti, 2003), protein microarrays (MacBeath, 2002; Espina et al., 2003; Liotta et al., 2003), and imaging mass spectrometry (Caprioli et al., 1997; Chaurand et al., 1999; Stoeckli et al., 2001). On the genomic side, there are equally powerful platforms for biomarkers discovery, which use polymerase chain reaction (PCR) (Datta et al., 1994; Krismann et al., 1995), serial analysis of gene expression (SAGE) (Velculescu et al., 1995), and DNA microarrays (Young, 1995; Ramaswamy and Golub, 2002). These techniques and relevant references are fully covered in Chapter 2. Brief description of certain aspects of these technologies together with other elements relevant to biomarkers discovery is described in the following sections of this overview.

1.2. CANCER BIOMARKERS

Broadly speaking, cancer biomarkers can be divided into three categories: (a) Diagnostic (screening) biomarkers are used to detect and identify a given type of cancer in an individual. This type of biomarkers is expected to have high levels of diagnostic sensitivity and specificity, especially if it is used in large-screening trials. (b) Prognostic biomarkers are commonly used, once the disease status has been established. They are expected to predict the likely course of the disease, its recurrence, and thus they have an important influence on the aggressiveness of the therapy. (c) Stratification (predictive) biomarkers are often DNA based and serve to predict the likely response to a drug before starting treatment, classifying individuals as "responders" or "nonresponders." This type of biomarkers is the result of recent advances in genetic research, which made it possible to predict clinical outcome from the molecular characteristics of the patient's tumor (Van de Vijver et al., 2002). Such predictive classification is of a major importance in designing clinical drug trials to define an intended use for the drug under investigation. In my opinion, the dividing line between screening and prognostic markers is rather flexible. In other words, there is no valid reason as to why a screening marker cannot be used as prognostic marker and vice versa. The role of a chosen marker does not end once the target cancer has been diagnosed. For example, the expression levels of a protein can be exploited as a biomarker and at the same time for assessment of the therapeutic response and recurrence following the diagnosis of the disease. Regardless of which definition is used, cancer biomarkers can be DNA, mRNA, metabolites, or processes such as apoptosis, angiogenesis, or proliferation that can be associated with a given type of cancer and can be measured quantitatively or qualitatively by an appropriate assay or technique. These markers can be found in a wide range of specimen, including body fluids (plasma, serum, urine, saliva, etc.), tissues, and cell lines. If the source of the biomarker is not the tumor itself then the term remote media is used. Such term refers to body fluids, lavages, detached cells, biopsies of nonmalignant tissues, and so on.

Tremendous amount of work in the area of biomarkers made it abundantly clear that the efficacy of a given biomarker assay is determined by its sensitivity and specificity. Both the terms take on precise meanings in the development of biomarker tests for population-based screening or for clinic-based surveillance of high-risk population. The clinical sensitivity of a biomarker can be simply defined as the proportion of individuals with confirmed disease who test positive for the biomarker assay, whereas the specificity refers to the proportion of control subjects (individuals without the disease) who test negative for the biomarker assay (Sullivan Pepe et al., 2001). A receiver-operating curve is commonly used to evaluate the efficacy of an assay. This is a graphical representation of the relationship between sensitivity and specificity. The ideal graph is the one giving the maximum area under the curve.

At present, there are many clinical situations in which tumor biomarkers are already being used; these existing markers are still the focus of further research efforts to increase their specificity, optimize them, and gain further information relevant to future generation of these or other new classes of markers. A commonly cited marker is the prostate-specific antigen (PSA), which is commonly used to screen male patients for prostate cancer (Stamey et al., 1987; Hudson et al., 1989; Thomson et al., 2004). Despite the tremendous impact of this marker on many aspects of the management of prostate carcinoma, the fact remains that this marker lacks specificity resulting in false-positives as high as 30%. In other words, almost a third of the patients with an elevated level of PSA do not necessarily suffer from this form of cancer. The most thoroughly assessed ovarian cancer biomarker is Carcinoma-associated glycoprotein antigen (CA-125), which was first identified by Bast et al., (1981; 1983). This biomarker and PSA are discussed in more details in Chapter 3.

Another serum-based marker is carcinoembryonic antigen (CEA), which was first reported by Gold and Freedman (1975). Levels of this protein are normally used to monitor disease progression and response to therapy in patients with colorectal cancer. One of the main limitations of this marker is that only a proportion of colorectal cancers tend to express elevated CEA levels at the time of diagnosis (Benson et al., 2000). This biomarker is a representative case of how initial findings regarding a promising biomarker are not always reproducible. CEA was initially purported to be nearly 100% sensitive and specific for colorectal cancer screening (Thomson et al., 1969). Subsequent studies have demonstrated that such levels of sensitivity and specificity were rather too optimistic (Read et al., 1995). The failure to reproduce the initial results was, in large part, due to the fact that individuals who were initially studied had an advanced stage of this disease, whereas individuals who were later studied had less extensive asymptomatic cancer in which CEA levels did not experience the expected increase. It has to be pointed out that this negative experience with CEA had some positive influence on methods development and rules of validation by which diagnostic tests are judged today.

Various forms of cancer are currently the target of major interdisciplinary efforts aiming at elucidating the molecular mechanisms governing disease pathogenesis, discovering new biomarkers for diagnosis, prognosis, and response to therapy. Within these interdisciplinary efforts, protein- and DNA-based technologies are expected to play a key role in the understanding and treatment of various human disorders including cancer.

1.3. PHASES OF BIOMARKERS DEVELOPMENT

The surge in research to develop cancer-screening biomarkers prompted the establishment of the Early Detection Research Network (EDRN) by the National Cancer Institute (Srivastava and Kramer, 2000). The aim of the EDRN is to coordinate research among biomarker-development laboratories, biomarker validation laboratories, clinical repositories, and population-screening programs. With the goals of EDRN in mind, Sullivan Pepe et al. (2001) proposed five phases that a biomarker needs to pass through to become a useful population-screening tool. These five phases can be summarized as follows: (i) The first phase is based on preclinical exploratory studies, comparing tumor with nontumor specimen. The aim of this phase is to identify unique characteristics of tumor that might lead to ideas for clinical tests capable of detecting the cancer. In this phase, various techniques can be employed, including immunochemistry, western blots, gene-expression profiles, protein-expression profiles, and levels of circulating antibodies against thousands of cancer antigens. (ii) Phase two involves the development of clinical assays preferably using specimen, which can be obtained noninvasively. A protein uniquely expressed by tumor and measured with serum antibody can be considered as an example of such assays. The main aim of this phase is to establish the true-positive rate (the proportion of case subjects who are biomarker positives), false-positive rate (the proportion of control who are biomarker positive), and the receiver operating characteristics. The authors of these guidelines noted that since the case subjects in this phase have established disease, such phase does not determine if the same disease can be detected early with the same biomarker. (iii) Comparison of clinical specimens collected from subjects with cancer before their clinical diagnosis and compared with those from control subjects (subjects who have not developed the disease) can provide initial evidence on the capability of the biomarker to detect disease in the preclinical phase. In this phase, retrospective longitudinal repository studies are conducted. The aim of this phase is to evaluate, as a function of time before clinical diagnosis, the capability of the biomarker to detect preclinical disease, and to define criteria for a positive screening test in preparation for phase 4. In other words, if the levels of the biomarker in case subjects measured at a time close to clinical diagnosis show little deviation from those in control subjects, the biomarker has little promise for screening. On the contrary, if the levels in case subjects demonstrate distinct differences from those in control subjects months or years before the appearance of clinical symptoms, then the biomarker's potential for screening is enhanced. (iv) This final phase estimates the reduction in mortality of a given type of cancer as a result of screening tests employing a selected biomarker for that type of cancer. This phase has to address a number of difficulties before its findings can be truly related to the benefits of screening. Some of these difficulties have been pointed out within the guidelines by Sullivan Pepe et al. (2001) and include the following: (a) ineffective treatment for screen-detected tumors, (b) poor compliance with the screening program, (c) prohibitive economic costs of screening itself and of the diagnostic work-up of subjects who falsely screen positive for the disease, and (d) the overdiagnosis.

Before considering the applicability of such guidelines, it is relevant to take into account the indications by The World Health Organization (WHO) regarding early detection and disease control. These indications can be summarized as follows: first, the disease must be common and associated with serious morbidity and mortality. Second, screening tests must be able to accurately detect early-stage disease. Third, treatment after detection through screening must have been shown to improve prognosis relative to usual diagnosis. Fourth, evidence must exist that the potential benefits outweigh the potential harms and costs of screening (Winawer et al., 1995).

To appreciate the difficulties in constructing practical and reliable screening tests, it is sufficient to consider existing screening tests, which are in use for a number of cancers. For example, in the case of colorectal cancer screening, many guidelines recommend sigmoidoscopy (hollow tube inserted into the rectum for imaging the lower part of the colon and rectum) or colonscopy (similar to sigmoidoscopy, but examines the entire length of the colon), which are both expensive and above all are not well accepted in terms of time required, discomfort involved, and the risk of adverse outcome. Biomarkers that are in use for prostate-cancer screening, the PSA carries a substantial risk of overdiagnosis due to its poor specificity. A similar situation is found in ovarian cancer, where the use of CA-125 as a biomarker can result in false-positive rates that lead to unacceptably high number of surgeries to confirm the disease; the same biomarker fails to detect many early-stage cancers.

1.4. NEW APPROACH TO BIOMARKERS DISCOVERY

The early approaches to discover and identify cancer biomarkers were mainly based on preliminary clinical or pathological observations. A representative example of such approaches is the overexpression of the CEA, which was first reported by Gold and Freedman (1975). The isolation and purification of PSA (Wang et al., 1979), which is currently the only biomarker for prostate cancer is another example. A simple comparison between the methods used to discover these well-established cancer biomarkers and those currently employed in the search for new biomarkers reveals an unmistakable new approach for such discovery. Such comparison underlines two apparent differences: first, there is a clear shift in investigative strategy from an orderly inquiry into biological mechanisms toward a "brute force" approach that can be described as "collect the set, generate, and mine data." Furthermore, present methods attempt to identify distinguishing pattern (s) and/or multiple markers rather than a single one. Second, the discovery is conducted by different techniques even within the same laboratory. It is reasonable to suggest that such changing approach to biomarkers discovery is the direct result of technical advancement and newly acquired knowledge of the biology and molecular basis of various forms of cancer. Human genome sequencing, the discovery of oncogenes, tumor-suppressor genes, and tremendous advances in DNA-based and proteomic-based analyses have started to have a tangible impact on the landscape of biomarkers discovery. So, how such emerging technologies together with the newly acquired knowledge of cancer biology have influenced the discovery of cancer biomarkers? A partial answer to this question can be postulated through the following considerations:

1.4.1. New and Powerful Technologies

The past 10 years have witnessed an impressive growth in the field of large scale and high-throughput biology, resulting in a new era of technology development and the accumulation of new knowledge, which highlighted a number of challenges, including the need to elucidate the function of almost every encoded gene and protein in an organism and to understand the basic cellular events mediating a host of complex processes and their possible role in various diseases. Such newly acquired knowledge made it clear that a comprehensive analysis of the molecular basis of cancer and other disease states requires the integration of the distinct, but complementary information gained from genomics and proteomics. A number of emerging approaches have been used to tackle this prohibitive task, including large-scale analysis of genes and proteins. Over the past few years, miniaturized and parallel assay systems have already demonstrated a part of their potential in large-scale and high-throughput biological analysis. Today, the expression of thousands of genes can be simultaneously assessed under different conditions, including disease state and treatment. Powerful technologies including PCR, SAGE, single nucleotide polymorphism analyses, and microarrays can target almost any DNA, RNA, or protein sequence. These microarrays have been used for the detection of sequence variations and for mapping the targets of transcription factors (Lyer et al., 2001; Heller, 2002; Horak et al., 2002). A drawback of DNA microarrays is their unsuitability for protein analysis. There are two experimentally demonstrated reasons behind such limitation: first, there is little correlation between mRNA and protein expression levels (Anderson and Seihamer, 1997; Gygi et al., 1999). Second, proteins are often derived from different alternative spliced RNAs, and/or contain posttranslational modifications, which result in distinct functions and activities (Harada et al., 2004; Rammensee, 2004).

Although it is still too early to compare the success of protein microarrays with that already achieved by their DNA counterparts, there is no doubt that the first type of microarrays has made substantial progress in terms of construction and applications, including the area of biomarkers discovery. In recent years, there have been considerable achievements in preparing microarrays containing over 100 proteins and even an entire proteome (Madoz-Gúrpide et al., 2001; Cahill and Nordhoff, 2003; Michaud et al., 2003; Haab, 2005). Different array formats have been developed, including tissue, living cells, peptides/small molecules, antibody/antigen (s), protein, and carbohydrate arrays, which are described in more details in Chapter 2. The capability of these formats to provide simultaneous assessment of expression/interaction of 100s and even 1000s of proteins can be considered one of the emerging developments, which is paving the way to new and more powerful strategies in biomarkers discovery.

Mass-spectrometry-based methods for proteomic analysis have been improved on various fronts; new generation of mass spectrometers allows higher mass accuracy, higher detection capability, and shorter cycling times, allowing higher throughput and more reliable data. Two-dimensional chromatography coupled to MS/MS is getting more acceptance as a powerful tool for the analysis of complex protein mixtures. Recent improvements on the chromatography side included high-pressure LC systems and smaller diameter packing material allowing shorter analysis times and higher detection limits. With regard to data analysis, there are now several data mining tools for analyzing global protein expression data generated by this approach. Several publicly available software packages are currently in use to map proteomic data sets generated by searching peptide collision induced dissociation spectra against one or more major protein databases such as Swiss-Prot/TrEMBL, International Protein Index (IPI), and the National Center for Biotechnology Information (NCBI). Other relevant Web sites are given in Table 1.1.

In the emerging field of systems biology, accurate quantification of proteins and their changing patterns represent an important component. Recently, MS-based quantitative proteomics has become an important component in biological and clinical research. Over a number of years, multidimensional chromatography coupled to tandem MS has demonstrated its capability to identify hundreds to thousands of proteins within complex mixtures. However, the same platform fails short of routinely providing accurate quantitative analysis of proteins in complex media such as serum or cell lysate. A number of strategies have been devised to enhance the potential of this approach for protein quantification, including some posttranslational modifications. Many of these modifications, such as phosphorylation and glycolysation, have well-documented roles in signal transduction, regulation of cellular processes, clinical biomarkers, and therapeutic targets. A limited number of recent

TABLE 1.1. Some links relevant to DNA- and protein-based analysis available for public use.

ArrayExpress: http://www.ebi.ac.uk/arrayexpress
Biocarta: http://www.biocarta.com
Biomolecular Interaction Database:
http://www.blueprint.org/bind/bind.php
CaCORE: http://ncicb.nci.nih.gov/core
Cancer Biomedical Informatics Grid:
http://cabig.nci.nih.gov
Cancer Genome Anatomy Project: http://cgap.nci.nih.gov/
Cytoscape: http://www.cytoscape.org
Database of Interacting Proteins:
http://dip.doe-mbi.ucla.edu/
Cancer Genome Anatomy Project: http://cgap.nci.nih.gov
Cancer Genome Project: http://www.sanger.ac.uk/CGP
dbEST: http://www.ncbi.nlm.nih.gov/dbEST
Gene Expression Omnibus: http://www.ncbi.nlm.nih.gov/geo
Human Cancer Genome Project:
http://www.ludwig.org.br/ORESTES
IMAGE Consortium: http://image.llnl.gov
Mitelman Database of Chromosome Aberrations in Cancer:
http://cgap.nci.nih.gov/Chromosomes/Mitelman
SAGE Genie: http://cgap.nci.nih.gov/SAGE
SAGEmap: http://www.ncbi.nlm.nih.gov/SAGE
Spectral Karyotyping/Comparative Genomic Hybridization
Database: http://www.ncbi.nlm.nih.gov/sky
Access to this interactive links box is free online.
ExPASy: http://ca.expasy.org
Gene Ontology Consortium: www.geneontology.org
Swiss-Prot/TrEMBL: http://ca.expasy.org/sprot
International Protein Index(IPI): www.ebi.ac.uk/IPI/IPIhelp.html
National Center for Biotechnology Information (NCBI): www.ncbi.nlm.nih.gov
MouseSpec: http://tap.med.utoronto.ca/~posman/mousespec
Protein families data base: www.sanger.ac.uk/Software/Pfam
InterPro: www.ebi.ac.uk/interpro
PSORT II: http://psort.ims.u-tokyo.ac.jp
TreeView: http://jtreeview.sourceforg.net
GenMAPP; www.GenMapp.org

strategies have demonstrated the potential for large-scale analysis of phosphorylated and glycosylated proteins. MacCoss et al. (2002) have described what they termed "shotgun" approach for the identification of various forms of protein modifications (including phosphorylation) in complexes and in lens tissue. To digest the investigated protein mixtures, the authors used three different enzymes, one that cleaves at a specific site, whereas the other two cleave at nonspecific sites. The mixture of the resulting peptides was separated by multidimensional liquid chromatography and analyzed by tandem mass spectrometry. This approach has been applied to simple protein mixture, Cdc2p protein complexes isolated by affinity tag, and to lens tissue from a patient with congenital cataracts. These results yielded various sites of phosphorylation, acetylation, methylation, and oxidation.

In two relatively recent articles, which appeared in the same issue, two independent groups (Kaji et al., 2003; Zhang et al., 2003) described similar strategies for the identification and quantification of N-linked glycoproteins. The first group used a strategy that combines hydrozyde chemistry, stable isotope labeling, and mass spectrometry, whereas the second group used lactin affinity capture in combination with isotope-coded tagging and mass spectrometry. The approach by Kaji et al. (2003), termed as isotope-coded glycosylation-site-specific-tagging (IGOT), is based on the lactin column-mediated affinity capture of glycopeptides generated by tryptic digestion of protein mixtures, followed by peptide-N-glycosidase-mediated incorporation of stable isotope tag ^{18}O specifically into the N-glycosylation site. The tagged peptides are then identified by multidimensional LC coupled to mass spectrometry. This approach was tested on N-linked high-mannose and and/or hybrid-type glycoproteins derived from an extract of *Caenorhabditis elegans*. The authors reported the identification of 250 glycoproteins, including 83 putative transmembrane proteins, with the simultaneous determination of 400 unique N-glycosylation sites. To demonstrate the potential of IGOT strategy for protein quantification, the authors processed two peptide aliquots differentially labeled with ^{18}O and ^{16}O and the mixed preparation was examined by LC/MS. Although the isotope distribution of the two-tagged peptides partly overlapped owing to the natural isotopic abundance, both spectra were good enough to permit relative quantification of ^{16}O- and ^{18}O-tagged peptides.

Visible-coded affinity tag (VICAT) is a tagging reagent which allows absolute quantification of protein(s) in a complex biological sample (Lu et al., 2004). This tagging procedure can be considered a variant of the well-established ICAT procedure. VICAT reagents target thiol groups of Cys or thioacetylated amino groups and introduce into the tryptic peptide a biotin affinity handle, a visible moiety for tracking the chromatographic location of the target peptide by a detection method other than mass spectrometry. Initial capability of this reagent was demonstrated by the absolute determination of human group V phospholipase A_2, in eukaryotic cell lysates.

Another approach for high-throughput quantitative analysis has been recently reported by Zhang et al. (2005). This approach is designed to simplify the analysis of serum and allow targeted quantification of proteins, which happen to have relatively low concentration. This method is based on the selective isolation of those peptides from serum proteins that are N-linked glycosylated in the native protein and the use of LC/MS and LC/MS–MS to analyze the peptide mixture of the deglycosylated forms of these peptides. This method has two apparent advantages: first, a dramatic reduction in the total number of peptides, and second, a reduction in the complexity of the acquired spectra due to the removal of oligosaccharides that contribute significantly to the peptide pattern heterogeneity. The potential of this method was demonstrated by generating peptide patterns, which could distinguish the serum proteome of cancer bearing mice from genetically identical normal mice.

A method, termed stable isotopelabeling with amino acids in cell culture (SILAC), has recently gained popularity for its ability to compare the expression levels of hundreds of proteins in a single experiment (Everley et al., 2004). SILAC is based on the use of ^{12}C- and ^{13}C-labeled amino acids added to the growth media of separately cultured cell lines, giving rise to cells containing either light or heavy proteins. Lysates collected from these cells are then mixed, separated on SDS-Page, separated bands are excised and digested, and can be injected into a tandem mass spectrometer for protein identification/quantification.

1.4.2. Promising Sources for Biomarkers

1.4.2.1. DNA Methylation. The past few years have seen a substantial advance in our understanding of the functional consequences of DNA methylation and its interaction with chromatin structure and the transcriptional machinery (Laird, 2003). First insights into what causes DNA methylation patterns to undergo changes in cancer cells have also been acquired (Di Croce et al., 2002; Song et al., 2002). From a clinical perspective, DNA methylation changes in cancer represent a highly attractive therapeutic target, as epigenetic alterations, including DNA methylation are, in principle, more readily reversible than genetic events (Karpf and Jones, 2002). However, the great strength of DNA methylation in clinical applications promises to be in the areas of molecular diagnostics and early detection. The introduction of a highly sensitive methylation specific PCR (MSP) procedure by Herman et al. (1996) rendered DNA methylation a fertile ground for biomarkers research. The main advantage of the MSP assay is its sensitivity and capability to detect methylation in the presence of contaminating normal tissue or cells. The same assay can be conducted directly on tissue sections (in situ MSP) to identify clonality of the gene silencing in tumors and premalignant lesions (Nuovo et al., 1999). Recent improvements in the sensitivity of this assay are: A more sensitive assay called methylLight capable of detecting methylated alleles in the presence of 10^4-fold excess of unmethylated alleles has been described by Eads et al. (2000). This is a high-throughput assay capable of quantitative determination of a particular pattern of DNA methylation. A further improvement in MSP assays has been introduced by Palmisano et al. (2000). This improved assay was designated nested MSP and had the capability to detect a single methylated allel in ~50,000 unmethylated alleles. The list of cancer-associated methylated genes detected by this type of assays is expanding (see Chapter 4, Tables 4.4 and 4.5).

1.4.2.2. Mitochondrial DNA Mutations. This is another area targeted as a possible source for the identification of cancer biomarkers. Mitochondria dysfunction was proposed to be involved in cancer over 50 years ago (Warburg et al., 1967). Mitochondria are believed to be more susceptible to exogenous mutagens and also have less efficient DNA-repair mechanisms. There is accumulating evidence suggesting that mitochondria regulate several cellular processes that are linked to apoptosis, which include electron transport and energy metabolism. They are also the storage site for a number of soluble proteins that mediate apoptosis, including cytochrome c. Many of the signals that elicit apoptosis converge on the mitochondria, which respond to apoptotic signals by releasing cytochrome c (Verma et al., 2003). Information gained recently on the connection between mitochondrial dysfunction, deregulation of apoptosis, and tumorigenesis together with an increasing knowledge of proteins that are involved in cancer progression may lead to a new class of markers for the early detection and possible prevention of certain types of cancer.

1.4.2.3. Phosphatidylinositol-3 Kinases (PI3Ks). PI3Ks constitute a lipid kinase family characterized by their ability to phosphorylate inositol ring 3′-OH group in inositol phospholipids to generate the second messenger phosphatidylinositol-3,4,5-triphosphate (PIP3) at the inner side of the cell membrane (Cantley and Neel, 1999; Cantley, 2002). PIP3 in turn contributes to the recruitment and activation of a wide range of downstream targets, including the serine–threonine protein kinase Akt (also known as protein kinase B). The PI3K-Akt signaling pathway regulates many normal cellular processes including cell proliferation, survival, growth, and motility, processes that are critical for tumorigenesis. In the last decade, much of the cancer research has focused on the central role of *RAS*, the first identified oncogene, in neoplastic transformation. Extensive biochemical and genetic studies of the signaling components upstream and downstream of this small GTPase in model organisms led to the model of mitogenic signaling by receptor tyrosine kinases (RTKs) through RAS and motigen-activated protein kinases (MAPKs). The central importance of this pathway in neoplastic cell proliferation in humans has been strongly supported by the clinical success of therapeutics that target tyrosine kinases.

In recent years, a second pathway downstream of RTKs that involves phosphatidylinositol-3 kinase and Akt has come onto the scene and is proving to be an important regulator of mammalian cell proliferation and survival. Indeed, the role of this pathway in oncogenesis has been extensively investigated and altered expression or mutation of many of its components has been implicated in various forms of human cancer (Vivanco and Sawyers, 2002).

Currently, there are a number of therapeutic strategies in development, which target this pathway. Quantification of signaling throughput in the PI3K to Akt pathways has the potential of providing prognostic information to distinguish clinically important subsets of cancer. For example, a number of findings have specifically linked this pathway to prostate cancer. *PTEN* inactivation or loss of heteozygosity is common in prostate cancers, especially metastatic carcinoma (Suzuki et al., 1998; Sansal and Sellers, 2004), and targeted deletion of *PTEN* in mouse prostate activates Akt and induces prostate carcinoma (Wang et al., 2003). In a xenograft model for progression of the androgen-dependent (or androgen-sensitive) LNCaP cell line to androgen independence, Akt activity (but not expression) was elevated and correlated with Ser^{473} phosphorylation (Graff et al., 2000). Introduction of constitutively activated Akt into these cells permitted androgen-independent growth. Progression from normal prostate epithelium to prostatic intraepithelial neoplasia or carcinoma is associated with elevated Akt phosphorylation (Paweletz et al., 2001; Malik et al., 2002). These studies reported that mitogen-activated protein kinase activation monitored with phospho-extracellular signal-regulated kinase (ERK) antibodies was enhanced in prostatic intraepithelial neoplasia but reduced in carcinoma.

1.4.2.4. Profiling Tyrosine Phosphorylation. Over the last two decades, it has become clear that tyrosine phosphorylation plays a central role in a variety of important signaling pathways in multicellular organisms. Such role has been recently enforced by the success of specific tyrosine kinase inhibitors in cancer treatment (Druker, 2002). Functional profiling of the tyrosine phosphoproteome is likely to lead to the identification of novel targets for drug discovery and provide exciting and novel molecular diagnostic approaches. A major challenge in this direction is to develop the means to rationally control and manipulate the cellular tyrosine phosphorylation state. It is reasonable to state that the detection, identification, and quantification of phosphoproteins, and mapping of their phosphorylated sites, are the main objectives of phosphoproteomics. Over the past few years, a number of approaches have shown the potential to achieve some of these objectives. These include MS-based phosphoproteomic approaches (Conrads et al., 2002; Mann et al., 2002), two-dimensional gel electrophoresis with and without ^{32}P labeling (Immler et al., 1998; Larsen et al., 2001; Yoshimura et al., 2002), immunoaffinity-based methods (Pandey et al., 2002; Steen et al., 2002), and western blotting (Nollau et al., 2001; Machida et al., 2003).

1.4.2.5. Proteins Expression. Extensive activities dealing with protein profiling and analyses have generated a tremendous amount of data on the expression/modification of proteins under disease conditions, including various forms of cancer. Subsequent interpretation and assignment of biological roles of a part of such data allowed the identification of families which have the potential to be translated into biomarkers capable of an early detection of some forms of cancer. Three of these families are considered in the present text, two of which are briefly introduced below. These two families together with kallikerins are described in more details in Chapter 4.

Extensive research activities over the last 30 years have shown that HSPs and their close constitutively expressed relatives are in effect molecular chaperons. Following exposure to proteotoxic stressors, the cells in most tissues dramatically increase the production of this group of proteins. Various studies have proposed diverse roles for chaperones as sensors and regulators of stress-induced apoptosis. The molecular pathways that mediate apoptosis are tightly regulated by a series of positive and negative signals, the balance of which determines whether or not cells commit suicide. Increasing evidence suggests that HSPs can influence this process through direct interaction with key components of the apoptotic machinery. In other words, these proteins serve as cellular safeguards to protect the network of protein–protein interactions that sense stress signals and relay them to the apoptotic machinery (Mosser and Morimoto, 2004). These authors suggested that the ability of HSPs proteins to influence a cell's fate through modulation of numerous control points endows these proteins with the unusual capacity to contribute in a decisive way and at multiple points in the process of tumorigenesis.

Molecular cloning and biochemical characterization of 14-3-3 proteins have revealed seven homologous isoforms in mammalian cells, which were designated with the Greek letters β, γ, ε, η, σ, τ (sometimes referred to as θ), and ζ (Ichimura et al., 1988; Fu et al., 2000). Most of these isoforms are expressed in all human tissues, although the σ form expression is restricted to epithelial cells (Leffers et al., 1993). This family of proteins is implicated in the regulation of numerous cellular signaling circuits that are involved in the development of various forms of cancer. These proteins have attracted interest because they are involved in important cellular processes such as signal transduction, cell-cycle control, apoptosis, stress response, and malignant transformation. These different roles are in part due to their capability to bind more than 100 different binding partners. Of all the 14-3-3 genes, 14-3-3σ has been most directly linked to cancer. Inactivation of this gene occurs at many levels, and the high frequency of its inactivation suggests that it has a crucial role in tumor formation. This role has been consolidated by a number of recent investigations. Osada et al. (2002) have demonstrated frequent and histological-specific inactivation of 14-3-3σ in human lung cancer. The loss of 14-3-3σ expression in breast carcinoma was attributed to methylation silencing (Umbricht et al., 2001). Another report along these lines was given earlier by Ferguson et al. (2000). By using SAGE analysis the authors reported that the expression of this gene was 7-fold lower in breast carcinoma cells compared with normal breast epithelium.

1.5. INITIATIVES RELEVANT TO BIOMARKERS DISCOVERY

The complexity of the process of biomarkers discovery and validation together with the tremendous research activities involved in such process have underlined the urgent need for various initiatives, both at the national and international levels, to facilitate scientific collaboration and access to data generated by various research groups working in the field of biomarkers discovery. Some of these initiatives are briefly discussed in the sections below, and a more detailed description of these and other initiatives will be given in the latter part of this book.

1.5.1. Initiatives of the Human Proteome Organization (HUPO)

The HUPO formed in 2001 has launched several major initiatives. These are focused on the plasma proteome, the liver proteome, brain proteome, protein standards/bioinformatics, and certain technologies, including large scale antibody production. The aim of these initiatives is to foster organized international efforts in the field of proteomics, including more effective strategies for early disease detection (Hanash, 2004). The initial planning meetings around this initiative have drawn some sort of a checklist to be followed by various research groups. Regarding the plasma proteome project, interdisciplinary groups of experts have proposed a pilot phase to address the following issues: assessment of the sensitivity of various techniques; guideline on all aspects of specimen collection and handling; methods of depleting or prefractionation of the most abundant proteins; comparison of advantages and limitations associated with plasma versus serum; enumeration and categorization of visualized and identified proteins, with particular attention to their posttranslational modifications and tissue of origin; separation of intact proteins versus separation of their digested peptides, comparing gel-based methods with multidimensional liquid chromatography, and assessment and advancement of specific labeling chemistry. Looking at this list of issues, it is not difficult to realize that such pilot stage is an attempt to address some of the drawbacks and limitations, which have been highlighted by previous and more recent proteomic analyses generated by various techniques. Whether such list of issues can be rigorously implemented is difficult to predict. Regardless of such prediction, such initiative will no doubt contribute to more reproducible data and the identification of the most suitable platform(s) to handle the complexity of plasma or serum proteome.

1.5.2. Data Mining in Cancer Research

The complexity of tumors biology renders the use of interdisciplinary approaches a necessity rather than a choice. Furthermore, the tremendous amount of data generated by a wide and diverse proteomic and genomic approaches underlined the need for the creation of easily accessible repositories to allow the interrogation of databases and other tools. Currently, there are a number of data repositories containing enormous amount of data on gene expression in normal and cancer cells. These data are the result of initiatives such as the Cancer Anatomy Project and the Director's Challenge Initiative, funded by the National Cancer Institute (NCI). There are other resources available for data mining. One of these resources, GoMiner, was developed by Zeeberg et al. (2003), a program package that organizes lists of genes, such as down- and overexpressed genes from a range of microarray experiment(s). This program package provides quantitative and statistical output files and two different visualizations. Genes displayed in GoMiner are linked to major public bioinformatics resources. The NCBI at the National Institute of Health was created almost 20 years ago to develop information systems for molecular biology. In addition to maintaining the GenBank nucleic acid sequence database, to which data are submitted by the scientific community, NCBI provides data retrival systems and computational resources for the analysis of GenBank data and a variety of other biological data. Wheeler et al. (2006) described the major resources of NCBI, which are available on a home page at the Web site, http://www.ncbi.nlm.nih.gov, and download of bulk data underlying these resources is also available through a link ftp.ncbi.nih.gov from the NCBI home page.

There are resources relevant to data mining in cancer, including the Kyoto Encyclopedia of Genes and Genomes (KEGG) (Kanehisa et al., 2002). This publicly available web database contains more than 150 pathways with emphasis on well-defined metabolic pathways. Gene Microarray Pathway Profiler (GenMAPP) is a freely available program for viewing and analyzing expression data on microarray pathway profiles representing biological pathways or other functional grouping of genes (Doniger et al., 2003). Other useful links are listed in Table 1.1.

1.6. CONCLUDING REMARKS

The material in this book is an attempt to underline tremendous advances in the technologies and current knowledge of the biology and molecular basis of cancer. The same material, however, leaves no doubt that despite such advances there are still a number of challenges before many forms of cancer can be defeated. The correlation between what has been already achieved and what is remained to be done has been elegantly described by the following phrase; "One day, we imagine that cancer biology and treatment at present, a patchwork quilt of cell biology, genetics, histopathology, biochemistry, immunology, and pharmacology will become a science with a conceptional structure and logical coherence that rivals that of chemistry or physics" (Hanahan and Weinberg, 2000). Before moving to the next chapter it is useful to have in mind a number of general considerations, which in part illustrate some of the difficulties still facing research scientists in their efforts to discover biomarker(s) for the various forms of cancer:

- Despite enormous proteomic and genomic efforts, in only a few tumor diseases have relevant markers been established that can be used for early diagnosis or improved therapy in cancer. We are still facing the dilemma where in many cases cancer is not diagnosed and treated until cancer cells have already invaded surrounding tissues and metastasized throughout the body. No one disputes the fact that serum-based markers such as CA125 and PSA have saved many lives, yet both markers suffer two well-recognized limitations: The first is their low specificity, which in turns results in a high rate of false-positives, whereas the second limitation is associated with what can be considered an unacceptable time-lag between the detection and the in situ state of the disease. In other words, an elevated level of these markers seems to manifest at an advanced stage of the malignancy. To address certain limitations there have been some recent attempts to emphasize the utility of multiple markers rather than relying on "one-at-a-time" approach. On the proteomic side, for example, the use of serum-based proteomic patterns analysis started to gain more momentum. The SELDI analysis is a representative example of such emerging approach, where patterns containing a number of different molecular ions are used to distinguish between healthy and diseased samples. The basic principle of such analysis is not substantially different from that applied in two-diminsional gel analysis, where alteration in protein expression is monitored for multiple rather than for single proteins. The same principle is also applied in some DNA-based analysis, where, for example, patterns of DNA methylation are sought rather than the methylation of a single entity.
- The inefficacy of some existing serum-based cancer markers particularly regarding their capability for the early detection of the disease cast some shadow on the search strategies used to discover them. These strategies have been criticized on the ground that the media in which the target markers are detected (e.g., serum) do not necessarily reflect the in situ situation of the tumor. In other words, data delivered by these strategies are not easily traced back to the

biological properties or the heterogeneity of the tumor itself. Such criticism can be partially justified if we consider the in situ complexity of many forms of cancer. Such complexity has been underlined by Hanahan and Weinberg (2000) in an article entitled "The hallmarks of cancer." The authors suggested that the vast catalog of cancer genotypes is a manifestation of six essential alterations in cell physiology that collectively dictate malignant growth: self-sufficiency in growth signals, insensitivity to growth-inhibitory (antigrowth) signals, evasion of programmed death (apoptosis), limitless replicative potential, sustained angiogenesis, and tissue invasion and metastasis. The same authors went on to propose that all six capabilities are common to most and perhaps to all types of human tumors. The complexity of many forms of cancer is further aggravated by the absence of a comprehensive knowledge on the signaling pathways within a cell, which are more and more mimicking complex electronic integrated circuits, where transistors are replaced by proteins and electrons by phosphates and lipids. If we apply the same principle to the signaling pathways in cancerous cells, then we can appreciate that an elevated level of a single protein in serum may represent a useful marker for a given type of cancer, yet at the same time we have to accept that such marker is going to have a number of limitations.

- Over the last 20 years, both proteomic and genomic activities have begun to make extensive use of products of human origin. This new trend has raised many ethical and social issues, particularly those involving the individual rights, including issues of consent. This means that researchers should carefully consider several aspects when designing studies in which samples of human origin are required. These aspects will surely include the extent of risk for human volunteers, biosafety in particular when international collaboration is needed, rules on data acquisition and storage that also has to be carefully assessed particularly when using computerized databases. Presently, the ethical and regulatory framework for using human tissues in various areas of research is still vague and lacks precise guidelines. Besides these ethical and social problems, research scientists looking for new disease markers are still facing hurdles related to approvals by Food and Drug Administration (FDA) and European Agency for the Evaluation of Medicinal Products (EMEA). As far as proteomic-based tests are concerned, neither of the two regulatory bodies has an official guideline on how such tests should be submitted. However, it has to be said that the FDA has taken an important step toward defining a policy designated "Multiplex Tests for Heritable DNA Markers, Mutations, and Expression Patterns which mainly focuses on DNA tests, including DNA microarrays." Although such draft guideline could also cover some proteomic tests, it is hoped that a similar guideline, which may cover a wider range of proteomic tests, is something that can materialize in the near future. The situation is more complicated in the case of EMEA, which is not in charge of evaluating diagnostic tests, yet it does monitor developments in the field, which may impact on pharmaceuticals.

- Continuing the discussion on the theme of serum-based markers, it is worth considering a current point of contention regarding the relevance of establishing the identity of such markers. For example, if we use mass spectrometry-based method to analyze healthy and diseased samples, would differences in the pattern of unidentified MS peaks be sufficient for use as a diagnostic tool? The answer to this question strongly depends on the person who gives it. A research scientist would argue that the identification of each peak is important for current and future attempts to decipher the complex biology and signaling pathways associated with cancer. Another line of thought advocates that a pattern of unidentified MS peaks, which has been tested on extremely large number of samples, might be more than sufficient to satisfy doctor–patient perspective. In other words, if we have a reliable and selective marker for a given type of cancer, then its identity is not the top priority of either the patient or his physician. Leaving aside the difference between the two opinions, it is not difficult to spot a common objective, which is called discovering new and reliable markers for a class of devastating diseases.
- High-throughput analysis techniques raise the question of overfitting of data generated in discovery-based research. Such danger can be encountered when large amount of data are generated and analyzed for discriminatory patterns to use in diagnosis or prognosis (Stears et al., 2003). For example, RNA expression levels of thousands of genes from a cancer specimen can be analyzed for patterns that predict a patient's prognosis or response to therapy. Similarly, thousands of peaks generated by mass spectrometry of serum sample can be analyzed for protein/peptide patterns that discriminate between a healthy person and a patient. Ransohoff (2004) underlined the problem of overfitting by a simple yet efficacious example, which is worth considering. According to this author, overfitting can occur when large numbers of potential predictors are used to discriminate among a small number of outcome events. This scenario has been exemplified by imagining 10 people with cancer and 10 without who are screened using 20,000 features with no relation to cancer, such as the type of films they watch or the number of times they chew their food. The author commented that if enough predictors are examined, even if nonsensical and random, a pattern could be found to discriminate among the group of individuals derived from a training set, but it would not discriminate in an independent validation set. Biostatistics and empirical assessment has also demonstrated how overfitting can occur in RNA expression analysis (Ambroise and McLachlan, 2002; Simon et al., 2003). Simon and collaborators constructed a group of imaginary individuals, 10 with and 10 without cancer, along with expression data for 6000 genes. They then applied different methods of cross-validation, in a manner highly representative of real experiments, to discover discriminatory patterns. The authors reported that using one common method, 98% of the models fit perfectly in the training set, indicating how frequently overfitting can occur. Such overfitting has to be carefully assessed in approaches that use multivariable analysis such as artificial neural networks (Selaru et al., 2002), genetic algorithms (Petricoin et al.,

2002), boosted decision-tree analysis (Qu et al., 2002), and metagenes (Huang et al., 2003) are commonly used in discovery–based research.

REFERENCES

Aebersold, R., Goodlett, D. R. (2001) Chem. Rev. **101**, 269.
Aebersold, R., Mann, M. (2003) Nature **422**, 198.
Ambroise, C., McLachlan, G. J. (2002) PNAS. USA **99**, 6562.
Anderson, L., Seihamer, J. (1997) Electrophoresis **18**, 533.
Bast, R. C., Jr, Freeney, M., Lazarus, H., et al. (1981) J. Clin. Invst. **681**, 1331.
Bast, R. C., Jr, Klug, T. L., St John, E., et al. (1983) N. Engl. J. Med. **309**, 883.
Cahill, D. J., Nordhoff, E. (2003) Adv. Biochem. Engin. Biotechnol. **83**, 177.
Cantley, L. C. (2002) Science **296**, 1655.
Cantley, L. C., Neel, B. G. (1999) PNAS. USA **96**, 4240.
Caprioli, R. M., Farmer, T. B., Gile, J. (1997) Anal. Chem. **69**, 4751.
Chaurand, P., Stoeckli, M., Caprioli, R. M. (1999) Anal. Chem. **71**, 5263.
Conrads, T. P., Issaq, H. J., Veenstra, T. D. (2002) Biochem. Biophys. Res. Commun. **290**, 885.
Datta, Y. H., Adams, P. T., Drobyski, W. R., et al. (1994) J. Clin. Oncol. **12**, 475.
Di Croce, L., Raker, V. A., Corsaro, M., et al. (2002) Science **295**, 1079.
Doniger, S., Salomonis, N., Dahlquist, K., et al. (2003) Genome Biol. **4**, R7.
Druker, B. J. (2002) Cancer Cell **1**, 31.
Eads, C. A., Danenberg, K. D., Kawakami, K., et al. (2000) Nucleic Acids Res. **28**, e32.
Espina, V., Mehta, A., Winters, M. E., et al. (2003) Proteomics **3**, 2091.
Everley, P. A., Krijgsveld, J., Zetter, B. R., et al. (2004) Mol. Cell. Proteomics **3.7** 729.
Ferguson, A. T., Evron, E., Umbricht, C. B., et al. (2000) Proc. Natl. Acad. Sci. USA **97**, 6049.
Fu, H., Subramanian, R. R., Masters, S. C. (2000) Annu. Rev. Pharmacol. Toxicol. **40**, 617.
Gold, P., Freedman, S. O. (1975) JAMA. **234**, 190.
Graff, J. R., Konicek, B. W., McNulty, A. M., et al. (2000) J. Biol. Chem. **275**, 24500.
Gygi, S. P., Rochon, Y., Franza, B. R., et al. (1999) Mol. Cell. Biol. **19**, 1720.
Haab, B. B. (2003) Proteomics **3**, 2116.
Hamdan, M., Righetti, P. G. (2003) Mass Spectrom. Rev. **22**, 272.
Hanahan, D., Weinberg, R. A. (2000) Cell. **100**, 57.
Hanash, S. (2004) Mol. Cell. Proteomics **3.4**, 298.
Harada, K., Yewdell, J. W., Young, J. C. (2004) Nature **427**, 252.
Heller, M. J. (2002) Ann. Rev. Biomed. Eng. **4**, 129.
Herman, J. G., Graff, J. R., Nelkin, P. D., et al. (1996) PNAS. USA **93**, 9821.
Horak, C. E., Luscombe, N. M., Qian, J., et al. (2002) Genes Dev. **16**, 3017.
Huang, E., Cheng, S. H., Dressman, H., et al. (2003) Lancet **361**, 1590.
Hudson, M., Bahnson, R., Catalona, W. (1989) J. Urol. **142**, 1011.
Hutchens, T. W., Yip, T. T. (1993) Rapid Commun. Mass Spectrom. **7**, 576.

Ichimura, T., Isobe, T., Okuyama, T., et al. (1988) PNAS. USA **85**, 7084.
Immler, D., Gremm, D., Kirsch, D., et al. (1998) Electrophoresis **19**, 1015.
Kaji, H., Saito, Y., Yamauchi, T., et al. (2003) Nat. Biotecnol. **21**, 667.
Kanehisa, M., Goto, S., Kawashima, S., et al. (2002) Nucleic Acids Res. **30**, 42.
Klose, J., Kobalz, U. (1995) Electrophoresis **16**, 1034.
Krismann, M., Todt, B., Schroder, J., et al. (1995) J. Clin. Oncol. **13**, 2769.
Laird, P. W. (2003) Nat. Rev. Cancer **3**, 253.
Larsen, M. R., Sorensen, G. L., Fey, S. J., et al. (2001) Proteomics **1**, 223.
Lu, Y., Bottari, J., Stemman, O., et al. (2004) Anal. Chem. **76**, 4104.
Karpf, A. R., Jones, D. A. (2002) Oncogene **21**, 5496.
Leffers, H. P., Madsen, H. H., Rasmussen, B., et al. (1993) J. Mol. Biol. **231**, 982.
Link, A. J., Eng, J., Scvhieltz, D. M., et al. (1999) Nat. Biotechnol. **17**, 676.
Liotta, L. A., Espina, V., Mehta, A., et al. (2003) Cancer Cell **3**, 317.
Lyer, V. R. (2001) Nature **409**, 533.
MacCoss, M. J., McDonald, W. H., Saraf, A., et al. (2002) PNAS. USA **99**, 7900.
MacBeath, G. (2002). Nat. Genet. Supl. **32**, 526.
Machida, K., Mayer, B. J., Nollau, P. (2003) Mol. Cell. Proteomics **2**, 215.
Madoz-Gúrpide, J., Wang, H., Misek, D. E., et al. (2001) Proteomics 1, 1279.
Malik, S. N., Brattain, M., Ghosh, P. M., et al. (2002) Clin. Cancer Res. **8**, 1168.
Mann, M., Ong, S. E., Gronborg, M., et al. (2002) Trends Biotechnol. **20**, 261.
Michaud, G. A., Salcius, M., Zhou, F., et al. (2003) Nat. Biotechnol. **21**, 1509.
Mosser, D. D., Morimoto, R. I. (2004) Oncogene **23**, 2907.
Nollau, P., Mayer, B. J. (2001) PNAS. USA **98**, 13531.
Nuovo, G. L., Plaia, T. W., Belinsky, S. A., et al. (1999) PNAS. USA **96**, 12754.
O'Farrell, P. H. (1975) J. Biol. Chem. **250**, 4007.
Osada, H., Tatematsu, Y., Yatabe, Y., et al. (2002) Oncogene **21**, 2418.
Palmisano, W. A., Divine, K. K., Saccomanno, G., et al. (2000) Cancer Res. **60**, 5954.
Pandey, A., Blagoev, B., Krachmarova, I., et al. (2002) Oncogene **81**, 8029.
Paweletz, C. P., Charboneau, L., Bichsel, V. E., et al. (2001) Oncogene **20**, 981.
Petricoin, E. F., Ardekani, A. M., Hitt, B. A., et al. (2002) Lancet **359**, 572.
Qu, Y., Adam, B-L., Yasui, Y., et al. Clin. Chem. **48**, 1835.
Ramaswamy, S., Golub, T. R. (2002) J. Clin. Oncol. **20**, 1932.
Rammensee, H-G. (2004) Nature **427**, 203.
Ransohoff, D. F. (2004) Nat. Rev. Cancer **4**, 309.
Read, M. C., Lachs, M. S., Feinstein, A. R. (1995) JAMA 274, 645.
Sansal, I., Sellers, W. R. (2004) J. Clin. Oncol. **22**, 2954.
Selaru, F. M., Yan, X., Yin, J., et al. (2002) Gastroenterology **122**, 606.
Simon, R., Radmacher, M. D., Dobbin, K., et al. (2003) J. Natl. Cancer Inst. **95**, 14.
Song, J. Z., Stirzaker, C., Harrison, J., et al. (2002) Oncogene **21**, 1048.
Sporn, M. B. (1996) Lancet. **347**, 1377.
Srivastava, S., Kramer, B. S. (2000) Lab. Invest. **80**, 1147.

Stamey, T. A., Yang, N., Hay, A. R., et al. (1987) N. Engl. J. Med. **317**, 909.
Stears, R. L., Martinsky, T., Schena, M. (2003) Nature Med. **9**, 140.
Steen, H., Kuster, B., Fernandez, M., et al. (2002) J. Biol. Chem. **277**, 1031.
Stoeckli, M., Chaurad, P., Hallahan, D. E., et al. (2001) Nat. Med. **7**, 493.
Sullivan Pepe, M., Etzioni, R., Feng, Z., et al. (2001) J. Natl. Cancer Inst. **93**, 1054.
Suzuki, H., Freije, D., Nusskern, D. R., et al. (1998) Cancer Res. **58**, 204.
Tang, N., Tornatore, P., Weinberger, S. R. (2004) Mass Spectrom. Rev. **23**, 34.
Thomson, I. M., Pauler, D. K., Tangen, C. M., et al. (2004) N. Engl. J. Med. **350**, 2239.
Thomson, D. W., Krupey, J., Freedman, S. O., et al. (1969) PNAS USA **64**, 161.
Umbricht, C. B., Gabrielson, E., Ferguson, A., et al. (2001) Oncogene **20**, 3348.
Van de Vijver, M. J., He, Y. D., van 't Ver, L. J., et al. (2002) N. Engl. J. Med. **347**, 1999.
Velculescu, V. E., Zhang, L., Vogelstein, B., et al. (1995) Science **270**, 484.
Verma, M., Kagan, J., Sidransky, D., et al. (2003) Nature Rev. **3**, 789.
Vivanco, I., Sawyers, C. L. (2002) Nat. Rev. Cancer **2**, 489.
Wang, S., Gao, J., Lei, Q., et al. (2003) Cancer Cell **4**, 209.
Wang, H., Hanash, S. M. (2003) J. Cromatogr. B. **787**, 11.
Wang, M. C., Valenzuela, L. A., Murphy, G. P., et al. (1979) Investig. Urol. **17**, 159.
Warburg, A., Geissler, A. W., Lorenz, S. (1967) Physiol. Chem. **348**, 1686.
Washburn, M. P., Wolters, D., Yates, J. R., 3rd. (2001). Nat. Biotechnol. **19**, 242.
Wheeler, D. L., Barrett, T., Benson, D. A., et al. (2006) **34**, D173.
Winawer, S. J., St. John, D. J., Bond, J. H., et al. (1995) Bull. World Health Organization. **73**, 7.
Young, R. A. (1995) Cell **102**, 9.
Yoshimura, Y., Shinkawa, T., Taoka, M., et al. (2002) Biochem. Biophys. Res. Commun. **290**, 948.
Zeeberg, B. R., Feng, W., Wang, G., et al. (2003) Genome Biology **4**, R28.
Zhang, H., Li, X-J., Martin, D. B., et al. (2003) Nat. Biotechnol. **21**, 660.
Zhang, H., Yi, E. C., Li, X-J., et al. (2005) Mol. Cell. Proteomics **4**, 144.

2

PROTEOMIC PLATFORMS FOR BIOMARKERS DISCOVERY

Currently we have a number of cancer markers, which are routinely used for screening, diagnosis, prognosis, and for monitoring response to therapeutic treatments. Most of these markers have been discovered using immunological and monoclonal antibody assays. With the completion of the human genome project and the unprecedented progress in proteomic and bioinformatics, the balance has shifted toward the use of high throughput proteomic and genomic approaches in the search for new cancer biomarkers. On the proteomic side, we have a number of platforms, which are applied in the area of markers discovery including, surface enhanced laser desorption ionization (SELDI) (Hutchens and Yip, 1993; Dayal and Ertel, 2002; Tang et al., 2004), mass spectrometry combined with two-dimensional liquid chromatography (Link et al., 1999; Wall et al., 2000; Washburn et al., 2001; Wang and Hanash, 2003) or two-dimensional gel electrophoresis (O'Farrel, 1975; Klose and Kobalz, 1995; Aebersold and Goodlett, 2001; Aebersold and Mann, 2003; Hamdan and Righetti, 2003), and protein microarrays (MacBeath, 2002; Espina et al., 2003; Liotta et al., 2003). Imaging mass spectrometry (Caprioli et al., 1997; Chaurand et al., 1999; Stoeckli et al., 2001) has also shown the potential to become one of the future tools in the search for cancer biomarkers. These techniques and some of their applications in the search for potential cancer biomarkers have been described in this chapter.

2.1. SURFACE ENHANCED LASER DESORPTION IONIZATION

The principle of separating proteins from crude mixtures on chip surfaces can be easily compared to the classical adsorption/desorption principle used in column

Cancer Biomarkers: Analytical Techniques for Discovery, by Mahmoud H. Hamdan

chromatography. Unlike column chromatography, SELDI uses small flat surfaces (chips) to perform initial fractionation, and the desorption is effected through laser irradiation. Although this approach is commonly perceived as separation-free, it should not be forgotten that the series of surfaces (chips) designed to capture selected protein population can be looked upon as a serial or a cascade fractionation device; that is, a given protein population adsorbed onto a cation-exchange surface can be subsequently eluted and further purified onto an anion-exchange surface, to be further manipulated onto a metal affinity surface, or a reversed phase. The concept of SELDI was first demonstrated over 10 years ago (Hutchens and Yip, 1993), while the ionization method used in this approach was demonstrated few years earlier (Tanaka et al., 1987; Karas and Hillenkamp, 1988). The impressive speed by which the SELDI approach has gained a leading role in biomarkers discovery can be attributed to a successful combination between affinity technology, well tested and highly versatile ionization technique, and highly advanced bioinformatics. In the following sections, an attempt has been made to describe the basic elements of this strategy, its current limitations, and its future potential.

2.1.1. Some Basic Considerations

The SELDI arrangement consists of three main components: protein-chip arrays, mass spectrometer, and the software for data acquisition, manipulation and interpretation. The first component can be considered the heart of this approach and it is the single element that distinguishes the technology from more traditional MS-based strategies. As it currently stands, SELDI analysis requires pretreatment of a small amount ($\sim$0.5–1 μL) of biological specimen such as serum, plasma, urine, intestinal fluid, cell lysates or cellular secretion products. Such pretreatment is commonly conducted on surfaces with 8 or 16 spots containing specific chromatographic or biologically functionalized surfaces. Currently available chromatographic surfaces are based on hydrophobic, ion-exchange, metal affinity, normal or reverse-phase chromatography, whereas the biological surfaces rely on the immobilization of specific reagents such as antibodies, receptors, DNA, and so forth. Following the pretreatment step, the surfaces are washed to remove unbound components and a suitable laser absorbing matrix is added and allowed to crystallize with the captured proteins. The remaining steps including laser irradiation, ion analysis, and data acquisition are identical to steps commonly used in conventional MALDI-TOF mass spectrometry. A schematic representation of the various steps in SELDI analysis is given in Figure 2.1. Over the last 5 years, a number of excellent reviews have described the principle and applications of SELDI (Merchant and Weinberger, 2000; Issaq et al., 2003; Dayal and Ertel, 2002; Tang et al., 2004; Diamandis, 2004). However, the rapid diffusion of this technique and its increasing application in the search for potential biomarkers provide sufficient new material to be evaluated by current and possibly future users of SELDI technology.

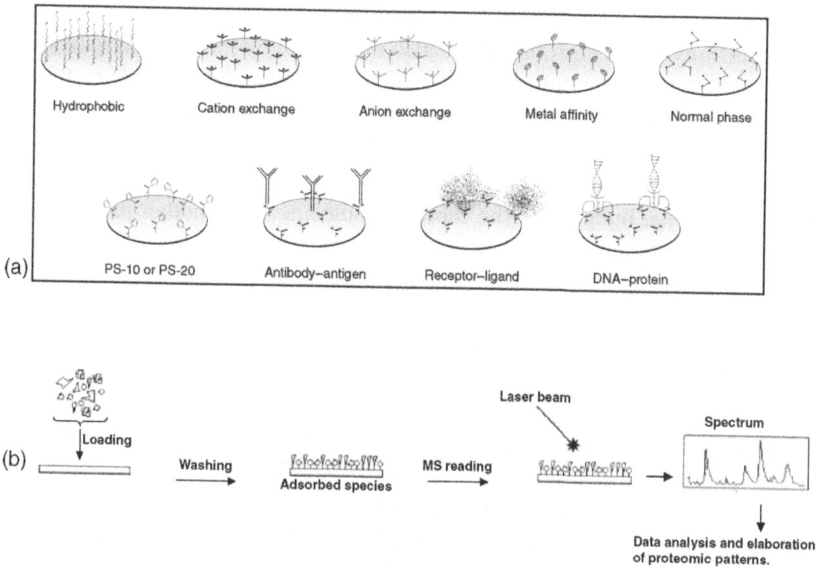

Figure 2.1. (a) Commercially available SELDI chips comprising chemically (upper row) and biochemically modified (lower row) surfaces. (b) The main steps in a typical SELDI analysis.

2.1.2. Protein Capture Surfaces

As it has been already pointed out at the onset of this chapter that the capture surfaces array is a central component in SELDI technology, which merits some detailed comments. Ideally, when a complex biological sample is deposited onto a surface, such surface would only bind the component(s) whose properties match perfectly its binding characteristics. However, before achieving such a tantalizing objective a number of difficulties have to be resolved. Some of these difficulties and possible solutions have been discussed by Tang et al. (2004). The following considerations are mainly based on some of the arguments advanced by these authors.

- The goal of developing a wide-range of binding surfaces can be compromised by nonspecific binding reactions, which can be extensive within complex clinical samples. These undesired nonspecific reactions can be provoked by different causes including the scaffold chemistry of the supporting substrate, which can be responsible for hydrophobic, van der Waals, hydrogen bonding, and electrostatic interactions.
- The capacity of an affinity surface can generally be defined as the total number of a given class of molecules, which can be specifically bound by the surface. Such parameter is of particular importance in SELDI analyses, which handle biological samples containing both extremely low- and high-abundance proteins. Finding high capacity affinity surfaces will no doubt influence our capability to detect

low-abundance proteins in a complex medium. It is needless to say that the capacity of a given affinity surface/probe will be also influenced by the characteristics of the molecule to be captured and the way by which such characteristics are matched by the functional groups attached to the surface. The heterogeneity and widely diverse protein abundances within a biological sample are sufficient to render such a task fairly daunting.

- Preserving the integrity of the captured molecule(s) is another parameter that can determine the suitability and efficacy of the capturing surface. This is particularly critical for biomolecules in general and proteins in particular. In other words, the pH and organic solvent contents associated with various immobilization steps should be carefully assessed to avoid disruption of the native structure and/or degradation of the target species.
- The search for potential biomarkers in biofluids most often takes a nonbiased approach in which the aim is to characterize as many components as possible. This approach has a price because such biofluids are generally dominated by high abundance proteins. For instance, in the case of plasma and serum, only 22 proteins account for 99% of their protein contents. These include albumin, transferrins, immunoglubulins, and complement factors. These proteins have all been well characterized; therefore, what the SELDI analysis should aim at is the remaining 1%, which is made up of lower abundance circulatory proteins as well as proteins that are shed or excreted not only by healthy cells but also by cells of the cancer itself or its microenvironment. To facilitate the detection of this interesting part of the proteome, it became more evident that we need various forms of enrichment/fractionation prior to sample deposition on the capture surfaces. Some of these approaches are briefly described in the following sections.

2.1.3. Enrichment/prefractionation Prior to SELDI Analysis

Existing literature describes several methods for the removal of high-abundance proteins, most of which derive from classical approaches in the domain of affinity chromatography. For example, albumin can be removed by using an immobilized dye named Cibacron blue-F3GA (Leatherbarrow and Dean, 1980). Most but not all, serum antibodies of immunoglobulin G (IgG) class can be removed by using an immobilized protein A (Lindmark et al., 1983). The main drawback of these separation protocols is that a large number of other species are co-adsorbed on the column. This has been recently reported by Guerrier et al. (2005) while attempting to separate serum proteins in several fractions for further analysis. The authors observed that the IgG fraction isolated by protein A affinity adsorption contained 794 polypeptides of low-concentration and the albumin fraction isolated on immobilized Cibacorn Blue dye contained 835 low-concentration polypeptides. Currently, there are a number of strategies, which can be applied to attenuate concentration differences between the components within complex biological samples. All such strategies can be used in MS-based analysis; however, some of them have more potential use in SELDI analysis.

2.1.3.1. Combinatorial Affinity. Interestingly, one of the most promising approaches for reducing the immense differences between low- and high-abundance proteins in a serum sample prior to SELDI analysis is based on the principle that has been tested in combinatorial chemistry for the synthesis of extensive libraries of small organic molecules. This approach is based on the pioneering work of Merrifield (Merrifield, 1963) in solid-phase synthesis, and using the "split, couple, recombine" method, libraries of potentially billions of different amino acid ligands can be created. Briefly, such synthesis is performed in parallel-sequential chemical reactions: the first step is a batch of millions of microscopic resin beads divided into different reaction vessels, where the first building block, a protected amino acid, is coupled to the resin. The beads are then mixed together and washed extensively, the amino group of the coupled amino acid is deprotected, and the beads are distributed randomly into a second set of reaction vessels and coupled with the next set of building blocks. This process is repeated until a ligand of the desired length is obtained (Furka, et al., 1991; Lam et al., 1991). This means that using the 20 natural amino acids and making six reaction steps, a library of linear, hexamer ligands containing a minimum of 20^6 different members can be made. Without going into further details regarding this principle, it is not difficult to appreciate the potential of such tremendous ligand-diversity in capturing almost every protein, peptide, antibody, and so forth, present in a complex biological sample. The basic steps in applying this approach to reduce concentration difference in samples prior to SELDI analysis are depicted in Figure 2.2. Preliminary data on the application of this protocol to reduce concentration differences between various proteins in serum sample prior to SELDI analysis has been described in a recent article (Righetti et al., 2005). The authors hypothesized that when a complex protein mixture of a first variance, such as human

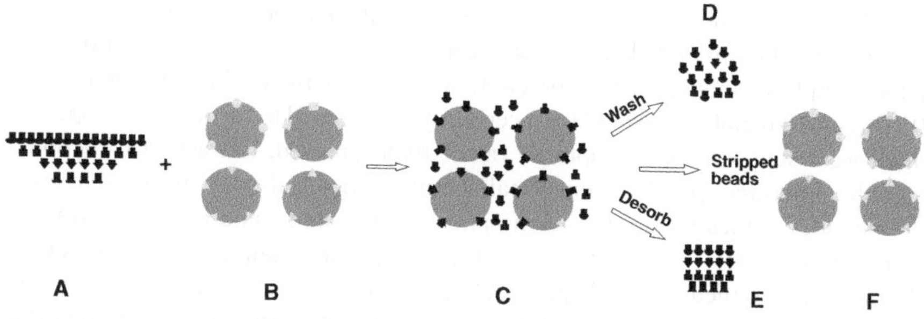

Figure 2.2. Representation of the principle of reducing the concentration difference between species mixed within the same solution. A solution composed of a mixture of four species "A" at different concentrations is loaded on a combinatorial solid-phase ligand library "B" under large overloading conditions. Each bead captures the corresponding protein partner until saturation, leaving the excess in the supernatant "C." A wash follows eliminating thus all proteins "D." Resulting beads are then treated with appropriate solutions so that adsorbed species are collected "E." This latter contains all considered species but at different concentration compared to the initial sample "A." From Righetti et al. (2005) with permission.

serum, is incubated with such library, representatives of each component within the mixture will bind to these individual ligands. Under large overloading conditions, high-abundance proteins saturate their specific affinity ligand and excess is removed during the washing step, whereas low-abundance proteins continue to concentrate on their specific affinity ligands. After processing, representatives of all the original components can be eluted to produce a sample with a second variance in which each representative is still present, but in different concentrations. High-abundance proteins are significantly diluted, whereas the low-abundance species are concentrated. It is interesting to note that the authors (Righetti et al., 2005) presented preliminary SELDI mass spectra with and without the application of the above protocol. These preliminary spectra showed that the application of the above procedure resulted in a higher number of peaks, particularly in the low and middle molecular mass range.

2.1.3.2. Magnetic Beads. An emerging approach to simplify samples for analysis by MADI-MS including SELDI is the use of reversed phase magnetic beads (Vallanueva et al., 2004; Zhang et al., 2004). The first group used reversed-phase (C-8) magnetic beads to investigate their capability of capturing the maximum number of peptides in serum before and after the removal of the most abundant proteins within the serum. The removal of these abundant proteins prior to the use of beads was effected using different methodologies including, ethanol precipitation, membrane ultra filtration, or albumin depletion using Cibacorn blue-derivatized beads. Surprisingly, MALDI-TOF-MS analysis displayed a much lower number of peptides compared with those observed in the mass spectrum of the whole serum. This unexpected result can be partially explained by the following considerations: First, during the depletion of albumin using Cibacorn blue-derivatized beads, a relatively large number of other species are co-adsorbed and therefore also removed. As it has been already pointed out, Guerrier et al. (2005) reported that over 800 polypeptides were removed with albumin on immobilized Cibacron Blue column. This may explain the reduced number of peptides captured by the magnetic beads following the depletion of albumin on Cibacron Blue beads. Second, the additional manipulation of the sample including ethanol precipitation or centrifugal ultra filtration over a 30 kDa cut-off membrane would have certainly contributed to further loss of species, which were supposed to be captured rather than removed. Extraction of peptides from blood plasma using C-8-functionalized magnetic beads was also used to compare peptide profiles in samples from patients with asthma and healthy volunteers (Zhang et al., 2004). Prior to peptides extraction, plasma samples were diluted two-fold in an unspecified buffer. Peptides eluted from the beads were examined with MALDI-TOF-MS and the acquired data were interpreted using pattern recognition software.

2.1.3.3. Stacked Sorbents. Another promising approach for sample fractionation is the use of several stacked columns, equilibrated with a single binding buffer for adsorption and for elution of the captured proteins. This concept was recently demonstrated by Guerrier et al. (2005). This approach uses a set of different solid-phase chemical entities serially connected in a stack mode, equilibrated in the same binding physiological buffer. As the sample crosses the different adsorbent layers, proteins within are sequentially trapped by the various components of the stack. Once the loading and capturing is achieved, the sequence of columns is disassembled, and each column containing different complement of proteins is eluted separately in a single step and under optimal elution conditions. When compared to classical single-chemistry fractionation based on, for example, anion-exchange and pH stepwise elution, this approach shows much lower protein overlap between fractions, and therefore, greater resolution. This results in a larger number of detectable species including low-abundance species. By the choice of orthogonal chemistries, their sequence, the loading volume amount, and by proper selection of the binding buffer, it is expected that the bound proteins will be quite different in each of the capturing blocks. With this chromatographic separation approach, redundancy (the same protein found in different fractions) is almost eliminated. Such an approach is expected to be further optimized and will certainly prove highly useful in analysis aimed at discovering low abundance components within a protein mixture. This is because the method generates a limited number of fractions and can be operated extremely rapidly even with a large number of samples. Furthermore, species that are relatively diluted are captured only by one column section and are consequently concentrated. The main steps in the stacked sorbents procedure are schematically represented in Figure 2.3.

2.1.3.4. Organic Solvent Extraction. Exploitation of differences in solubility is one of the oldest and easiest methods for the separation of various components in a chemical mixture. Precipitation of blood plasma proteins with ethanol was demonstrated almost 60 years ago (Cohn et al., 1946). A number of recent publications have used a similar approach to reduce the complexity of some biofluids before submitting them for SELDI analysis. Certov et al. (2005) have used acetonitrile containing 0.1% trifluoroacetic acid to extract peptides and low molecular weight proteins from serum under denaturing conditions. The authors also observed that the presence of some ion-pairing agents contributed to a more efficient dissociation of peptides and small proteins, which happen to be somewhat associated with large, abundant proteins. The same authors reported that the use of this extraction approach together with SELDI allowed the detection of some proteins, including apoliprotein A-II, which were down-regulated in mice with B cell lymphoma. The use of acetonitrile for the extraction of peptides and small proteins has also been reported in other recent studies (Alpert and Shukla, 2003; Merrell et al., 2004).

2.2. BIOINFORMATICS IN SELDI

Most MS-based strategies use bioinformatics to acquire calibrated m/z values, which with the help of certain computer algorithms in combination with various database searches can provide fairly reliable information on the identity of various proteins within a given mixture. Bioinformatics support to SELDI analysis has two minor differences from the above approach: First, it is not the objective of such analysis to

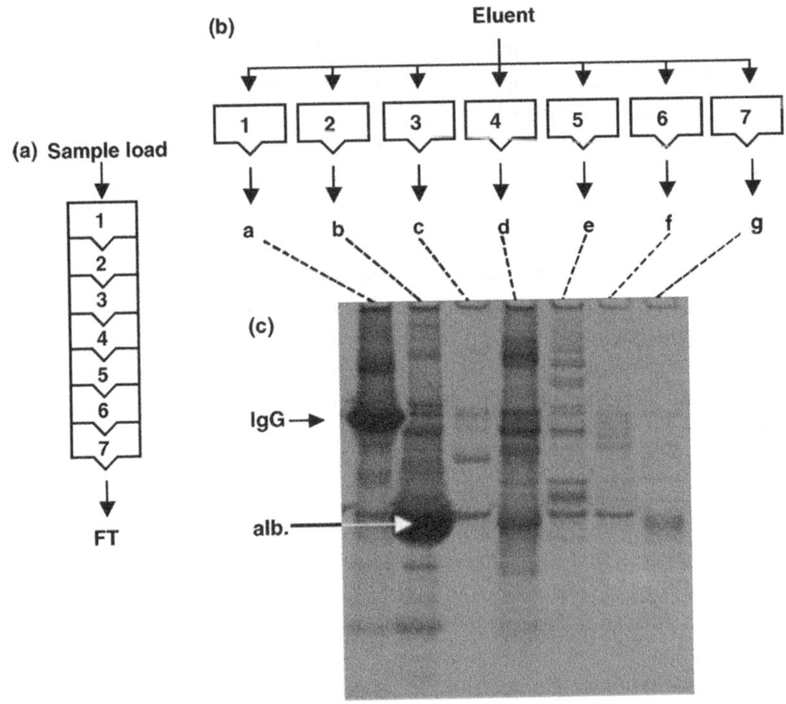

Figure 2.3. Scheme of fractionation using stacked columns. (a) Assembled stack of seven capture columns (from 1 to 7) filled with different sorbents. FT refers to the flow-through fraction containing species that are not captured. (b) Disassembled construct ready for elution of the captured proteins. (c) SDS-PAGE analysis of the seven eluates from single disassembled columns. Adsorption buffer was tris-phosphate (for all columns); elution was operated by means of either a mixture of trifluoroacetic acid–water–acetonitrileisopropanolor of ammonia–wateracetonitrile–isopropanol. The position of IgG (fraction 1) and of alb. (human serum albumin) (fraction 2) is highlighted. From Righetti et al. (2005) with permission.

identify the observed m/z values but to identify pattern(s) containing a number of these values, which can be used to differentiate two sets of measurements (e.g. control versus diseased). This can be implemented by using a fraction of the clinical samples as a "training set" to deduce the interpretation algorithm, and the remaining samples used as a "test set." Second, traditional and well-established MS-based strategies for protein identification rely on the use of MS-MS data, which renders this type of data almost free from interferences associated with the ionization source and/or impurities within the investigated sample. On the contrary, most SELDI measurements published so far use MS data, which encapsulate the complexity of clinical samples and undesired interferences by the ionization process itself. This means that the generated mass spectra can comprise tens of thousands of m/z values, hence analysis is rarely straightforward and usually requires substantial preprocessing before the data can be further analyzed by statistical or machine-learning methods. The last few years have witnessed an ever increasing number of bioinformatic tools designed to handle the complexity of the SELDI data (Ball et al., 2002; Qu et al., 2002; Yasui et al., 2003a, b; Mian et al., 2003; Sorace and Zhan, 2003). These tools have been given various names such as "Boosted decision tree" (Qu et al., 2002), "Artificial neural network algorithms" (Mian et al., 2003; Ball et al., 2002). Machine-learning techniques seek to semi automatically build and validate mathematical models of data that can then be used for classification or regression and for examining which part of the data were relevant and in what way. There are over 40 different machine-learning terms, which have been published so far (Listgarten and Emili, 2005). Regardless of the name of the tool, most if not all involve four main steps: first, positing a class of appropriate mathematical or statistical models, which are not known a priori; second, learning which particular model in the assigned class is most suitable for the data; third, validation of the model by use of a test set, cross validation, or similar method; and fourth the application of the final model to the new data. This seemingly straightforward procedure involves extensive number of steps such as, peak alignment in different spectra; the construction of the correct classifier; width of the window encompassing the diagnostic peak(s), and so forth. It is needless to state that each mathematical and/or statistical model/algorithm has its strong and weak points, and therefore, it would be wise to compare the output of more than one model before coming to final conclusions regarding the robustness of one model or another. This cautious and scientifically meaningful approach would of course go against the trend of developing and testing high throughput strategies for biomarkers discovery. Recent literature has already warned against coming to hasty conclusions regarding SELDI analysis (Sorace and Zhan, 2003; Diamandis, 2004; Baggerly et al., 2005), a warning, which in my modest opinion should be taken seriously. The above statement, which may sound too overcautious, can be supported by the following considerations: Qu et al. (2002) pointed out that one of the concerns in the construction and use of learning algorithms is the possibility of over-fitting the data. Furthermore, it is still too difficult to establish how robust these algorithms will be when used at different times, or on different sets of clinical samples. A representative example of this possible difficulty in the use of SELDI for discriminating renal cell carcinoma from controls has been pointed out by Rogers et al. (2003). Proteomic profiling of urinary proteins in renal cancer was investigated by SELDI. Initial data resulted in sensitivity and specificity values of 98–100%. Repeating similar measurements 10 months later resulted in sensitivity and specificity values in the range 41–76%.

2.3. SOME REPRESENTATIVE SELDI APPLICATIONS

Most of the applications described here refer to the use of SELDI approach in the search for potential markers associated with prostate and ovarian cancer. However, when appropriate application of the same technology to search for biomarkers for other forms of cancer will be cited. After surveying the literature for prostate and ovarian cancer diagnostics, I came across a number of works that use SELDI approach in the search for potential biomarkers associated with both forms of cancer. Most if not all of such investigations are based on the principle of comparing healthy and diseased samples to identify a proteomic pattern which can distinguish the two sets of analyses. Early SELDI investigations (Petricoin III, et al., 2002; Qu et al., 2002) related to prostate and ovarian cancers had a number of shortcomings, which latter studies have attempted to address (Zhang et al., 2004 ; John- Semmes et al., 2005).

The frequency of publications describing the use of SELDI in the search for potential biomarkers makes it prohibitive to include all of them in a single table. I have chosen a number of such publications, which give a reasonable indication of the potential of the technology, its advantages and drawbacks in the search for disease biomarkers. Although the publications listed in Table 2.1 cover a relatively short period of time; it is not difficult to note that within such a brief period of time, there have been serious attempts to improve the performance of this relatively young analytical technique. These relatively early investigations used different chip types and different discriminating m/z values, which rendered the evaluation of reproducibility, and the assignment of biological significance of the generated numbers by other research groups rather problematic. Before describing more recent SELDI investigations, which have addressed some of the drawbacks of earlier studies it is worth considering some comments related to those earlier investigations. Such comments by various research groups have underlined a number of drawbacks which latter investigations have attempted to address. Some investigations listed in Table 2.1 have generated considerable controversy concerning the suitability of SELDI as a profiling approach, most notably by Diamandis (2004). Other concerns regarding this approach have been raised by other research scientists (Sorace and Zhan, 2003; Baggerly et al., 2004). Without going into too many details, the following points give a fair summary of the main concerns:

(i) Diamandis (2003, 2004) doubted whether SELDI in its current state is capable of capturing putative proteomic changes provoked by small and localized tumors. The author argued that modification of the serum will be too insignificant to be captured by this approach. The same author emphasized that such serum modifications will require ultra sensitive techniques capable of measuring concentrations in the range of 10^{-12} mol/L, which is far lower than the current SELDI capability. The bias of the SELDI approach toward high-abundance molecules within the serum was another concern raised by the same author. He argued that the capturing surfaces in current use are not specific enough for any type of protein. Given that the serum contains an extensive array of high abundance molecules (e.g. albumin) together with very low abundance molecules, it will be highly unlikely that such highly informative, low abundance molecules will be captured efficiently. For example, in serum, the PSA concentration in healthy males is ~1 μg/L whereas the total protein concentration is in the order of 8×10^6 μg/L. When proteins are exposed to the chip, each PSA molecule (or other molecules of similar abundance) will encounter competition for binding to the same surface by millions of irrelevant molecules. It would thus seem very unlikely that molecules with very low-abundance will be detected by this method. Failure to identify well-established cancer biomarkers, reluctance to identify the discriminatory peaks, and disagreement between peaks generated by different research labs were also pointed out as some of SELDI limitations.

TABLE 2.1. References, capture surfaces, and brief description of a number of SELDI-MS investigations related to some forms of cancer.

References	Capture surface	Description
Petricoin III, et al. (2002)	Hydrophobic C16.	This was one of the early investigations to use SELDI-MS to detect serum proteomic patterns associated with prostate cancer. Mass spectra covering m/z range, 0–20,000 derived from prostate cancer and benign conditions were compared. The authors claimed that within m/z window from 2750–4500 there were 7 discriminatory peaks.
Adam, et al. (2002)	IMAC-CU. (Immobilized metal affinity capture-copper array).	SELDI-MS in combination with an artificial intelligence learning algorithm was used to differentiate prostate cancer from noncancer cohorts. Serum from 167 patients, 77 with benign prostate hyperplasia, and 82 age-matched unaffected healthy subjects were used to train and develop a decision tree classification algorithm that used a nine protein m/z pattern to classify the investigated samples. The authors claimed 97% specificity at 83% sensitivity.
Banez et al. (2003)	IMAC3-CU, WCX2 (weak cation exchange array).	SELDI combined with decision tree algorithm classification were used to analyse samples from prostate cancer patients and healthy subjects. A total of 106 patients and 56 controls were randomly allocated to a training set and a test set. The training set was used to build a decision tree algorithm, while the test set was used in a blind fashion to validate the decision tree.
PetricoinIII, et al. (2002)	Hydrophobic C16.	SELDI mass spectra were generated from 50 unaffected women and 50 ovarian cancer patients. The acquired spectra were analyzed by iterative searching algorithm, which according to the authors identified a proteomic pattern, which completely discriminated cancer from noncancer. This pattern was then used to classify an independent set of 116 masked serum samples. The authors reported 100% sensitivity, 95% specificity, and 94% positive predictive value.
John-Semmes et al. (2005)	IMAC3-CU.	Assessment of SELDI reproducibility in the analysis of prostate cancer. This was carried out by measuring mass accuracy, resolution, signal-to noise ratio, and normalized intensity of three m/z values present in standard pooled serum sample. The assessment of these parameters was carried out on a number of instruments within a single laboratory and compared with results generated across various laboratories.

(continued)

TABLE 2.1. (*Continued*)

References	Capture surface	Description
Zhang et al. (2004).	IMAC3-CU, SAX2 (strong cation exchange), H50, WCX2.	This study involved over 500 women including patients with invasive epithelial ovarian cancer, benign pelvic masses, other forms of ovarian cancers, as well as a number of healthy women. Serum proteomic expressions were analyzed in a five centre case-control study, where data from patients with early stage ovarian cancer and healthy women at two centres were analyzed independently and the results cross-validated to look for potential biomarkers. These results were validated using the samples from the remaining centres. This was one of the few SELDI studies to identify a number of proteins considered by the authors to be potential biomarkers for early stage ovarian cancer. These identified proteins were used in a multivariate model, which included the well established CA-125 biomarker to assess the sensitivity and specificity of this combination.
Grizzle et al. (2004)	IMAC3-CU	This article describes The Early Detection Research Network (EDRN) validation study for SELDI analysis of prostate cancer. In a three-stage study, the probability and reproducibility of the technique was assessed. A predictive algorithm was refined in a multi-institutional case-control population; followed by ultimate validation in the context of a prospective trial with the aim of complete disease ascertainment. An interesting aspect of this study is the use of two groups of cancer cases: high-risk group (Gleason score ≥ 7) and low-risk group (Gleason score ≤ 6).
Malyarenko et al. (2005).	Blank, IMAC-CU, Hydrophobic normal phase (NP20)	In this study the authors used three types of capture surfaces, blank, hydrophobic and IMAC-CU incubated with standard peptide mixtures or pooled serum. Spectra were recorded after one or more laser shots at different positions on the surface. Newly developed algorithms were applied to the SELDI data, which according to the authors compensated for electronic noise, removed baseline caused by charge accumulation, detected and corrected peak jitter, enhanced signal at high masses, and improved resolution through the use of a deconvolution filter.

TABLE 2.1. (*Continued*)

References	Capture surface	Description
Le et al. (2005)	Reverse phase (H4), IMAC	The use of SELDI for protein profiling of sera from prostate cancer patients with and without bone metastases. The authors reported the identification of a cluster of proteins in sera of patients with bone metastases, which was identified as isoforms of serum amyloid A.
Kozak et al. (2003)	SAX2	Sera derived from ovarian cancer patients, benign tumor patients, and healthy donors were examined by SELDI in combination with univariate and multivariate statistical models. The authors reported the discovery of three protein panels, that when used together could distinguish healthy controls from patients with either benign or malignant ovarian neoplasia.
Diamandis (2004)	None	This review gives a fairly comprehensive and informative evaluation of a number of exiting SELDI investigations associated with prostate and ovarian cancers. The author did not limit himself to pointing out the merits and limitations of this technology, but he proposed possible solutions for a number of limitations.
Baggerly et al. (2005)	None	Evaluation of reproducibility of published SELDI analysis of ovarian cancer. This is an attempt to draw attention to possible bioinformatics artefacts in some published SELDI analysis. The authors examined the reproducibility of the proteomic patterns across data sets. The authors concluded that patterns, which have been used for ovarian cancer classification, were biologically implausible. The same authors added that noise peaks within such studies could achieve perfect classification of normals and patients.
Yip et al.(2005)	IMAC-CU(II), weak cation exchange (CM10). SARS and control sera were first fractionated using anion-exchange Q-hyper D ceramic resin.	The use of SELDI for serum profiling in patients suffering from severe acute respiratory syndrome (SARS). The authors reported nine significantly increased and three significantly decreased peaks in the sera of the SARS patients. One of these peaks was further examined with MS/MS and identified as serum amyloid A.

(*continued*)

TABLE 2.1. (*Continued*)

References	Capture surface	Description
Rogers et al. (2003)	WCX2	The authors examined the robustness and clinical utility of the combination of neural network modelling and SELDI profiling using urine samples from patients with renal cancer, initially examining the ability of such models to discriminate between normal and malignant disease, and also the potential to differentiate from various benign urological conditions. The authors draw the attention to some experimental results which may manifest in this type of analysis. The authors observed that the sensitivity/specificity of a test for discriminating renal cell carcinoma from controls was initially 98–100%. However when the same procedure was used 10 months later in a new set of patients, the sensitivity dropped to 41%. This dramatic drop in performance was tentatively attributed to sample stability, laser parameters, or chip variability.

(ii) The data published by Petricoin et al. (2002) on the use of proteomic patterns in serum to identify ovarian cancer received particular attention by various research groups. A number of reservations were related to possible bioinformatic artefacts during the phase of data interpretation. For example, it has been suggested that m/z values below 2000 may derive from matrix and/or background and chemical noise (Banez et al., 2003). It is worth recording that two out of five discriminatory peaks reported by Petricoin III et al. (2002) were below 1000 m/z. In a recent analysis of the original raw mass spectrometric data on ovarian cancer, Sorace and Zhan (2003) identified a number of peaks that could be used to discriminate between normal and cancer patients. According to these authors such discriminating peaks had no biological sense. Similar conclusions were reached by Baggerly et al. (2004) who have also shown that background peaks can reach a high level of discrimination between normal and cancer patients. Concerns regarding the influence of bioinformatic tools on the choice of the discriminatory peaks have been also raised regarding prostate cancer SELDI analysis. The two works in Table 2.1 (Adam et al., 2002; Qu et al., 2002) were published by the same research group but were examined by two different bioinformatic approaches. Although the same patients were analyzed, which generated a single set of data, the analysis of this single set by two different bioinformatic tools resulted in two different sets of discriminating peaks.

2.3.1. Addressing Reproducibility in SELDI Analysis

Most recent analyses of SELDI data on ovarian, prostate, and other forms of cancer leaves no doubt that one of the main drawbacks of this approach is the difficulty to reproduce the discriminating peaks by different research groups. Of course, the question of reproducibility is not limited to this relatively young technique. Older and well engrained techniques have gone through a similar phase. The most obvious example is two-dimensional gel electrophoresis, which was introduced almost 30 years ago (O'Farrel, 1975). The question of reproducibility remained unresolved until the introduction of commercial immobilized pH gradients (Bjellqvist et al., 1982), which had a decisive influence on both the reliability and reproducibility of this technique. Even this major development was not sufficient on its own to allow full reproducible analyses; however, more recently further success in this direction has been achieved, where the comparison of two samples can be conducted on a single rather than on multiple gels. This version of gel analyses is referred to as differential in gel electrophoresis (DIGE), first introduced by Ünlü et al. (1997). I am not saying that we have to wait for another 20 years before having reproducible SELDI analyses, though what I am sure of is that further efforts are needed to address this central aspect of SELDI analyses. It is encouraging to note that over the last 2 years or so, a number of SELDI studies have been reported, where the question of reproducibility represented a central theme in their design. To support the above statement, it is worth considering some recent examples involving ovarian, prostate, and other forms of cancer. In a recent study by Zhang et al. (2004), the authors have used SELDI analysis to look for potential markers associated with early stage ovarian cancer. This study involved over 500 women including patients with invasive epithelial ovarian cancer, benign pelvic masses, other forms of ovarian cancers, as well as a number of healthy women. Serum proteomic expressions were analyzed in a five-centre case-control study, where data from patients with early stage ovarian cancer and healthy women at two centres were analyzed independently, and the results cross-validated to look for potential biomarkers. These results were validated using the samples from the remaining centres. At this point it is worth asking how this study differs from earlier ones. This question can be partially answered by the following considerations: First, this study was conducted in five different centres, where the acquired data could be cross-validated; such a step adds confidence to such data and allows reasonable assessment of reproducibility. Second, unlike earlier SELDI studies, the authors have identified a number of proteins, which they maintained, were relevant to their objective of discovering potential biomarkers. These proteins included apoliprotein A1, a truncated form of transthyretin, and a cleavage fragment of inter-α-trypsin inhibitor heavy chain H4. Identification of at least some of the proteins in SELDI analysis has been always one of the points of contention in this type of analyses (Diamandis, 2004). Such identification is an important element in the assessment of the biological significance of such proteins and can be also relevant if such measurements are to be assessed by different research groups who might wish to use different technology. Third, the use of the three proteins together with a well-established marker, CA-125 in a multivariate model is another new element, which addresses the concern that earlier SELDI measurements did not take well-established cancer biomarkers into account. SELDI reproducibility was also recently assessed by John-Semmes et al. (2005). This report described the first stage of a National Cancer Institute/Early Detection Research Network-Sponsored multi-institutional evaluation and validation of SELDI for the detection of prostate cancer. In such study, two sequential experimental phases were conducted to establish interlaboratory calibration and standardization of SELDI instrumental and assay platform output. One of the initial steps was to ensure that mass accuracy, resolution, signal-to-noise ratio, and normalized intensity for three m/z values present in pooled serum sample could be standardized in all six sites involved in the study. A set of SELDI spectra generated at the indicated centres is given in Figure 2.4.

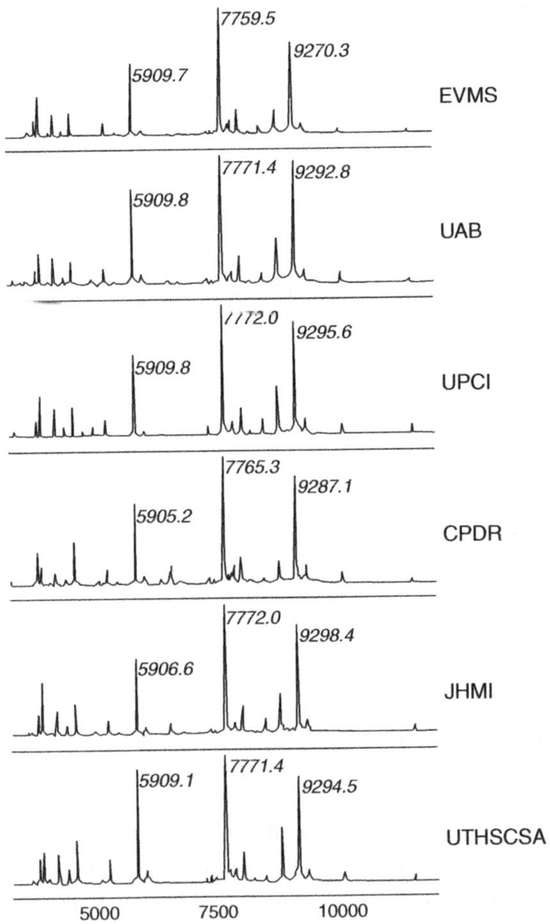

Figure 2.4. SELDI-TOF mass spectra obtained using IMAC3-Cu surface for the quality control serum from each validation laboratory site. The spectra were processed with baseline subtraction and normalization. The three peaks used for instrument synchronization are labeled. EVMS, Eastern Viginia Medical School, UAB, University of Alabama at Birmingham; UPCI, University of Pittsburgh Cancer Institute; CPDR, Walter Reed Army Medical Center, Center for Prostate Disease Research; JHMI, Johns Hopkins Medical Institutions; UTHSCSA, University of Texas Health Science Center at San Antonio. From John-Semmes et al. (2005) with permission.

Although this study has not been completed yet, the data generated in phase I are sufficient to make a number of comments: First, this is one of the few SELDI studies to set a clear guideline regarding instrumentation and sample handling in six different sites. Such guidelines can be considered a step forward toward addressing the question of reproducibility and reduction of artefacts, which in the early studies were responsible for scarce reproducibility, even within the same laboratory. Second, the same three diagnostic peaks were identified at multiple sites and were sufficient to differentiate case/control samples at all sites. These results can be considered as the first step to address the question of inconsistent discriminating peaks, which is commonly cited as one of the weaknesses of the SELDI approach. The authors, however, did not reveal the identities of m/z values within the diagnostic proteomic pattern because, according to them, phase II was required to remain blind to the identities of such peaks. This means that the readers cannot evaluate the biological sense of such peaks. This is of course a temporary drawback of such a study, which is likely to be addressed once the study is completed.

2.3.2. Limitations and Other Open Questions Regarding Current SELDI

Most of the works listed in Table 2.1 and the references included seem to agree that SELDI technology has the potential to become one of the tools for the discovery and identification of new biomarkers for various forms of cancer. The same works raise a number of concerns and list various limitations of the technique, which have to be resolved before such technology can deliver on its promise. Some of these limitations are either intrinsic to the technology or they can be artefacts, which manifest in the various phases of sample preparation, data acquisition, and elaboration. This statement can be better comprehended by the following considerations.

- The first intrinsic limitation is associated with the ionization method, which is a central component in SELDI analysis. Assessment of up- or downregulation of a given set of m/z values in a SELDI spectrum is one of the main elements of such an analysis. Such assessment in effect is a form of quantification, which has a number of inherent limitations. Setting aside the limited dynamic range of MALDI ionization, there are other elements, which can influence the reliability of such an assessment. For instance, let us assume two likely scenarios in SELDI analyses: First, the analyzed chip contains a simple homogenous population of proteins/peptides, and the second and more likely scenario is that the same chip can capture a fairly complex and heterogeneous population. Regarding the first situation, it is reasonable to expect similar ionization efficiency of the components captured by the chip. Having said that, existing literature contradicts such an expectation. It has been known for a number of years that MALDI ionization is biased toward species that contain argenine residues over those that contain lysine (Krause et al., 1999). The same study reported that arginine-containing peptides were detected with 4–18-fold more intensity than lysine-containing peptides. The increase in intensity is commonly attributed to a higher condensed-phase basicity and/or gas-phase basicity of the arginine-containing peptides. Existing studies in proteome analyses leave no doubt that the ionization efficiency of a given peptide within a tryptic mixture has a direct impact on the quantification of this peptide. Such information has been exploited by Hale et al. (2000), who proposed a derivatization procedure to facilitate de novo sequencing of lysine-terminated tryptic peptides through the guanidination of the ε-amino group. This reaction converts lysine residues into the more basic homoarginine without modification of the amine terminus of the same residue. This means that the presence of two informative molecules within a sample, which happen to differ by a single amino acid (e.g. Lys instead of Arg), will result in two different intensities and consequently misleading assessment due to different ionization efficiency rather than a biological effect. In the second scenario, where the chip contains a complex mixture of various components having a wide range of concentration, an informative molecule of a low concentration has to compete with numerous components, which may have a higher concentration and possibly higher proton affinity. This means that the ionization efficiency of such an informative molecule is likely to suffer and will no doubt influence its eventual quantification.

- Another limitation which is hardly mentioned in existing SELDI literature is the capability of this technique to detect and possibly quantify protein posttranslational modifications. Although the identification of proteins in complex mixtures is becoming routine, such identification alone provides only a limited insight into protein functions and signaling pathways. Covalently modified proteins provide an important component of protein regulation and protein functions. These modifications (co- or posttranslational) cannot be obtained from protein sequences deduced from nucleotide sequences. Over 200 different modifications have been described (Krishna and Wold, 1993). Many of these modifications such as phosphorylation and glycolysation, have well documented roles in signal transduction, regulation of cellular processes, clinical biomarkers, and therapeutic targets. These and other protein modifications continue to challenge well-established analytical techniques and will certainly represent a particular challenge for SELDI analyses. The question of protein modifications is discussed later in the text, however, for the sake of the present discussion, I would like to refer briefly to two representative examples of such modifications: First, protein glycolysation has long been recognized as one of the central biochemical alterations associated with various forms of disease. From the analytical point of view, glycolysation is one of the most challenging tasks, simply because carbohydrates are the most abundant and structurally diverse compounds found in nature. Unlike the linear polymers such as proteins and nucleic acids, oligo and polymeric carbohydrates can form branched structures because the linkage of the constituent monosaccharides can occur at a number of positions. It has been calculated (Laine, 1994) that for a simple hexasaccharide, there are excess of 1.05×10^{12} possible isomeric structures. Luckily, the highly specific biosynthetic pathways, which are limited by the available glycosyltransferases, render the above number far less prohibiting. Typically, carbohydrates are linked to the side chains of serine or threonine residues (O-linked glycosylation) or of asparagine residues (N-linked glycosylation). Despite the fact that glycoproteins represent an attractive target in the search for biomarkers and new therapeutic targets, an indication of the difficulties associated with the characterization and quantification of glycoproteins is reflected in the limited number (only 172) of experimentally confirmed human glycoproteins listed in the current Protein Information Resources Protein Sequence Database (http://pi.georgetown.edu/pirwww/search/textpsd.shtml) (Kaji et al., 2003).

Second, the phenomenon of protein ubiquinitation has been known for almost 30 years (Goldknopf et al., 1975). The involvement of ubiquinitation in processes as diverse as cell cycle regulation, DNA repair and receptor mediated endocytosis demonstrate the biological significance of such phenomenon (Ciechanover et al., 2000; Finley, 2001). Although such biological significance is perceived, its study however remains more challenging than other forms of modifications, and its study is inherently difficult because the modification is large (~8 kDa) and the turnover of ubiquitinated proteins is very rapid, so that steady-state conjugate levels are characteristically low. In two reports, Marotti et al. (2002) and Peng et al. (2003) have described the use of mass spectrometry for the identification of protein ubiquitination sites. The work by Marotti et al. (2002) was the first example of the use of mass spectrometry for the direct identification of an in vivo ubiquinitation site. The same work was a rare example of a G protein-signaling component that undergoes ubiquinitation. Peng et al. (2003) have described a proteomics approach to enrich, recover, and identify ubiquitin conjugates from *Saccharomyces cerevisiae* lysate. This method exploits a characteristic of trypsin proteolysis of ubiquitine-conjugated proteins, which produces a signature peptide at the ubiquitination site containing glycine–glycine residues derived from the C-terminus of ubiquitin and that is still covalently attached to the target lysine residue. In the same study, ubiquitin conjugates from a strain expressing 6 × His-tagged ubiquitin were isolated, proteolized with trypsin, and analyzed by multidimensional liquid chromatography coupled to tandem mass spectrometry. The authors claimed the identification of 1075 proteins containing 110 specific ubiqutination sites present in 72 ubiquitin-protein conjugates. Furthermore, ubiquitin itself was found to be modified at seven lysine residues. The authors observed that the detected conjugates reported in their study could only represent a subset of the total ubiquitine conjugates; according to the authors such partial identification can be attributed to two possible causes: First, mass spectrometry is a concentration dependent detection method and therefore bias toward more abundant conjugates. Second, fast protein degradation following ubiquinitation resulted in the absence of known, short-lived regulators of the cell cycle. One way to facilitate the detection of such fast degrading proteins is to use some form of chemical or genetic stabilization, neither of which was applied in the study by Peng et al. (2003).

Some may argue that to expect SELDI in its current state to deliver quantitative information on post-translational modifications is like asking the technique to run before it can walk. In my opinion though, if SELDI technology aspires to become a method of choice for the profiling of serum proteins, then detecting both modified and nonmodified proteins should be within the type of information expected by the user. To be able to furnish this type of information, SELDI analysis has to be conducted on mass spectrometers, which can perform MS-MS measurements. Furthermore, extensive efforts have to be dedicated to derivatization and tagging chemistry, and new technical developments to allow the interrogation of functional protein microarrays by SELDI-MS.

2.3.3. Other Open Questions

In a recent article (Diamandis, 2004), the author pointed out a number of aspects, which have to be addressed before SELDI technology can deliver on its promise. Some of these questions have been already mentioned; however, I find it useful to underline some of these questions before expressing my own opinion regarding likely solutions for some of them: (a) Disagreement between different research groups regarding the discriminating m/z values associated with the same disease. (b) SELDI technology so far has not demonstrated its capability to detect well-established serum cancer markers such as PSA and CA-125. (c) Absence of highly specific capture surfaces means the failure to capture informative and generally low-abundance molecules within the investigated sera. (d) Sample preparation and handling differs from one group to another, which renders reproducibility assessment a difficult task. (f) Scarce correlation exists between discriminatory molecules and known cancer biology. (e) Mass spectrometry is a poor quantitative technique rendering the correlation between measured peak height and actual concentration within the sample rather ambiguous.

A closer look at the above points reveals that some of them are intrinsic to the technology, whereas others are because of the implementation and experimental design. Most of us would agree that intrinsic limitations are difficult to address and generally require a drastic action such as replacing the technology or giving it a radical face-lifting. Luckily, only point (e) can be considered a truly intrinsic limitation, which is associated with the mass spectrometry side of the technology. Considering that no one including myself would dare to eliminate mass spectrometry from SELDI technology, then let us look deeper at point (e) and see whether such limitation is indeed fully intrinsic. It is worth reminding ourselves that the mass spectrometry component within a SELDI setup encompasses sensitivity, mass resolution, and sequencing information. If the analysis is conducted on a low performance mass spectrometer, then we have to expect rather disappointing results. It has to be said that most existing SELDI measurements have been acquired on low-performance mass spectrometers, which may explain why sensitivity and mass accuracy are commonly cited among the limitations of this approach. I am more than convinced that performing the same measurements on ICR mass spectrometer or on a hybrid mass spectrometer (e.g. TOF–TOF; linear ion trap-TOF) will present a different picture. Having said that, we have to bear in mind that in a SELDI setup, the MS component cannot be evaluated in isolation from the sample analyzed and the bioinformatics. The author is well aware that current MS sensitivity cannot compete with methods based on radioisotope labeling and multiphoton detection, yet there is a limited number of reports, which demonstrated that TOF-MS equipped with crygenic detector could detect a single protein ion (Twerenbold et al., 1996, 2001). In another report (Belov et al., 2000 a,b), the authors used FT-ICR with enhanced ion transmission facility to investigate proteins at low zeptomol level. At this point, one is entitled to ask why such high sensitivity mass spectrometers are not used in current SELDI analysis. The answer to this question can be simply attributed to the following: First, mass spectrometers equipped with cryogenic detectors are not commercially available. Second, the relatively high cost (compared to a simple TOF) and the high technical skills needed to maximize their output rendered FT-ICR unpopular for SELDI analyses.

2.3.4. Outlook

Over the last 5 years, SELDI technology has demonstrated part of its potential as a high-throughput analytical tool for protein profiling of sera and other biofluids. Given the scientific as well as the social impact of such technology in the area of biomarkers discovery, it was bound to attract both enthusiasm and scepticism. Most recent literature regarding this relatively young approach suggests that regardless of which side you are on, there is a common agreement that for this technology to deliver on its promise it has to address a number of current limitations. Addressing such limitations not only will answer the concerns of those who believe in this technology, but above all will convince those who have strong doubts about the validity of this approach. Some of these limitations have been pointed out in the previous sections; however, I find it relevant to underline some of these limitations, which will no doubt influence the future direction of the technology. It is also reasonable to suggest that the outlook for this technology is not only related to overcoming some of its current limitations, but also in further optimization of what we call today the successful facets of this technology.

- It is becoming more evident that the use of m/z patterns generated on low-performance mass spectrometers is not reliable enough to distinguish between cancer and noncancer patients. Generating these patterns on high-resolution and high-sensitivity mass spectrometers not only will reduce interference by the matrix ions, but will also enhance the reproducibility of such patterns and the eventual identification of protein/peptide associated with the individual m/z values within such patterns. Accurate mass spectra on their own are of little use unless they are accompanied by the identification of the species associated with the discriminating peaks. It would be highly desirable if such identification is carried out by different technologies, for example, those based on immunochemistry.
- Capture surfaces (chips) available at the time of writing this text do not have the binding capacity or specificity to capture specific protein subpopulation, which happen to have widely differing abundances. This means that prefractionation/enrichment prior to sample deposition on the SELDI chips is rather mandatory. A number of strategies for prefractionation have been described in Section 2.1.3. of this chapter. In my opinion, a more frequent use of such methods prior to SELDI analyses will no doubt facilitate the detection of low-abundance and possibly more informative species buried within a complex medium.
- There is a wide range of commercial SELDI chips, however none of these chips gives an indication of the possible number of charges on such surface, which can react with their opposite charges within the deposited sample. In other words, if a user can estimate the number of amino acids and hence the number of expected charges

within the the sample, then it is relevant to know if all or part of such charges will be captured by the surface. This information may sound trivial, yet they can provide a clear guidline on the requirements for prefractionation and/or enrichment of the sample before its deposition. A crude method to get this kind of information is to use a standard solution of a known protein and perform SELDI analyses at different molarity followed by data analyses to establish the range of molarity at which the signal of a set of m/z values derived from the investigated protein reaches a plateau. I am sure that the manufacturer of these chips may devise a more sophisticated approach to give the future user of these chips some indication on the number of charges, which can be captured by various surface types.

- Another central component within the SELDI strategy is the bioinformatics support, which needs to be further improved in terms of reproducibility, particularly when such tools are used to interpret data related to the same disease but are generated from different samples and even by different research groups. In most published SELDI works, fraction of the clinical samples are used as a "training set" to establish the interpretation algorithm, whereas the remaining samples are used as a "test set." Such approach has the risk of overfitting the data and even including peaks originating from the matrix as being discriminating peaks (Diamandis, 2004; Qu et al., 2002).
- Future SELDI analyses will certainly benefit from the HUPO initiatives relative to clinical proteomics (Hanash, 2004). These initiatives include three major projects: Plasmaproteome Project (PPP), Liver Proteome Project (LPP), and Brain Proteome Project (BPP). The first proteome project to be implemented will be the PPP, which is currently in its piloting phase. The objectives of this project include the following: (i) comprehensive analysis of plasma protein constituents in normal humans in large cohorts of subjects; (ii) determination of the extent of variation in plasma proteins within populations in various countries and across various populations from around the world; and (iii) identification of biological sources of variation within individuals over time and assessment of the effect of age, sex, diet, and lifestyle, as well as common medications and common diseases. These objectives will no doubt impact on the way future SELDI analyses will be conducted. Adhering to the guidelines to achieve such objectives will indirectly address some of the points of contention regarding published SELDI analyses, particularly reproducibility, and identification of proteins within the discriminating proteomic patterns.

2.4. TWO-DIMENSIONAL POLYACRYLAMIDE GEL ELECTROPHORESIS

Since its introduction over 30 years ago (O'Farrell, 1975), two-dimensional polyacrylamide gel electrophoresis (2-D PAGE) has experienced three main developments, which rendered it a major player in today's efforts to find potential biomarkers for various pathologies: (a) the introduction of immobilized pH gradients gels (IPG), which eliminated gradient instability and poor sample loading capacity associated with carrier ampholyte pH gradients (Bjellqvist et al., 1982).

The availability of commercial precast IPG gels in a variety of narrow and broad pH ranges was a determining factor in rendering 2-D PAGE a reproducible high-resolution method for protein separation. (b) The second significant development was the introduction of differential in gel electrophoresis (DIGE), first described by Ünlü et al. (1997). This method relies on pre-electrophoretic labeling of the investigated samples with one of three CyDyes (C2, C3, and C5), which are mass- and charge-matched and possess distinct excitation and emission spectra, allowing multiplexing of samples on a single gel, which translates into shorter analyses times, higher reproducibility, and relatively less sample compared to conventional 2-D PAGE. (c) The third and possibly the most significant development is the increasing use of mass spectrometry and associated bioinformatics in the characterization of the separated proteins. Prior to the development of routine MS and MS/MS protein identification methods, many 2-D PAGE studies failed to identify the proteins that changed in their expression between samples. Several reports characterizing protein differences between normal and diseased prostate belong to this category (Grover and Resnick, 1995; 1997; Partin et al., 1993). Whether the comparison of protein expression in two samples is conducted on a single or on multiple gels, there are a number of basic steps and experimental conditions, which have to be implemented in both forms of analyses. Some of these steps and how they influence the final outcome of the analysis, limitations, and advantages of this approach, its current role in the search for potential cancer biomarkers are the main elements discussed in the following sections. To avoid unnecessary confusion in the use of terminology, I shall refer to the relatively older 2-D PAGE version as 2-DE, whereas the relatively more recent version as 2-D-DIGE.

2.4.1. Sample Preparation

Sample preparation prior to any electrophoretic separation has a major influence on the final outcome of 2-D maps. This step involves solubilization, denaturation, reduction, and alkylation. Such a procedure is designed to break the interactions between proteins, remove nonprotein sample components such as nucleic acids (Rabiloud, 1996) and prevent reformation of disulfide bridges during the separation process, which can produce smears and even spurious bands in the final 2-D map (Herbert et al., 2001). Ideally, one would achieve complete sample solibilization in a single step to avoid excessive handling and unnecessary sample losses. Current solubilzation mixtures are likely to contain 8 M urea, 2 M thiourea, 40 mM Tris, 3 mM tributyl phosphine or 50 mM DTT, 2% detergents, and 10–100 mM an alkylating agent (such as iodoacetamide, 2- or 4-vinylpyridine or acrylamide). The role of the individual components within the solibilization cocktail and how to optimize their combination has been described in a number of works (Rabilloud, 1996, Herbert et al., 1998, Hamdan and Righetti, 2005). Having said that, the question of the composition of the solubilization mixture and the sequence of the various steps in sample preparation is continuously evolving. It is sufficient to remember that many membrane proteins as well as proteins from tissues that are highly resistant to denaturation, still pose a challenge because of their poor solubility in existing solubilization mixtures. Although a detailed discussion of the question of sample preparation and protein solubility is not the subject of this text, I do think that the potential reader of this text would be interested to know about some caveats and possible artefacts, which can be avoided or even eliminated during the phase of sample preparation. To underline the importance of proper implementation of the sample preparation step prior to any electrophoretic step, and how it may influence the final two-dimensional map of a given analysis, it would be helpful to consider the influence of certain components within the solubilization mixture.

- The identity of the alkylating reagent, when and how to effect alkylation is a crucial step in sample preparation. Incomplete alkylation can result from an insufficient reaction time, the nature of the reagent and its possible scavenging by certain components within the solubilization medium. For example, Iodoacetamide (IAA), which is one of the most commonly used alkylating reagents has sulphur atom that is known to react with thiourea as efficiently as the ionized, free -SH groups of protein–bound Cys at alkaline pH values. As a result of this reaction, free IAA is quickly depleted by thiourea, via the formation of an intermediate adduct, which is rapidly deamidated to form the cyclic compound thiazolinidone monoamine (Galvani et al., 2001a). On the basis of MALDI-TOF-MS analysis, it became evident that such a reaction strongly competes with the direct addition of IAA onto the -SH group in proteins, resulting in a very poor alkylation of the same proteins. In fact, the same MS data have shown that the depletion of IAA in the solubilization medium took less than 5 min to complete. Although the use of acrylamide as an alternative reagent eliminated the scavenging problem, a second limitation has emerged, which showed that substantial alkylation required more than 6 h, a period that is longer than the times recommended by widely adapted protocols for sample preparation. The influence of the identity of the alkylating reagent on the efficiency of the reaction has been described in a number of articles (Glavani et al., 2001b; Hamdan and Righetti, 2002), which have demonstrated that the failure to achieve complete alkylation renders certain -SH groups vulnerable to attack by nonpolymerized chemicals within the gel and protein–protein interaction with the inevitable formation of aggregates. It has been also shown that poor alkylation may result in a considerable loss of spots particularly in the alkaline gel region, a phenomenon that could manifest when certain proteins, at their pH, regenerate their disulfide bridges with concomitant formation of macro-aggregates, which become entangled within the polyacrylamide gel fibers, thus quenching their transfer to the subsequent SDS-PAGE. The formation of aggregates in the alkaline pH region has been demonstrated by Herbert et al. (2001). These authors compared 2-D maps of human globin chains, with and without alkylation, prior to the first dimension and the resulting spots were examined by MALDI- TOF-MS. These results showed that the alkylated sample generated a two-dimensional map containing a single string of spots at 16 kDa associated with the monomers of these chains, whereas in the case of the nonalkylated sample a number of strings in the range 16–70 kDa were observed. The MALDI-MS analyses have revealed that these unexpected high molecular weight spots were simply homo- and heteroaggregates of α- and β-globin chains. If we are to consider that such simple polypeptides where one (α-) contains a single Cys residue, whereas the other (β-) contains two Cys residues, can generate such large aggregates, then we should take appropriate steps to prevent their formation, particularly when dealing with complex mixtures such as total cell lysates.

• Existing protocols for protein separation in two-dimensional gels include a 2% concentration of surfactants such as SDS (when interfacing the first with the second dimension or CHAPS, prior to the first dimension). These detergents contribute to protein solubilization and prevent hydrophobic interactions that can be caused by the exposure of protein hydrophobic domains induced by chaotropes (Rabilloud, 1996; Jones, 1999). The hydrophobic tails and the polar head groups of these detergents both play an important role in protein solubilization. The surfactant tail binds to hydrophobic residues to allow the dispersal of these domains into aqueous medium, and the polar head groups can disrupt ionic and hydrogen bonds to contribute to the overall dispersion. Galvani et al. (2001b) used MALDI-TOF-MS to probe the influence of SDS and CHAPS on the alkylation efficiency. The data in Figure 2.5 were obtained from three solutions that contained 50 pmol/ mL of reduced α-lactalbumin (LCA) that was incubated for 1 h with 260 mM iodoacetamide at pH 8.8 in the presence of 6 M urea, together with 2% SDS or 2% CHAPS. The kinetics of the complexation of LCA with eight IAA molecules (considering that the 8 Cys residues are the most likely sites of alkylation) (Bordini et al., 1999) shows that the presence of SDS has a substantial negative influence on the observed alkylation efficiency. This observation is not surprising, in view of a model proposed by Lundahl et al. (1992) and Ibel et al. (1994), which has been called the "protein-decorated, micelle model." On the basis of a small-angle neutron-scattering study, this model proposes that adjacent, protein-decorated, spherical SDS micelles are formed; the polypeptide chain is buried within such micelles, and only short segments (4–5 amino acids) are found outside these domains. In other words, owing to the high hydrophobicity of Cys, it is likely that such residues will be buried inside the SDS globules, thus greatly hampering their interaction with the alkylating agent.

2.4.2. Reducing Sample Complexity

Even in simple organisms, Proteins heterogeneity in terms of the ranges of their isoelectric points, molecular masses and abundances can challenge most powerful and high resolution techniques including 2-DE. It is sufficient to recall that in a standard human cell, the most abundant protein is often actin, which is present at $\sim 10^8$ molecules per cell compared with some cellular receptors or transcription factors, which are probably present at 100–1000 molecules per cell (Rabilloud, 2002). The situation can be even worse in sera, where albumin is at 40 mg/mL and cytokines at pg/mL levels. This means that enrichment or prefractionation strategies are needed if low-copy proteins are to be reached. Traditional biochemical separation methods still offer the possibility of reducing the complexity from thousands of proteins within a whole cell sample to hundreds of proteins in each subcellular fraction. The first step in these methods is to separate nuclei and unbroken cells from cytoplasmic organelles by differential sedimentation at low centrifugal force to obtain the postnuclear

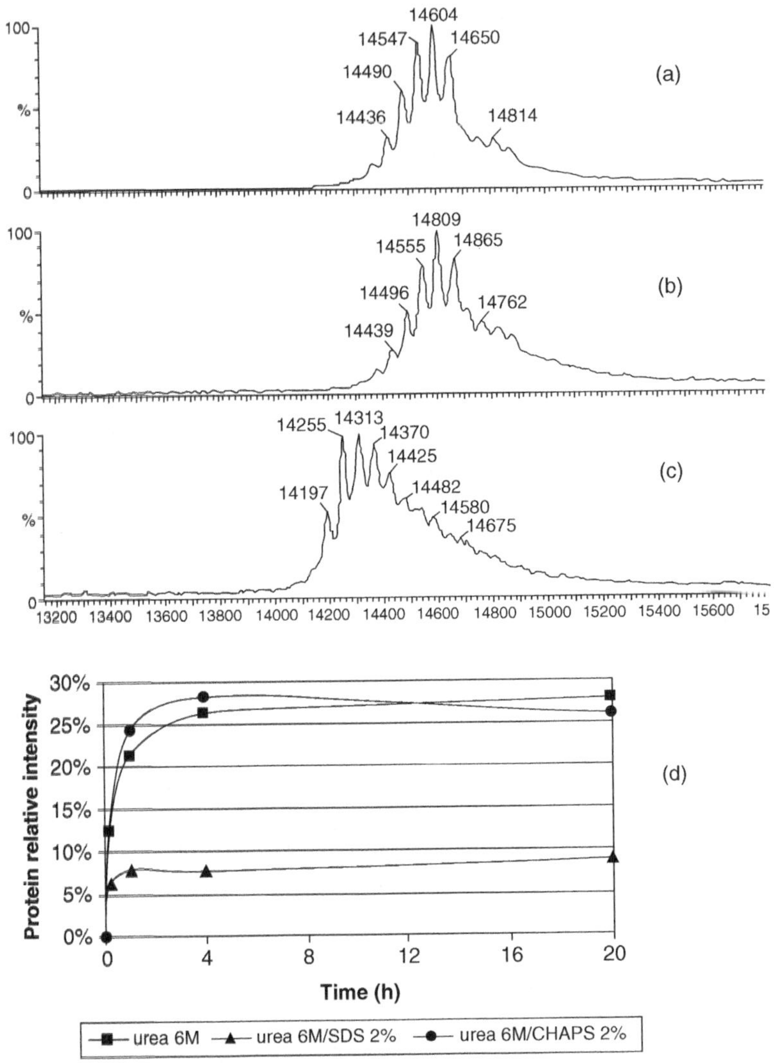

Figure 2.5. MALDI-TOF mass spectra of reduced α-lactalbumin incubated for 1 h with (a) 260 mM IAA and 6 M urea; (b) 6 M urea + 2% CHAPS, and (c) 6 M urea + 2% SDS; (d) kinetics of the reactionat pH 9 over 20 h reaction time. From Galvani et al. (2001b) with permission.

supernatant. This supernatant is then subjected to various density gradients to isolate specific organelles such as mitochondria or lysosomes (Huber et al., 1996).

Righetti et al. (1989) described a multicompartment electrolyser where each compartment is separated by a polyacrylamide gel membrane with a specific pH. This method relies on isoelectric membranes, fabricated with the same acrylic monomers adopted in IPG fractionations. This approach has a number of advantages:

(i) it offers a method that is fully compatible with the subsequent first-dimensional separation in 2-DE maps, a focusing step based on Immobiline technology. Thus, protein mixtures harvested from the various chambers of this apparatus can be loaded onto IPG strips without any need for further treatment, in that they are isoelectric and isoionic; (ii) it permits harvesting a population of proteins having pI-values precisely matching the pH gradient of both narrow and wide IPG strips; (iii) as a corollary of the previous point, much reduced chances of protein precipitation will occur. In fact, when an entire cell lysate is analyzed in a wide gradient, there are fewer risks of protein precipitation; on the contrary, when the same mixture is analyzed in a narrow gradient, massive precipitation of all nonisoelectric proteins could occur; (iv) owing to the fact that only proteins co-focusing in the same IPG interval will be present, much higher sample loads can be operative, permitting detection of low-abundance proteins.

Another electrokinetic apparatus, called the Gradiflow, has been described by Corthals et al. (1997). This is a multifunctional electrokinetic membrane apparatus that can be used to process and purify protein solutions based on differences in mobility, pH, and size. Its interfacing with 2-DE analysis was demonstrated by Corthals et al. (1997), who adapted this apparatus for prefractionation of native human serum and enrichment of protein fractions.

Protein tagging (labeling) can be another way to reduce sample complexity and to enhance the sensitivity of a subpopulation within a complex cell lysate. Surprisingly, recent literature reveals a paucity of attempts to incorporate such strategies in two-dimensional gel electrophoresis. A brief consideration of a limited number of such applications reveals a potential, which is yet to be realised. A representative example of such potential can be found in strategies, which rely on protein biotinylation. The high affinity and specificity of avidine-biotin interactions have been exploited in diverse applications including immunology, histochemistry, in-situ hybridization, affinity chromatography, and other areas (Wilbur et al., 1999; Wilchek and Bayer, 1999; Diamandis and Christopoulos, 1991; Bayer and Wilchek, 1990; Chapman-Smith and Cronan, 1999). Biotinylation reagents can provide the "tag" that renders poorly detectable proteins into probes, which can be recognized by a labeled detection reagent. The combination of biotinylation with two-dimensional gels has been demonstrated by Hewett (2001), where protein identification was largely based on the use of specific antibodies. In latter studies, a global surface protein biotinylation strategy coupled with mass spectrometry was used to examine protein profiles in *Helicobacter pylori* (Sabarth et al., 2002). A similar strategy was used by Shin et al. (2003) for global protein profiling of a variety of cancer cell types. The approach used by these authors consisted of the biotinylation of plasma membrane proteins of freshly isolated cells, primary cultures or cell lines, followed by their comprehensive profiling and identification. As all tagging strategies, biotinylation requires certain precautions to ensure its effectiveness. First, to selectively label only those lysine residues that are extracellular in orientation, nonmembrane permeable biotin reagents need to be utilized to prevent the entry of biotin into the cell. Sulfo-NHS-LC biotin is water-soluble; thus, it is not permeable across hydrophobic lipid bilayers and can be utilized for the selective labeling of surface membrane proteins (Hurely et al., 1985; Busch et al., 1989). Furthermore, to retain both biological activities and ligand-binding properties and to facilitate subsequent identification of the tagged proteins by mass spectrometry, it is necessary to avoid extensive biotinylation. Second, it is not always possible to use the most efficient reagents for solobulization because they tend to interfere with the capture and purification of the tagged proteins when captured on a monomeric avidin column. This means that some of biotinylated proteins will not adequately solubilize to be effectively captured and identified. Thus, further refinement of existing biotinylation strategies, such as by partial digestion of biotinylated proteins, which do not solubilize efficiently in the present solubilization mixtures, may further increase the repertoire of biotinylated proteins detected on the cell surface.

2.4.3. Various Nomenclatures In-gel Analysis

Two dimensional-gel analysis, as we know it today, was first introduced over 30 years ago and designated 2-D PAGE (O' Farrell, 1975). Current applications of this technique in proteomic analysis show that the basic strategy has remained unchanged, yet the same strategy has been given different names. These different nomenclatures are meant to distinguish labeling and detection approaches associated with the initial strategy. In the course of this chapter it will become clear that the central theme within these different nomenclatures is the comparison of protein profiles in two samples (e.g. control and disease) to detect and possibly quantify changes in such expression, which can be possibly related to the disease state. The following sections give a brief description of a number of these protocols as well as a section on Laser capture microdissection (LCM). The inclusion of the latter in the following part of this chapter is mainly motivated by the frequent use of LCM as part of the sample preparation for gel analysis of various forms of cancer.

2.4.3.1. Multiple-gels Two-dimensional Analyses. Prior to the introduction of what I would call two-dimensional analyses on a single gel (see below), most if not all 2-DE analysis of a single sample used to be conducted on a number of replicates. The generation of a number of replicates is designed to enhance confidence in spot reproducibility and to minimize possible artfactual spots. During the 1980s and 1990s, this approach has been used to investigate various forms of cancer including, brain (Hanash et al.,1985), thyroid (Lin et al.,1995), breast (Wirth et al.,1987; Giometti et al.,1997; Rasmussen et al.,1997), colon (Ward et al., 1990), kidney (Sarto et al., 1997), and bladder (Celis et al.,1996). More recent applications of the same approach are described within the following sections. The experimental design to compare protein profiles in control and pathological samples can be summarized as follows: A minimum number (4–5) of two-dimensional map replicates is generated for each sample. Following staining and optical density scanning the remaining work is handled by one of the commercially available statistical packages such as Melanie, PD Quest, Z3, Z4000, Phoretix, and Progenesis. These analyses generate two master two-dimensional maps, one for the normal and another for the pathological samples. Subsequent analysis by the same software would compare these two maps to

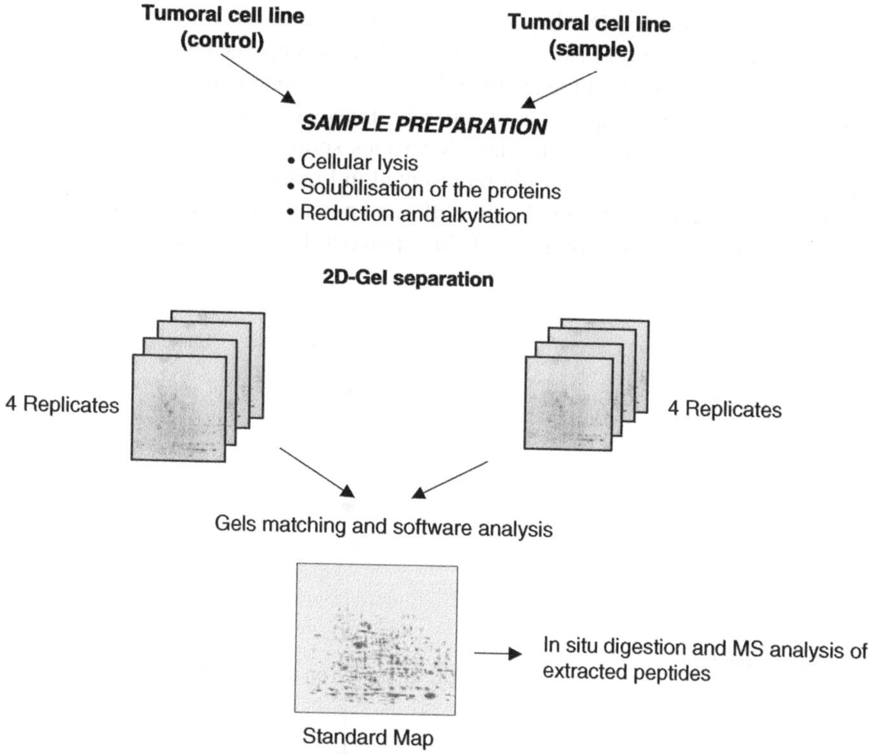

Figure 2.6. The main steps in comparing protein expression in two different samples by 2-DE in combination with mass spectrometry.

identify proteins which have experienced measurable up- or downregulation in their intensity. A schematic representation of the main steps in such protocol is given in Figure 2.6.

2.4.3.2. Two-dimensional DIGE Analysis. Protein labeling prior to electrophoretic separation is a central step in the DIGE (Differential- in gel-electrophoresis) protocol. Currently, there are two forms of labeling: *N*-hydroxy succinimidyl ester reagents for low-stochiometry labeling of the ε-amine groups of lysine side chains (minimal labeling), and maleimide reagents for labeling cystine sulfhydyls to saturation (saturation labeling) (Ünlü et al., 1997; Tonge et al., 2001; Lilley et al., 2002; Shaw et al., 2003; Wheeler et al., 2003). The most commonly used members of this CyDye family are designated C2, C3, and C5. In the minimal labeling procedure, care is taken to ensure that only 2–3% of the total number of lysine residues is labeled. Maintaining a low dye/protein ratio is necessary to avoid multiple dye additions, which would severely interfere with protein quantification, and result in heterogeneity in electrophoretic mobility. In the second labeling protocol, the CyDyes have an additional thiol-reactive maleimide group and carry no intrinsic charge. Regardless of the labeling chemistry, DIGE protocol for comparative protein expression consists of four main steps: (a) Prior to any electrophoretic step, the individual samples are labeled with one of the three cyanine dyes; (b) electrophoretic co-separation of the labeled samples; (c) acquisition of separate images for each sample on the gel; (d) software analysis of images to identify various spots and difference in protein abundances. These are schematically represented in Figure 2.7. The following brief comments are meant to highlight the strengths and potential limitations of the DIGE approach: First, one of the strengths of this approach lies in the fact that multiple

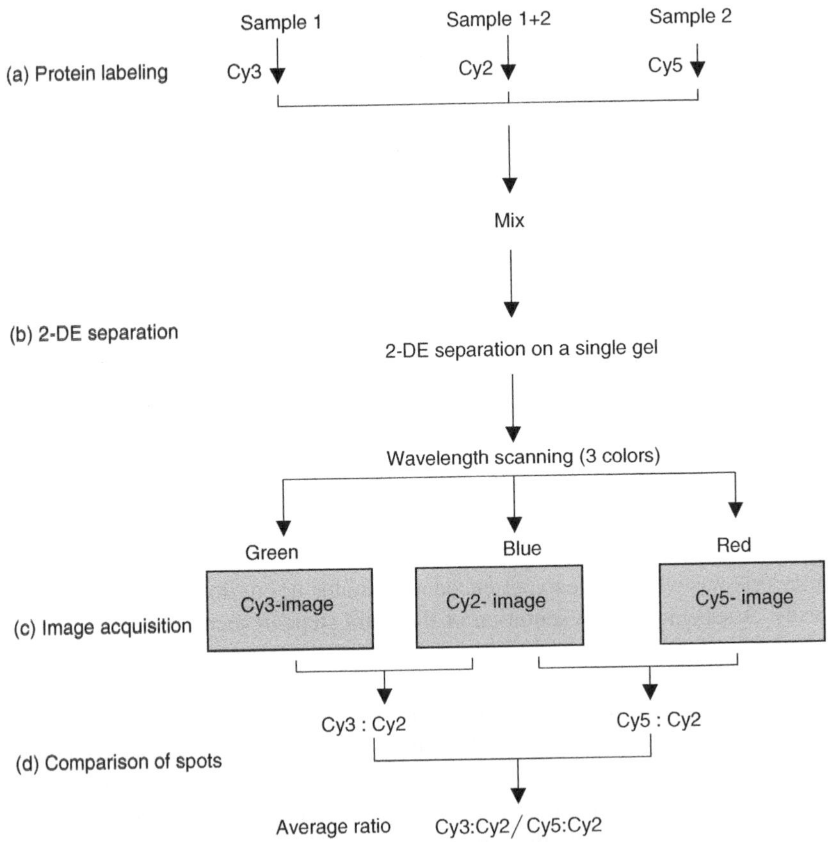

Figure 2.7. Schematic representation of the 2D-DIGE workflow for the differential analysis of protein abundance in two different samples: (a) one of the samples is labeled with Cy3, while the other with Cy5. Equal amounts of both samples are mixed and labeled with Cy2 to be used as an internal standard. (b)The three labeled samples are then mixed and run into the same gel. (c) For each of the three Cydyes, an image is acquired by successive scanning of the gel at specific excitation and emission wavelengths associated with the three fluorophores. (d) During intragel comparison (analysis of parallel gels), for each spot the ratios of normalized volumes between samples and internal standard are determined. Differences in protein abundance are then determined by calculating the average ratios (Cy3:Cy2):(Cy5:Cy2).

samples can be run on a single gel. This is rendered possible by the use of three spectroscopically different dyes allowing separate images pertaining to each sample to be detected, and thus eliminating gel-to-gel variations, enhancing reproducibility, and reducing the required sample. Second, although CyDyes have an excellent detection limit (~0.25–0.95 ng) and almost 4 orders of magnitude linear range, such benefits are severely hampered in the minimal labeling protocol, where more than 97% of the total protein remains unlabeled. This necessitates the use of a poststaining reagent to visualize such proteins. Sypro Ruby, which is commonly used for poststaining has a linear range comparable with CyDyes but a lower detection limit (~1–2 ng). Third, labeling with any of the three CyDyes will inevitably increase the molecular mass of the protein by approximately 500 Da (assuming a single addition). Under minimal labeling conditions, only less than 3% of a specific protein is labeled and this minor fluorescent population is generally shifted to a slightly higher molecular mass position. Therefore, the position of the bulk amount of unlabeled proteins would be shifted about one spot diameter down. This represents a problem for the subsequent spot excision to perform MS analysis. On the contrary, if minimal labeling is not strictly conducted, then serious effects are likely to manifest including poor solubility, and more seriously, extensive labeling would interfere with trypsin action on the modified lysines, thus resulting in missed cutting sites, which would translate into larger peptides and surely less efficient database search.

2.4.3.3. Multiphoton Detection Imaging. The measurement of radiolabled proteins by scintillation counting has long been one of the most reliable methods for accurate, quantitative measurements. This method has been applied within proteomioc research (Gygi and Aebersold, 1999; Gygi et al., 2000) and does offer gains in absolute sensitivity and dynamic range. Multiphoton detection (MPD) imaging of proteins separated by 2-DE is the latest addition to the terminology used under the umbrella of two-dimensional gel analysis. The principle on which this method is based can be easily compared with 2-D DIGE, the main difference lying in the labeling procedure, where MPD uses radioisotope labeling, while DIGE uses fluorescent dyes. The application of MPD for the quantification of gel-separated proteins has been recently demonstrated by Kleiner et al. (2005). Basically, sample A is labeled with ^{125}I and sample B with ^{131}I; the two samples are combined and run on the same gel. The resulting spots are then scanned by the MPD imager, which has two position-sensitive photomultiplier detectors coupled with a scintillating array. This type of analysis would yield three simultaneous images associated with ^{125}I, ^{131}I, and, an image representing the difference between these two images. This experimental setup was tested on acidic/phosphorylated and total proteins from 60 ng of HeLa cells proteins. On the basis of this analysis the authors made a number of claims, which merit further comments: first, MPD is linear over 6–7 orders of magnitude, which means it could be used to quantify radiolabeled proteins in the concentration range, zeptomoles-femtomoles. If this dynamic range can be reproduced in future studies, then it would address one of the main limitations of 2-DE because of its limited dynamic range (~10^4). Second, it is assumed that the labeling is optimal and only a single iodine atom per protein is introduced. If this is the case, then such a labeling scheme will not have detrimental effects on the electrophoretic behavior of the separated proteins. Even such an optimistic assumption is not without limitations. It is known that iodination changes the pK of tyrosine and, therefore, such a method will not be useful for highly basic proteins.

2.4.3.4. Stable-isotope Labeling with Amino Acids in Cell Culture (SILAC). Another emerging approach, which can be used either in 2-D gel analysis or in multidimensional chromatography is designated SILAC (Ong et al., 2002). These authors described the in vivo incorporation of specific amino acids into the mammalian proteins. Basically, mammalian cell lines are grown in a medium lacking a standard yet essential amino acid but supplemented with a nonradioactive, isotopically labeled form of that amino acid. On the basis of cell morphology, doubling time, and the ability to differentiate, the authors claimed that when using deuterated lucine (Leu-d3), the growth of cells was not different from those cells which were grown in a normal medium. Figure 2.8 compares the main steps in SILAC and ICAT labeling protocols. This approach with slight modification in the labeling procedure was used by Everly et al. (2004) to investigate prostate cancer cells with varying metastatic characteristics.

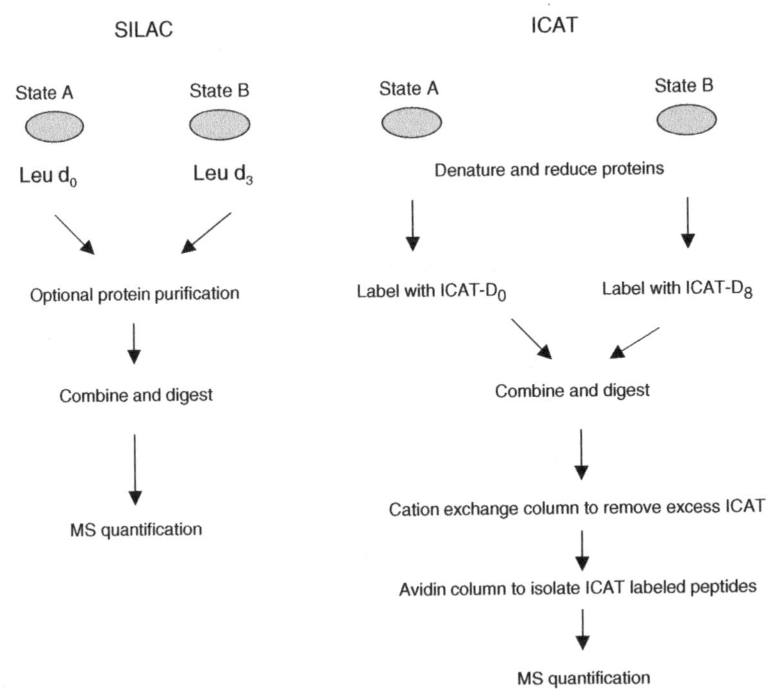

Figure 2.8. A schematic comparison of SILAC (using leucine labeling) and ICAT labeling strategies. For SILAC, the cell culture has been adapted to normal leucine (Leu-d0) or Leu-d3 media at the start of the experiment. Adapted from Ong et al. (2002) with permission.

Basically, two different cell lines are grown in media containing in one case ^{12}C labeled amino acids, whereas the second medium contains ^{13}C labeled amino acids. These growth media give rise to two protein populations, which can be described as "light" or "heavy." Lysates collected from the two media are then mixed in equal ratios, subjected to SDS-PAGE separation, individual bands are then excised, digested, and subjected to MS analysis, which can provide relative quantification on the labeled proteins.

2.5. LASER CAPTURE MICRODISSECTION

The use of laser capture microdissection to procure selected human cell population from a section of complex, heterogenous tissue is an efficient way to reduce the complexity of the sample prior to its gel analysis. It is generally argued that attempts to localize specific alterations in tumor DNA, RNA, and proteins can be hampered by our inability to adequately isolate specific cell types from pathologic specimens. Over the years, a host of techniques have been employed for including cell scraping and affinity column purification (Franzen et al., 1995), and manual microdissection of tissues (Berger, 1980; Radford, 1983). While these techniques could provide pure cell lines for evaluation of intracellular processes, the same techniques can result in the loss of relevant information regarding elements, which may confer a unique phenotype upon an individual tumor. For example, the tumor microenvironment of a carcinoma consists of not only the malignant epithelial component, but also the surrounding stroma and normal tissue. These distinct microcompartments use receptors, cell junctions, and inter- and intra-cellular signaling molecules to allow tumor cells to communicate with their surroundings and play an active role in their own control or progression (Liotta and Kohn, 2001). Removing a subpopulation of these cells for growth in an in vitro system interrupts potentially important cell–cell and cell–matrix interactions that may affect tumor behavior, thus giving the researcher a misleading impression of the actual in vivo tumor composition and physiology. The advent of laser capture microdissection (LCM), first introduced by Emmert-Buck et al. (1995), has enhanced our ability to remove specific subpopulations of cells from frozen or ethanol-fixed tissues under direct microscopic visualization.These tissues can be microdissected either stained or unstained. In fact, coupling rapid immunohistochemical staining techniques with LCM may allow for more accurate microdissection of cell subsets (Fend et al., 1999). LCM under direct microscopic visualization permits rapid one-step procurement of selected human cell populations from a section of complex, heterogeneous tissue. In this technique, a transparent thermoplastic film (ethylene vinyl acetate polymer) is applied to the surface of the tissue section on a standard glass histopathology slide; a CO_2 laser pulse then specifically activates the film above the cells of interest. Strong focal adhesion allows selective procurement of the targeted cells. The last few years have witnessed an exponential increase in the use of LCM in combination with two-dimensional gel electrophoresis in the analysis of a wide range of cancer forms. Having said that, LCM like other techniques has a number of limitations, particularly when used with

2-DE analysis: First, as no amplification technique can be applied to proteins (as PCR for DNA/RNA), it is necessary to start any proteomic analysis with a significant amount of material. For example, a minimum of 100,000 cells is necessary to perform two-dimensional gel analysis in a way that allows subsequent characterization of proteins by mass spectrometry (Wulfkuhle et al., 2002). Of course, alternative "off-gel" methods, such as isotope-coded affinity tagging (Aebersold and Mann, 2003), certainly allows one to start with a lower cell number, but tens of thousands of cells are still required. Although such a number of cells represent the upper limit of the theoretical capabilities of current microdissection techniques, the preparation of such a number of cells require extensive work with a constant problem in the choice of which cells to select. Indeed, cell morphology alone is not totally reliable to recognize cell types (especially cancer *versus* normal). The second problem of using microdissection for proteomic analysis is related to the biochemical quality of the dissected material. For instance, the use of fixatives should be kept to a minimum or even avoided as the requisite compounds create artificial boundaries between amino acid residues. In addition, because of the time necessary to perform microdissection, protein modifications such as degradation or dephosphorylation are likely to take place (Hondermarck, 2003). This last point is related to a more general concern about working with tissue biopsies; the collecting procedure of the samples must be standardized to a maximum in order to avoid differential protein modification and/or degradation. Standardization of procedures and tractability of the samples are crucial points to take into serious account. This issue is certainly more significant with proteomic analysis than it is with genomics because, in contrast to RNA that can easily be extracted from frozen tissues, protein resolubilization can be hampered after freezing. Ideally, protein extraction should be performed from fresh tissue to allow maximum solubilization and recovery of the proteins in SDS; after that, samples can be frozen before performing proteomic analysis. When considering the setup of large-scale analyses of cancer proteomes, underway in several international programs, with tumor resection and protein extraction being performed in different clinical centres, the issue of standardization is obviously important. Finally, it is worth pointing out that analyses of pure tumor cells may facilitate the detection of very low-abundance potential biomarkers. However, this clear advantage may also result in the loss of valuable information compared with analyses of whole-tumor samples. This information may be related to the cross-talk between tumor cells and surrounding stroma, which has been demonstrated to be important in breast and melanoma biology (Wernert, 1997; Liotta and Kohn, 2001).

2.6. MS ANALYSIS OF GEL-SEPARATED PROTEINS

Two-dimensional maps, image analysis, and software tools can provide apparent mass associated with each spot, its pH, and the relative protein content within such spot. For over a decade, this information was used to investigate protein expression in control and in disease samples associated with various forms of cancer. In this approach, protein expression in two different samples (e.g. one normal and the other pathological) is compared by generating a number of two-dimensional map replicates for each sample. Following staining, optical density scanning, and software analyses, two master two-dimensional maps, one for the normal and another for the pathological samples are acquired. Subsequent analysis by the same software would compare these two maps to identify proteins which have experienced measurable up- or downregulation in their intensity. For the last 10 years most 2-DE analyses have been conducted in conjunction with mass spectrometry to provide more reliable identification of the separated proteins. MALDI-MS has been the traditional choice for the analysis of gel-separated proteins. When comparing two samples (control versus disease), the two-dimensional maps generated by these samples are subjected to image analysis to establish the spots, which have experienced substantial alteration in their intensity. The second step is to use mass spectrometry to identify the protein(s) within each of the spots of interest. In this procedure, the spot of interest is excised from the gel, distained, enzymatically digested, and the tryptic peptides are extracted. The extracted peptides are then loaded onto a target plate by mixing 1 μL of each solution with the same volume of a suitable MALDI matrix, which upon dryness can be introduced into the MALDI ion source. The process of spot excising, cutting, digestion, and deposition on various sample plate formats including 96-well plates are now fully robotized. More details on the transfer of gel-separated proteins into mass spectrometers have been given in a number of recent works (Mann et al., 2001, Hamdan and Righetti, 2005). The combination of 2-DE with MS can provide various types of information depending on the MS analyser. For example, MALDI-TOF equipped with a reflectron can provide fairly reliable masses of the investigated peptides. Such data is commonly termed peptide mass fingerprinting (Berndt et al., 1999; Perkins et al., 1999, Eriksson et al., 2000). The same type of analyser can provide information on the amino acid sequence of the individual peptides through the use of postsource decay (Kaufmann et al., 1996; Stahl-Zeng et al., 1996). More modern MALDI instruments such as quadrupole-Time-of-flight (Q-TOF) or TOF–TOF can furnish MS–MS data at low and high collision energy, respectively. These data can be searched against a number of Web-available databases. It is worth emphasizing here that the various steps to introduce a given sample to the ion source is one of the important aspects for the use of mass spectrometry in proteomic analyses. Moving and manipulating small quantities of proteins from the laboratory bench to the mass spectrometer should be conducted in a way which minimizes sample losses. In most two-dimensional gel analyses, protein purification starts with a whole-cell lysate and ends with a gel-separated protein band or spot. The MALDI-MS analysis of such band/spot is carried out on peptides obtained after enzymatic digestion. In special cases, the intact proteins are analyzed to gain more accurate determination of the molecular weight. In principle, any of the classical separation methods such as centrifugation, column chromatography, and affinity-based procedures can precede the final step of gel electrophoresis. As long as the proteins of the lysate can be adequately resolved, it is best to minimize the number of separation steps. Generally, silver-stained amounts are required for successful MS identification of proteins (5–50 ng or 0.1–1 pmol for 50 kDa protein) but this does not mean that higher sensitivities cannot be achieved. In this kind of analyses, it is important to minimize contamination caused by the presence of keratins, which are introduced by dust, chemicals, handling without gloves, and so forth, as the keratine peptides can easily dominate the spectrum. In MALDI-based analyses, most detergents and salts are eliminated in the gel washing procedure. Nevertheless, the protein should be as concentrated as possible in gel, to avoid excessive background signals in the MS analysis. Pooling of spots is not necessarily advantageous as both protein and background signals will be increased. Cross-linkers and harsh oxidizing agents should be avoided as they interfere with the extraction of peptides from the gel. It is also highly recommended to reduce and if possible to fully alkylate the investigated proteins prior to any electrophoretic step; the failure to do so may result in a number of artifacts including the formation of large aggregates, which hamper an efficient transfer of proteins from the first to the second (SDS) dimension. To enhance the quality of analysis, peptides are often desalted and concentrated through the use of short reversed phase columns (ZIPTIP).

2.7. REPRESENTATIVE APPLICATIONS OF 2-DE FOR BIOMARKERS DISCOVERY

Table 2.2 gives a summary of some examples in which 2-DE with and without mass spectrometry has been used to examine the variation in protein expression, which can be attributed to various forms of cancers. In my opinion, the potential reader of this text is not interested in simply going through an endless list of publications on this argument. On the contrary, the same reader is interested to know some answers to a number of questions including, what kind of information can this technology provide, how suitable the technology is for biomarkers discovery and, what are its current limitations and its future promise. At least partial answers to some of these questions can be found in some applications of this technology in the search for potential biomarkers in cancer research.

- The use of 2-DE to investigate prostate cancer had been reported over 10 years ago (Grover and Resnick, 1995; Partin et al., 1993). These early works reported protein expression differences in samples from normal individuals and patients affected by benign prostate hyperplasia and prostate cancer. The same studies have identified a number of proteins, which were differentially expressed in the investigated samples. However, the absence of MS data in those studies rendered protein assignments far from certain. More recent 2-DE studies that used MS methods to identify differentially expressed proteins reported that surgically resected metastatic prostate tumors contained significantly higher levels of a number of proteins including heat-shock protein 90, superoxide dismutase, oncoprotein 18, and elongation factor 2 when compared with benign prostate hyperlysia (BPH), while cytokeratin 18 and tropomysins-1 and -2 were downregulated in prostate cancer samples (Alaiya et al., 2000; 2001). Considering published works on the use of 2-DE methods to look for potential prostate cancer biomarkers in clinical samples reveal a number of difficulties, which are associated with some inherent characteristics of gel analysis.

First, such analyses fail to detect hydrophobic, insoluble membrane and membrane-associated proteins. Therefore, this class of proteins tends to be significantly under-represented in 2-DE studies (Gygi et al., 2000). Membrane proteins still give disappointing results in 2-DE separation. This is normally attributed to poor solubility,

TABLE 2.2. Brief comments and references of some 2-DE analyses with and without MS detection relevant to biomarkers discovery.

Reference	Comments
Ornstein et al. (2000)	Specific populations of normal and malignant epithelium from three radical prostatectomy tissue specimens were procured by LCM and examined by 2-D PAGE. Six proteins were only present in malignant cells, while two proteins were only present in benign epithelium. One of these proteins was identified by Western blot analysis as PSA. No mass spectrometry analyses were conducted.
Somiari et al. (2003)	Protein profiles from four different stages of infiltrating ductal carcinoma of the breast were individually compared with the protein profile from non-neoplastic tissue from a female donor with no personal or family history of the disease. 2-D DIGE in combination with MALDI-MS were used to effect the comparison and identify some differentially expressed proteins. Difference in regulation was found to be dependent on the stage of the disease, and ranged from ~15% to 31%. Proteins identified included, gelsolin, vinculin, lumican, α-1-antitrypsin, heat shok protein 60, cytokeratin-18, transferring, enolase-1, and β-actin.
Lexander et al. (2005)	Two-DE combined with MALDI-MS were used to compare protein profiles in three different anatomical zones (peripheral, transition, and central) of the prostate. The aim of the study was to gain insight into the functional differences between these zones, which may be reflected in their respective protein profiles.
Shekouh et al.(2003)	Analysis of normal and malignant pancreatic epithelial cells. Samples were procured by LCM and analyzed by DIGE and MALDI-MS-MS. Comparison of protein profiles in both samples revealed nine protein spots that were consistently differentially regulated, including S100A6, calcium-binding protein. To support some of these findings, sections from pancreas cancer tissue array containing 174 duplicate normal and malignant samples were analyzed by immunohistochemistry.This analysis showed the absence or weak presence of S100A6 in normal epithelial cells.
Gharbi et al.(2002)	This was one of the early studies to apply 2-D DIGE in combination with MALDI-MS to assess protein expression in a model breast cancer cell. Specifically the authors investigated ErbB-2 mediated transformation in a model cell line system comprised of an immortalized luminal epithelial cell line and a derivative stably overexpressing ErbB-2 at a similar level reported in breast carcinomas. The authors identified a number of differentially expressed proteins resulting from ErbB-2 receptor tyrosin kinase (also known as neu/HER2) overexpression.

(*continued*)

TABLE 2.2. ***(Continued)***

Reference	Comments
O'Neil et al.(2003)	This work describes a method development for the characterization of membrane and membrane associated proteins present in human breast cell lysates. This method profiles the differential expression of membrane proteins present in normal and mutated lines derived from MCF-10 breast epithelial cell line. The novelty of this work is mainly in the separation and sample preparation prior to MALDI-MS analysis. Intact membrane proteins are separated by hydrophobicity using nonporous reversed phase LC. The resulting fractions are further separated on SDS-PAGE. This hybrid liquid phase/gel phase help to avoid protein precipitation encounterd in traditional 2-DE, particularly when dealing with membrane proteins.
Shen et al.(2004)	Two-dimensional gel electrophoresis and MS to identify differentially expressed proteins in six cases of pancreatic adenocarcinoma, two normal adjacent tissues, seven cases of pancreatitis, and six normal pancreatic tissues. The authors reported the identification of forty proteins, a number of which had been associated previously with pancreatic disease in gene expression studies. The identified proteins include antioxidant enzymes, chaperones and/or chaperone-like proteins, calcium-binding proteins, proteases, signal transduction proteins, and extracellular matrix proteins. Among these proteins, annexin A4, cyclophilinA, cathepsin D, galectin-1, 14–3-3 ξ, α-enolase, peroxiredoxin I, TM2, and S100A8 were specifically over expressed in tumors compared with normal and pancreatitis tissues. Differential expression of some of the identified proteins was further confirmed by Western blot analyses and/or immunohistochemical analysis.
Huber et al.(2004)	The authors established an antiestrogen-resistant breast cancer cell line as an in vitro model system to analyze differentially expressed genes and proteins. The human cancer cell line, T47D and its antiestrogen-resistant derivative, T47D-r were examined by Affymetrix DNA chip hybridizations on a commercially available arrays and 2-DE analyses. The authors reported that 38 proteins were found up- or down-regulated by more than two-fold in T47D-r compared to T47D. Comparison of the 2-DE data with the differential mRNA data revealed that 19 of these proteins were up- or downregulated in parallel with the corresponding mRNA molecules, among which are the protease cathepsin D, GTPases Rab 11a. The athors concluded that the combination of mRNA data with 2-DE data provided a more detailed picture on how breast cancer cells are altered in their antiestrogen-resistant form compared to antiestrogen-snsitive state.
Meehan et al.(2002)	Two-dimensional maps of normal and malignant prostate were compared. Samples were talken from 34 radical prostatectomy cases. Differentially expressed proteins were identified using MALDI-MS and N-terminal sequencing. This comparison revealed 20 proteins which were lost in malignant transformation, including PSA and α-1-antichemotrypsin. One of the main findings of this study is the expression of NEDD8, calponin, and the follistatin-related protein in normal prostate tissues and their loss or reduced expression in prostate malignancy.

TABLE 2.2. ***(Continued)***

Reference	Comments
Castagna et al.(2004)	Two-DE combined with MALDI-TOF-MS has been used to investigate protein expression in cervix squamous cell carcinoma cell line A431 and its cisplatin-resistant subline, A431/Pt. The innovative feature of this study is in the experimental design of the investigation. The investigation was not limited to two-way comparison of control and drug-resistant cell lines, but also an acute cisplatin treatment of both cell lines, leading to a four-way comparison.
Cecconi et al.(2003)	A pancreatic adenocarcinoma cell line (PaCa44) was treated with a chemotherapeutic agent, 5-aza-2' and examined with 2-DE and MALDI-TOF-MS. Protein profiling of treated and control cell lines revealed 32 downregulated and 13 up-regulated proteins. Among the major changes in DAC-treated cell lines; cofilin and profilin were silenced; coactosin, peptidyl-propyl cis-trans-isomerase A and cystatin B down-regulated by 22, 16 and 15-fold respectively.
Kim et al.(2002)	The use of 2-DE in combination with MALDI-TOF-MS to compare protein profiles in normal and in human liver tumor tissue.The authors reported sixteen protein spots corresponding to 11 proteins were found significantly altered, they were identified by peptide mass fingerprinting.

precipitation during first dimension IEF employing immobilized pH gradients, and inefficient transfer of hydrophobic proteins from the first to the second dimension gel. Another reason for the poor separation of this class of proteins is their charge heterogeneity due to numerous glycosylated extracellular domains, which results in their focusing over a wide pH range during the IEF phase. There have been a number of works, which demonstrated that some difficulties associated with membrane proteins can be bypassed through various modifications to the conventional 2-DE analyses. For example, Simpson et al. (2000) used an experimental approach to investigate proteins from an enriched membrane preparation of the human colorectal carcinoma cell line LIM1215. Following fractionation by sodium dodecyl sulfate-polyacrylamide gel electrophoresis (SDS-PAGE), the unstained gel slices were subjected to in-gel tryptic digestion and the resulting peptide mixtures were examined by reversed phase LC coupled to mass spectrometry. The acquired MS–MS spectra were used to interrogate various databases for protein identification.

In another proteomic study Adam et al. (2003) investigated the protein content of breast tumor cell membranes. To enhance the possibility of obtaining the highest number of protein identifications, the authors took a number of steps. First of all, to simulate clinical heterogeneity of breast cancer the authors used multiple cell lines with different molecular pathologies. In addition, tumor-derived cell lines were used to ensure enrichment for cancer cell-specific plasma membrane proteins. Membrane preparations were resolved on one rather than two-dimensional gels to avoid discrimination against insoluble hydrophobic membrane proteins. The separated proteins were digested and the resulting peptide mixtures were examined by MALDI-TOF and MS/MS analysis. Second, invariably due to limited dynamic range of the gel method, the most-abundant soluble proteins are typically visualized and detected by 2-DE methods. Unfortunately, normal and diseased prostate epithelia are heterogeneous tissues that lack "pure cell populations" (Gleason, 1992). The use of laser capture microdissection (LCM) to procure homogenous cell population is one way to mitigate the question of heterogeneity. Although LCM has been used with great success in procuring pure populations in prostate cancer tissues (Emmert-buck et al., 1995; Macintosh et al., 1998), this technology is not without limitations when used in conjunction with 2-DE (see Section 2.5). Wright et al. (2005) have argued that if a 25 kDa potential biomarker is expressed at 1000 copies/cell and subjected to 2-DE, it would amount to running approximately 2.9 pg (120 amol) of the target biomarker into the gel. The limit of detection for a silver-stained protein spot in a 2-DE gel is around 1 ng. Thus, the biomarker would be ~2–3 orders of magnitude below visual detection by 2-DE staining methods. Even optimistically assuming very small sample losses during the complex 2-DE process, and assuming the biomarker was detectable in the gel, detecting 120 amol using a standard mass spectrometer would be very challenging. The very broad, dynamic protein expression range of serum samples has severely limited the use and success of 2-DE methods for detecting potential prostate cancer biomarkers. In contrast, 2-DE methodology has and will continue to be a powerful platform for identifying potential biomarkers that replicate prostate cancer progression using *in vitro* and xenograft model systems (Thompson et al., 2000). These experimental systems can generate abundant amounts of protein, which is typically not the case when clinical samples are utilized.

- The capability of 2-DE to separate thousands of proteins from a complex mixture made it an attractive tool in the analyses of breast cancer. These analyses were pioneered before the emergence of more powerful proteomic and genomic tools. Prior to the proteomic era, proteins separated by 2-DE could be identified by comigration, specific antibody labeling or chemical microsequencing. These approaches are known to be time-consuming and difficult to implement, yet they did produce sufficient data to establish 2-DE protein profiles of breast tumors as opposed to normal breast tissue. The same approaches have also identified a number of molecular indicators associated with tumorigenesis (see Table 2.3.). A period extending between the mid nineties and lasting to present day has witnessed the contribution of large-scale 2-DE analysis, which resulted in the establishment of 2-DE databases of human breast epithelial cell proteins [(http://www.anl.gov/BIO/PMG/projects/index_hbreast.html), Giometti et al., (1997); (http://www.ludwig.edu.au/jpsl/databases/MDA-MB231.asp), Rasmussen et al., (1997); (http://www.bio-mol.unisi.it/2d/2d.html.), Bini et al., (1997)]. Over the last 10 years, the 2-DE approach has generated a considerable amount of valuable data for breast cancer research; however, the translation of these data to clinical practice is still to be demonstrated (see Table 2.3). This rather disappointing deduction has a number of good reasons. One in particular is the fact that only a limited number of individual protein modifications commonly divide normal cells from their cancerous counterparts. This subtle

TABLE 2.3. Some protein biomarkers identified by 2-DE in breast cancer. Adapted from Hondermarck et al. (2001) with permission.

Proten (SwissPROT accessionNo.)	Function	Method of identification	Reference
Cathepsin D (P07339)	Protease	Antibody	Westley and Rochefort (1980); Augereau, et al. (1988).
Breast Carbonic anhydrase (mouse P16015)	Metabolic enzyme	Antibody	Ring et al. (1989)
Cytokeratins K8(P05787), K18 (P08727)	Cytoskeleton	Antibody	Trask et al. (1990); Moll et al. (1982) Cell **31**, 11.
Cytokeratin K5	Cytoskeleton	Antibody	Trask et al. (1990); Moll et al. (1982).
Tropomysin 1 (P09494)	Cytoskeleton	Antibody	Bhattacharya et al. (1990).
Tropomysin 2 & 3 (P12324)	Cytoskeleton	Comigration	Franzen et al. (1996a).
Nuclear matrix proteins (1-6)	Skeleton of the nucleus	Subcellular fractionation	Khanuja et al. (1993); Samuel et al. (1997).
Heat Shock Proteins HSP60 (P10809), HSP90 (P08238), calreticulin (P27797).	Molecular chaperones	Matching with published maps	Giometti et al. (1997); Franzen et al. (1996).
PCNA (P12004)	DNA replication and reparation.		Franzen et al. (1996b).
14-3-3σ (P31947)	Molecular chaperone	Mass spectrometry	Vercoutter-Edouart et al. (2001).

difference is by no means limited to proteomic analysis. A similar situation is encountered at the genomic level, where a limited number of molecular modifications, affecting oncogenes and suppressor genes, are required to transform normal cells into cancerous ones (Haber, 2000). This limited number of molecular modifications has two direct consequences: First, so far, no technique has been able to provide unmistakable marker for the early detection of this pathology. Secondly, finding drugs, which specifically target breast cancer cells, has not been achieved yet, and therefore patients continue to receive treatments which are known to have heavy side effects.

As it has been pointed out, the use of 2-DE to investigate protein expression in tissues and in biofluids has yielded a wide range of proteins, which showed a certain promise. The σ isoform of 14-3-3 can be consided an interesting case. Vercoutter-Edouart et al. (2001) have shown that the sigma form of 14-3-3 is easily detectable in 2-DE gels of normal breast epithelial cells using low-sensitivity Coomassie staining, whereas the spot was undetectable in breast cancer cell protein profiles. Nevertheless, the same isoform was also present in breast cancer cells, but due to very low levels, more sensitive silver staining was necessary for its detection. In the same study, the distribution of this protein was investigated in breast cancer biopsies. On the basis of the data gathered in cell culture, the analysis of breast tumors with 2-DE was set up with a narrow-range pH gradient to provide an optimal view of the gel area containing 14-3-3 σ. It was shown that the level of 14-3-3 σ was systematically down-regulated in tumor biopsies. At the mRNA level, it was also shown that gene expression of 14-3-3 σ is 7–10 times lower in breast cancer cells than in normal breast due to the high frequency of hypermethylation of the 14-3-3 σ locus (Ferguson et al., 2000). Interestingly, the mRNA for 14-3-3 σ was undetectable by Northern blot analyses in 45 of 48 primary breast carcinomas studied; in contrast, the same protein was detected in 30 of 35 primary tumor samples. From the clinical point of view, it has been reported that the downregulation of 14-3-3σ gene expression is an early event in breast carcinogenesis (Umbricht et al., 2001). In an earlier study, Nurcombe et al. (2000) reported that the overexpression of this isoform in cancer cells reverses the proliferative phenotype. The biological relevance of this and other isoforms of 14-3-3 family are discussed in more detail in Chapter 3.

- Pancreatic cancer is the fourth leading cause of cancer death in the United States (Greenlee et al., 2001; Jemal et al., 2002). Over the last few years, 2-DE in combination with mass spectrometry has been used to look for potential biomarkers through the investigation of protein profiles in tissues, pancreatic juice, and in various cell lines before and after pharmacological treatment. Shekouh et al. (2003) used 2-DE to study differentially regulated proteins in pancreatic cancer. The authors compared the 2-DE protein profiles from nonmalignant and malignant microdissected ductal epithelial cells and reported nine protein spots that were consistently differentially regulated. Five of these protein spots showed increased abundance in cancer cells, whereas the other four protein spots showed diminished abundance. To identify these proteins, the authors matched the position of the spots of interest from microdissected samples to the pattern obtained from large quantities of whole tissue sample, and sequenced the corresponding spot from whole-tissue samples. The limited amount of material directly from the 2-DE gel of microdissected sample was not sufficient for mass spectrometric identification. Only one protein spot was identified; the other eight differentially regulated protein spots remained unidentified. The identified protein, S100A6, belongs to the S100 protein family. Several members of S100 family have been reported to be overexpressed in pancreatic cancer both by mRNA and immunohistochemical analysis, and thus may be important in pancreatic cancer. Another study by Shen et al. (2004) identified 40 differentially-expressed proteins using 2-DE to analyze whole pancreatic cancer tissue. A considerably higher number of differentially expressed proteins were identified in the study because it used whole cancer tissue, thus a larger amount of sample was available for 2-DE analysis. Five of the 40 proteins identified had been previously associated with pancreatic disease in gene expression studies. Among the other proteins, annexin A4, cyclophilin A, cathepsin D, galectin-1, 14-3-3ξ, α-enolase, peroxiredoxin I, TM2, and S100A8 were specifically overexpressed in tumors compared with normal and pancreatitis tissues.

In a series of articles, Reghetti's group have used 2-DE in combination with mass spectrometry to investigate protein changes in pancreatic adenocarcinoma cell lines before and after treatment with a number of chemotherapeutic agents. In the first study, a pancreatic adenocarcinoma cell line (Paca44) was treated with trichostatin-A (TSA), an inhibitor of histone deacetylases (Cecconi et al., 2003a). The authors reported up- and downregulation in over 50 spots, which were characterized by MALDI-TOF fingerprinting. Among these proteins, of particular interest are the two downregulated proteins nucleophosmin and translationally controlled tumor protein. The upregulated proteins included stathmin (oncoprotein 18), programmed cell death protein 5 (also known as TFAR19). In a second study, the same cell line was treated with 5-aza-2′-deoxycytidine (DAC) and the two protein profiles before and after treatment were compared using 2-DE in conjunction with MALDI-TOF peptide fingerprinting. The authors reported substantial alterations in 45 proteins (Cecconi et al., 2003b). Among the major changes in DAC treated cells cofilin and profilin were silenced, while coactasin peptidyl-trans isomerase A and cystatin B were downregulated by 22, 16, and 15-fold respectively. In the third study, the same group (Cecconi et al., 2005) have treated human pancreatic adenocarcinoma cell line, T3M4 with two chemotherapeautic agents, gemcitabine (2′,2′-difluorodeoxycytidine), which is known to interfere with DNA methylaytion together with trichostatin A (a drug known to interfere with histone acetylation). The analyses were conducted by 2-DE and altered proteins were characterized by MALDI-TOF-MS peptide fingerprinting or ESI-MS/MS. Protein profiling was studied via a four-way comparison, comparing control cells and cell populations treated individually with the two drugs or via a simultaneous addition of both drugs. The authors set a threshold value, for differential profiling, at 2.5-fold instead of the 2-fold threshold commonly used in this type of analyses.These analyses identified a total of 81 polypeptide chains differentially expressed, and of those, 56 could be characterized by MALDI-TOF and electrospray MS-MS analyses (see Fig. 2.9). The authors attempted to draw pathways of functional association among all proteins found to be modulated by the single or combined treatment of the two drugs. This was done through the use of a software tool, called pathway assist (version 2.01). These analyses indicated that most of these associations are localized around four major biological processes, namely apoptosis, cell death, proliferation, and mitogenesis (see Fig. 2.10).

2.8. PROTEIN MICROARRAYS

Advances in genomics and proteomics have underlined an urgent need for maniaturized and robust platforms for high throughput analysis of proteins. Microarrays, generated by spotting biomolecules in an ordered arrangement on a solid surface at high spatial density, offer such possibility by allowing the simultaneous investigation of thousands of spotted targets. Historically, protein-detecting microarrays relied on the development of immunoassays. In fact, as early as 1929, antibodies were used in serology to precipitate antigen for subsequent quantification (Heidelberger and Kendall, 1929). After the impressive success of DNA microarray technology

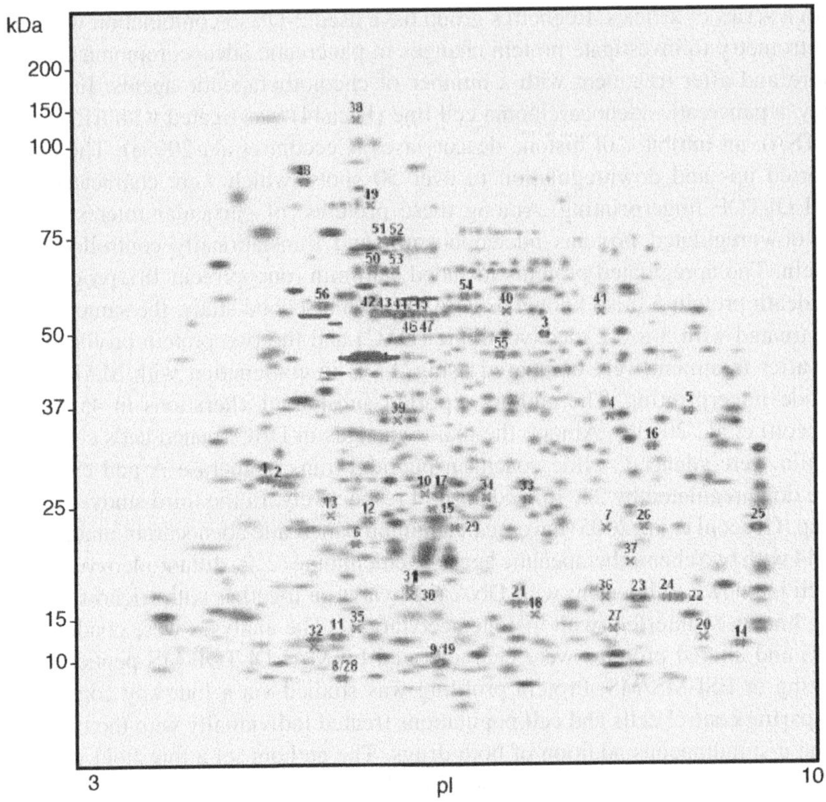

Figure 2.9. Reference map of the T3M4 adenocarcinoma pancreatic cell line. The identified proteins are marked with the corresponding gene name based on human gene nomenclature database (http://www.gene.ucl.ac.uk/nomanclature/). From Cecconi et al. (2005) with permission.

demonstrated in the late 1990s, attention has been shifted toward the construction and application of protein-based microarrays. Such shift has been encouraged by the realization that gene expression analysis does not give reliable protein abundance nor does it provide information about protein modifications and protein functions (Gygi et al., 1999; Anderson and Seilhamer, 1997). The early enthusiasm to simulate the success registered by DNA array technology was hindered by a number of intrinsic characteristics of proteins compared to DNA. For example, unlike the simple hybridization chemistry of nucleic acids, proteins possess a staggering variety of chemistries, affinities, and specifities. Second, there is no equivalent amplification process like PCR that can generate large quantities of proteins. Third, Protein purification is a demanding task and does not always guarantee the preservation of the functional integrity of the purified protein. Furthermore, many proteins are highly unstable, which raises concern about protein microarray shelf life. Over the past few years, two main categories of protein-based microarrays have emerged:

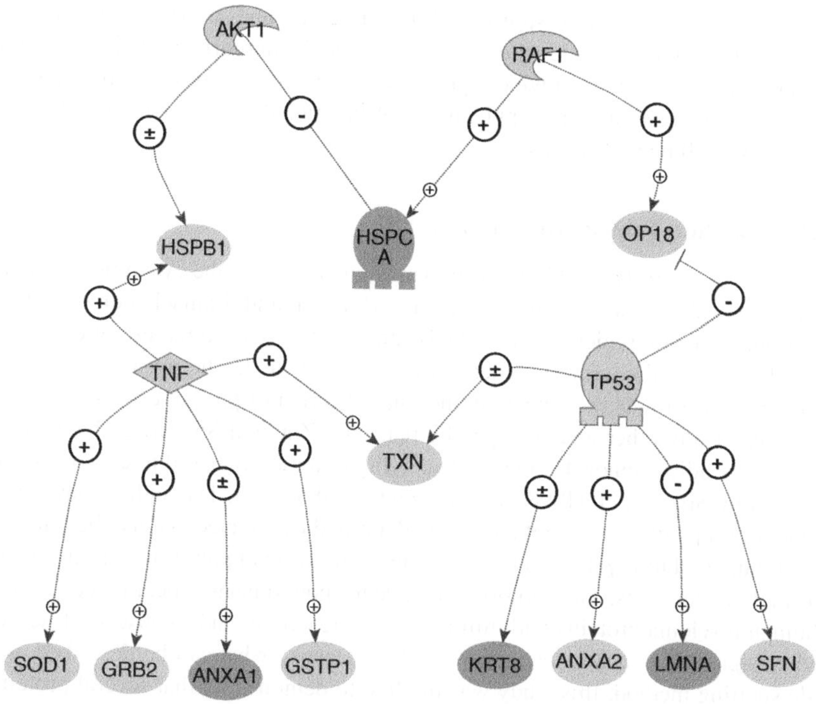

Figure 2.10. Functional associations of some differentially expressed proteins in T3M4 cells. Significantly regulated proteins were examined by pathway Assist (version 2.01), which resulted in functionally related network of proteins (genes). The major biologic processes (such as death, apoptosis, proliferation, mitogenesis, RNA localization, regulation of signal transduction, etc.) related to these proteins are represented as rectangles. From Cecconi et al. (2005) with permission.

analytical and functional. In analytical arrays, different types of ligands, including antibodies, antigens, DNA or RNA aptamers, carbohydrates or small molecules, with high affinity and specificity, are spotted onto a derivatized surface. These surfaces (chips), which can be used for assessing protein expression, probe for differences in cell types or varying conditions as well as for monitoring disease states, and are thus expected to be highly suitable for the identification and monitoring of potential markers of various diseases. On the contrary, functional protein arrays involve the high-density deposition and analysis of a set of proteins, or even an entire proteome. Such arrays can be used as screen for biochemical and enzymatic activities, interaction with proteins, lipids, or small molecules. Functional protein arrays are predicted to be very useful for elucidating the biochemical functions of proteins, developing protein interaction networks, screening for different disease markers, and for the discovery and development of novel drugs (MacBeath and Screiber, 2000; Ramachandran et al., 2004; Zhu et al., 2001; Boutell et al., 2004). Both types of microarrays share a number of basic elements associated with the surface chemistry, delivery systems, probing strategies, and so forth. Some of these elements are discussed under the title of analytical microarrays, which over the past few years have shown a wider application in the area of protein profiling and biomarker investigation. Some applications of functional and reverse phase protein microarrays will be also discussed.

2.8.1. Analytical Protein Microarrays

Currently, there are two additional nomenclatures to describe Analytical protein microarrays: abundance-based microarrays (LaBaer and Ramachandran, 2005), and protein-detecting microarrays (MacBeath, 2002). To avoid the use of excessive terms, I shall use the term analytical throughout this text. The most common use of this type of array is the detection and quantification of protein expression using antibodies or any other analyte–specific reagents (Zhu and Snyder, 2001). One of the earliest studies demonstrating the feasibility of antibody-based microarrays was published by Silzel et al. (1998) and Knezevic et al. (2001). The authors have used a standard inkjet printer to spot monoclonal antibodies directed against the four human immunoglobulin-γ (IgG) subclasses onto a thin sheet of polystyrene film to form 200 μm diameter spots. The authors were able to show subclass-specific recognition of human myeloma proteins with minimal cross reactivity and to observe dose-dependent signals for each investigated subclass. Despite what can be described as a crude spotting method, this study was the first to demonstrate that several proteins could be detected and quantified in parallel fashion using antibody microarrays.

2.8.2. Substrates and Protein Attachment Methods

The key elements in the construction of a protein array are considered to be the following: the generation and isolation of an extensive repertoire of recognition molecules with which the target proteins will interact; the means by which these recognition molecules may be immobilized in an array format; and the detection strategy of the bound target proteins. High quality substrates with well-defined and reproducible surface properties and optimized surface chemistry are essential for homogenous and functional conformation of protein immobilization. Given the extensive diversity of target proteins, which extends from known binders such as antibodies to the more delicate to immobilize, water insoluble membrane proteins, different substrates and different immobilization chemistries have to be considered.

Ideally, we would like the surface chemistry to meet a number of characteristics, including, protein immobilization which is highly specific and does not require prior purification; the surface which is inherently inert and resists nonspecific adsorption; the attachment process that allows the control of protein orientation; and the surface that contains uniform functional groups for the facile immobilization of the proteins. Soft substrates such as polystyrene, poly vinylidene fluoride (PVDF), and nitrocellulose membranes have been traditionally used to attach proteins in biochemical analyses (Cahill, 2001; Ge, 2000; Walter et al., 2000). However, early experience with these surfaces revealed that the spotted solutions tend to spread on the surface with the direct result of a limited spot density. Other limitations of this class of surfaces are the inability to control protein conformation and orientation, and their low signal-to-noise ratio due to high auto fluorescence background. These and other limitations have encouraged the use of glass substrates, which allow smaller spots, higher density arrays and low background noise. Glass, of course, must be chemically modified to allow protein attachment. The simplest modification approach, which has been well tested for cDNA microarrays, is to coat the glass with poly-L-lysine, a positively charged protein polymer that adsorbs to the negatively charged silica surface at an appropriate pH (Joos et al., 2000; Ge, 2000; MacBeath and Schreiber, 2000). This modification mode has been demonstrated by Haab et al. (2001). The authors investigated whether antibody microarrays could quantitatively detect antibody-antigen interactions. They chose 105 well-studied antibody-antigen pairs to set up both antigen and antibody microarrays, and found that about 30% showed a linear relationship. To allow more specific and stronger protein attachment, which can resist stringent wash conditions, attempts have been made to create reactive surfaces on glass that can covalently cross-link target proteins. A functionalized silane reagent is often used to create such sites for protein attachment. In general, a bifunctional silane cross linker is used to form a self-assembled monolayer, which has one functional group that reacts with the Si—OH groups on the glass surface, and another free one that can either directly react with primary amine groups of proteins or can be further chemically modified to enhance specificity (Mendoza, et al., 1999; MacBeath and Schreiber, 2000; Rowe et al., 1999). This cross-linking approach has different variants. For example, a gold-coated glass surface in which the self-assembled monolayer is formed through the use of a bifunctional thioalkylene, which has an SH-group that reacts with gold, and another free one that reacts with the capture molecules. The advantage of using gold-coated surfaces is that both mass spectrometry and surface plasmon resonance (SPR) can be used for detection as well as for monitoring the dynamics of the reaction and possibly to identify the captured molecules (Houseman et al., 2002; Rich et al., 2001). In another variation, the silane reagent is chosen with a hydrophobic tail, which allows the adsorption of biotinylated BSA, followed by attachment of streptavidin and biotinylated antibodies (Mooney et al., 1996). This approach has been demonstrated by Rowe et al. (1999), where avidin was directly cross-linked to glass, followed by binding biotinylated antibodies.

Over the last few years, both oriented and nonoriented modes of protein immobilization have been tried. Immobilization of proteins in an oriented manner provides a better spatial accessibility of active binding sites than that in nonorientated fashion. In the covalent cross-linking approach, where some reactive ligands can also exist in the side chains of the immobilized proteins, it is likely that those proteins can attach to the surface in a random fashion, which may alter their native conformation, and thus reduce their activity or make them inaccessible to probes. To favour the orientation of immobilized proteins, attempts have been made to attach proteins through highly specific affinity interactions. In this approach, proteins are fused with a high-affinity tag; these tagged proteins are then attached to the substrate surface via this tag. This capture method has the advantage that proteins are uniformly oriented at a distance from the array surface. However, activities that require a free amino or carboxyl terminus may be adversely affected depending on the location of the tag. The power of this method was demonstrated by Zhu et al. (2000; 2001) who managed to attach 5800 fusion proteins containing the 6xHis affinity tag on nickel-coated glass slides for screening experiments. The authors screened for calmodulin and lipid-binding proteins, leading to several novel calmodulin-binding proteins and defining a consensus calmodulin binding motif. They also identified 150 lipid binding proteins in the yeast proteome of which 52 were previously uncharacterized. The importance of selective immobilization of proteins with orientational control has been demonstrated in a number of studies, which used different approaches including the use of fusion proteins (Hodneland et al., 2002) and immobilized protein A or G, which binds to the Fc portion of antibodies (Turkova, 1999; Kanno et al., 2000), mRNA –protein hybrids (Weng et al., 2002). In a relatively recent study (Cha et al., 2005), the authors have described an interesting strategy for oriented protein microarrays using recombinant polyhistidine tags and surface-chelated metal ions. This approach is based on the well-established immobilized metal ion affinity chromatography (IMAC) (Porath et al., 1975). Basically, a chelate group, for example iminodiacedic or nitrolotriacetic is attached to a solid support. It can bind strongly to a metal ion such as Ni^{2+} or Cu^{2+} via three or four binding sites, leaving three or two vacant sites for coordination to imidazole groups of polyhistidine units on the tagged protein. A polyhistidine tag can be engineered to either the C- or N-terminus of the protein, and it is generally assumed that such tagging does not interfere with the structure or function of the tagged protein. Cha et al. (2005) used a high-density PEG-coated Si (111) surface for the immobilization of polyhistidine-tagged proteins. These Authors have used these surfaces to assess the influence of orientation on the activities of the immobilized molecules. Catalytic activities of immobilized enzymes were compared with and without molecular orientation. It was concluded that oriented protein molecules selectively immobilized on PEG-coated surface via polyhistidine tags faithfully reflected their activity in solution phase, whereas those randomly oriented did not.

2.8.3. Detection Strategies

One of the key elements to ensure the success of protein microarrays is the method by which such arrays are probed. In other words, how can minute quantities of bound proteins be detected, bearing in mind that there is no equivalent of PCR for the amplification of proteins. The most obvious method is to use an ELISA-based assay in which the probes are tagged with enzymatic or fluorescent molecules. At present, most detection methods to probe protein microarrays require some form of fluorescent or radiolabeling. Such a labeling step imposes extra time and cost demands and can in some cases, interfere in the processes which are designed to monitor/measure. For example, fluorescent labels are invariably hydrophobic, and in many screens, background binding can be significant. These and other limitations have encouraged the use of techniques which do not require labeling. Some of these techniques together with some labeling techniques are considered below.

TABLE 2.4. Manufacturers of biosensors and their Web addresses. Adapted from Cooper (2002) with permission.

Manufacturer	Web Address	Technology
Affinity sensors	www.affinity.sensors.com	Resonant mirror
Artifitial sensing instruments	www.microvacuum.co/research/memcos	Waveguide
Aviv instruments	www.avivinst.com	Grating- coupled SPR
Biacore	www.biacore.com	SPR
Farfield sensors	www.farfield-sensors.com	Waveguide
HTS biosystems	www.htsbiosystems.com	Grating-coupled SPR
IBIS	www.ibis-spr.nl	SPR
Luna analytics	www.lunaanalytics.com	Grating-coupled fiber optics
Nippon lasers	www.rikei.com/spr	SPR
SRU biosystems	www.srubiosystems.com	Guided-mode resonant filter
Prolinx	www.prolinx.com	SPR

2.8.3.1. Surface Plasmon Resonance (SPR). Optical biosensors that exploit surface plasmon resonance are now widely used to analyse biomolecular interactions. The main attraction of these sensors is their capability to provide real-time information on a wide range of biological problems, without the need for molecular tagging or labeling. Since their commercial introduction in the late 1980s, their use in research and development has been described in over 4000 scientific publications that cover many disciplines in pharmaceutical and diagnostic research. A list of manufacturers who produce this type of sensor together with their Web addresses is given in Table 2.4. SPR is one of the optical biosensors, which exploits evanescent-wave phenomenon to characterize interactions between receptors attached to the biosensor surface and ligands within a solution above such a surface. These waves are caused by the total internal reflection of light at a solid-solution interface. A simplified scheme of the main components of SPR biosensor is given in Fig. 2.11. Basically, binding of molecules in solution to surface-immobilized receptors alters the refractive index of the medium near the surface. This change can be monitored in real time to measure accurately the amount of bound analyte, its affinity for the receptor and the association and dissociation kinetics of the interaction. An extremely wide range of molecules can be analysed, from low-molecular-mass drugs to multiprotein complexes and bacteriophage, as well as their interaction affinities (Cooper, 2002). Most importantly, binding affinities and kinetics can be determined using very low amounts of compound without the need for prior chemical or radio-labeling. The interface between the sensor surface and the chemical or biological systems to be studied is a key component of optical biosensors. Receptors must be attached to some form of solid support, while retaining their native conformation and binding activity. This attachment must be stable over the course of a binding assay, and in addition, sufficient binding sites should be presented to the solution phase to interact with the analyte. Most crucially, the support should be resistant

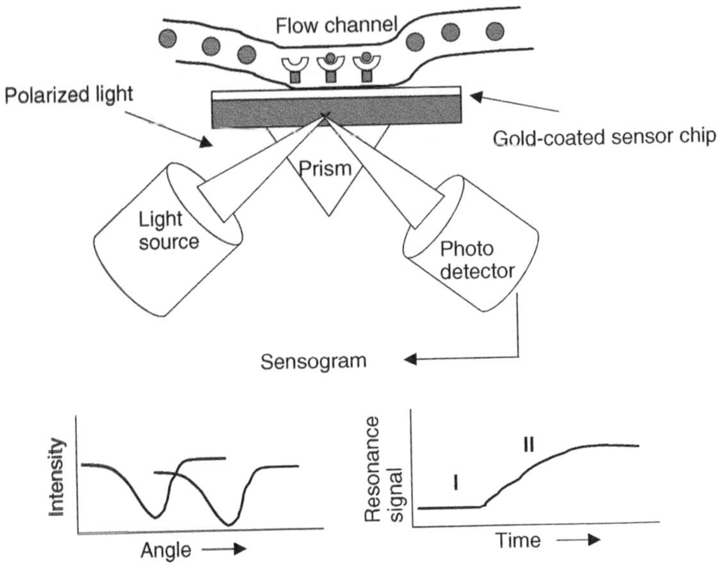

Figure 2.11. A typical set-up for an SPR biosensor. Surface plasmon resonance (SPR) detects changes in the refractive index in the immediate vicinity of the surface layer of a sensor chip. SPR is observed as a sharp shadow in the reflected light from the surface at an angle that is dependent on the mass of material at the surface. The SPR angle shifts (from I to II in the lower left-hand diagram) when biomolecules bind to the surface and change the mass of the surface layer. This change in resonant angle can be monitored noninvasively in real time as a plot of resonance signal (proportional to mass change) versus time. From Cooper (2002) with permission.

to nonspecific binding of the sample, which can mask the specific binding signal. One of the strengths of SPR is that it can be operated in situ; in other words, it does not require substrate rinsing and drying prior to analysis, which renders the technique highly suitable for obtaining kinetic information on binding interactions. This characteristic is highly important in the quantification of low-affinity protein–protein interactions that would normally not be stable under rinsing and drying conditions. The kinetics of binding of antibodies immobilized into a solid support is of a key interest for immunodiagnostics. Over the last 15 years, a number of biosensors that can measure such binding have proved to be a valuable tool in the characterization of binding of antigen to immobilized antibodies. Since its commercial introduction in the early 1990s, SPR (Rich and Myszka, 1999) has found its way to a wide range of biophysical and biochemical applications. This type of detector allows the analysis of receptor-ligand interactions over a wide range of molecular masses extending from a few hundreds of Daltons (Davis and Wilson, 2000) to whole cell binding (Quinn et al., 2000); it has an effective affinity range from nano to micromolar, and it can provide reliable binding rates. Regarding protein microarrays, SPR has been used to study the kinetics of antigen-antbody interactions and receptor-ligand interactions over a wide range of molecular masses, affinities and binding rates (McDonnell, 2001). Another promising feature of SPR is its ease of coupling with mass spectrometry. Several groups have already demonstrated the integration of these powerful techniques for affinity-based capture and characterization of ligands (Williams and Addona, 2000; Nelson et al., 2000). The strength of the combination of SPR and MS is that the first component could be used to detect, capture and subsequently digest and deliver nanomolar to femtomolar levels of ligand for MS analysis. Although commercially available SPR chips have a fairly limited number of channels, Myszka and Rich (2000) described a sensor surface with 64 individual immobilization sites in a single flow cell. It is worth noting here that SPR is one of the two techniques which can measure protein binding in real time, the other technique being fluorescence planar waveguide (Pawlak et al., 2002). However, unlike SPR the latter technique requires labeling.

2.8.3.2. Atomic Force Microscopy (AFM). A simple yet highly realistic description of AFM has been given in an article by Engel et al. (1999), where the authors compared the working principle of this technique to a blind person exploring his path with a stick. Deflections of the stick are recorded and assembled into an image in the brain. In the case of AFM, the sample is explored by a cantilever whose deflections are recorded by a computer. The breakthrough in biological applications of the AFM came with the development of the liquid cell, which permits the investigation of proteins and nucleic acid molecules under physiological conditions (Engel et al., 1999). This technique uses a sharp micron-scale tip to scan and amplify surface features, resulting in exceptionally detailed topographical information with amplification in the order of 10^6. AFM is used extensively in computer and semiconducting industries; however, the last few years have witnessed an increased use of this technology in biological sciences. This detector can work in different modes depending on the type of information the user seeks. In one of these modes, known as contact-force mode, a layer of proteins, or other biological polymer, is either adsorbed to the substrate or linked to it. When the tip and the substrate are brought together and then withdrawn, one or more molecules can adsorb to the tip. Variation in the distance between the tip and the substrate induces molecular extension and a restoring force which causes the deflection of special cantilever. The bending of the latter causes a deflection of a laser beam, which is then measured by a photodetector. The signal of this detector can be related to the angle of the cantilever and therefore to the applied force. This readout together with the known mechanical characteristics of the cantilever can be used to obtain spatial information within the nanometre range. A schematic representation of AFM in the force-extension mode is given in Figure 2.12. This detection approach has been exploited in different ways. AFM has been used to detect binding through the determination of the adhesive strengths of antigen-antibody interactions in which the sample is modified with an immobilized antibody and the tip is modified with an immobilized antigen and vice versa (Dammer et al., 1996; Allen et al., 1997). In other variations, AFM has been used to measure the change in height that results from ligand-receptor binding. Height changes of 3-4 nm have been observed as a consequence of adsorption of antigenic IgG to a surface,

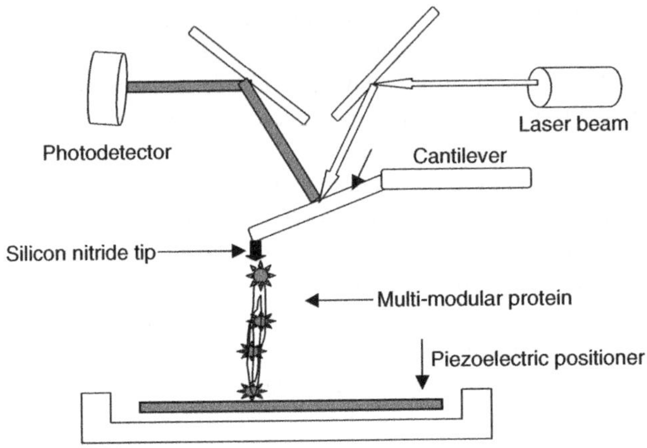

Figure 2.12. Representation of the force–extension mode of the atomic force microscope (AFM). When pressed against a layer of protein attached to a substrate, the silicon nitride tip can adsorb a single protein molecule. Extension of a molecule by retraction of the piezoelectric positioner results in deflection of the AFM cantilever. This deflection changes the angle of reflection of a laser beam striking the cantilever, which is measured as the change in output from a photodetector. From Fisher et al. (1999) with permission.

followed by a similar change in height upon antibody-antigen binding (Browning-Kelley et al., 1997; Morgan et al., 1995). The combination of AFM topographic imaging capability with a compositionally patterned array of immobilized antigen rabbit IgG on gold to perform immunoassays has been demonstrated by Jones et al. (1998). In an interesting study by Gad et al. (1997), functionalized tips were used to measure specific forces of interaction between ligand-receptor pairs and to map the location of polysaccharides on a living microbial cell surface. These measurements were carried out as part of developing an antigen-sensitive probe, which can be used to identify its target molecules within a heterogeneous molecular population on the surface. Other more recent studies have used AFM to investigate DNA and DNA-protein complexes immobilized on different surfaces (Liu et al., 2005; Zhang et al., 2005). A novel application of AFM, which combines its surface profiling capabilities with fixed immunocapture using antibodies immobilized in a nanoarray format, has been evaluated (Huff et al., 2004). The idea behind such evaluation is to develop a platform for direct, label-free detection and characterization of viral particles and other pathogens.

2.8.3.3. Enzyme-linked Immunosorbent Assay (ELISA). Traditional protein analysis methods have relied heavily on the detection of chemiluminescent products of enzymatic reactions. Detection strategies based on ELISA methodology have been developed and applied to probe various proteins both on filter (Joos et al., 2000; Büssow et al., 1998) and on glass (Mendoza et al., 1999; Arenkov et al., 2000) microarrays. In this approach, an immobilized capture antibody is used to bind the protein of interest, and a second detection antibody conjugated to biotin binds to a second epitope on the captured protein to create an antibody sandwich. Visualization of the protein antibody complex can be achieved using streptavidin conjugated either to a fluorescent molecule such as cyanine dyes (e.g., C3 or C5) or to horseradish peroxidase together with the activation of multiple copies of a dye or a hapten-labeled tyramide derivative, to enhance the detection of low-abundance analytes (Barry and Soloviev, 2004). These analyses are commonly carried out in 96-well format with one, or more, capture reagents specific to a particular analyte being mixed together in each well. In this type of assay, a number of wells are commonly reserved for controls. This sandwich-type ELISA protein array has a number of advantages and limitations. Regarding the first aspect, such assay uses well-established protocols, which allow the generation of quantitative data based on internal standard curves; it does not require labeling, which can mask recognition epitopes; the use of two antibodies per analyte increases the specificity of the assay. One of the main limitations of this type of assay is the fact that more than one antibody is needed per protein, each being required to recognize a separate epitope leading to high costs, time consumption and rather complex procedure.

2.8.3.4. Radio Isotope Labeling. Isotope labeling, in which a radioactive isotope is incorporated into a molecule, has been used to probe protein microarrays. Zhu et al. (2000) have used this method of labeling to investigate 119 protein kinases using 17 different substrates. The yeast kinases were radiolabeled with ^{33}P-γ-ATP, and the phosphorylated substrates were probed with phosphoimager. Radio isotope detection has a low-background signal and extremely high sensitivity because of the fact that radioactive signal can be easily integrated. Another advantage of this approach is in the labeling protocol, which does not require bulky molecules, and which can interfere in the activity and specificity of the labeled protein. On the contrary, this approach has its limitations including costly reagents, lengthy preparation, and the extra precautions required when working with radioactive labels. Another limitation, which this approach shares with ELISA, is that only one signal channel is normally available, meaning that it is not possible to directly measure binding of more than one specie per spot. Furthermore, the amount of capture material in each spot has to be tightly controlled to enable quantitative comparison between spots.

2.8.3.5. Fluorescence Detection. High sensitivity, high resolution, simplicity, and safe use have made fluorescence detection the preferred method in protein analyses. Fluorescence detection can be implemented in various modalities, depending on the sample/medium to be probed; some of these modalities are considered below. A potentially general method to detect protein–protein interactions uses fluorescence resonance energy transfer (FRET) between fluorescent tags attached to the interacting proteins (Wouters et al., 2001). Basically, interaction between two proteins can be imaged by detecting FRET between donor and acceptor fluorophores. Following the excitation of the first fluorophore, FRET is detected either by emission from the second fluorophore using suitable filters or by alteration of the fluorescence lifetime of the donor tag. Several fluorescent detector molecules suitable for use in protein microarrays have been described (Wise, 2003). FRET commonly uses two variants of green fluorescent protein (GFP): cyan fluorescent protein (CFP) and yellow fluorescent protein (YFP). The obvious attraction of FRET is that the measurements can be conducted in living cells, which allows the detection of protein interactions at the location(s) where they normally occur. This means that relevant physiochemical conditions and spatial restrictions to possible interactions, as imposed by compartmentalization, are maintained. Inducible interactions have been demonstrated, such as the binding of Grb2 to activated epidermal growth factor receptors (Sorkin et al., 2000) and the hormone-induced binding of coactivator proteins to nuclear receptors (Llopis et al., 2002). Although the potential of FRET for high throughput screening of binding can be considered as an optical analogue to genome-wide-yeast-two-hybrid, till date technological difficulties have restricted the use of FRET on a large scale. These difficulties include extremely short working range (typically 2–6 nm), unfavourable orientations of the donor-receptor pair and relatively high false-negative detection of interactions.

Intrinsic time-resolved UV fluorescence has been recently proposed as an approach to probe protein interactions without the need for prior labeling (Striebel et al., 2004). This approach measures changes in the fluorescence decay time in the intrinsic fluorophores tryptophan and tyrosine of proteins due to protein–protein interaction. This detection method is based on the comparison between measured fluorescence lifetimes of immobilized proteins before and after interaction. This method assumes that the lifetime of the excited electronic states of molecules is sensitive to their environment. For example, such environment can be changed by protein–protein interaction, which results in changes in the fluorescence lifetime. The excitation of the intrinsic fluorescence of tryptophan and tyrosine could be achieved using a laser beam ($\lambda = 280$ nm), provided the pulse duration is much shorter than the monitored fluorescence lifetime. This label-free approach offers two advantages: first, modifications of proteins are minimal, which reduces possible perturbation of protein–protein interactions. Second, low-abundance proteins are not diminished further by labeling and associated purification steps. Compared with measuring absorption or fluorescence intensity changes upon binding, time resolved methods tend to be less sensitive to alteration in concentrations of the substrate as well as the binding protein. This approach has been tested for proteins in solution toward protein partners immobilized on protein micrroarrays. Other measurements within the same study investigated different binding states of protease/protease-substrate pairs and tubulin/kinesin systems.

Current literature demonstrates that protein labeling with fluorophores followed by fluorescence detection is the most common approach for microarrays probing. Immunoassay strategies used in protein-detecting microarrays include the following: (a) sandwich immunoassay in which the capture antibodies are immobilized on the solid support, and bound proteins are detected using a second, labeled detection antibody; (b) antigen capture assay in which proteins are similarly captured by immobilized antibodies, but the captured proteins are detected directly. This is usually accomplished by chemically labeling the complex mixture of proteins before applying them to the array. In the two-color version of this assay, two samples are labeled independently with distinguishable fluorophores, and the samples are mixed before applying them to the array; (c) in a direct assay, the complex mixture of proteins is itself immobilized on the solid support, and specific proteins in that mixture are visualized using labeled detection antibodies. Each of these strategies has its advantages and limitations. For instance, two-color antigen capture methods are less demanding in terms of sample preparation and spotting because protein quantification is based on ratio comparison. On the contrary, and as in the case of direct assay, the two-color approach is limited by the specificity of the affinity reagents. Furthermore, protein labeling may introduce other complications such as preferential labeling of antigenic epitopes of some proteins leading to the loss of their ability to be captured by their affinity reagents (MacBeath, 2002). The only source of signal amplification in the two-color assay is that associated with the fluorescent dyes, which may explain a rather limited sensitivity (~ng/mL). Attempts to address this limitation included enzyme-catalyzed chemiluminescence (Moody et al., 2001) and rolling-circle amplification (Lizardi et al., 1998; Schweizer et al., 2002). Unlike the antigen capture method, the sandwich assay does not require protein labeling, which simplifies sample preparation and increases the throughput. Furthermore, and more importantly, the sandwich assay addresses the question of specificity, which is considered as one of the main obstacles confronting detection strategies used to probe protein microarrays. The superior specificity of this assay compared with direct assay and the antigen capture assay has been elegantly described by MacBeath (2002). The author compared the latter assays to a western blot in which each lane has been compressed into a single band, and therefore, all crossreacting proteins will contribute to the measured signal. By contrast, a sandwich assay is rather like an immunoprecipitation reaction coupled with a compressed western blot. In this case, one is far more likely to observe a single band on a western blot when the product of an immunoprecipitation reaction is analyzed than when total cellular lysate is analyzed. The main drawback of the sandwich assay is that two noncompeting affinity reagents are needed for each protein.

In one of the earliest and most extensive attempts to examine the feasibility of a two-color labeling approach, Haab et al. (2001) assembled 115 antibodies and their protein ligands. The authors prepared defined mixtures of the antigens, labeled them with activated Cy5 dye and combined each mixture with a reference mixture labeled with Cy3 dye. The proteins were then applied to antibody microarrays comprising the 115 antibodies spotted at high density on slides coated with poly-L-lysine. In this study, it was found that only 20% of the arrayed antibodies provided specific and accurate measurements of their target antigens at a concentration of 1.6 μg/mL or less, but that some antibodies could detect ligands at concentrations of less than 1 ng/mL and at partial concentrations of 1 part in 1,000,000. The same two-color approach was applied to the proteomic analysis of LoVo colon carcinoma cells in response to treatment with ionizing radiation (Sreekumar et al. 2001). Proteins in lysates from untreated cells and cells collected 4 h after irradiation were labeled with Cy3 and Cy5 dye, respectively. The samples were mixed and applied to a microarray composed of 146 distinct antibodies directed against proteins involved in stress response, cell-cycle progression and apoptosis. In addition to observing the upregulation of five apoptotic proteins already known to be induced by radiation, these authors discovered six other proteins that were up-regulated and one that was downregulated; most of these results were confirmed subsequently by immunoblotting.

2.8.4. Functional Protein Microarrays

These microarrays are produced by printing the proteins of interest on the array using methods designed to maintain the integrity and the activity of the proteins, allowing hundreds to thousands of target proteins to be simultaneously screened for function (Zhu and Snyder, 2003; MacBeath, 2002; Mitchell, 2002). This type of microarray can be used to examine target protein interactions with other molecules, such as drugs, antibodies, nucleic acids, lipids or other proteins. Furthermore, the same microarrays can be interrogated to find substrates for enzymes (Cahill and Nordhoff, 2003; Jona and Snyder, 2003). Possible applications of this class of microarrays may include the following: Screening a class of enzymes with a candidate inhibitor to assess binding selectivity; probing a range of enzymes with a drug candidate to discover undesired binding targets, which may contribute to unexpected toxicities (LaBaer and Ramachandran, 2005). A number of studies have demonstrated the potential of this type of microarray in various areas. In a relatively recent study, function-based protein microarrays were generated by printing complementary DNAs into glass slides, and then translating target proteins with mammalian reticulocyte lysate (Ramachandran et al., 2004). Epitope tags fused to the proteins allowed them to be immobilized in situ. This approach offered a number of advantages including bypassing protein stability problems associated with the storage; it allowed the capture of sufficient proteins to perform functional studies, and above all, it obviated the need to purify proteins. This experimental design was used to map pairwise interactions among 29 human DNA replication initiation proteins, recapitulate the regulation of Cdt1 binding to select replication proteins and map its geminin-binding domain.

High spatial density is one of the perquisites of functional protein microarrays. Two examples of achieving such objective have been given by MacBeath and Schreiber (2000) and Zhu et al. (2001). In the first example, a high-precision robot designed to manufacture cDNA microarrays was used to spot proteins onto chemically derivatized glass slides at extremely high spatial densities (1600 spots /cm^2). Although the spotted proteins were covalently attached to the slide surface, the authors demonstrated that these proteins have retained their ability to interact specifically with other proteins or with small molecules in solution. These microarrays were used to demonstrate three different applications: screening for protein–protein interactions, identification of the substrates of protein kinases and identification of protein targets of small molecules.

Zhu et al. (2001) constructed a yeast proteome microarray containing approximately 80% yeast proteins and screened it for a number of biochemical activities. To prepare the proteome chips, the authors cloned 5800 open reading frames, overexpressed and purified their corresponding proteins, which were then printed in duplicate onto glass slides using a commercially available microarrayer. In their initial experiments, the authors used aldehyde-treated microscope slides (MacBeath and Schreiber, 2000) in which fusion proteins were attached to the surface through their primary amines at their NH_2-termini. In subsequent studies, proteins were spotted onto nickel-coated slides, in which the fusion proteins were attached through their HisX6 tags. Comparing the performance of the two types of slides, it was concluded that nickel-coated slides gave a better signal for various protein preparations. The efficiency of fusion protein attachment was assessed by probing the surfaces with antibodies to GST (glutathione S-transferase), whereas the reproducibility of such attachment was assessed by comparing the signals associated with each pair of duplicated spots. Using these functional microarrays, the authors were able to identify many new calmodulin and phospholipids interacting proteins; a common potential binding motif for many of the calmodulin-binding proteins.

Despite a notable success of the way to construct and apply functional microarrays, it is not always easy to produce and purify the many proteins needed for protein spotting microarrays. Purification of mammalian proteins from Escherichia coli is a representative case of such difficult task. Some recent studies have suggested more pragmatic approaches to circumvent such difficulties. In these approaches, it was attempted to concentrate on a family of closely related proteins using the same protocol, which would allow the use of a single protocol to purify. This was recently demonstrated by Boutell et al. (2004), where a function-based microarray was constructed and used for parallel quantification of the effects of mutations and polymorphisms on the DNA-binding function of the p53 oncoprotein. Investigation of protein domains rather than the full-length proteins is another alternative to simplify purification. This approach was demonstrated by Espejo et al. (2002), where over 200 domains were expressed and purified and microarrayed onto nitrocellulose slides. These domains included SH3, SH2 and PDZ, and were screened with labeled peptides. The binding of these peptides in a predicted pattern was interpreted as an indication of the stability and accessibility of the immobilized domains.

2.8.5. Reverse-phase Protein Microarrays

Reverse-phase protein microarrays can be considered as newcomers to the field of protein microarrays technology. This class of microarrays can use tissue or cell samples to perform either functional or analytical tasks, which explains their increasing use as a tool for the discovery of potential tumor biomarkers. Unlike the earlier generation of protein microarrays in which immobilized capture molecules are directed against certain target proteins (e.g. an antibody), reverse phase microarrays immobilize the whole repertoire of sample proteins associated with tissue, or cell line, lysate. This type of array is characterized by high linearity and good sensitivity; furthermore, there is no need for a labeling procedure. To underline the potential role of this type of microarray in the search for potential cancer biomarkers, some applications are worth considering. High-density reverse-phase lysate microarrays have been constructed and applied in proteomic profiling of the NCI-60 cancer cell lines (Nishizuka et al., 2003). The authors examined 60 human cancer cell lines used by the National Cancer Institute to screen compounds for anticancer activities. Each glass slide microarray included 648 lysate spots representing the cell lines and controls. One of the main findings of this study was the identification of two pathological markers, which could potentially distinguish colon from ovarian adenocarcenomas. A second interesting finding by this study is the correlation between protein expression and the mRNA levels obtained for the same genes. The latter data were obtained by cDNA and oligonucleotide microarrays. The authors reported that cell structure-related proteins showed a high correlation between mRNA and protein levels across the NCI-60 cell lines, whereas noncell structure-related proteins showed poor correlation. This correlation has been previously assessed within single cell types (Anderson and Seilhamer, 1997; Gygi et al., 1999); however, the characteristics of reverse-phase format and the diversity of the investigated cell lines allowed different types of assessment across different cell types for each protein. This meant that the authors could evaluate whether different classes of proteins could result in different correlation between mRNA and protein levels.

The combination of laser capture microdissection and cDNA microarray has been used by Paweletz et al. (2001) to construct reverse-phase microarrays, which were used for longitudinal analysis of prosurvival checkpoint proteins from patients with matched histologically normal prostate epithelium to prostate intraepithelial neoplasia (PIN) and then to invasive prostate cancer. A high degree of sensitivity, precision and linearity was achieved, making it possible to quantify the phosphorylated status of signal proteins in subpopulations of human tissue cells. Basically, histopathological cell polpulations were microdissected, lysed in a suitable buffer, and approximately 3 nL of the lysate were arrayed onto glass-backed nitrocellulose slides. This arraying resulted in 250–350 μm each containing the entire cellular repertoire corresponding to a given pathologic state that has been captured. Subsequently, each slide was probed with a suitable antibody that could be detected by fluorescent, colorimetric, or chemiluminescent assays. In this experimental arrangement, over a 1000 individual cellular lysates could be accommodated on a 20×30 mm slide using 1 μL of lysate. On the basis of this investigation, the authors reported that cancer progression was associated with increased phosphorylation of the serine/threonine kinase Akt, suppression of apoptosis pathways and decreased phosphorylation of extracellular signal-regulated kinase (ERK). At the transition from histologically normal epithelium to intraepithelial neoplasia, a statistically significant surge in phosphorylated Akt was observed, together with a concomitant suppression of downstream apoptosis pathways preceding the transition into invasive carcinoma.

The combination of laser capture microdissection, reverse-phase tissue microarrays and phosphor-specific antibodies was also used by Wulfkuhle et al. (2003) to examine the activation status of several key molecular gates involved in cell survival and proliferation signaling in human ovarian tumor tissues. Frozen ovarian tumor tissues were collected at a family cancer research institute and the National Ovarian Cancer early detection program clinic. A total of 40 cases encompassing major histotypes and stages of ovarian cancer were examined. Approximately 100 protein lysate arrays were printed from 40–50 μL of starting material for each case. These arrays were then probed with antibodies against proteins representing key nodes in a wide variety of signaling cascades. Antibodies were first validated for specificity by western blotting against microdissected tissue lysates. The main deductions of this study were as follows: the levels of activated extracellular-regulated kinase (ERK1/2) varied considerably in tumors of the same histotype, but no significant differences between different histotypes were observed; advanced stage tumors had slightly higher levels of phosphorylated ERK1/2 compared with early stage tumors. A similar approach was used by Grubb et al. (2003) in which the authors used LCM in combination with reverse-phase microarrays to investigate the status of key points in cell signaling involved in pro-survival, mitogenic, apoptotic, and growth regulation pathways in the progression from normal prostate epithelium to invasive prostate cancer. Both this study and the one by Wulfkuhle et al. (2003) were among a number of studies which first demonstrated the utility of reverse phase tissue microarrays for the multiplexed analysis of signal transduction in cells procured directly from human prostate and ovarian tumor specimens. Some of the deductions in these studies suggest that patterns in signal pathway activation may be patient specific rather than type or stage specific. Of course, further data are needed to give a stronger support for such suggestions.

Numerous protein posttranslational modifications are generally difficult to capture using either recombinant proteins or antibodies that do not distinctly recognize specific forms of a protein. Cell- and tissue-based reverse-phase microarrays offer the possibility to perform a comprehensive analysis of proteins in their modified forms. A study along these lines has been conducted by Madoz-Gúrpide et al. (2001). In this approach, various separation modes including reverse-phase and ion-exchange LC were used to fractionate cell and tissue lysates followed by arraying and probing of the individual fractions. A similar approach but for a different objective has been reported by Yan et al. (2003). Two-D liquid phase separation composed of pI-based fractionation (pH 4–7) is followed by nonporous LC of each pI fraction. The resulting cellular protein fractions were arrayed onto nitrocellulose slides and used to investigate humoral response in prostate cancer cell line, LnCAP. This cell line was fractionated using a gradient at 0.2 pI intervals. Each of these fractions was collected and separated in the second dimension by nonporous silica reversed-phase LC. The authors estimated that the sample was fractionated into as many as 1400 protein bands, some of which contained multiple proteins. Proteins eliciting immune response were identified by various mass spectrometric analyses.

2.8.6. Future Prospects

Over the past few years, microarray technology has become a central component for the large-scale and high-throughput biology. This technology allows fast, simple, and parallel interrogation of thousands of addressable elements in a single experiment. Protein microarrays are poised to become one of the most powerful tools in the field of large-scale biology. Despite its recent introduction, it has already demonstrated a great potential in basic research, diagnostics, and biomarkers discovery. However, the full impact of this emerging technology on proteomics, medical research, and drug and biomarkers discovery is yet to be fully realized. Given the relatively short period of time during which this technology has demonstrated a part of its potential, it is reasonable to expect that such technology will experience further developments and improvements, which will no doubt render it one of the most powerful tools in the field of large-scale biology. Some of these expected developments can be underlined by the following considerations:

- Label-free detection methods such as biosensors and mass spectrometry are expected to assume a major role in future design of protein microarrays. These methods offer two obvious advantages: first, modifications of proteins are kept to the minimum, which in turn contribute to the conservation of the native states of the arrayed proteins. Second, minute amounts of interesting proteins are not diminished further by labeling reactions and purification steps. At present, most screens that are used to probe protein microarrays require some type of fluorescent or radiolabeling. This labeling step imposes extra time, costs, and in some cases interferes with the monitored reaction. We also have to bear in mind that most fluorescent compounds are hydrophobic, which results in a high background noise with the inevitable result of false positives (Cooper, 2002). Several groups have already demonstrated the integration of these powerful techniques for affinity-based capture and characterization of ligands (Nelson et al., 2000, Williams and Addona, 2000). The strength of the combination of SPR with MS is that the first component could be used to detect, capture, and subsequently digest and deliver nanomolar to femtomolar levels of ligand for MS analysis.
- Preparing a large number of soluble, pure, and stable recombinant proteins still represents a bottleneck for extensive application of protein microarrays particularly for functional studies. Widening of this bottleneck has also to be accompanied by the production of high-quality substrates with well-defined and reproducible surface properties and optimized surface chemistries to immobilize the capture proteins homogeneously and in a functional conformation. Currently, most arrays are created on glass or silicon slides treated with various agents to immobilize the proteins. There are promising attempts to increase the density of the microarrays, which would allow the analysis of more samples than is currently possible. A promising approach in this direction uses photolithography to etch miniature wells on silicon surfaces. The immobilized proteins or antibodies are located in the flow chambers on the chip, so that they are always kept in aqueous solution and do not denature through drying out (Mitchell, 2002). In a relatively recent study, Lee et al. (2002) applied photolithography to construct nanoarrays with sample features of 0.1 μm. To probe these high-density microarrays, it is necessary to use high resolution and fast-scanning readers, which are still not commercially available.
- The preparation of arrays with a diverse set of protein families is still challenging (Lee and Mrksich, 2002). The usual approach involving cloning, expressing, and purifying can be used to prepare hundreds of proteins, yet it is not possible to assay the activity of each protein, and therefore an unknown proportion of the products will not be active in the array. This situation can be further aggravated by some posttranslational modifications including phosphorylation, proteolysis, and glycosylation, which can alter the activities of the proteins, requiring the preparation of

multiple forms of each protein. This means that mammalian expression systems will be required to prepare proteins that have specific posttranslational modifications. In one innovative approach (Ziauddin and Sabatini, 2001), seeded mammalian cells on a surface having a patterned array of cDNA found that cells were transfected by the DNA, leading to production of the encoded proteins.

- Future protein microarrays have also to address an increasing demand for chips that contain arrays of functional integral membrane proteins, which are key participants in signal transduction events and represent the site of action of approximately half of the approved drugs (Drew, 2000). The feasibility of constructing this type of array has been demonstrated by Fang et al. (2002). The authors described the fabrication of arrays of G-protein-coupled receptors and assays for screening of ligands of these arrays.

- Emerging microarrays including carbohydrates and peptides are expected to assume a stronger role in the investigation of protein activities. Immobilized 9-peptide substrates on a gold-coated glass surface were used to form a high-density peptide microarray (Houseman et al., 2002). These microarrays were probed with SPR, fluorescence, and phosphoimaging to study the substrate of the nonreceptor tyrosine kinase *c-Src*. The same study provided preliminary quantitative evaluation of the effect of three known inhibitors of the kinase. Zhu and Snyder (2003) also reported the design of twenty 17-mer peptide substrates covalantely immobilized to epoxy-activated glass surface and used them to screen 120 yeast kinases for their preferred substrates.

2.9. MULTIDIMENSIONAL LIQUID CHROMATOGRAPHY COUPLED TO MS

Traditionally, 2-DE has been the method of choice for resolving complex protein mixtures derived from biological samples such as sera and tissues (Klose and Kobalz, 1995). This powerful analytical technique is still considered a central component in the search for disease biomarkers. A number of limitations including limited automation, relatively narrow dynamic range, and poor separation of insoluble, highly basic, highly acidic, low (<10 kDa), and high(<150 kDa) molecular mass proteins have encouraged the search for more efficient strategies for protein mixtures analyses. Over the last decade, liquid chromatography coupled to tandem mass spectrometry (LC/MS–MS) has emerged as an effective method for the identification/quantification of proteins from complex mixtures (Eng et al., 1994; Yates III et al., 1996; Link et al., 1997). The fact that this strategy deals with peptides rather than intact proteins results in a number of advantages and eliminates a number of drawbacks encountered in separations on gel. On the chromatographic side, these strategies employ a bibasic column with a section of reversed-phase material flanked by strong cation exchange (SCX) resin. In other variation, a third component of reverse phase material can be added to allow online desalting of the sample. The evident advantages of this strategy are higher automation, capability to analyze membrane proteins, and reduction of sample complexity through successive steps, which allow the mass spectrometer to identify most if not all the components reaching it. Recently, this strategy has been enhanced through the introduction of various labeling schemes. Isotope-coded affinity tag (ICAT) (Gygi et al., 1999) and stable isotope labeling by amino acids in cell culture (SILAC) (Ong et al., 2002) are two of such schemes, which have gained popularity. The incorporation of stable isotopes into proteins allows for the simultaneous identification and quantification of proteins associated with two cellular states. Whereas on one hand, the SILAC method is not suitable for the analysis of clinical samples because the isotope labeling is achieved through metabolic route; on the other hand, the ICAT and other chemical labeling methods use post-extraction isotope labeling, which render them suitable for the profiling and quantification of proteins derived from clinical samples.

Strategies using liquid chromatography coupled to mass spectrometry have four main components each of which has its impact on the final outcome of the analysis: liquid chromatography, mass spectrometry, labeling schemes, and bioinformatics, which naturally encompass database searches. These components together with representative examples on the application of this technology in the search for potential cancer biomarkers are discussed below.

2.9.1. Protein Labeling

Labeling proteins to facilitate their quantification in two different samples (e.g., control and disease) can be easily compared with the pioneering work by Schena et al. (1995). In that approach, mRNA species in control and experimental samples were labeled with different fluorescent dyes, the two samples were mixed, individual mRNA species were isolated by hybridization, and the relative degree of change in mRNA concentration between the two samples was determined by fluorescent ratio measurements. Current labeling strategies for protein quantification can be divided into two main groups, each of which has within it a number of variations. In all MS-based strategies, the first and may be the most critical step is to modify the molecular mass of a specific amino acid so that it can be distinguished from its unlabeled counterpart in the detection phase. This can be done in various ways: In one approach, stable isotope labeling is used without changing the chemical identity of the amino acid as in the case of introducing heavy atoms of H, C, N, and O within various functional groups. In a second approach, chemical modification with or without stable isotope labeling is used. Alkylating Cys is an example of the first case, whereas guanidination which transfers C-terminal Lys to homoarginine is an example of the latter.

2.9.2. Labeling a Specific Amino Acid

The ICAT approach was first introduced by Gygi et al. (1999; 2000) where light/heavy chemical reagents were used to alkylate Cys residues. The ICAT reagent consists of three components: the first is a thiol-specific reactive group for alkylating Cys residues; the second component is simply a polyether linker to allow the replacement of 8 H-atoms with their corresponding D-atoms; and the third component is a biotin group which is used to isolate ICAT-labeled peptides during the chromatography phase. In this strategy, proteins from two different cell states are harvested, denatured, reduced, and labeled at cysteines, one of the states with light, whereas the other with heavy ICAT. The samples are then combined and digested with trypsin. ICAT-labeled peptides are isolated by biotin-affinity chromatography and then analyzed by online LC–MS, and MS–MS. The measured ratio of the ion intensities of an ICAT labeled pair of peptides can be used for quantifying the relative abundance of their parent proteins present in the two samples. Furthermore, the MS–MS data can provide amino acid sequences, which can be used for database searches and for the identification of the detected proteins.

More recently, two new variations of the ICAT approach have been reported. The first is designated absolute quantification of proteins (AQUA) (Gerber et al., 2003), whereas the second is dubbed (VICAT), which stands for visible isotope-coded affinity tag (Lu et al., 2004). In the VICAT approach, the labeling reagent tags the thiol groups of Cys or thioacetylated amino groups and introduces into the tryptic peptide a biotin affinity component, a visible moiety for tracking the chromatographic location of the tagged peptide by methods other than mass spectrometry. In the AQUA method, short synthetic peptides that are chemically identical to the native target peptide but labeled with stableisotope tags serve as internal standards to precisely and accurately quantify the absolute levels of the protein after proteolysis using selected reaction monitoring in a tandem mass spectrometer.

Over the last 5 years, the principle on which the ICAT approach is based has been replicated by different research groups. Various amino acids were chosen for the labeling, whereas the basic steps were almost identical to the ICAT method. In a recent report, Kuyama et al. (2003) have described a method for relative protein quantification based on labeling of tryptophan residues through the reaction with light and heavy 2-nitrobenzenesulfenyl chloride (NBSCI-$^{12}C_6$ or -$^{13}C_6$). The two versions of this chemical reagent differ by 6 Da, which can be used in MS analysis to distinguish and quantify two identical sequences present in two different protein mixtures. It is not difficult to note that the underlying principle of this approach is identical to that applied in ICAT. The main difference of course is in the chemistry of labeling and in the choice of the amino acid to be labeled. The authors reported that the NBS reagent exhibited selectivity for the indole ring of tryptophan, and could incorporate a sixfold stable isotope labeled benzene nucleus. The main steps in the experimental procedure can be summarized as follows: The tryptophan residues in a denatured protein sample derived from cell state 1 are modified with isotopically light NBS, and the equivalent residues derived from cell state 2 are modified with the isotopically heavy NBS. The two samples are combined and passed through a Sephadex LH-20 column to remove excess reagent and other small molecules. Disulfide bridges in the eluting mixture were reduced and alkylated and the mixture was subjected to enzymatic digestion. Peptides containing labeled tryptophan were enriched using sephadex LH-20 column and were analyzed by MALDI–TOF–MS or LC–ES–MS. This approach was tested for fairly simple standard mixtures of peptides to verify the reliability of the quantification and the influence of labeling on the elution times of light and heavy-labeled peptides. Biologically derived samples were also spiked with known proteins and examined by the same approach.

Guanidination to convert C-terminal Lys to homoarginine was used by Cagney and Emili (2002), who introduced the term mass coded abundance tagging (MCAT) and demonstrated its use in combination with LC/MS–MS for the relative quantification of proteins in mixtures. As in the case of ICAT, this approach is based on the principle of modifying a specific amino acid residue and of following its course through the use of chromatography coupled to mass spectrometry, and database searching to identify and possibly quantify the precursor protein(s) associated with the two different states. When comparing the initial work on ICAT with MCAT, two qualitative differences can be identified: The first chooses Cys for modification through an alkylation reaction, whereas MCAT uses guanidination to convert C-terminal Lys to homoarginine. In the original ICAT approach, the reaction is affected on intact proteins, whereas the MCAT applies the reaction on the tryptic peptides. In the ICAT approach, the mass difference between two identical peptides (one modified) is 8 Da, whereas the reaction used for the MCAT approach results in 42 Da difference (in both cases the difference refers to singly charged ions). Apart from these differences, the two approaches adopt a very similar procedure for the relative quantification of proteins present in two different states. In the MCAT approach, the two tryptic digests before and after guanidination are combined with a predetermined ratio, separated by reversed-phase capillary LC, and introduced into an ESI source for MS and MS/MS analysis. The resulting full-scan mass spectra are used to obtain the relative abundance of correlated peptides by comparing the intensities of their reconstructed single-ion chromatograms.

The esterification of carboxyl groups of aspartic and glutamic acid residues and at the C-terminus of peptides is another method which has been tested for protein quantification. This method was first introduced to address the issue of in vitro labeling of Cys-free peptides, which the already existing ICAT approach could not handle (Goodlett et al., 2001). In this approach, protein mixtures to be compared are digested and the resulting peptides are methylated using either d_0- or d_3-methanol. Methyl esterefication converts carboxyl acids, such as those in the side chains of aspartic and glutamic acids as well as the carboxyl terminus, to their corresponding methyl esters. The two-peptide mixtures marked with d_0- or d_3- are combined with a known ratio and then examined by micro-capillary liquid chromatography coupled to tandem mass spectrometry. There are a number of good reasons as to why this approach can be considered complementary to the ICAT technique; not all proteins within a mixture contain Cys residues, some proteomes will have much lower total protein coverage than the 92% reported for yeast. Some Cys residues may contain posttranslational modifications prior to isolation making them unavailable for alkylation with Cys-specific ICAT reagents, and while one advantage of the in vivo labeling approaches is that an internal standard for all peptides is produced, not all systems of biological interest are amenable to tissue culture growth conditions.

2.9.3. Stable Isotope Incorporation

Several groups have used stable isotope-labeled internal standards (e.g., $^{2}H_{1}$ ^{13}C, ^{14}N, ^{18}O, etc.) to obtain global quantitative protein profiles. Stable isotopes have been introduced to proteins via metabolic labeling using heavy salts or amino acids (Conrads et al., 2001), enzimatically via transfer of ^{18}O from water to peptides (Oda et al., 1999; Mirgorodskaya et al., 2000; Yao et al., 2001). In principle, isotope tags can be incorporated into proteins during cell growth or after cell lyses. For example, Oda et al., (1999) used this approach in conjunction with LC/MS and SDS–PAGE/MS to identify and quantify a number of proteins in yeast culture. In another study, Smith et al. (2002) have used capillary LC combined with accurate mass measurements to obtain quantitative changes in protein abundances. In their study, D. radiodurans was cultured in two media, the first containing the natural abundance of ^{14}N and ^{15}N, whereas the second medium was chemically enriched with ^{15}N (>98%). The two cultures were mixed, processed, and the combined proteome sample was analyzed by capillary LC coupled to high-resolution FT–ICR mass spectrometer.

Yao et al. (2001) described stable isotope labeling, which could provide quantitative information as well as comparison between individual proteins from two entire proteome pools or their sub-fractions. Basically, two ^{18}O atoms are incorporated universally into the carboxyl termini of all tryptic peptides during the proteolic cleavage of all proteins in the first pool. Proteins in the second pool are cleaved analogously with the carboxyl termini of the resulting peptides containing two ^{16}O. The two peptide mixtures were pooled for fractionation, separation, and examined by high-resolution mass spectrometry, which yielded the accurate masses and the isotope ratios of peptide pairs, which differ by 4 Da.

Stable isotope labeling with amino acids in cell culture, or SILAC, has recently gained popularity for its ability to compare the expression levels of hundreds of proteins in a single experiment. SILAC makes use of ^{12}C- and ^{13}C- labeled amino acids added to the growth media of separately cultured cell lines, giving rise to cells containing either "light" or "heavy" protein populations (Ong et al., 2002; Blagoev et al., 2003). In this approach, protein populations from experimental and control samples can be directly mixed, digested, and subjected to mass spectrometry analysis. A comparison between SILAC and ICAT labeling and analysis protocols is given in Figure 2.8. This approach was recently applied to compare the expression levels of over 400 proteins in the microsomal fractions of prostate cancer cells with varying metastatic potential. The authors reported that in the highly metastatic cells, 60 proteins have experienced 3-fold up-regulation, whereas another 22 proteins were found to be 3-fold down-regulated.

2.9.4. Limitations of Labeling

Each of the labeling protocols cited above have demonstrated its capability in delivering samples suitable for the comparison of protein levels derived from two different samples. Having said that, none of these schemes on its own can be considered sufficient to handle the complexity of the proteome within complex biological samples.

This is simply because of limitations associated with the labeling chemistry and the subsequent purification, separation, and detection. Setting aside the efficiency and specificity of the labeling, which are never 100%, other limitations include: (i) labeling a single amino acid would by definition exclude the detection and quantification of proteins, which do not contain such amino acid or happen to contain a modified version. The ICAT strategy, which targets Cys, is a representative example of such situation. Existing literature suggests that almost 7% of known proteins do not contain Cys in their sequence (Vuong et al., 2000). To capture this percentage, it is evident that a different or a complementary strategy has to be considered. This situation is almost the opposite of that encountered in Lys-labeling with cyanine dyes (Cy2, Cy3, and Cy5), where excessive labeling is one of the major limitations of the DIGE strategy (Ünlü et al., 1997). (ii) When dealing with a complex tryptic digest of a total cell lysate, one might want to use different chromatographic procedures. For example, isotopically labeled peptides could be separated by ion-exchange chromatography followed by a reversed-phase column, or by reversed-phase chromatography followed by ion-mobility separation. It turns out that if labeled peptides are separated through a reversed-phase column, then a well-known isotope effect takes place, by which the deuterated peptide elutes earlier than its nondeuterated counterpart. Such separation causes a variation in isotope ratio across the two differently eluting profiles of the isoforms, an effect which becomes more pronounced with small peptides and can influence the reliability of the quantification. (iii) Optimal performance of most if not all exiting labeling strategies cannot be achieved unless such strategies are accompanied by some form of prefractionation and enrichment of low-abundance proteins. Labeling and quantification of serum proteins is a representative example of such analyses. There are some recent examples to suggest that the enhancement of the performance of certain labeling strategies can be achieved through the exploitation of the physical properties of a given subclass of proteins. For example, a glycocapture method was developed so that isotope labels were introduced into the glycopeptides after glycocapture, which allowed a simultaneous protein identification and quantification by LC/MS–MS (Zhang et al., 2003). This method has great potential and several obvious advantages in biomarker discovery and validation. In this method, albumin, the predominant protein component in serum, is effectively left behind by the glycoprotein purification step, which effectively removes a large contaminant from the sample and increases the chance of detecting low-abundant proteins in the samples. Furthermore, isotopic labeling of glycopeptides could facilitate biomarker quantification in patient samples under different disease states. The glycocapture method could also be applied to microdissected tissue specimens because many cell-surface proteins also undergo N-linked glycosylation. For example, prostate-specific antigen (PSA) is N-linked glycosylated, which should be detectable and quantifiable using this methodology. Of course, glycosylation is not the only modification that the above strategies have to deal with. For example, some of the in vitro pre- and postdigestion labeling procedures, which target a specific amino acid, are not suitable for the quantification of phosphorylation. On the contrary, in vivo labeling has such potential, which has been demonstrated by Oda et al., 1999. The authors demonstrated the concept of using isotopically labeled media for the quantification of various phosphorylation sites. For some of the existing labeling strategies to be able to address the question of protein phosphorylation, such strategies have to be complemented with various steps such as specific chemical reaction methods for the phosphate group, and metal affinity columns or antibodies (MacCoss et al., 2002).

2.10. CHROMATOGRAPHIC SEPARATION

In chromatographic approaches, the proteins, in most cases, are digested into peptides prior to separation. The advantage is that peptides, particularly from membrane proteins, are more soluble in a wide variety of solvents and hence easier to separate than the intact proteins. The disadvantage is a tremendous increment in the number of components within such a mixture, which have to be fractionated prior to entering the mass spectrometer. To address the question of efficient fractionation of complex peptide digests prior to their detection, a number of research groups have proposed various combinations of chromatographic and electrophoretic approches. Considering existing configurations, it is not difficult to realize that they have two general characteristics in common: first, the combination is made of orthogonal separation methods. In other words, such methods are based on different separation principles. The application of the principal of orthogonal combination of two separation methods to handle complex protein mixtures has been demonstrated almost 30 years ago (O'Farrell, 1975). The combination of isoelectric focusing, which separates proteins according to their charge with SDS–PAGE which, separates the same proteins according to their mass, facilitates the separation of thousands of proteins within the same run. Another example demonstrating the power of orthogonality is the combination of reverse-phase LC and capillary zone electrophoresis (CZE). The first technique exploits the hydrophobicity of various molecular entities, whereas the latter relies on electrophoretic mobility (mass-to-charge ratio). Second, the combined methods have compatible buffers, amenable to automation, and easy to couple to various detection methods particularly mass spectrometry.

In the following sections, I shall avoid reviewing various works, which have been described in a number of recent reviews (Issaq, 2001; Issaq et al., 2002; Wagner et al., 2002). Instead, I shall discuss in some detail a number of specific works, which will give the reader a reasonable feel for the capability and future potential of multidimensional chromatography combined with mass spectrometry. This potential will be better illustrated by considering a number of applications of this technology in the area of biomarkers discovery.

2.10.1. Three Dimensional Separation

The first truly three-dimensional separation arrangement was first described by Moore, Jr and Jorgenson (1995). The authors used size exclusion chromatography (SEC) as the first dimension to separate components according to their mass. The

SEC effluent was repetitively sampled online into a rapid two-dimensional reversed-phase LC/CZE system. The capability of this system was tested using a digest of hen ovalbumin, and the separated components were detected by laser-induced fluorescence. The acquired results demonstrated that the various components of the digest were located in three-dimensional space by their size, hydrophobicity, and electrophoretic mobility. This early example of three-dimensional separation has underlined a number of experimental observations, which provided clear indications of the advantages and pitfalls of such experimental arrangement: (i) the major advantage of this three-dimensional arrangement is the gain in peak capacity. Geddings (1987) has shown that the peak capacity of a two-dimensional chromatographic method is roughly the product of the peak capacities of the one-dimensional component. The same work showed that this rule is equally applicable to a multidimensional system of a number of dimensions *n*. Thus, the final peak capacity in a three-dimensional system can be roughly defined as the product of the three values associated with the three separating components. The authors reported that the addition of SEC as a third dimension resulted in a substantial increase in the overall peak capacity; (ii) the authors rightly pointed out that highly concentrated samples were used to overcome dilution factors associated with the physical interface between SEC and reversed-phase LC. This observation implies that sample-limited situations would challenge such experimental arrangement; (iii) to obtain reliable and reproducible elution/migration times in 3-D, rigorous control of temperature in each of the dimensions is needed. Failure to do so would result in time drifts due to temperature fluctuations; and (iv) analysis time in the first dimension (SEC) was in the order of a few hours, whereas the remaining 2-D (LC/CZE) took only a few minutes. This huge disparity is associated with the fact that each analysis in the two-dimensional system represents only a single point in the first dimension. Thus, to take full advantage of the potential of such arrangement, the analysis times of the two-dimensional portions have to be much shorter than those in the first dimension to allow complete sampling of the components separated in the first dimension.

In another and much earlier work, Geddings (1991) reported the coupling of three different chromatographic columns to perform three-dimensional analysis. This approach involved some form of fractions collection in the first dimension, which was then reinjected into the remaining two-dimensional system. Strictly speaking, such an arrangement cannot be considered a comprehensive three-dimensional system comparable to that described by Moore, Jr and Jorgenson (1995). The high price in terms of sample requirements, long analysis times, and fairly labor-intensive procedures associated with three-dimensional arrangement have convinced many research groups that two-dimensional chromatography coupled to mass spectrometry is a more attractive alternative for high-throughput proteomic analyses.

2.10.2. Two-dimensional Chromatography

The simplest method to perform online two-dimensional separation is to directly couple in series two columns with orthogonal separation mechanisms. However, similar results in terms of resolution and peak capacity can be obtained by using

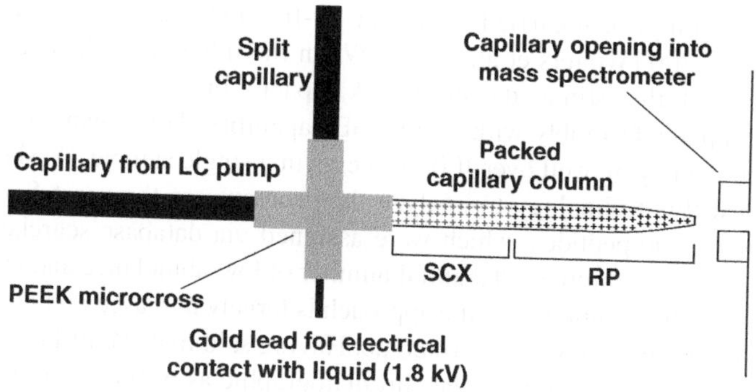

Figure 2.13. A diagram of MudPIT column and electrospray interface. Adapted from Wolters et al. (2001) with permission.

a single capillary column packed with two different stationary phases, as shown in Figure 2.13. The most prominent work that follows this experimental arrangement is dubbed multidimensional protein identification technology (MudPIT), developed by Yates and colleagues (Link et al, 1999; Washburn et al., 2001; Wolters et al., 2001). In this experimental arrangement, the microcapillary column is a packed fused silica capillary with a pulled tip at the outlet end for direct coupling to MS instrumentation. The end of the capillary nearest to the pulled tip is packed with several centimeters of reversed-phase material, followed by several centimeters of strong cation-exchange (SCX) material. The column is connected to a PEEK microcross which splits the flow from a conventional quaternary LC pump, reducing it to 0.1–0.2 μL/min, and also provides a connection for the voltage needed to operate the electrospray ionization source.

2.10.3. Basic Considerations Regarding MudPIT

In a typical MudPIT analysis, denatured and reduced protein complex is first digested to generate a mixture of peptides. The acidified complex peptide mixture is applied to a strong cation-exchange (SCX) chromatography column, and a discrete fraction of the absorbed peptides are displaced onto a reversed-phase chromatography column. To elute fractions of the sample from the SCX portion of the column onto the reverse phase portion, a step gradient of buffer with increasing ionic strength is used. Peptides are retained on the RP column, but contaminating salts and buffers are washed away and diverted to waste. The peptides are then eluted from the RP column into the mass spectrometer using a gradient of increasing acetonitrile. Finally, the RP column is reequilibrated in preparation for absorbing another fraction of peptides from the SCX column. An iterative process of increasing salt concentration is then used to displace additional fractions of peptides from the SCX column onto the RP column. Each simplified fraction is eluted from the RP column into the mass spectrometer. The capabilities of MudPIT as a separation method for proteomics analysis have been assessed by several studies. The peak capacity of this experimental arrangement carried out in one 15-fraction analysis was estimated as approximately 3200 (Wolters et al., 2001). When the inherent peak capacity of the MS was included, the estimated overall peak capacity increased to 23,000, an estimate that compares favorably with 2-D PAGE capability. This means that MudPIT is capable of identifying well over 1000 proteins in a single sample. Washburn et al. (2001) applied this technology to analyze the proteome of the yeast *S. cerevisiae* identified over 5000 peptides, which were assigned via database searches to 1484 proteins. Among these were a substantial number of low-abundance and trans-membrane proteins, which implies that the approach is largely unbiased and gives a fairly representative picture of the yeast proteome. Having said that, MudPIT has its limitations, which are associated with the chromatographic as well as with the MS side of the technology. Paradoxially, the chromatographic limitations are mainly due to the simplicity of the setup, which is mainly composed of a quaternary gradient LC pump, packed microcapillary column, and the minimum number of interfacing components. Such a simple arrangement renders the system rather inflexible. Both the two separation modes coupled in tandem must utilize gradient elution. Furthermore, the gradient in the first dimension must always be run in a stepwise fashion in order to elute concise fractions of sample onto the second column. The resolution contributed by the first dimension is limited by the number of fractions transferred, which is equal to the number of steps used in the first-dimension gradient. Each step requires substantial time, typically 100 min, which includes periods for reequilibration and washing the salts off the column. A maximum of 15 steps in the first-dimension salt gradient have been reported (Washburn et al., 2001), which may suggest that the first dimension is undersampled in spite of long run times. Notably, the only combination of separation modes that has been used so far in MudPIT is strong cation-exchange reversed phase, perhaps because this is the only arrangement where suitable gradients can be applied alternately to two different stationary phases in a single column in order to bring about an effective, orthogonal two-dimensional separation. On the MS side, there are still some technical as well as tintrinsic difficulties, which limit the potential of MudPIT technology: first, despite a relatively lengthy two-dimensional fractionation, which allows the capture of thousands of fragmentation events within a collision cell, the scan rates of most exiting mass spectrometers is still too slow to cope with all the peptides eluting into the ion source. Second, signal suppression in an ion source containing a wide range of peptides including those derived from posttranslationally modified proteins still represents one of the limitations of this strategy.

2.10.4. Mass Spectrometry and Data Analysis

As it has been stated earlier, mass spectrometry is one of the main components of MudPIT strategy. Mainly, electrospray ionization coupled with different mass analyzers have been used for this type of analysis. The eluting peptides are fed to the ion source at a flow rate of 0.1–0.2 μL/min. This source is generally coupled to a mass analyzer capable of providing tandem mass spectrometry data. These data are obtained by the isolation of an individual peptide and fragmenting it by collision with an inert gas (e. g., He or Ar). Although most of the early MudPIT measurements were mainly conducted on three-dimensional ion trap (Link et al., 1999; Washburn et al., 2001; McDonald et al., 2002), it would be useful to give a brief description of this and other types of mass spectrometers, which are highly suitable for this type of analysis.

The three-dimensional quadrupole ion trap has a number of characteristics, which made it one of the main instruments in current proteomic analysis. These characteristics include robustness, sensitivity, and relatively low price. The same trap, however, has a number of limitations associated with its relatively low trapping efficiency for externally generated ions (<10%) and space-charge effects due to a relatively small volume to which the ions are injected. The isolation of a particular *m*/*z* value to perform MS/MS measurements can be accomplished through a combined manipulation of the DC and RF potentials to bring the ion to an apex of the stability diagram where all other *m*/*z* values will be unstable (Dawson et al., 1969; Mather et al., 1978). To perform collision-induced dissociation (CID), the amplitude of the resonance signal can be adjusted to induce collisions with the He damping gas rather than ejection from the ion trap (Louris et al., 1987). Multiple low-energy collisions are sufficient to transfer enough energy into a peptide to cause its fragmentation in a manner similar to that obtained in CID in a triple quadrupole. While performing CID, the amplitude of the RF signal, which sets the stability of the isolated ion, has to be carefully chosen to eject all ions below a specific *m*/*z*, a working condition which also limits the acquired fragmentation information associated with low *m*/*z* values. This means that to obtain a complete set of b- and y-type ions, it is necessary to use multiple stages of analysis (MS^n).

The linear or two-dimensional ion trap is a relatively recent development, which addresses some of the limitations associated with the three-dimensional ion trap. There are a number of advantages associated with ion trapping in a linear ion trap; this type of trap has a high ion beam acceptance since there is no quadrupole field along the axis of the beam. Ions admitted into a pressurized linear quadrupole undergo multiple collisions, which reduce their axial energy prior to encountering the end electrodes thereby enhancing trapping efficiency. The larger volume of the pressurized linear trap relative to the three-dimensional conventional ion trap means that more ions can be stored before the onset of the charge space effects. Radial containment of ions within a linear trap results in a strong focusing effect along the center of the trap, a behavior different from that encountered in the conventional ion trap, where the focusing is achieved at a specific point.

In a recent article, Hager (2002) described two experimental configurations based on the ion path of a traditional triple quadrupole. In the first arrangement, a pressurized collision cell placed between two conventional quadrupole mass filters was used as the linear trap. In such configuration the product ion scans are accomplished by selecting the precursor ion using the first RF/DC transmission quadrupole. Fragment ions are generated by accelerating the precursor ion into the pressurized collision cell where the fragment ions and the remaining precursor ions are trapped and then axially scanned out of the ion trap through an RF-only quadrupole to the detector. This configuration bears a close resemblance to conventional triple quadrupole instrument with the additional capability of ion trapping. This means that scan functions such as multi-reaction monitoring and constant neutral loss scans can be performed, functions that cannot be performed on the conventional three-dimensional ion traps. The second arrangement described by the same author (Hager, 2002; Hager and Yves Le Blanc, 2003) uses the final quadrupole as the linear trap. In such configuration the product ion spectra are obtained by using the first quadrupole to select the precursor ion, which is then accelerated to the pressurized transmission collision cell. Ions emerging from this cell are trapped in the linear ion trap and subsequent mass is selectively scanned out of the trap to the detector.

There are other two trapping configurations, which are still to make their impact in MudPIT technology: Fourier transform ion cyclotron resonance (FT-ICR) (Marshal et al., 1998) and orbitrap mass analyzers (Makarov, 2000; Hardman and Makarov, 2003). In the first technique the ions are trapped by the combined effect of an electric field and a strong magnetic field. Among its strengths are high resolution, high mass accuracy, and high sensitivity. Despite its enormous potential, the expense, operational complexity, and low peptide fragmentation efficiency has limited its routine use in proteomics research. It has been proposed (Makarov, 2000) that the orbitrap mass analyzer could potentially become a feasible alternative to FT-ICR. The orbitrap mass analyzer employs the trapping of pulsed ion beams in an electrostatic quadro-logarithmic field. This field is created between two electrodes, where stable ion trajectories are produced by the combination of rotation around one of the electrodes with harmonic oscillations along the same electrode. Whether this type of analyzer can be an alternative to FT-ICR is still to be demonstrated.

The abbreviation QqTOF is commonly used to describe a hybrid analyzer, where Q refers to a mass resolving quadrupole, q refers to an RF-only quadrupole or hexapole collision cell, and TOF refers to the time-of-flight section. The development of this configuration followed closely the development of the so-called ESI-TOF arrangement, where the principle of orthogonal ion injection to a TOF analyzer was applied. The QqTOF can be regarded either as the addition of RF/DC resolving quadrupole to an ESI-TOF or a triple quadrupole in which the third mass filter is replaced by a TOF analyzer. The principle and other detailed theoretical treatment of orthogonal TOF have been given a number of previous reviews (Guilhous et al., 2000; Yefchak et al., 1990). In the MS/MS mode, the quadrupole is operated in the mass filter mode to transmit only the parent ion of interest, typically selecting a mass window from 1 to 3 Da wide, depending whether the transmission of the full isotopic cluster is required. The ion is then accelerated to an energy between 20 and 200 eV before it enters the collision cell, where it undergoes CID. After the first few collisions with neutral gas molecules (usually argon or nitrogen), the resulting fragment ions as well as the remaining parent ions are collisionally cooled and focused into the TOF section of the instrument, where they are detected in reflector mode.

2.10.5. Data Analysis and Interpretation

Multiple steps MudPIT analysis can generate thousands of MS/MS spectra, which pose a real challenge in terms of interpretation and database interrogation. Strategies for automated data analysis, streamlined filtering, and extracting meaningful data out of enormous lists of numbers are therefore required. Several search algorithms are currently available to interpret MS/MS data generated by MudPIT analysis, the two most prominent algorithms being SEQUEST (Eng et al., 1994) and Mascot (Perkins et al., 1999). Centroided MS data are submitted to Mascot in the form of peak lists, optionally with associated intensity values. In the case of MS/MS data, peak detection is also required in the chromatographic run, so that multiple spectra from a single peptide can be summed and spectra from the baseline can be discarded. The fundamental step in this approach is to calculate the probability that the observed match between the experimental data set and each sequence database entry is a chance event. The match with the lowest probability is reported as the best match. Whether such match is significant depends on the size of the database. The authors who first introduced this algorithm (Perkins et al., 1999) adopted a convention often used in sequence similarity searches, and reported a score which is "$-10\log_{10}(P)$," where P is the probability. This means that the best match is the one with the highest score, and a significant match is typically of the order of 70.

The SEQUEST algorithm converts the character-based representations of amino acid sequences in a protein database to a fragmentation pattern that can be used to match fragment ions in a given MS/MS spectrum. The algorithm initially identifies amino acid sequences in the database that match the measured mass of the peptide ion and predicts the fragment ions expected for each sequence. A score is calculated for each amino acid sequence by matching the predicted ions to the ions observed in the acquired tandem mass spectrum, and the highest scoring amino acid sequences are then reported. Matching peptide sequencing by these algorithms is commonly obtained through the interrogation of various databases among which are Swiss-Prot/TrEMBL (http://ca.expasy.org/sport), International Protein Index (www.ebi.ac.uk/IPI/IPIhelp.html), and National Center for Biotechnology Information (www.ncbi.nlm.nih.gov).

2.10.6. Application of Multidimensional Chromatography/MS

To underline the potential of MudPIT technology in the area of biomarkers discovery, a number of recent applications are described below:

Tirumalai et al. (2003) described the application of MudPIT for the profiling of low molecular mass proteins in human serum. Briefly, standard human serum was purchased from the National Institute of Standards and Technology. High-molecular-mass proteins were removed by centrifugal ultrafiltration using solvent conditions to disrupt protein–protein interactions and to release low-molucular-mass species that might be bound to large prorteins. Following filtration the low-molecular-weight serum proteome was digested with trypsin, fractionated by SCX chromatography, and analyzed by microcapillary-reversed-phase chromatography coupled to mass spectrometry. The acquired MS/MS spectra were searched against a human protein database using SEQUEST algorithm. This study has identified over 340 human serum proteins and demonstrated that serum albumin was efficiently removed from the sample. The findings of this study were later used by

Anderson et al. (2004) to compile a list of 1175 distinct gene products based on different methodologies including MudPIT in the investigation of human plasma proteome. A two-dimensional chromatograph (SCX/RP–LC) coupled to a linear ion trap mass spectrometer was also used by Fujii et al. (2005) to investigate a limited number of samples from healthy and lung adenocarcinoma patients.

The combination of LCM, ICAT labeling, and 2-dimensional liquid chromatography/tandem mass spectrometry was used to conduct a fairly large-scale study to investigate the qualitative and quantitative proteoms of heptocellular carcinoma (Li et al., 2004). Proteins extracted from microdissected Heptacellular Carcino (HCC) and from non-HCC hepatocytes were examined by 2-D-LC-MS/MS with or without the attachment of a photocleavable ICAT label. The authors reported qualitative identification of 644 proteins and unambiguos quantification of 261 proteins.

Detection and quantification of various forms of protein posttranslational modifications remain a challenge for most existing analytical technologies including MudPIT. Current litertature contains a few examples, which, nevertheless, have demonstrated the potential of such technology. MacCoss et al. (1998) used differential proteolysis in combination with MudPIT for mapping a sizable number of posttranslational modifications (PTMs). The use of three different enzymes resulted in a large series of overlapping peptides with the direct result of increasing the sequence coverage used to identify various proteins within the investigated mixtures. The authors reported the impressive number of 73 PTMs among 270 identified proteins. These modifications included phosphorylation, methylation, acetylation, and oxidation.

Another example demonstrating the success of this approach has been recently reported (Pan et al., 2005). The method is based on the selection and chemical synthesis of isotopically labeled reference peptides that uniquely identify a particular protein and the addition of a panel of such peptides to the sample mixture consisting of tryptic peptides from the proteome under investigation. Peptides were first separated by reversed-phase LC, spotted on a MALDI plate, and subjected to MALDI–MS–MS analysis. This experimental arrangement was tested for the identification and quantification of glycopeptides in human serum. The innovative feature of this study is the capability to perform high-throughput analysis of selected proteins and to obtain their absolute abundance. Future studies based on this strategy may provide accurate information on the expression of targeted proteins in tissue and serum samples. Such anticipated investigations are likely to prove powerful tools for the discovery of potential biomarkers for various forms of cancer.

2.10.7. Outlook for Multidimensional LC/MS

Using multidimensional LC to resolve peptides through orthogonal dimensions prior to MS analysis continue to gain a broader acceptance as an alternative to the well-established 2-DE. One of the reasons for this new trend lies in the fact that peptides are more uniformly soluble and much easier to handle both outside and inside the mass spectrometer. Recently, multidimensional chromatography has been enhanced by the more frequent use of stable isotope labeling such as ICAT and SILAC, which substantially facilitate quantification of proteins in a wide range of samples including clinical samples. These isotopic-labeling approaches have been pioneered using in vitro cell culture systems (Aebersold and Mann, 2003); however, their application to clinical samples has yet to be fully tested. For example, the ICAT method has been proposed to quantify differentially expressed cell-surface proteins in normal and in tumor tissues derived from colon and lung cancers (Domon and Broder, 2004). It is becoming more evident that no single technology platform can handle the complexity of even a simple proteome. However, each of the existing platforms have particular strengths, which if fully exploited can respond to many unresolved proteomic challenges. Mutidimensional chromatography in combination with mass spectrometry and certain labeling strategies is certainly in a position to be a key player in the search for disease biomarkers. However, before such a platform can deliver on its promise, a number of aspects have to be further improved or extended including: (i) an increased use of prefractionation to deplete highly abundant proteins (e.g., in serum) would certainly enhance the capability of this approach to reveal and even quantify low-abundant and generally more informative proteins/peptides. A number of works along these lines have been cited earlier in the text; however, I find it useful to point out two other works, which underline the importance of abundant protein depletion. Adkins et al. (2002) at Pacific Northwest National Laboratory used microcapillary liquid-phase chromatography, digestion to peptides, and ion trap mass spectrometry on samples in which immunoglobulins were depleted using binding proteins (proteins A/G) and reported 490 distinct proteins. Pieper et al. (2003) fractionated serum proteins by immunoaffinity chromatography to remove eight highly abundant proteins (albumin, haptoglobin, transferrin, transthyretin, alpha-1 antitrypsin, alpha-1 acid glycoprotein, hemo-pexin, and alpha-2macroglobulin), followed by sequential anion-exchange and size-exclusion chromatography, before 2-DE. They resolved about 3700 protein spots and identified 1800 by MS, which were recognized as 325 distinct proteins after sequence homology and similarity searches to eliminate redundancies. (ii) Further optimization and increased use of strategies which target specific class of proteins. Preliminary validation of this approach has been demonstrated by Zhang et al. (2003). To simplify the investigated mixture, the authors applied what they called the "glycocapture method," which allows for the selective enrichment of glycosylated proteins in serum, cells, or tissues. This strategy was adapted so that isotopic labels were introduced into the glycopeptides after glycocapture, which allowed the simultaneous protein identification and quantification by LC–MS/MS. Two other methods to reduce sample complexity and to allow accurate protein quantification have been designated as "AQUA" (Aebersold, 2003; Gerber et al., 2003) and "VICAT" (Lu et al., 2004). These methods have in common the mass spectrometer that is focused on the analysis of the targeted analyte and the process ignores the complex matrix of peptides that are unrelated to the targeted protein. Wright et al. (2005) suggested that such methods that also have the potential to determine the absolute quantity of an analyte may become increasingly important to monitor the levels of previously discovered biomarkers during prostate cancer tumorigenesis and in oncology in general. In the AQUA method, short synthetic peptides that are chemically identical to the native target peptide but labeled with stable isotope tags serve as internal standards to precisely and accurately quantify the absolute levels of the protein after proteolysis using selected reaction monitoring in a tandem mass spectrometer. For example, AQUA peptides representing a specific biomarker can be added to resected normal, benign prostate hyperplasia, and prostate cancer tissues and the abundance of the corresponding native protein can be monitored during prostate cancer tumorigenesis. In principle, the VICAT method is very similar to the AQUA method, except the VICAT reagent reacts with cysteine-containing peptides and thus provides another level of enrichment and quantification of proteins.

2.11. IMAGING MASS SPECTROMETRY

Mass spectrometry is not normally conceived as a technique which can provide spatially resolved information. This wrong conception was first challenged by Casting and Slodzian (1962) over 40 years ago. In their work the authors generated ions by bombarding solids with few tens of kiloelectron particles and monitored the ejected ions by a magnetic analyzer. These types of analyses formed the basis for today's secondary ion mass spectrometry (SIMS). The same authors argued that it should be possible to build an ion-optical collection system, analogous to a lens used in a light microscope to preserve the spatial relationship of the desorbed ions from sample to detector. The various aspects and phases of development of imaging using mass spectrometry have been reviewed by Packolski and Winograd (1999). The modern version of imaging mass spectrometry and its application to investigate the distribution of proteins and drug candidates in brain tissues had been pioneered in the late 1990s by Caprioli's group at Vanderbilt University (Caprioli et al., 1997; Chaurand et al., 1999; Stoeckli et al., 2001).

This technology requires matrix assisted laser desorption ionization (MALDI) to generate the ions, sample plates large enough to accommodate whole tissue samples and can be moved in the x and y direction with sufficient accuracy, the diameter of the laser beam and its localization on the sample should be within the micrometer range, and of course adequate software tools to acquire, interpret, and archive the enormous amount of data generated by this type of analysis. The outcome of the analysis is profiles and two dimensional ion density maps of molecules. It is reasonable to state that this technology is still going through a phase of technical optimization related to sample preparation, operational characteristics of the laser beam, and above all the software packages to support such a technology. Although there is still a paucity of applications of this approach in the area of biomarkers discovery, the same technology has demonstrated a huge potential in various areas of proteomics, and it will be a matter of time before such technology becomes a key player in the search for biomarkers for various pathologies. Indeed imaging mass spectrometry has already demonstrated its utility in the profiling of normal and diseased tissues. These studies included the spatial distribution of proteins in xenographs of human gliomas (Stoeckli et al., 2001), both normal and cancerous mouse prostate (Masumori et al., 2001), healthy and cancerous mouse colon tissue

(Chaurand et al., 2001), mapping of neuropeptides in the rat brain (Fournier et al., 2003), and in human lung tumor biopsies (Yanagisawa et al., 2003). The following sections describe some technical aspects, applications, and future potential in the area of biomarkers discovery.

2.11.1. Tissue Preparation and Matrix Application

The crystallization of samples within a suitable matrix compound plays a crucial role in obtaining high-quality mass spectra from complex biological samples. Over the last decade, various protocols have been developed for various samples. Limiting the present discussion to the analysis of tissues, it can be reasonably stated that one of the first and possibly the most important elements in this strategy is the preparation of the tissues including matrix application prior to MS analysis. Tissue preparation methods have to be carefully designed to preserve the integrity and the spatial arrangement of the investigated molecules. Schwartz et al. (2003) proposed that following the surgical removal of the tissue, they should be loosely wrapped in aluminum foil and lowered gently into liquid nitrogen. The authors cautioned that rapid plunging of the tissue into the liquid nitrogen may result in cracking and brittle edges of these tissues. Furthermore, placing freshly excised tissues into plastic tube prior to freezing can cause sticking to the sides of the tube and can modify the shape of the stored slice.

Obtaining an optimal tissue thickness is still considered as a compromise by the users of this technology. Thick sections are easier to manipulate but take longer to dry and may result in poor electrical conductivity inside the mass spectrometer, leading to poor quality mass spectra. On the contrary, very thin sections are more difficult to handle and can tear easily. The current compromise is to choose a thickness of the order of 10–20 μm, which is of the order of the diameter of a mammalian cell, which implies that the majority of the cells within the investigated tissue are cut open, exposing the intercellular contents. The frozen tissues are generally cut into thin sections in a cryostat, which allows accurate sectioning, while minimizing sample contamination. Transfer of the tissue slice to the MALDI plate can be accomplished in different ways: The preferred method involves a cold plate that is prepared by simply placing it in the cryostat chamber held at −15 °C while sectioning. Following careful positioning of the tissue on the cold plate, it can be quickly warmed to thaw the mounted tissue. The advantage of this method is the preservation of water-soluble proteins. The second method uses a room temperature plate, which can be placed over the frozen section and, as the slice thaws, it adheres to the plate. This method can result in the loss of soluble proteins in the ice left behind, and it may cause protein delocalization during the thawing procedure. One additional practical consideration, which may add some advantages to imaging analysis, is the material of the MALDI plate. Once the plate is inserted into the ion source, it would be very helpful to be able to distinguish the surface of the plate from that of the tissue. Such differentiation would contribute to a more accurate localization of the laser beam during acquisition, which has a direct impact on the reproducibility of the analysis. For example, a polished, gold-coated surface provides high contrast to the dull surface of the tissue, thus enhancing tissue outline recognition. Additionally, differences in cell morphology can frequently be distinguished in sections on a gold surface. This property is useful for obtaining high-quality digital photographs of tissue sections. Stainless steel plates, on the contrary, do not provide this advantage. One of the important aspects of the investigation of disease tissue is the ability to compare histologic features obtained from stained sections using light microscopy with MS images. Previously, this was accomplished using two separate adjacent sections, one for histology and the other for mass spectrometry. Often visual registration between both sections was difficult because of differences in morphology or because of preparation-induced physical changes. Furthermore, it is difficult to locate tissue features, which may not always be visible, particularly when tissue sections are mounted on different opaque plates. To perform both an optical evaluation and MS analysis on the same tissue section, attempts have been made to mount tissue sections on optically transparent and conductive sample plates. The conductivity is necessary to avoid surface-charge accumulation once the plate is introduced into the ion source, which would interfere with the accelerating voltage and subsequent calibration of the acquired mass spectra. Novel conductive glass slides and MALDI-friendly staining protocols have been developed by Caprioli's group as an alternative to metal plates (Chaurand et al., 2004a,b). These slides allow one tissue section to be first stained for histological classification and then analyzed by imaging mass spectrometry. This approach offers the advantage of being able to directly analyze the stained section by MALDI, whereas metal plates require collecting an adjacent section of tissue on a microscope slide for staining to correlate morphological regions with the mass spectrometric analysis.

Once the initial tissue preparation steps have been completed, matrix solution is deposited on the tissue surface prior to analysis. Choosing the best matrix and optimizing application parameters are necessary to obtain high-quality mass spectra directly from tissue samples as well as maintaining the spatial integrity of the tissue surface. Although matrix selection and concentration are generally fixed for a given analysis, variations in solvent composition can be manipulated to favour the detection of a class of molecules in the tissue sample. Sinapinic acid (3,5-dimethoxy-4-hydroxycinnamic acid, SA) is routinely used for high-molecular mass proteins, whereas α-cyano-4-hydroxycinnamic acid (HCCA) is a more common matrix for lower molecular mass molecules such as peptides or drugs. Because of the widely varying surface properties of tissue sections, the matrices used must effectively crystallize over the entire area of the tissue to yield high-quality spectra. Matrix crystallization and the quality of the generated mass spectra from mouse liver sections were evaluated by comparing the performance of three commonly used matrices: SA, HCCA and DHB (2,5-dihydroxy benzoic acid) (Schwarz et al., 2003). Droplets of matrix (20 mg/ mL of each matrix in 50:50 acetonitrile/ 0.1% trifluoroacetic acid in water) were deposited on the surface of a mouse liver section and analyzed by imaging mass spectrometry. On the basis of these analyses, the authors concluded that both SA and HCCA formed dense crystals on the tissue, whereas DHB showed poor crystallization. Furthermore, SA showed a superior quality mass spectrum, containing over 300 individual signals, good signal-to-noise ratio and a low base line. On the contrary, HCCA showed a less number of signals, poor signal-to-noise ratio and a disturbed base line. In imaging MS analysis, the method of matrix application can also influence the outcome of the analysis. In tissue analysis by imaging MS, there are various methods for applying matrix to tissue sections: These include the addition of matrix in individual droplets, spraying by nebulization, acoustic waves and electrospray deposition. The first method targets a specific region in the investigated tissue and uses small droplets of matrix to cover the targeted region. This method is useful when the aim of the investigation is to compare protein profiling in two regions of the same tissue (e. g. normal vs diseased or treated vs untreated). Other methods of matrix deposition are designed for whole tissue coating (Chaurand et al., 2004 a,b; Schwarz et al., 2003). In the nebulization approach, a nebulizer is used to deposit the matrix directly onto the tissue surface. The primary goal is to maximize solubilization of proteins within the tissue in order to enhance co-crystallization of proteins and matrix molecules as well as the final protein profile, while minimizing protein delocalization. Increasing the duration for which proteins are exposed to large quantities of solvent also increases the chance the molecules can physically move over the tissue surface. To minimize the latter effect, full matrix coverage over the entire tissue sample is carried out in short-time cycles in which small volumes of matrix solution are deposited during each cycle. The sample plate is held vertically a few centimetres away from the sprayer nozzle. As matrix is nebulized from the sprayer, the device is moved parallel to the target to evenly apply matrix and just barely wet the sample surface to avoid the undesired deposition of a large quantity of solvent in any one region of the tissue. The sample is allowed to dry before the next coating cycle is performed. An average of about ten cycles of coating and drying should be performed to achieve an even, dense crystal field. Cycling small volumes of matrix spray allows a slow development of a dense matrix coat while minimizing the time the sample remains damp. Since different types of tissues can have different surface properties, the number of coating cycles can vary depending on the sample.

2.11.2. MS Acquisition

Most published data on MS imaging have been conducted on commercially available MALDI-TOF instruments operated in linear mode. The ions are generated by a 337 nm nitrogen laser, which can be operated at repetition rates of 3 or 20 Hz. The diameter of the laser beam, which has a direct influence on the resolution of the acquired image, is in the range of 50–80 μm. We have to bear in mind that the laser spot is halfway between circular and elliptical shape. This type of instrument generates MS data, and if operated in reflector mode, can generate MS/MS data through the use of post source decay analysis. More recent analyses have used hybrid instruments (e.g. qudrupole/time-of-flight) capable of generating MS/MS data (Reyzer et al., 2003). This type of analysis is almost mandatory when the objective of the study is the imaging of small molecules (>1000 Da) which can suffer from interference by matrix clusters and/or indigenous molecules within the investigated tissue. Ion image acquisition is generally controlled by custum software, that interfaces the instrument controller and acquisition software. This software controls acquisition over a predetermined area and reconstructs ion density maps or images by plotting measured signal intensities over such area. This is generally done by using the instrument control software to set a data-acquisition grid that defines a discrete cartesian pattern across the sample surface. This pattern has a fixed center-to-center distance between spots, in the order of 50–150 µm depending on the dimension of the section and on the imaging resolution required. Mass spectra are then acquired using this grid pattern and a predetermined number of laser shots per spot. Image construction is done by dedicated software, which integrates signal intensity at the desired m/z values across the data set. The acquisition time of each image is directly related to the area of the tissue to be analyzed, resolution, the monitored m/z range and the number of laser shots per spot. High resolution imaging measurement can generate data files in the range of 1–2 GB. High resolution imaging is highly desirable in the analysis of tumor biopsies, where the differentially expressed MS signals may furnish valuable information on the architectural arrangement of various cell types and likely, molecular interactions between the tumor and its surrounding stroma. Before these types of analyses become routine, a number of technical difficulties have to be resolved. Such difficulties are directly related to laser characteristics, tissue preparation and software and data elaboration: First, Laser-beam size and its frequency are essential for high spatial resolution and for reasonable analysis times. Reducing the beam size to 1 µm or less would certainly facilitate spatial resolution of the desorbed ions particulariy from tissue structures and single cells which are less than 10 µm. Currently, laser beams on commercial instruments have a diameter of 50–100 µm. Increasing imaging speed through the use of high repetition rate lasers (kHz range) in combination with fast data acquisition systems would certainly reduce the current analysis times. Improvement on the bioinformatics side are also needed to handle storage, processing and mining of an enormous amount of data generated by high-resolution imaging of tissue sections. It is sufficient to remind ourselves that even at medium image resolution, thousands of spots are irradiated per image. Normalization of these spectra and their comparison with those from other sections, mixed pattern recognition as well as other data related tasks, represent a formidable challenge, which has to be addressed. There are other minor shortcomings associated with the current capability of commercially available MALDI instruments. For example, at the present time and even under delayed extraction conditions, the mass resolution required for this type of analysis starts to deteriorate above m/z 50,000 (Bahr et al., 1997). Above this value, peaks become broader resulting in inevitable low mass accuracy. Most MALDI-TOF instruments use microchannel plates for ion detection, which are known to be more biased toward smaller ions, which have a higher velocity than their heavy counterparts present within the same ion beam.

2.11.3. Some Representative Applications of Imaging MS

Considering existing literature on imaging MS, it is reasonable to state that the full potential of this approach for biomarkers discovery is yet to be realized. Having said that, the technique has already demonstrated such potential as a powerful tool for the analysis of both proteins and drugs in tissues. Some of these applications and the type of information provided are considered in the following section.

In a recent study, which can be described as a combination of imaging mass spectrometry and SELDI, Yanagisawa et al. (2003) described the use of MALDI-MS to generate proteomics patterns from fresh frozen lung-tumor tissues, which were used to classify and predict histological groups as well as nodal involvement and survival in resected non-small-cell lung cancer (NSCLC). The describtion of the study as a combination of the two technologies derives from the fact that mass spectra were directly generated from fresh frozen tissues (typical of Imaging MS), but rather than constructing images, the authors used the data together with statistical packages to generate proteomics patterns (typical of SELDI analysis). To establish proteomics patterns in lung tumors, the authors assessed protein expression profiles of 50 tissue samples in a training cohort, which included surgically resected lung tumors, pulmonary metastases of previously resected NSCLC, and pulmonary carcinoid. The authors reported various numbers of MS signals, which could distinguish normal lung from tumor and from primary NSCLC. The latter was also differentiated from cancer metastatic to the lung, adenocarcinoma from squamous-cell carcinoma and squamous-cell from large cell carcinoma. It has to be pointed out that the identities of most of the proteins within these distinguishing profiles were largely unidentified. In another study (Schwarz et al., 2004), protein patterns in primary brain tumors have been assessed in tissue by imaging mass spectrometry. The authors analyzed twenty consecutive human tumor and nontumor tissues and concluded that the acquired MS data provided reliable identification of normal and neoplastic tissues and could be used to discriminate between tumors of different grades. Given the extremely low number of patients, such conclusions have to be considered as preliminary, and further studies involving a more substantial number need to be conducted to support such findings.

The capability of imaging mass spectrometry to map both small and large molecules in tissues has been described by Rohner et al. (2005). The authors used commercial MALDI-TOF instrument equipped with a YAG laser, which could be pulsed up to 300 Hz together with a fast digitizer board. One of the interesting analyses was the imaging of amyloid beta (Aβ) peptides of 4-5 kDa molecular mass. These peptides are considered as one of the pathological features of Alzheimer's disease. Brain sections from transgenic mice overexpressing human amyloid precursor protein were prepared and coated with a matrix solution containing insulin as an internal standard for m/z calibration and to normalize signal levels on the whole area of the brain section. The acquired images demonstrated that Aβ-(1-40) and Aβ-(1-42) were the most abundant amyloid peptides. This result is in reasonable agreement with earlier findings showing that Aβ-(1-40) is the major component in *aqueous* cerebral cortical extracts in patients suffering from Alzheimer's disease (Mori et al., 1994), whereas the insoluble Aβ-(1-42) peptide was primarily in senile plaque cores (Miller et al., 1993; Roher et al., 1993). Within the same study, the authors demonstrated the power of imaging MS/MS when dealing with low molecular masses, which are *vulnerable* to interferences by *indigenous* ions and/or ions associated with the matrix. This approach has the advantage of selecting a target molecule within a complex medium and monitors its fragments. This type of analysis can provide additional information, which well-established techniques such as whole-body autoradiography fail to provide. Although autoradiography provides high resolution images, it does not distinguish between a parent molecule and its metabolite simply because it measures the radioactivity, whereas imaging MS–MS does have the capability to distinguish between a molecule and its metabolite and/or its fragment ion(s).

Troendle et al. (1999) used MALDI in conjunction with a quadrupole ion trap equipped with a laser microprobe to detect the drug paclitaxel (MW 853) from rat liver and human ovarian tumor tissue. The liver tissue was incubated with a solution of paclitaxel, whereas the ovarian tumor tissue was obtained from an animal dosed with paclitaxel. The concentration of drug was approximately 50 mg/kg in each tissue of interest. In both cases, the presence of the drug was unambiguously determined by comparison of the MS–MS spectra from the tissue with those obtained from the standard drug. This investigation, however, did not provide information on the localization of the drug within the investigated tissue. Reyzer et al. (2003) also used MS–MS imaging to analyze drug candidates in tissue. A hybrid instrument (qudrupole-TOF) equipped with a MALDI source was used to analyze tissue of mouse tumor dosed with antitumoral compounds in the phase of development.

The last few years have witnessed increasing efforts to characterize various forms of cancer through the use of noninvasive techniques prior to deciding on surgery. Magnetic resonance imaging (MRI) combined with magnetic resonance spectroscopy has used brain metabolite information as a classifier of brain tumor (Simonetti et al., 2003). Cross-correlating this approach with a classification based on imaging mass spectrometry would certainly improve the eventual diagnosis, and may even allow a more reliable classification. A fairly recent study (Crecelius et al., 2005) described the development of three-dimensional visualization of proteomics data correlated to anatomical features in the brain. As an example, the authors chose to model the distribution of myelin basic protein (MBP) in the corpus callosum (CC). The mouse brain was chosen because of its small size and because an atlas already exists which can be used as a reference for the methodology development. To create the three-dimensional model of MBP within the CC, optical, histologic and MALDI IMS images were acquired and processed according to a fairly elaborate workflow (for full details see Crecelius et al., 2005). In such an approach, a number of registration steps were required to produce the final three-dimensional protein distribution. First, optical micrographs of unstained sections were aligned to a reference atlas. Next, the optical images and MALDI MS images were registered to each other on the basis of landmarks visible in both imaging modalities. Finally, a three-dimensional surface of the anatomical feature was created from the aligned optical images, and the created 2-D MALDI MS images were inserted into the three-dimensional surface model according to their z coordinate position. The resulting three-dimensional model provided a unique correlated view of MBP distribution within the CC of the mouse brain. These initial data suggest that this methodology can be used to map the distribution of any protein visible in the ion image spatially correlated with anatomical features observed in optical micrographs.

2.11.4. Current Limitations and Potential Developments

The present analysis of blots or tissue sections by MALDI-MS typically reveals about 300 distinct mass signals in the m/z range up to about 100,000, with the majority of these signals under m/z 20,000. Such limit is directly related to the operating characteristics of commercially available instruments. Most, if not all, such instruments use microchannel plates for ion detection. This type of detector detects electrons generated by ions impacting on its surface, which renders the detection velocity dependent. Within m/z 100,000, there is a wide range of velocity, where the lighter and faster ions are detected more efficiently than the heavier and slower ions. Over the last decade, there has been increasing interest in cryogenic detectors; however, we are still waiting to see their use on commercial instruments. Cryogenic detectors have been developed since the early 1980s for applications in particle physics, astrophysics, and also for material analysis. There are a number of reasons behind their use in these diverse fields. These include high resolving power, low energy threshold and large sensitive absorbing areas. The ability of these detectors to detect slow-moving massive ions with 100% efficiency and the additional information gained from measuring the energy of the same ions have attracted a number of attempts to use them in TOF mass spectrometry. This type of detector has a number of characteristics, which make its use in TOF-MS highly attractive. On paper, cryogenic detectors should not exhibit any decrease in sensitivity for large masses. Any impacting particle will create a signal as long as the energy deposited exceeds the low noise level of the detector, typically a small fraction of an electron volt. Cryogenic detectors are, therefore, expected to detect large slow-moving ions or neutrals with 100% efficiency. The energy resolution provided by these detectors may also be used to differentiate between charge states. This is because a doubly charged ion would have twice the kinetic energy of its singly charged counterpart, and thus would result in doubling the pulse height. Such a feature can be used, for example, to distinguish between a doubly charged dimer and its singly charged monomer, even though they both have the same flight time. The same characteristic can be used to discriminate between fragments and their precursors. Further details on the characteristics and operation principles of cryogenic detectors have been given in a number of works (Frank et al., 1999; Kraus, 2002; Towerenbold et al., 1996; Towerenbold, 1996; Towerenbold et al., 2001).

Another important aspect of imaging mass spectrometry is the speed of data acquisition. Commercially available instruments are generally equipped with low frequency (up to 20 Hz) nitrogen lasers, limiting the speed of image acquisition. At such frequency, the number of laser shots per spot and the time necessary to reposition the sample plate for the subsequent acquisition can amount to 2–3 s for movements as small as 50 µm. An improvement in this direction has been reported in a recent article (Rohner et al. 2005). The authors described the use of ND: YAG laser, which could be pulsed at 300 Hz, an almost 15-fold increase in speed compared with most commercial instruments. Data download and processing time is another element, which is in continuous evolution. Currently there are transient recorders operating at 1 kHz, which can average 50 laser shots per spot at a laser frequency of 200 Hz, giving a cycle time of ~0.25 s. The total time needed to acquire an image is, of course, dependent on the dimensions of the tissue section to be imaged and the spot or image resolution. For example, a 5 × 5 mm tissue section imaged with steps of 50 μm and laser frequency of 20 Hz would require 10,000 data points, or about 5 h to acquire. This means that for practical high-resolution images, cycle times in the order of 0.1–0.2 s would be required.

For the most part, using sample preparation protocols described in the literature for direct tissue analysis by imaging MS, mainly soluble proteins and noncovalently surface bound proteins are detected. This is because the matrix together with other organic solvents do not sufficiently distrupt cell membranes to render membrane-bound proteins available for ionization. Solubilization of membrane and other highly hydrophobic tissue proteins is not a problem limited to MS imaging. For example, to enhance the solubilization of this class of proteins prior to 2-DE separation, a fairly complex solubilzation mixture has to be used that contains high molar components such as urea and thiourea. Unfortunately, such components are known to suppress MS signals rendering them unsuitable for sample preparation for MS imaging. New methods to allow single step solubilization of hydrophobic proteins in tissue are still to be developed.

REFERENCES

Adam, B-A., Qu, Y., Davis, J. W., et al. (2002). Cancer Res. **62**, 3609.

Adam, P. J., Boyd, R., Tyson, K. L., et al. (2003) J. Biol. Chem. **278**, 6482.

Adkins, J. N., Varnum, S. M., Auberry, K. J., et al. (2002) Mol. Cell. Proteomics **1**, 94.

Aebersold, R., Goodlett, D. R. (2001) Chem. Rev. **101**, 269.

Aebersold, R. (2003) Nature **422**, 115.

Aebersold, R., Mann, M. (2003) Nature **422**, 198.

Alaiya, A., Roblick, U., Egevad, L., et al. (2000) Anal. Cell Path. **21**, 1.

Alaiya, A., Oppermann, Langridge, J., et al. (2001) Cell. Mol. Life Sci. **58**, 307.

Allen, S., Chen, X., Davies, J., et al. (1997) Biochemistry **36**, 7457.

Alpert, A. J., Shukla, A. K. (2003) Association of Biomolecular Resource Facilities, Denver, Feb. 10–13.

Anderson, L., Seilhamer, J. (1997) Electrophoresis **18**, 533.

Anderson, N. L., Palanski, M., Pieper, R., et al. (2004) Mol. Cell. Proteomics **3.4**, 311.

Arenkov, P., Kukhtin, A., Gemmell, A., et al. (2000) Anal. Biochem. **278**, 123.

Augereau, P., Garcia, M., Mattei, M. G., et al. (1988) Mol. Endocrinol. **2**, 186.

Baggerly, K. A., Morris, J. S., Edmonson, S. R., et al. (2005) J. Natl. Cancer Inst. **97**, 307.

Bahr, U., Stahl-Zeng, J., Gleitsmann, E., et al. (1997) J. Mass Spectrom. **32**, 1111.

Ball, G., Main, S., Holding, F., et al. (2002) Bioinformatics **18**, 395.

Barry, R., Soloviev, M. (2004) Proteomics, **4**, 3717.

Bayer, E. A., Wilchek, M. (1990) J. Chromatogr. **27**, 3.

Belov, M. E., Gorshkov, M. V., Udseth, H. R., et al. (2000a) J. Am. Soc. Mass Spectrom. **11**, 19.

Belov, M. E., Gorshkov, M. V., Udseth, H. R., et al. (2000b) Anal. Chem. **72**, 2271.

Banez, L. L., Prasanna, P., Sun, L., et al. (2003) J. Urrulogy **170**, 442.

Berger, S. J. J. (1980) Biol. Chem. **255**, 3128.

Berndt, P., Hobohm, U., Langen, H. (1999) Electrophoresis. **20**, 3521.

Bhattacharya, B., Prasad, G. L., Valverius, E. M., et al. (1990), Cancer Res. **50**, 2105.

Bini, L., Magi, B., Marzocchi, B., et al. (1997) Electrophoresis. **18**, 2832.

Bjellqvist, B., Ek, K., Righetti, P. G., et al. (1982) J. Biochem. Biophys. Meth. **6**, 317.

Blagoev, B., Kratchmarova, I., Ong, S. E., et al. (2003) Nat. Biotechnol. **21**, 315.

Bordini E, Hamdan M, Righetti PG. (1999) Rapid Commun. Mass Spectrom. **13**, 2209.

Boutell, J. M., Hart, D. J., Godber, B. L. J., et al. (2004) Proteomics **4**, 1950.

Browning-Kelly, M. E., Wadu-Mesthrige, K., Hari, V., et al. (1997) Langmuir **13**, 343.

Büssow, K., Cahill, D., Nieffeld, W., et al. (1998) Nucleic Acids Res. **26**, 5007.

Busch, G., Hoder, D., Reutter, W., et al. (1989) J. Immunol. Methods **50**, 257.

Cagney, G., Emili, A. (2002). Nat. Biotechnol. **20**, 163.

Cahill, D. J. (2001) J. Immunol. Methods. **250**, 81.

Cahill, D. J., Nordhoff, E. (2003) Adv. Biochem. Eng. Biotechnol. **83**, 177.

Caprioli, R. M., Farmer, T. B., Gile, J. (1997) Anal. Chem. **69**, 4751.

Castagna, A., Antonioli, P., Astner, H., et al. (2004) Proteomics. **4**, 3246.

Casting, R., Slodzian, G. (1962) J. Microsc. **1**, 395.

Cecconi, D., Scarpa, A., Donadelli, M., et al. (2003a) Electrophoresis. **24**, 1871.

Cecconi, D., Astner, H., Donadelli, M., et al. (2003b) Electrophoresis. **24**, 4291.

Cecconi, D., Donadelli, M., Milli, A., et al. (2005) J. Proteome Res. In press.

Celis, J. E., Ostergaard, M., Basse, B., et al. (1996) Cancer Res. **56**, 4782.

Certov, O., Simpson, J. T., Biragyn, A., et al. (2005) Expert Rev. Proteomics **(2.1)**, 139.

Cha, T., Guo, A., Zhu, X-Y. (2005) Proteomics **5**, 416.

Chapman-Smith, A., Cronan, J. E., Jr (1999) Trends Biochem. Sci. **24**, 359.

Chaurand, P., Stoeckli, M., Caprioli, R. M. (1999) Anal. Chem. **71**, 5263.

Chaurand, P., Dague, B. B., Pearsall, R. S., et al. (2001) Proteomics **1**, 1320.

Chaurand, P., Schwarz, S. A., Bilheimer, D., et al. (2004a) Anal. Chem. **76**, 1145.

Chaurand, P., Sanders, M. E., Jensen, R. A., et al. (2004b) Am. J. of Pathology **165**, 1057.

Ciechanover, A., Orian, A., Schwartz, A. L. (2000) Bioassays. **22**, 442.

Cohen, E. J., Strong, L. E., Hughes, W. L., et al. (1946) J. Am. Chem. Soc. **68**, 459.

Conrads, T. P., Alving, K., Veenstra, T. D., et al. (2001) Anal. Chem. **73**, 2132.

Cooper, M. A. (2002) Nat. Rev. **1**, 515.

Corthals, G. L., Molloy, M. P., Herbert, B. R., et al. (1997) Electrophoresis **18**, 317.

Crecelius, A. C., Cornett, D. S., Caprioli, R. M. (2005) J. Am. Soc. Mass Spectrom. **16**, 1093.

Dammer, U., Hegner, M., Anselmetti, D., et al. (1996) J., Biophys. **70**, 2437.

Davis, T. M., Wilson, W. D. (2000) Anal. Biochem. **284**, 348.

Dawson, P. H., Hedman, J., Whetten, N. R. (1969) Rev. Sci. Instrum. **40**, 1444.

Dayal, B., Ertel, N. H. (2002) J. Proteome Res. **1**, 375.

Diamandis, E. P., Christopoulos, T. K. (1991) Clin. Chem. **37**, 625.

Diamandis, E. P. (2003) Clin. Chem. **49**, 1271.

Diamandis, E. P. (2004) Mol. Cell. Proteomics. **3.4**, 367.

Domon, B., Broder, S. (2004) J. Proteome Res. **3**, 253.

Drew, J. (2000) Science **287**, 1960.

Emmert-Buck, M. R., Vocke, C. D., Pozzatti, R. O., et al. (1995) Cancer Res. **55**, 2595.

Eng, J. K., McCormack, A. L., Yates III, J. R. (1994) J. Am. Soc. Mass Spectrom. **5**, 976.

Engel, A., Lyubchenko, Y., Müller, D. (1999) Trends in Cell Biol. **9**, 77.

Eriksson, J., Chait, B. T., Fano, D. (2000) Anal. Chem. **72**, 999.

Espejo, A., Cote, J., Bednarek, A., et al. (2002) Biochem. J. **367**, 697.

Espina, V., Mehta, A., Winters, M. E., et al. (2003) Proteomics **3**, 2091.

Everly, P. A., Krijgsveld, J., Zetter, B. R., et al. (2004). Mol. Cell. Proteomics **3.7**, 729.

Fang, Y., Frutos, A. G., Lahiri, J. (2002) J. Am. Chem. Soc. **124**, 239.

Fend, F., Emmert-Buck, M. R., Chiquita, R., et al. (1999) Am. J. Pathl. **154**, 61.

Ferguson, A. T., Evron, E., Umbricht, C. B., et al. (2000) Proc. Natl. Acad. Sci. USA **97**, 6049.

Finley, D. (2001). Nature. **412**, 285.

Fisher, T. E., Oberhauser, A. F., Carrion-Vazquez, M., et al. (1999) Trends in Biochem. Sci. **24**, 379.

Fournier, I., Day, R., Salzet, M. (2003) Neuroendocrinol. Lett. **24**, 9.

Frank, M., Labov, S. E., Benner, W. H. (1999) Mass Spectrom. Rev. **18**, 155.

Franzen, B., Hirano, T., Okuzawa, K., et al. (1995). Electrophoresis. **16**, 1087.

Franzen, B., Linder, S., Urju, K., et al. (1996a) Br. J. Cancer **73**, 909.

Franzen, B., Linder, S., Alaiya, A. A., et al. (1996b) Br. J. Cancer **74**, 1632.

Fujii, K., Wakano, T., Kanazawa, M., et al. (2005) Proteomics **5**, 1150.

Furka, A., Sebestyen, F., Asgedom, M., et al. (1991) Int. J. Peptide Prot. Res. **37**, 487.

Gad, M., Itoh, A., Ikai, A. (1997) Cell Biol. Int. **21**, 697.

Galvani, M., Rovatti, L., Hamdan, M., et al. (2001a) Electrophoresis **22**, 2066.

Galvani M, Hamdan M, Herbert B, et al. (2001b) Electrophoresis. **22**, 2058.

Garbi, S., Gaffney, P., Yang, A., et al. (2002) Mol. Cell. Proteomics **1.2**, 91.

Ge, H. (2000) Nucleic Acids Res. **28**, e3.

Geddings, J. C. (1987) J. High Resolut. Chromatogr **10**, 319.

Geddings, J. C. (1991) Unified Separation Science (1991) John Wiley Sons, New York.

Gerber, S. A., Rush, J., Stemman, O., et al. (2003) Proc. Natl. Acad. Sci. USA. **100**, 6940.

Giometti, C. S., Williams, K., Tollaksen, S. L. (1997) Electrophoresis **18**. 573.

Gleason, D. F. (1992) Hum. Pathol. **23**, 273.

Goldknopf, I. L., Taylor, C. W., Baum, R. M., et al. (1975) J. Biol. Chem. **250**, 7182.

Goodlett, D. R., Keller, A., Watts, J. D., et al. (2001) Rapid Commun. Mass Spectrom. **15**, 1214.

Greenlee, R. T., Hill-Harmon, M. B., Murray, T., et al. (2001). CA. Cancer J. Clin. **51**, 15.

Grizzle, W. E., John-Semmes, O., Bsler, J., et al. (2004) Urol. Oncol. **22**, 337.

Guerrier, L., Lomas, L., Boschetti, E. (2005) J Chromatogr. A. **1073**, 25

Grover, P. K., Resnick, M. I. (1995). Prostate **26**, 12.

Grover, P. K., Resnick, M. I. (1997). Electrophoresis **18**, 814.

Grubb, R. L., Calvert, V. S., Wullfkuhle, J. D., et al. (2003) Proteomics. **3**, 2142.
Guilhous, M., Selby, D., Mlynski, V. (2000) Mass Spectrom. Rev. **19**, 65.
Gygi, S. P., Rochon, Y., Franza, B. R., et al. (1999) Mol. 7 Cell. Biol. **19**, 1720.
Gygi, S. P., Aebersold, R. (1999). Methods Mol. Biol. **112**, 417.
Gygi, S. P., Rist, B., Gerber, S. A., et al. (1999) Nat. Biotechnol. **17**, 994.
Gygi, S. P., Corthals, G. L., Zhang, Y., et al. (2000). Proc. Natl. Acad. Sci. USA **97**, 9390.
Haab, B. B., Dunham, M. J., Brown, P. O. (2001) Genom Biol. 2-R4.
Haber, D. N. (2000). N. Engl. J. Med. **343**, 1566.
Hager, J. W. (2002). Rapid Commun. Mass Spectrom. **16**, 512.
Hager, J. W., Yves Le Blanc, J. C. (2003) Rapid Commun. Mass Spectrom. **17**, 1056.
Hale, J. E., Butler, J. P., Knierman, M. D., et al. (2000) Anal. Biochem. **287**, 110.
Hamdan, M., Righetti, P. G, (2002) Mass Spectrom. Rev. **21**, 287.
Hamdan, M., Righetti, P. G. (2003) Mass Spectrom. Rev. **22**, 272.
Hamdan, M., Righetti, P. G., "Proteomics Today" (2005) Wiley Interscience, New York.
Hanash, S. M., Gagnon, M., Seeger, R. C., et al. (1985) Prog. Clin. Biol. Res. **175**, 261.
Hanash, S. (2004) Mol. Cell. Proteomics **3**, 298.
Hardman, M., Makarov, A. A. (2003) Anal. Chem. **75**, 1699.
Heidelberger, M., Kendall, F. E. (1929) J. Exp. Med. **50**, 809.
Henderson, E. (2004) J. Biomol. Screening **9**, 491.
Herbert, B. R., Molloy, M. P., Gooley, B. A. A., et al. (1998) Electrophoresis **19**, 845.
Herbert, B. R., Galvani, M., Hamdan, M., et al. (2001) Electrophoresis **22**, 2046.
Hewett, P. W. (2001). Int. J. Biochem. Cell. Biol. **33**, 325.
Hodneland, C. D., Lee, Y. S., Min, D-H, et al. (2002) Proc. Natl. Acad. Sci. USA **99**, 5048.
Hondermarck, H. (2003) Mol. Cell. Proteomics. **2.5**, 281.
Houseman, B. T., Huh, J. H., Kron, S. J., et al. (2002) Nat. Biotechnol. **20**, 270.
Huber, L. A., Pasquali, C., Gagescu, R., et al. (1996). Electrophoresis **17**, 1734.
Huber, M., Bahr, I., Krätzschmar, J. R., et al. (2004). Mol. Cell. Proteomics **3.1**, 43.
Huff, J. L., Lynch, M. P., Nettikadan, S., et al. (2004) J. Biomol. Screening **9**, 491.
Hurely, W. L., Finkelstein, E., Holst, B. D. (1985) J. Immunol. Methods **85**, 195.
Hutchens, T. W., Yip, T. T. (1993) Rapid Commun. Mass Spectrom. **7**, 576.
Ibel, K., May R. P., Sandberg, M., et al. (1994) Biophys. Chem. **53**, 77.
Issaq, H. (2001) Electrophoresis **22**, 3629.
Issaq, H., Conrads, T. P, Janini, G. M., et al. (2002) Electrophoresis **23**, 3048.
Issaq, H., Conrads, T. P., Prieto, D. A., et al. (2003) Anal. Chem. **75**, 149A.
Jemal, A., Thompson, A., Murray, T., et al. (2002). CA. Cancer J. Clin. **52**, 23.
John-Semmes, O., Feng, Z., Adam, B-A, et al. (2005) Clin. Chem. **51**, 102.
Jona, G., Snyder, M. (2003) Curr. Opin. Mol. Therapy **5**, 271.
Jones, V. W., Kenseth, J. R., Porter, M. D., et al. (1998) Anal. Chem. **70**, 1233.
Jones, M. N. (1999) Int. J. Pharm. **177**, 137.
Joos, T. O., Schrenk, M., Höpfl, P., et al. (2000) Electrophresis. **21**, 2641.
Kaji, H., Saito, Y., Yamauchi, Y., et al. (2003) Nat. Biotecnol. **21**, 667
Kanno, S., Yanagida, Y., Haruyama, T., et al. (2000) J. Biotech. **76**, 207.

Karas, M., Hillenkamp, F. (1988) Anal. Chem. **60**, 2299.

Kaufmann, R., Chaurand, P., Kirsch, D. (1996). Rapid Commun. Mass Spectrom. **10**, 1199.

Khanuja, P. S., Lehr, J. E., Soule, H. D., et al. (1993) Cancer Res. **53**, 3394.

Kim, J., Kim, S. H., Lee, S. U. (2002) Electrophoresis **23**, 4142.

Kleiner, O., Price, D. AS., Ossetravo, N., et al. (2005) Proteomics. **5**, 2322

Klose, J., Kobalz, U. (1995) Electrophoresis **16**, 1034.

Knezevic, V., Leethanakul, C., Bichsel, V. E., et al. (2001) Proteomics. **1**, 1271.

Kraus, H. (2002) Int. J. Mass Spectrom. **215**, 45.

Kozak, K. R., Amneus, M. W., Pusey, S. M., et al. (2003). Proc. Natl. Acad. Sci. USA **100**, 12343.

Krause, E., Wenschcuh, H., Jungblut, P. R. (1999) Anal. Chem. **71**, 4160.

Krishna, R. G., Wold, F., (1993) Adv. Enzymol. Relat. Areas Mol. Biol. **67**, 265

Kuyama, H. Watanabe, M. Toda, C., et al. (2003) Rapid Commun. Mass Spectrom. **17**, 1642.

LaBaer, J., Ramachandran, N. (2005) Curr. Opin. Chem. Biol. **9**, 14.

Lam, K. S., Salmon, S. E., Hersh, E. M., et al. (1991) Nature **354**, 82.

Laine, R. A. (1994) Glycobiology **4**, 759.

Le, L., Chi, K., Tyldesley, S., et al. (2005) Clin. Chem. **51**, 695.

Leatherbarrow, R. J., Dean, P. D. G. (1980) Biochem. J. **189**, 27.

Lee, K. B., Park, S. J., Mirkin, C. A. (2002) Science. **295**, 1702.

Lee, Y-S., Mrksich, M. (2002) Trends Biotechnol. **20** (suppl.) S14.

Lexander, H., Franzén, B., Hirschberg, D., et al. (2005) Proteomics **5**, 2570.

Li, C., Hong, Y., Tan, Y-X., et al. (2004) Mol. Cell. Proteomics **3.4**, 399.

Lilley, K. S., Razzaq, A., Dupree, P. (2002) Curr. Opin. Chem. Biol. **6**, 46.

Lin, J. D., Huang, C. C., Weng, H. F., et al. (1995). J. Chromatogr. B.

Lindmark, R., Thoren-Tolling, K., Sjoquist, J. (1983). J. Immunl. Methods **62**, 1613.

Link, A. J., Hays, L. G., Carmack, E. B. (1997) Electrophoresis **18**, 1314.

Link, A. J., Eng, J., Scvhieltz, D. M., et al. (1999) Nat. Biotechnol. **17**, 676.

LLopis, J., Westin, S., Ricote, M., et al. (2000) Proc Natl Acad Sci. USA. **97**, 4363.

Liotta, L. A., Kohn, E. C. (2001) Nature. **411**, 375.

Liotta, L. A., Espina, V., Mehta, A., et al. (2003) Cancer Cell **3**, 317.

Listgarten, J., Emili, A. (2005). Mol. Cell. Proteomics. **4.4**, 419.

Liu, Z., Zhou, L. H., Wei, G., et al. (2005) J. of Microscopy. **218**, 233.

Lizardi, P. M., Huang, X., Zhu, Z., et al. (1998) Nature Genet. **19**, 225.

Louris, J. N., Cooks, R. G., Syka, J. E. P., et al. (1987) Anal. Chem. **59**, 1677.

Lu, Y., Bottari, P., Turećek, F., et al. (2004) Anal. Chem. **76**, 4104.

Lundahl P, Watanabe Y, Takagi T. (1992) 1992. J. Chromatogr **604**, 95.

MacBeath, G., Schreiber, S. L. (2000) Science. **289**, 1760.

MacBeath, G. (2002). Nat. Genet. Supl. **32**, 526.

MacCoss, M. J., McDonald, W. H., Saraf, A., et al. (2002) PNAS. **99**, 7900.

Madoz-Gúrpide, J., Wang, H., Misek, D. E., et al. (2001) Proteomics **1**, 1279.

Makarov, A. A. (2000) Anal. Che. **72**, 1156.

Malyarenko, D., Cooke, W. E., Adam, B-A., et al. (2005) Clin. Chem. **51**, 65.

Mann, M., Hendrickson, R. C., Pandey, A. (2001) Ann. Rev. Biochem. **70**, 437.

Marotti, Jr, L. A. Newitt, R., Wang, R., et al. (2002) Biochemistry. **41**, 5067.

Marshall, A. G., Hendrickson, C. L., Jackson, G. S., (1998) Mass Spectrom. Rev. **17**, 1.

Masumori, N., Thomas, T. Z., Chaurand, P., et al. (2001) J. Cancer Res. **61**, 2239.

Mather, R. E., Waldren, R. M., Todd, J. F. J. (1978) Dyn. Mass Spectrom. **5**, 71.

McDonald, W. H., Ohi, R., Miyamoto, D. T., et al. (2002) Int. J. Mass Spectrom. **219**, 245.

McDonnell, J.M. (2001). Curr. Opin. Chem. Biol. **5**, 572.

Meehan, K. L., Holland, J. W., Dawkins, H. J. S. (2002). Prostate **50**, 54.

Mendoza, L. G., McQUARY, P., Mongan, A., et al. (1999). Biotechniques. **27**, 778.

Merchant, M., Weinberger, S. R. (2000) Electrophoresis **21**, 1164.

Merrifield, R. B. (1963). J. Am. Chem. Soc. **85**, 2149.

Merrell, K., Thulin, C. D., Southwick, K., et al. (2004) J. Bioml. Techniques. **15**, 11.

Mian, S., Ball, G., Hombuckle, J., et al. (2003) Proteomics **3**, 1725.

Miller, D.L., Papayannopoulos, I. A., Styles, J., et al. (1993) Arch. Biochem. Biophys. **301**, 41.

Mirgorodskaya, O. A., Kozmin, Y. P., Titov, M. I., et al. (2000) Rapid Commun. Mass Spectrom. **14**, 1226.

Mitchell, P. (2002) Nat. Biotechnol. **20**, 225.

Moll, R., Franke, W. W., Schiller, D. L., et al. (1982) Cell **31**, 11.

Moody, M. D., Van Arsdell, S. W., Murphy, K. P., et al. (2001) Biotechniques **31**, 186.

Mooney, J. F., Hunt, A. J., Mcintosh, J. R., et al. (1996) Proc. Natl. Acad. Sci. USA. **93**, 12287.

Morgan, H., Pritchard, D. J., Cooper, J. M. (1995). Biosens Bioelectron. **10**, 841.

Mori, H., Takio, K., Ogawara, M., et al. (1992) J. Biol. Chem. **267**, 17082.

Moore, A. W., Jorgenson, J. W. (1995) Anal. Chem. **67**, 3448.

Myszka, D. G., Rich, R. L. (2000) Pharm. Sci. Technol. Today **3**, 310.

Nelson, R. W., Nedelkov, D., Tubbs, K.A. (2000) Electrophoresis **21**, 1155.

Nishizuka, S., Charboneau, L., Young, L., et al. (2003) Proc. Natl. Acad. Sci. USA **100**, 14229.

Nurcombe, V., Smart, C. E., Chipperfield, H., et al. (2000) J. Biol. Chem. **275**, 30009.

Oda, Y., Huong, K., Cross, F. R., et al. (1999) Proc. Natl. Acad. Sci. USA. **96**, 6591.

O'Farrell, P. H. (1975) J. Biol. Chem. **250**, 4007.

Ong, S-E, Blagoev, B., Krachmarova, I., et al. (2002). Mol. & Cell. Proteomics. **1.5**, 376.

O'Neil, K. A., Miller, F. R., Barder, T. J., et al. (2003). Proteomics. **3**, 1256.

Ornstein, D. K., Gillespie, J. W., Paweletz, C. P., et al. (2000) Electrophoresis **21**, 2235.

Packolski, M. L., Winograd, N. (1999) Chem. Rev. **99**, 2977.

Pan, S., Zhang, H., Rush, J., et al. (2005) Mol. Cell. Proteomics **4.2**, 182.

Partin, A. W., Getzenberg, R. H., CarMicheal, M. J., et al. (1993). Cancer Res. **53**, 744.

Paweletz, C. P., Charboneau, L., Bichsel, V. E., et al. (2001) Oncogene **20**, 1981.

Pawlak, M., Schick, E., Bopp, M. A., et al. (2002). Proteomics **2**, 383.

Peng, J., Schwarz, D., Elias, J. E., et al. (2003) Nat. Biotechnol. **21**, 921.

Perkins, D. N., Pappin, D. J. C., Creasy, D. M., et al. (1999) Electrophoresis **20**, 3551.

Petricoin III, E. F., Ornstein, D. K., Paweletz, C. P., et al. (2002) J. Natl. Cancer. Res. **94**, 1576.

Petricoin III, E. F., Ali, M. A., Hitt, B. A., et al. (2002) Lancet. **359**, 572.

Pieper, R., Gatlin, C. L., Makusky, A. J., et al. (2003) Proteomics **3**, 1345.

Porath, J., Carlsson, J., Olsson, I., et al. (1975) Nature **258**, 598.

Qu, Y., Adam, B-A, Yasui, Y., et al. (2002) Clin. Chem. **48**, 1835.

Quinn, J. G., O'neil, S., Doyle, A., et al. (2002) Anal. Biochem. **281**, 135.

Rabilloud, T. (1996) Electrophoresis **17**, 813.

Rabilloud, T. (2002). Proteomics **2**, 3.

Radford, D. M. (1983) Cancer Res. **53**, 2947.

Ramachandran, N., Hainsworth, E., Bhullar, B., et al. (2004) Science **305**, 86.

Rasmussen, R. K., Ji, H., Eddes, J. S., et al. (1997) Elecrophoresis **18**. 588.

Reyzer, M. L., Hsieh, Y., Krfmacher, W. A., et al. (2003) J. Mass spectrom. **38**, 1081.

Rich, R. L., Myszka, D. G. (2000) J. Mol. Recog. **13**, 388.

Rich, R. L., Day, Y. S.Morton, T. A., et al. (2001) Anal. Biochem. **296**, 197.

Righetti, P.G., Castagna, A., Antonioli, P., et al. (2005) Electrophoresis **26**, 297.

Righetti, P. G., Wenisch, E., Faupel, M. (1989) J. Chromatogr. **475**, 293.

Ring, D. B, Kassel, J. A., Hsieh-Ma, S. T, et al. (1989) Cancer Res. **49**, 3070.

Rogers, M. A., Clarke, P., Noble, J., et al. (2003) Cancer Res. **63**, 6971.

Rohner, T. C., Staab, D., Stoeckli, M. (2005) Mechanisms of Aging & Development **126**, 177.

Roher, A. E., Lowenson, J. D., Clarke, S., et al. (1993) Proc. Natl. Acad. Sci. USA. **90**, 10836.

Rowe, C. A., Scruggs, S. B., Feldstein, M. J., et al. (1999) Anal. Chem. **71**, 433.

Sabarth, N., Lamer, S., Zimny-Arndt, U., et al. (2002). J. Biol. Chem. **277**, 27896.

Sarto, C., Marocchi, A., Sanchez, J. C., et al. (1997) Electrophoresis **18**. 599.

Samuel, S. K., Minish, T. M., Davie, J. R. (1997) J. Cell Biochem. **66**, 9.

Schena, M., Shalon, D., Heller, R., et al. (1995) Science **270**, 467.

Schwartz, S. A., Reyzer, M. L., Caprioli, R. M. (2003) J. Mass spectrom. **38**, 699.

Schwarz, S. A., Weil, R. J., Johnson, M. D., et al. (2004) Clin. Cancer Res. **10**, 981.

Schweizer, B., Roberts, S., Grimwade, B., et al. (2002) Nat. Biotechnol. **20**, 359.

Shaw, J., Rowlinson, R., Nickson, J., et al. (2003). Proteomics. **3**, 1181.

Shekouh, A. R., Thompson, C. C., Prime, W., et al. (2003) Proteomics **3**, 1988.

Shen, J., Person, M. D., Zhu, J, et al. (2004) Cancer Res. **64**, 9018.

Shin, B. K., Wang, H., Yim, A. M., et al. (2003) J. Biol. Chem. **278**, 7607.

Silzel, J. W., Cercek, B., Dodson, C., et al. (1998) Clin Chem. **44**, 2036.

Simonetti, A. W., Melssen, W. J., van der Graaf, M., et al. (2003) Anal. Chem. **75**, 5352.

Simpson, R. J., Connolly, L. M., Eddes, J. S., et al. (2000) Electrophoresis **21**, 1707.

Smith, R. D., Anderson, G. A., Lipton, M. S., et al. (2002) Proteomics **2**, 513.

Somiari, R. I., Sullivan, A., Russell, S., et al. (2003) Proteomics. **3**, 1863.

Sorkin, A., McClure, M., Huang, F., et al. (2000) Curr. Biol. **10**, 1395.

Sorace, J. M., Zhan, M. (2003) BMC Bioinformatics **4**, 24.

Sreekumar, A., Nyati, M. K., Varambally, S., et al. (2001). Cancer Res. **61**, 7585.

Stahl-Zeng, J., Hellenkamp, F., Karas, M. (1996) Eur. J. Mass Spectrom. **2**, 23.

Stoeckli, M., Chaurad, P., Hallahan, D. E., et al. (2001) Nat. Med. **7**, 493.

Striebel, H-M., Schellenberg, P., Grigaravicius, Greulich, K. O. (2004) Proteomics **4**, 1703.

Tang, N., Tornatore, P., Weinberger, S. R. (2004) Mass Spectrom. Rev. **23**, 34.

Tanaka, K., Waki, H., Ido, Y., et al. (1987) 2nd Japan-China Joint Symposium on Mass Spectrometry, Osaka, Japan.

Thompson, T. C., Timme, T. L., Park, S. H., et al. (2000) Prostate **43**, 248.

Tirumalai, R. S., Chan, K. C., Prieto, D. A., et al. (2003) Mol. Cell. Proteomics **2**, 1096.

Tonge, R., Shaw, J., Middleton, B., et al. (2001) Proteomics. **1**, 377.

Trask, D. K., Band, V., Zajchowski, D. A., et al. (1990) Proc. Natl. Acad. Sci. USA **87**, 2319.

Troendle, F. J., Reddick, C. D., Yost, R. A. (1999) J. Am. Soc. Mass Spectrom. **10**, 1315

Turkova, J. (1999). J. Chromatogr. B: Biomed. Sci. Appl. **722**, 11.

Twerenbold, D. (1996) Nuclear Instruments and methods. **A 370**, 253.

Twerenbold, D., Gerber, D., Gritti, D., et al. (2001) Proteomics. **1**, 66.

Twerenbold, D., Vuilleumier, J-L, Gerber, D., et al. (1996) Appl. Phys. Lett. **68**, 3503.

Umbricht, C. B., Evron, E., Gabrielson, E., et al. (2001) Oncogene **20**, 3348.

Ünlü, m., Morgan, M. E., Minden, J. S. (1997) Electrophoresis **18**, 2071.

Vallanueva, J., Philip, J., Entenberg, D., et al. (2004) Anal. Chem. **76**, 1560.

Vercoutter-Edourt, A. S., Lemoine, J., le Bourhis, X., et al. (2001) Cancer Res. **61**, 76.

Vuong, G. L., Weiss, S. M., Kammer, W., et al. (2000) Electrophoresis **21**, 2594.

Wagner, K., Miliotis, T., Marko-Varga, G., et al. (2002) Anal. Chem. **74**, 809.

Wall, D., B., Kachman, M. T., Gong, S., et al. (2000) Anal. Chem. **72**, 1099.

Walter, G., Büssow, K., Büssow, D., et al. (2000) Curr. Opin. Microbiol. **3**, 298.

Wang, H., Hnash, S. M. (2003) J. Cromatogr. B. **787**, 11.

Ward, L. D., Hong, J., Whitehead, R. H., et al. (1990) Electrophoresis. **11**, 883.

Washburn, M. P., Wolters, D., Yates, J. R., 3rd. (2001). Nat. Biotechnol. **19**, 242.

Weng, S., Gu, K., Hammond, P., et al. (2002) Proteomics **2**, 48.

Wernert, N. (1997). Virchows Arch. **430**, 433.

Westley, B., Rochefort, H. (1980) Cell 20, 353.

Wheeler, J. X., Wait, R., Stone, T., et al. (2003) Rapid Commun Mass Spectrom. **17**, 2563.

Wilbur, D. S., Pathare, P. M., Hamlin, D. K., et al. (1999) Biomol. Eng. **16**, 113.

Wilchek, M., Bayer, E. A. (1999). Biomol. Eng. **16**, 1.

Williams, C., Addona, T. A. (2000) Trends Biotechnol. **18**, 45.

Wirth, P. J., Egilsson, V., Gudnasen, V., et al. (1987) Breast Cancer Res Treat. **10**, 177.

Wise, R. (2003) Luminescence. **18**, 25.

Wolters, D. A., Washburn, M. P., Yates III, J. R. (2001) Anal. Chem. **73**, 5683.

Wouters, F. S., Verveer, P. J., Bastiaens, P. I. H., Trends in Cell Biology. **11**, 203 (2001).

Wright, M. E., Han, D. K., Aebersold, R. (2005). R., Mol. Cell. Proteomics **4.4**, 545.

Wulfkuhle, J. D., Aquino, J. A., Calvert, V. S., et al. (2003) Proteomics **3**, 2085.

Wulfkuhle, J. D., Sgroi, D. C., Krutzsch, H., et al. (2002) Cancer Res. **62**, 6740.
Yanagisawa, K., Shyr, Y., Xu, B. J., et al. (2003) Lancet **362**, 433.
Yan, F., Sreekumar, A., Laxman, B., et al. (2003) Proteomics **3**, 1228.
Yao, X., Freas, A., Ramirez, J., et al. (2001) Anal. Chem. **73**, 2836.
Yasui, Y., McLerran, D., Adam, B-L., et al. (2003a) J. Biomedicine 7 Biotechnol. **4**, 242.
Yasui, Y., Pepe, M., Thompson, M. L., et al. (2003b) Biostatistics. **4**, 449.
Yates III, J. R., McCormack, A. L., Link, A. J., et al. (1996) Analyst **121**, 65R.
Yefchak, G. E., Schultz, G. A., Allison, J., et al. (1990) J. Am. Soc. Mass Spectrom. **1**, 440.
Yip, T. T. C., Chan, J. W. M., Cho, W. C. S., et al. (2005) Clin. Chem. **51**, 47.
Zhang, Z., Bast, R. C., Yu, Y., et al. (2003) Proteomics **3**, 1725.
Zhang, Z., Bast, R. C., Yu, Y., et al. (2004) Cancer Res. **64**, 5882.
Zhang, F., Ji, L-N., Tang, L., et al. (2005). Acta Biochem. et Biophys. Sinica **37**, 113.
Zhang, X., Leung, S. M., Morris, C. R., et al. (2004) J. Biomol. Tech. **15**, 167.
Zhu, H., Klemic, J. F., Chang, S., et al. (2000) Nat. Gen. **26**, 283.
Zhu, H., Snyder, M. (2001) Curr. Opin. Chem. Biol. **5**, 40.
Zhu, H., Snyder, M. (2003) Curr. Opin. Chem. Biol. **7**, 55.
Zhu, H., Bilgin, M., Bangham, R., et al. (2001) Science **293**, 2101.
Ziauddin, J., Sabatini, D. M. (2001) Nature **411**, 107.

3

SOME EXISTING CANCER BIOMARKERS

3.1. INTRODUCTION

The first-known tumor biomarker was described over a century and a half ago, when in 1846 Henry Bence-Jones described the precipitation of a protein in an acidified urine from patients with multiple myeloma. Interestingly, the detection of the monoclonal light-chain immunoglobulin in this disease is still in use. Around the same period and more precisely in 1853, J. Adams, a surgeon at The London Hospital, described the first case of prostate cancer, which he discovered by histological examination (Adams, 1853). Remarkably, a century and a half later, this form of cancer has become a leading cause for many deaths and the focus of impressive research activities. In today's world of biomarker discovery, it is commonly argued that two of the most important parameters in determining the efficacy of a cancer marker assay are the levels of sensitivity and specificity. The former simply means the minimal amount of the substrate that can be detected, and the later the percentage of assays that correctly distinguishes normal from cancerous conditions. Considering existing serum-based cancer markers, it is reasonable to state that there is always a trade-off between sensitivity and specificity. For example, high sensitivity might reduce specificity, leading to detection levels that do not correlate to the actual state of the disease, a situation, which commonly leads to false-positive results. On the contrary, a highly specific assay may not be sensitive enough to detect low but relevant levels of the marker, which may lead to false-negative results. This trade-off is still a focus of extensive research activities involving well-established serum-based markers such as prostate-specific antigen (PSA) for prostate

Cancer Biomarkers: Analytical Techniques for Discovery, by Mahmoud H. Hamdan

cancer and carcinoma-associated glycoprotein antigen (CA-125) for ovarian cancer. It is reasonable to state that within clinical settings such a trade-off is commonly implemented in a fairly pragmatic manner. For example, screening for colon cancer demands a high specificity because all individuals with a positive test result should subsequently undergo colonoscopy, a time-consuming and costly procedure. On the contrary, a breast cancer screening test could be acceptable at high sensitivity with low specificity. This is because such a test is commonly considered as a premammography test, and all positively tested individuals could subsequently be offered mammography, a noninvasive and simple procedure.

In this chapter, I have chosen two serum markers, which are well established for the detection and monitoring of two forms of cancer. The choice of these two markers is motivated by two considerations: First, both markers were identified over 20 years ago, a period sufficiently long to allow the accumulation of enough data to obtain a rational assessment of their respective performances. Such assessment encompasses the contribution of these markers in the fight against diseases, their advantages and shortcomings, efforts to improve their performance, the search for alternative markers, and the future prospects of these markers. Second, the impressive volume of literature dealing with both the markers is a valuable source of information in our endeavor to improve their performance and to draw lessons, which are indispensable in the search for more sensitive and above all more specific alternatives to these markers. The same literature is rich enough in both technical approaches and biological information to allow the present discussion to be extended (at least in specific cases) to other markers, which are used for other forms of cancer.

3.2. HISTORIC GLIMPSE AT PSA

The first serum marker to be extensively used in the diagnosis and management of prostate cancer was prostatic acid phosphotase (PAP), which was first detected in prostatic fluids and tissue (Dmochowski and Assenhajm, 1935; Kutscher and Worner, 1936). It has to be said that even today, the exact biological function of PAP and its role in prostate cancer pathogenesis is still unknown. A serum enzymatic assay for PAP was developed shortly after its discovery, and by the early 1940s it was being used for the diagnosis and monitoring of prostate cancer (Gutman et al., 1936). Despite subsequent refinement in enzymatic assays together with the introduction of increasingly sensitive detection methods including radioimmunoassay, the spread of PAP as a marker for prostate cancer was compromised by a number of limitations. These include diurnal variation in PAP serum levels, a lack of tissue specificity, cross-reactivity of the PAP assay with other acid phosphatases in serum, and its inherently low sensitivity particularly in localized disease (Sullivan et al., 1942).

One of the earliest investigations of tissue-specific antigens in human prostate was reported in 1970 (Albin et al., 1970 a,b). Prostatic antigens were later reported in seminal plasma (Hara et al., 1971; Li and Beling, 1973). While searching for a potential marker to be used in the investigation of rape crimes, Sensabaugh and Crim (1978) isolated and characterized PSA from human seminal plasma. However, the isolation, purification from prostate tissue, and its characterization as prostate specific in nature were first reported by Wang et al. (1979). PSA is a member of the kallikren gene family, which has 33 kDa molecular mass serine protease secreted by the prostatic epthilium and the epithelial lining of the periurethral glands. This antigen is recognized for its function in the liquefaction of the seminal coagulum to allow the release of spermatozoa (Lilja, 1985). It has to be pointed out that early investigators were hesitant to use PSA for screening purposes due to a substantial overlapping between measured values in individuals with or without carcinoma and the resulting poor test specificity (Guinan et al., 1987; Stamey et al., 1987; Barak et al., 1989). Indeed, recognizing the critical role of specificity in screening trials, Guinan et al. (1987) have proposed a PSA upper limit for normal (ULN) as 24 ng/mL. This value was based on a study of a group of control men who either had a histological diagnosis of benign prostatic hyperplasia (BPH) or had no genitourinary symptoms. However, none of the patients within the same study had undergone a prostate biopsy. Within the same year, another study by Stamey et al. (1987), which looked at over 2000 serum samples from about 700 men over a half of whom had prostate carcinoma concluded that ULN for PSA should be 2.5 ng/mL.

3.3. PROSTATE-SPECIFIC ANTIGEN

Prostate cancer is a major health burden throughout the world, although there is a large variation in its incidence. The highest rates are in the USA, Canada, Sweden, Australia, and France (Hsing, 2000). Although the causes of this variation are likely to be differences in screening methods, diet and health-related behaviors, clinical practice patterns, and environmental risk factors, the role of genetic differences is still an evolving argument. The observations that the risk of prostate cancer is higher among Japanese migrants to Hawaii (Kolonel, 1997), and the large variation of the incidence of prostate cancer among Chinese in different geographic locations (Parkin et al., 1997) suggest that diet and environmental differences play a major role in the development of the disease. There is, however, consistent evidence across different racial and ethnic groups that a family history of prostate cancer increases the risk that a man will get prostate cancer. Ever since its description in 1979 (Wang et al., 1979), PSA has gone through various phases before being accepted as a screening tool for prostate cancer. In 1986, the Food and Drug Administration (FDA) approved PSA testing for use in monitoring prostate cancer progression. Since then, measurement of PSA levels in serum has become the most common test for the diagnosis of prostate carcinoma. The use of this test over the last 20 years has resulted in an impressive increase in the rate of disease detection. Further measurements over the past 10 years or so led to the consensus that a PSA level of more than 4.0 ng/mL had a predictive value for the diagnosis of this type of cancer (Cooner et al., 1990). However, more recent data suggest that a PSA level of more than 2.5 ng/mL has a predictive value similar to that of a value of 4.0 ng/mL (Labrie et al., 1999; Krumholtz et al., 2002). The latter findings regarding the level of this marker have been strongly supported by the results of the prostate cancer prevention trial (Thompson et al., 2004). This study was sponsored by the National Cancer Institute and conducted by the Southwest Oncology Group. A total of 18,882 men were chosen through an eligibility criteria that included a serum PSA level of not more than 3.0 ng/mL, a normal digital rectal examination (DRE), and an age of at least 55 years. One of the main conclusions of this massive trial was that biopsy detected prostate cancer, including high-grade condition, is not rare among men with PSA levels of 4.0 ng/mL or less. Such a conclusion challenges the long-held notion that these levels are considered to be in the normal range. In an article describing the effect of verification bias on screening for prostate cancer by measurement of SPA (Punglia et al., 2003), the authors have argued that the sensitivity and specificity of such a test should be considered biased when disease status is not verified in all subjects and when the likelihood of confirmation depends on the test results itself. To clarify the above statement, it is worth considering further elements of this study. Over a 6-year period, almost 700 men underwent PSA-based screening. Of these men, 11% subsequently underwent biopsy of the prostate. A mathematical model was used to estimate adjusted receiver-operating characteristics curves (i.e., the overall diagnostic performance). The authors reported that adjusting for verification bias resulted in a significant improvement in the diagnostic performance of PSA for both below and above 60 years of age. The main deduction, however, was regarding the PSA level, where the authors reported that setting the threshold at 4.1 ng/mL would have resulted in missing 82% of prostate cancer in younger men and 65% in older men. The adjusted value of PSA levels for men with prostate cancer ranged from 2.1 to 3.9 ng/mL, depending on age and the results of DRE. These values were in complete disagreement with an earlier study (Morgan et al., 1996), where PSA levels in the range of 6.3–7.5 ng/mL were reported. On the contrary, the results reported by Punglia et al. (2003) were in agreement with those reported by Oesterling et al. (1993). The agreement and disagreement between the various studies regarding the correct threshold of PSA is still a point of contention between various leading experts in the field of prostate cancer and will therefore be further discussed in the following sections. This contention has been emphasized in a recent article (Stamey et al., 2004) in which the recommendation for lower PSA thresholds (2.6 instead of 4.1 ng/mL), to decide on biopsy advanced by Punglia et al. (2003), has been described as misguided because this is precisely the range of serum PSA for most men with BPH.

3.4. PSA AS A SCREENING MARKER

Since its approval by the FDA in 1986, PSA has probably become the most useful tumor marker in clinical practice today and is now routinely used in the diagnosis and monitoring of prostate cancer. There is a general agreement among clinicians that the PSA test has the highest predictive value for prostate cancer particularly when the malignancy is in its early stages. At the same time, there is a disagreement as to what level of PSA should prompt a prostate biopsy. This particular issue is passionately debated because of its profound impact on extremely difficult decisions. For example, the use of relatively high PSA thresholds risks missing the detection of cancer until it is too late for a cure. On the contrary, the use of low PSA levels will inevitably result in unnecessary biopsies or biopsies that identify clinically insignificant disease. The publication of numerous reports in which the need for a biopsy of the prostate was mainly based on the results of PSA tests has led to the consensus that a PSA level of more than 4.0 ng/mL had a predictive value for the diagnosis of this type of cancer (Cooner et al., 1990). Subsequent publications, however, suggested that a PSA level of more than 2.5 ng/mL has a predictive value similar to that indicated above (Labrie et al., 1999; Krumholtz et al., 2002). The question of the right PSA level to detect prostate cancer at the right time and at the right stage is still attracting an intense scientific and clinical debate, which is likely to continue for years to come. This debate had started almost a quarter of a century ago and judging from the most recent literature, it is likely to continue for a long time to come. One of the central points in such a debate is the level of PSA that should prompt a prostate biopsy. The answer to this question is not as simple as it might appear simply because such level is susceptible to a host of parameters including age, race, lifestyle, genetic mutations just to name a few. Some of these parameters will be discussed later. However, as things stand at the present time, it can be hypothesized that the current disagreement between various opinions stems from the following: The use of higher PSA thresholds risks missing the detection of significant cancer until it is too late for a cure, whereas the use of lower PSA thresholds will not only result in unnecessary biopsies but also increase the proportion of biopsies that identify insignificant stage of the disease. The above hypothesis can be partially justified by considering some recent opinions of leading scientists in the area of prostate cancer and the use of PSA for its detection. A recent article in *The New England Journal of Medicine* by Thompson et al. (2004) entitled "Prevalence of prostate cancer among men with PSA level ≤4.0 ng/mL" has attracted a number of comments by other eminent scientists in the field; however, before considering such comments, a brief summary of this article should be helpful. In Thompson's study, 18,882 men were enrolled in the prevention trial, almost half of them were randomly assigned to receive placebo and had an annual measurement of PSA and DRE. Within this half, 2950 men never had a PSA level of more than 4.0 ng/mL or an abnormal DRE, had a final PSA determination, and underwent a prostate biopsy after being in the study for 7 years. This study concluded that biopsy-detected prostate cancer, including high-grade cancers, is not rare among men with PSA levels of 4.0 ng/mL or less, levels generally considered to be in the normal range. Such conclusion by a leading scientist in the field has inevitably attracted a number of responses and comments by other fellow scientists. Before considering some of these comments, it would be useful to cite another recent study dealing with the preoperative role of the serum PSA. This study by Stamey et al. (2004) appeared in the *Journal of Urology* under the title "The PSA era in the USA is over for prostate cancer: What happened in the last 20 years?" This article deserves careful consideration for a number of reasons: First, it is written by an expert regarded as the father of PSA testing; second, he was one of the first persons to suggest that PSA level was directly proportional to increasing clinical stages of prostate cancer (Stamey et al., 1987, 1989); and third, in the two earlier articles the same author pointed out a number of serious limitations in the relationship between serum PSA, prostate cancer volume, and Gleason grade 4/5 cancer (Stamey et al., 2001, 2002). In their article, the authors have addressed the question of PSA reliability in the diagnosis of prostate cancer. A total of 1317 consecutive radical prostatectomies were divided into four 5-year periods between 1983 and 2003, and examined sequentially in 3-mm step sections by the same pathologist. The largest cancer and five other histological variables in each prostate were measured. Preoperative clinical stages were tabulated for each 5-year period. Means, Pearson correlation coefficients, percentage of change, and multiple regressions were used to assess a number of selected variables. The main conclusion of this study was that in the last 5 years, serum PSA has only been related to benign prostatic hyperplasia, and therefore, there is an urgent need for new serum markers that reflect the size and grade of this ubiquitous form of cancer. The authors went on to suggest that the original benefit of PSA testing was that a high level correlated well with the detection of large tumors. Today, however, PSA screening has become widespread and testing triggers biopsy at much lower scores. Recent suggestions to use lower thresholds (2.6 ng/mL) of PSA for recommending prostate biopsy will simply compound the problem and will add millions of men to the biopsy list.

At this point, it is worth considering some of the comments regarding both the articles by other scientists in the fields. In an editorial in the same issue in which Thompson's article has appeared, an editorial by Carter (2004) asked whether we should now recommend lowering the threshold of PSA, which prompts a biopsy? The negative answer to this question was supported by the following considerations: First, the author argued that it should not be surprising that 10–27% of the men with PSA values of 4.0 ng/mL or less, who ranged in age from 62 to 91 years, were found to have prostate cancer in the study by Thompson et al. (2004). On the basis of the results of 5250 autopsies reported in the U.S. literature, the prevalence of prostate cancer was 15–60% among men who were 60–90 years of age, and increased with age (Carter et al., 1990). Ninety percent of men at 50–90 years of age have PSA values of 4.0 ng/mL or less (Smith et al., 1996). Thus, quite a few men with PSA levels of 4.0 ng/mL or less must harbor prostate cancer. Second, prostate cancers detected at lower PSA levels are more likely to have a small volume (less than 0.5 mL) and to be low-grade (Carter et al., 1997) and are, thus, more likely to represent clinically insignificant disease, because cancer volume and grade are surrogates for biologic potential. McNeal et al. (1986) found that only cancers that were much larger than 1 mL in volume and poorly differentiated were associated with metastatic disease. Furthermore, prostate cancers with a volume of less than 1 mL do not usually result in PSA levels above 4.0 ng/mL (Brawn et al., 1991), so that the unexpected detection of cancer at lower PSA levels is more likely to identify disease for which treatment not only may be unnecessary but also may fail to improve survival (Vis et al., 2002). Third, in an earlier investigation similar to that of Thompson et al. (2004), it has been shown that men with baseline PSA levels between 1.0 and 4.0 ng/mL are at a significantly higher risk for a diagnosis of prostate cancer over the next 10 years than are men whose baseline PSA level is below 1.0 ng/mL (Gann et al., 1995). The authors found that a cut-off value of 3.3 ng/mL resulted in optimal sensitivity and specificity, but the gain was minimal as compared with that afforded by a cut-off value of 4.0 ng/mL. In addition, Morgan et al. (1996) have shown that the PSA cut-off value that results in 95% sensitivity is close to 4.0 ng/mL for men between the ages of 50 and 70 years and 2.5 ng/mL for men in the fifth decade of life. Because most of the variability in PSA levels is due to benign prostate enlargement that occurs with age, and men below the age of 50 years are unlikely to have such an enlargement, a threshold of 2.5 ng/mL seems reasonable for men below the age of 50 years.

Carter (2004) argued that the lifetime risk of death from prostate cancer is about 3%, whereas the lifetime possibility of prostate cancer diagnosis is about 16%, which implies that any approach that finds more cancers without identifying the clinical significance of the detected disease will only result in overdiagnosis and overtreatment. Furthermore, there is no convincing evidence that, with contemporary therapy, men who are treated when their cancers were detected at PSA levels at or below 4.0 ng/mL have better outcomes than men who are treated when the PSA is slightly higher than such a level. In short, detection of prostate cancer at a PSA threshold lower than 4.0 ng/mL has not been shown to improve the disease-free outcome. With a PSA level in the range of 2.6–6.0 /mL, younger men are more likely than their older counterparts to have curable prostate cancer (Carter et al., 1999) and a disease-free outcome (Khan et al., 2003). Such observations are probably driven by the fact that older men are more likely to have high-grade cancers. Hence, the weight of the evidence suggests that the detection of prostate cancer at younger ages should have a greater effect on the likelihood of being free from disease after treatment, than would the detection of prostate cancer at a PSA level of 4.0 ng/mL or less. A partial support for some of the above comments by Carter (2004) comes from a recent investigation by Welch et al. (2005) who examined the effect of lowering the PSA threshold to 2.5 ng/mL on the number of American men who would be labeled as having an abnormal level by a single PSA test. To conduct their investigation, the authors combined data obtained from two main sources: First, they obtained PSA data on a nationally representative sample of American men 40 years of age or older with no history of prostate cancer and no current inflammation or infection of the prostate gland ($n = 1308$) from the 2001–2002 National Health and Nutrition Examination Survey (NHANES). Second, the authors obtained data on the 10-year risk of prostate cancer death in the pre-PSA era from the National Cancer Institute (NCI), data generated by software, which calculates the probability of dying of cancer. On the basis of these analyses, the authors concluded that lowering the PSA threshold to 2.5 ng/mL would double the number of men defined as abnormal, to up to 6 million. In commenting on these figures, the same authors stated that until there is evidence that screening is effective, increasing the number of men recommended for prostate biopsy would be a mistake.

3.5. IMPROVING THE SPECIFICITY OF PSA

Since its inclusion in the early 1980s as a screening marker, the question of PSA specificity remains the focus of intense research activities. Several investigators have attempted to refine the performance of PSA in detection strategies to decrease the number of unnecessary biopsies while minimizing the number of missed cases of cancer. These strategies included PSA velocity (change in PSA level as a function of time) (Carter et al., 1992, 1995), PSA density (adjusting the measured levels according to the size of the prostate) (Benson et al., 1992; Brawer et al., 1993; Catalona et al., 1994), age-adjusted PSA levels to take account of age-related prostate growth (Oesterling et al., 1993; Gustafsson et al., 1998), PSA forms including its various isoforms and its complexation with other proteins (Mikolajczyk et al., 2001; Catalona et al., 2003), and the influence of ethnic variations on the measured levels of PSA (Moul et al., 1995; Eastham et al., 1998; Abdalla et al., 1999). To appreciate the potential influence of each of these parameters on the reliability of future PSA tests, it is worth considering a number of published works that have dealt with one or more of these parameters and their potential influence on the specificity of PSA as a marker.

3.5.1. Free/Complexed PSA

Although serum PSA testing has transformed the diagnosis and treatment of prostate cancer, the ability of the same test to distinguish between benign and malignant forms of the disease remains inadequate. PSA can exist as free or it can bind to several known serum proteins, primarily α-1-antichymotrypsin, a protease inhibitor present in high concentrations in the blood that inhibits PSA's proteolytic activity, and to a much lesser extent, PSA can also bound to α-2-macroglubulin (Lilja et al., 1991; Martinez et al., 2002; Partin et al., 2003). Existing PSA assays use monoclonal antibodies to qualitatively measure the levels of both free and complexed PSA present in serum. Currently, there are three commercial assays, which can be used to measure free, complexed, and total PSA levels, where the latter represents the oldest and most ubiquitous assay, which measures the sum of immunoreactive PSA (Parsons and Partin, 2004). Over the last decade, there have been a number of studies (Catalona et al., 1998; Allard et al., 1999; Okihara et al., 2001; Horinger et al., 2002; Partin et al., 2003), which suggested that measuring the complexed form of PSA, rather than its free form, may contribute to a more specific test and consequently may reduce the number of unnecessary biopsies. In a multicenter study by Horninger et al. (2002), an attempt was made to assess the clinical value of complexed *c*PSA in comparison to the total value *t*PSA, the free/total ratio and the *c*PSA/*t*PSA in early detection of prostate cancer in men with *t*PSA values in the range 2–4 ng/mL. The entire study population comprised 831 patients with a mean age of 63 years ranging between 30 and 92 years. Of these, 25.8% had *t*PSA levels between 2 and 4 ng/mL and 27% had suspicious DRE results. According to these authors, the data generated showed that *c*PSA provides a statistically significant increase in specificity for early prostate cancer detection in men with *t*PSA levels in the range 2–4 ng/mL at good sensitivity. Furthermore, the pathologic features in the evaluated patients indicated that most had clinically significant early-stage prostate cancer that was potentially curable. In another study along these lines, Partin et al. (2003) have compared the clinical performance of *c*PSA with *t*PSA. Over 800 patients scheduled for initial biopsy of the prostate were enrolled for the study, which was conducted by seven university centers and community-based urology practices. The authors concluded that the use of *c*PSA as a single test provided improved specificity over *t*PSA in all clinically relevant *c*PSA ranges (1.5–8.3 ng/mL) evaluated. At the same time, the use of the percentages of free and complexed PSA had little or no additional benefit in the differentiation between benign and malignant forms of disease compared with *c*PSA on its own. According to the same authors, a *t*PSA cutoff of 3.1 mg/mL yielded a sensitivity of 90% and specificity of 26%, and a comparable *c*PSA cutoff of 2.7 ng/mL yielded the same sensitivity but a much-improved specificity (40%). According to a recent article by Parsons and Partin (2004), complexed PSA is expected to contribute to three main areas associated with prostate cancer: First, one of the potential clinical uses of *c*PSA has been its capability to predict the pathological stage of the disease. One of the early studies in this regard (Sokoll et al., 2002) investigated preoperative serum samples from 420 men undergoing radical retropubic prostatectomy. Using univariate analysis, the authors have established that the capability of *c*PSA was equivalent to *t*PSA in predicting organ-confined disease and favorable pathological features. However, the main deduction of this study was that neither assay on its own could provide a reliable prediction of the final pathologic state. On the contrary, a more reliable prediction could be achieved in a multivariate logistic regression model composed of biopsy, Gleason score, clinical stage, and *c*PSA assay. Such preliminary data imply that nomograms incorporating PSA can choose to replace *t*PSA with *c*PSA without significantly affecting their prediction capability, a hypothesis that has to be supported by further experimental studies. Second, another proposed application of *c*PSA is its use in longitudinal protocols, which follow the evaluation of patients suffering from prostate cancer. Such evaluation may include surveillance for biochemical recurrence following definitive therapy for localized disease, and assessment of response to hormonal ablation and chemotherapy. In one of the early reports dealing with this aspect, Allard et al. (1999) compared serial *c*PSA levels with the disease course in 155 patients with prostate cancer. This number included men with both early- and late-stage disease. The authors reported that in 97% of cases, the *c*PSA levels correlated with the disease course, with *c*PSA decreasing after various forms of treatment including surgery, radiotherapy, hormonal ablation, or chemotherapy, whereas an increase in the level was found to correlate with the progression of the disease. Again, these preliminary data suggest that the use of *c*PSA rather than *t*PSA can be a valid tool for the longitudinal monitoring of the disease. As we have stated above, such a deduction needs to be consolidated through additional studies. The diagnostic performance of *t*PSA versus *c*PSA in various age groups has been assessed by a number of research groups. Veltri et al. (2002) conducted a study involving 3597 patients who underwent a biopsy procedure for either an abnormal DRE or elevated *t*PSA levels. Within such a study, the impact of increasing age on the cutoffs of both *t*PSA and *c*PSA was required to sustain a fixed sensitivity to detect cancer. The main conclusions of this relatively large study can be summarized as follows: In the cohort of men studied, the *c*PSA assay performed as well as or slightly better than the *t*PSA assay in the differentiation between benign and prostate cancer. The same study demonstrated that in a large contemporary referral population of men with a *t*PSA in the range 2.0–20.0 ng/mL, the cancer detection rate remained fairly constant over such a range. In another study, *t*PSA, free PSA, and PSA complexed to α_1-antichymotrypsin (α_1ACT) were measured in plasma and serum before and during treatment with finasteride (España et al., 2002). These measurements were performed on plasma and serum from a total of 40 men suffering from BPH (30 treated with finasteride and 10 with placebo). Based on these measurements, the authors concluded that finasteride treatment induced a significant reduction in *t*PSA, free PSA, and PSA-α_1ACT levels in plasma and in serum, while the ratios of complexed-to-total and free-to-total PSA remained constant in both groups.

In summary, the above works give an initial indication that the use of *c*PSA rather than *t*PSA and percentage of free PSA offers an improved specificity for prostate cancer detection. These initial data, however, need to be supported by further data particularly regarding the performance of *c*PSA assay for detection, staging, and longitudinal surveillance of the disease.

3.5.2. PSA Isoforms

Various research groups have attempted to demonstrate that the use of a precursor of PSA can be more specific as a serum marker for prostate cancer tests. Free PSA in serum is now known to exist in three distinct forms: One form has been identified as the proenzyme, or precursor of PSA (*p*PSA), and is associated with cancer (Mikolajczyk et al., 1997, 2000a). A second form of this specific antigen is designated BSPA, an internally cleaved or degraded form of PSA that is more highly associated with BPH (Mikolajczyk, 2000b). The third form appears to be composed of largely intact PSA similar to native, active PSA, except for some unknown conformational and/or structural changes, which renders it enzymatically inactive. The combination of immunosorption and MALDI-TOF mass spectrometry has been used to identify the various forms of truncated *p*PSA in the serum of prostate cancer patients (Peter et al., 2001). As well as the form containing all seven amino acids within the leading peptide (APLILSR) the same study revealed the presence of truncated forms, *p*PSA containing one, two, three, four, five, or six amino acids instead of the usual seven. A recent article by Mikolajczyk et al. (2002) described each of these forms and reviewed their potential clinical applications as serum markers. One of the relevant questions addressed in this article was whether the different isoforms of PSA can indeed offer a further insight into the cancer stage and/or grade. For instance, the authors asked whether the more stable and more truncated [−2] PSA form can be associated with a different stage or grade of cancer compared to its fully intact [−7] PSA form? This question was supported by a number of considerations. For example, the two forms have completely different biochemical properties; therefore, the absence of the truncated form in the seminal plasma suggests that the disruption of the secretory pathway in the cancer lesion may be responsible for the localized enrichment of the truncated form and their eventual release into the blood (Mikolajczyk et al., 2002). Currently, there are immunoassays specific for the truncated forms of PSA, which can be used for comparative studies between intact and truncated forms of PSA in cancer and in BPH. These relatively new assays for free PSA isoforms may contribute to a better understanding of a number of aspects associated with prostate cancer. Two representative examples, where such assays can have an additional value, are as follows: First, it is still not clear why free PSA rises in serum immediately after a biopsy or radical prostatectomy (Lilja et al., 1999). Second, the discrimination of prostatitis from BPH is another area, where these emerging assays are likely to provide further information (Van Iersel et al., 1995). In a study by Catolana et al. (2003), research immunoassays were used to measure native and truncated forms of *p*PSA. In their study, the authors used more than 1000 serum specimens derived from men enrolled in prostate cancer screening studies who had undergone prostate biopsy. The group was divided into PSA ranges of 2–4 ng/mL and 4–10 ng/mL, each group containing both benign and cancer cases. Within the total PSA range of 2–10 ng/mL, the authors have measured the sum of noncomplexed forms, [−2, −4, and, −5] *p*PSA. It was claimed that the acquired results demonstrated an increased detection of prostate cancer especially in the range less than 4 ng/mL, with the truncated [−2] *p*PSA form having the highest specificity for cancer detection. The same authors concluded that the measurement of *p*PSA forms provided a more cancer specific serum marker for prostate cancer than the combinations of free, complexed, or total PSA. They also added that the clinical usefulness of percentage of *p*PSA was most pronounced in the early detection PSA range of 2–4 ng/mL, although such percentage showed enhanced discrimination of benign from prostate cancer throughout the investigated range, 2–10 ng/mL. These findings are in line with those reported in an earlier study, which reported an improved cancer detection in the PSA range of 2.5–4 ng/mL (Sokoll et al., 2003). Other works, which lend partial support to the findings by Catolana et al. (2003) have been published earlier by Buting et al. (2002) and España et al. (2002). In the first work, it was shown that biological intra-individual variation of single assays for *c*PSA, *t*PSA, and free PSA was significantly greater than the ratios of these forms to each other. The same authors found that *c*PSA had the most biological variability, whereas the ratio *c*PSA/*t*PSA was the most stable. A conceptually similar example has been cited in the previous section (Section 2.5.1), which also showed that internal ratios of PSA forms were less susceptible to external perturbations. This was demonstrated with finasteride drug treatment, in which total serum PSA was found to decrease by about 50%, whereas the ratio of free-to-total PSA was found to be relatively constant (España et al., 2002).

3.5.3. Impact of Age, Race, and PSA Velocity

Data generated over the last 15 years on the use of PSA test have underlined three basic elements: First, the incidence of prostate cancer increases more rapidly with age than any other cancer type (Greenlee et al., 2001). Second, in men without evidence of prostate cancer, PSA levels increase with age, mainly because of an increase in prostate volume due to BPH, which led Oesterling et al. (1993) to develop age-specific reference ranges for white men from Olmsted County, Minnesota. Third, these age-specific reference ranges are influenced by the race (Oesterling et al., 1993). Various attempts have been made to improve the specificity of PSA test by compensating for changes in its level due to the age of the patient. In other words, lower limit for normal UNL should be used for younger men whereas a higher ULN may be used for older men. One of the early studies in this direction was reported by Oesterling et al. (1993) in a population of healthy men without clinical evidence of prostate cancer. The influence of patient age and prostatic size on the serum PSA concentration was assessed. This study was conducted between 1989 and 1991 and involved over 2000 healthy men aged between 40 and 79 years. On the basis of this study, the authors deduced that serum PSA concentration was correlated to the prostate volume, which in turn, was directly correlated with the patient's age; whereas, PSA density was only weakly correlated with age. The authors went on to suggest that, rather than rely on a single reference range for men of all age groups, it is more appropriate to have age-specific reference ranges. These age-specific reference ranges have the potential to make serum PSA a more discriminating tumor marker for detecting clinically significant cancers in older men. 3 years later, Partin et al. (1996) applied the same concept to men <60 years of age in a study involving 4600 men with clinically localized prostate cancer, and they detected 74 additional cancer cases by applying age-adjusted *t*PSA cut-offs. Reissigl et al. (1995) have also conducted a study along these lines, where an Austrian population of 21,000 men aged 45–75 years was involved. The authors reported that by the application of age- specific *t*PSA ranges, they were able to detect an additional 8% positive biopsies in men <59 years of age with normal DRE. The same study reported that in men >60 years of age and normal DRE, 21% fewer biopsies were performed, while missing 4% of organ-confined tumors. The additional value of age-specific *t*PSA was questioned by Catalona et al. (1994) who demonstrated that 44% of tumors were missed by applying such an approach. In a more recent study by Veltri et al. (2002), the impact of age on total and complexed PSA cutoffs was assessed. The diagnostic utility of *t*PSA and *c*PSA was evaluated using a diverse contemporary test population consisting of 3597 men (age range 45 to ≥80 years) who recently underwent a systematic biopsy by urologists in clinical practices throughout the United States. Part of the sera was obtained from men with no evidence of malignancy, and the other part from those having prostate cancer. The authors concluded that to maintain a given sensitivity level for *c*PSA and *t*PSA in the age ranges studied, continuous upward movement of the cutoffs was required as age increased. The same authors added that overall *c*PSA showed a marginal improvement in specificity versus *t*PSA across all age groups and at all sensitivity levels (80–95%) used in the study.

Over the last few years, there have been a number of works which indicate that the performance of PSA test can be influenced by the ethnic origins of the tested population. Investigators have reported higher PSA values among African American men (Moul et al., 1995; Eastham et al., 1998; Abdalla et al., 1999). Other works, however, have not supported such a correlation (McCammon et al., 1996). Individuals who identify themselves as African–American or black compose approximately 12% of the U.S. population. A number of studies have indicated a higher incidence of prostate carcinoma, less favorable stage at diagnosis, and higher mortality (Beahrs et al., 1992; Mettlin et al., 1995; Wingo et al., 1996). A hospital-based study by Fowler et al. (2002) involving a patient population with substantial black representation attempted to assess relation between race, PSA, and cancer detection with an initial prostate biopsy among men with a PSA level greater than or equal to 4.0 ng/mL and a normal DRE. The study was conducted over an 8-year period and involved a total of 2607 biopsy patients; 931 had a DRE that did not indicate the presence of cancer and a PSA ≥4.0 ng/mL. The main findings and limitations of this study can be summarized as follows: The PSA levels without detectable cancer were significantly greater in black men than in white men and that the differences were not caused by racial variability in prostate volume. This result was in accordance with earlier reports related to men with histologic BPH (Henderson et al., 1997; Fowler et al., 1999) and is consistent with the concept that equivalent volumes of benign prostatic tissue contribute more PSA to the circulating blood in black rather than in white men. The PSA with detectable cancer was found to be greater in black men than in white men and the difference remained significant when controlled for incremental Gleason score. This particular observation has been reported in other patient populations (Vijayakumar et al., 1992; Moul et al., 1995). Although it is difficult to quantify such characteristics, it is reasonable to hypothesize that the difference may be attributable in part to racial differences in serum PSA derived from benign prostatic tissue that coexists with malignancy. It is also reasonable to assume that the contribution of PSA to the circulating blood by equivalent volumes of histologically similar cancers may be greater in black men than in white men. Moul et al. (1995) have reported that PSA values in radical prostatectomy patients were greater in black men than in white men, when controlled for tumor volume and other relevant variations. Of course, this and other similar studies have their limitations, which have been pointed out by the same authors. First, although the authors were aware that needle biopsy was an imprecise means for the identification of local stage prostate carcinoma, many of the investigated subjects had not undergone repeated biopsy procedures. Second, the potential benefits of lower PSA reference ranges for black men younger than 60 years of age were not rigorously applied. Most of the patients within this study had no PSA history and underwent biopsy because the first determination of PSA was equal or greater than 4.0 mg/mL. This single measurement resulted in a wide variation in PSA levels, which meant that the acquired results could not be extended to longitudinal or clinical screening, in which the biopsies are prompted by a progressive increase in PSA levels, reaching or exceeding 4.0 ng/mL.

PSA density is commonly defined as the PSA level divided by prostate gland volume determined sonographically. The generally accepted normal cutoff value of PSA density is 0.1, above which the risk of harboring prostate cancer increases (Benson et al., 1992). To underline the influence of PSA density on the overall results of a PSA test, it is worth considering the results of two studies, a relatively recent one by Lam et al. (2003) and an earlier study by Abdalla et al. (1999). In the first study, a total of 404 consecutive Hispanic men and 341 consecutive white men with elevated serum PSA and/or abnormal DRE underwent transrectal ultrasound with lesion directed and systematic peripheral zone biopsies over a period of 5 years at a single institution. Before biopsy, all patients underwent volume measurements of the entire prostate. Based on this study, the authors came to a number of conclusions, which can be summarized as follows: At similar levels of total PSA, PSA density is higher in Hispanic men than in white men (0.13 versus 0.11). Furthermore, the PSA test was able to distinguish between malignant versus benign disease in Hispanic men, but not in white men. Before commenting on these findings, it would be helpful to consider two earlier studies by Abdalla et al. (1998; 1999). In the first investigation, the authors examined the influence of race or ethnicity on PSA levels and PSA density in a population of healthy men without clinically evident prostate cancer. This study was conducted over 5 years and involved 859 men composed of African Americans, Whites, and Hispanics. All men included in this study underwent a detailed clinical examination, including DRE, serum PSA determination, and transrectal ultrasound. One of the main conclusions of this study was that African Americans have higher serum PSA levels and higher PSA density than do Whites or Hispanics; such differences remained statistically significant after adjustment of age and prostate volume. Similar conclusions were reported by the same group and published a year later (Abdalla et al., 1999). These and other findings together with those reported by Lam et al. (2003) merit further considerations: First, the latter study reported that in groups of the investigated men with PSA between 2.5 and 10 ng/mL, a higher proportion of Hispanic men was found to have prostate cancer (35.1% versus 25.5% in white men, $p = 0.0147$). This finding is in complete contrast to those reported by Canto and Chu (2000), who reported incidence rates for prostate cancer in Hispanic men 30% lower than white men. On the contrary, Abdalla et al. (1998) found no difference in prostate cancer incidence between white men and Hispanic men in a study conducted in the Chicago area. Considering these findings, it is not difficult to note three different results in three different studies. This observation is by no means a criticism of any of these studies but to highlight the difficulty of assessing a single phenomenon, when such a phenomenon is susceptible to a multitude of parameters. Second, studies involving mixed races may run the risk of being biased toward one or other of the groups under study. For example, white patients might have been more heavily prescreened for prostate cancer than Hispanic patients. Populations that undergo an annual screening have shown that the chances of cancer detection on subsequent visits are significantly lower than at the initial screening (Carter et al., 1997). This may imply that the white patients in the study by Lam et al. (2003) had had more previous PSA testing than their Hispanic counterparts. Third, given the limited number of works and the limited number of samples within such works, it is reasonable to expect future studies involving a much larger number of samples to allow more rational assessment of some of the findings cited in the above works.

Several studies have reported that, when considered on its own, the rate of rise in PSA levels (PSA velocity) before the diagnosis of prostate cancer can predict tumor stage and grade (Carter et al., 1992; Goluboff et al., 1997; Egawa et al., 2000). Natural biological variations in PSA levels have been evaluated by Nixon et al. (1997). The authors investigated the daily variations in serum samples over a period of 2 weeks to determine the difference required between two consecutive PSA measurements that would indicate a significant elevation. These investigators have concluded that the degree of biological variation differs among patients, such that an increase between two consecutive PSA levels that is less than 20% to 46% may be due to biological and analytical variations alone. Similarly, Ornstein et al. (1997)

examined the biological variation of total-free and percentage-free PSA in a number of men older than 50 years. Testing for PSA was carried out on three occasions, each two weeks apart. This study showed a mean variation of ~15%. Year-to-year variation in PSA levels has also been investigated by Eastham et al. (2003).

The preoperative PSA velocity and the risk of death from prostate cancer after radical prostatectomy have been recently investigated by D'Amico et al. (2004). The authors have studied 1095 men with localized prostate cancer to assess whether the rate of rise in PSA during the year before diagnosis, the PSA level at diagnosis, the Gleason score, and the clinical tumor stage could predict the time of death from prostate cancer and death from any other cause after radical prostatectomy. The main conclusions of this study were as follows: First, an increase in PSA level at diagnosis, a Gleason score of 8, 9, or 10, and a clinical tumor stage of T2 could predict the time of death from prostate cancer. Second, men whose PSA level increases by more than 2.0 ng/mL during the year before the diagnosis of prostate cancer may have a relatively high risk of death from prostate cancer despite undergoing radical prostatectomy. Furthermore, for these men, who are otherwise in good health, watchful waiting may not be the best option. Instead, randomized clinical trials to identify possible treatment may improve their chances of survival. Having said that, an earlier study by Freedland et al. (2001) found no association between the preoperative PSA velocity and the postoperative pathological findings and PSA outcomes. It has to be said, however, that this study involved a small number (86 men) and a relatively brief follow-up period (2 years). Given the evidence that disease recurrence after surgery is often not predictive of death from prostate cancer (D'Amico et al., 2004) because of other possible competing causes of mortality and the long natural history of prostate cancer, long-term evaluation is pertinent.

3.6. LOOKING FOR OTHER SOLUTIONS

While PSA testing has resulted in an increase in prostate cancer detection, its routine use as a screening tool has been questioned because of its lack of specificity when levels are moderately elevated (4–10 ng/mL). A variety of methods have been suggested to improve the specificity of PSA testing, including age-specific PSA reference ranges and others (see Sections 2.5.1–2.5.3). Other attempts to achieve the same objective included proteomic approaches, investigation of genetic alterations, and the phosphorylated form of P13K/AKT signaling pathway. The application of various proteomic approaches in the search for alternative markers for PSA have been described in the previous chapter, whereas the other two approaches are considered below.

3.6.1. Genetic Alterations

There is accumulating evidence that family history represents a risk factor in prostate cancer. Approximately 10–15% of men with prostate cancer have at least one relative who is also affected (Hays et al., 1995; Whittemore et al., 1995). Furthermore, there are some indications that genetics is likely to play a role, at least for men with a positive family history. Such evidence comes from a variety of study designs, including case-control, cohort, twin, and family based (Bratt et al., 1999; Lichtenstein et al., 2000; Hemminki and Czene, 2002). Older age, ancestry, lifestyle-related factors, dietary-related factors, and a positive family history of prostate cancer have long been recognized as important risk factors (Nelson et al., 2003; Schaid, 2004). Over the last 20 years, various investigations have generated substantial data to support the role of genetics in this disease. These studies have provided a valuable insight into both hereditary and somatic genetic alterations associated with this form of cancer. The search for prostate cancer susceptibility genes by linkage studies offered early hope that finding genes would be as "easy" as finding genes for breast cancer and colon cancer susceptibilities. However, this hope has been dampened by the difficulty of replicating promising regions of linkage. Given the large body of evidence that prostate cancer is likely to have a strong genetic basis, and given the difficulty of finding inherited susceptibility genes, the evidence that prostate cancer is likely to be caused by multiple genes, possibly interacting in complex manners, and possibly interacting with environmental factors, continues to grow. Hereditary prostate cancer has been defined by Carter et al. (1993) as families that meet at least one of the following three criteria: (a) a cluster of three or more relatives affected with prostate cancer in any nuclear family; (b) the occurrence of prostate cancer in each of the three successive generations in either of the proband's paternal or maternal lineages; or (c) a cluster of two relatives, both affected with prostate cancer at 55 years of age or younger. This operational definition has been used in a number of studies, particularly those focused on linkage. However, this definition has been described as somewhat biased toward autosomal dominant transmission, and would likely to miss some families with autosomal recessive or X-linked transmission (Schaid, 2004). Studies in twins that compare the concordant occurrence of prostate cancer in monozygotic twins with that in dizygotic twins have consistently revealed a stronger hereditary component in the risk of prostate cancer than other type of human cancers (Gronberg et al., 1994; Lichtenstein et al., 2000). Furthermore, these studies together with case-control and family-based segregation studies have identified a number of prostate-cancer-susceptibility genes, which are summarized in Table 3.1.

Regarding somatic gene alterations in prostate cancer, a number of studies have indicated that human cancer cells typically contain somatically altered genomes, characterized by mutation, amplification, or deletion of critical genes (Fearon and Vogelstein, 1990). In addition, the DNA template from human cancer cells often displays somatic changes in DNA methylation (Fearon and Vogelstein, 1990; Baylin and Herman, 2000; Jones and Baylin, 2002). Over 10 years ago, Lee et al. (1994) reported that most human prostate cancers fail to express the π-class *GST*, and that regulatory sequences near the *GST1* gene, which encode the human π-class *GST*, appear to be commonly hypermethylated during prostatic carcinogenesis. Detecting hypermythelation of promoter regions of cancer-associated genes is being considered as an approach for finding more specific markers for various forms of

TABLE 3.1. Prostate cancer susceptibility and somatic genes, their alterations, and locations. On the basis of the data from Nelson et al. (2003) with permission.

Susceptibility gene alterations			Somatic gene alterations		
Gene	Location	Alteration	Gene	Location	Alteration
RNASEL	1q24-25	Base substitution leading to Met1Ile, Glu265X, and Arg462Gln alleles. Four-base deletion at codon 157 leading to premature protein truncation at codon 164.	*GSTP1*	11q13	CPG island hypermythelation (decreased expression).
			NKX3.1	8p21	Allelic losses (decreased expression).
			PTEN	10q23.31	Allelic losses, mutations, probable CPG Island hypermythelation (decreased Expression, function, or both).
ELAC2	17p11	Base insertion leading to premature Termination 67 amino acids after codon 157; base substitutions leading to Arg781His, Ser217 Leu, and Ala541Thr alleles.	*CDKN18*	12p12-13	Allelic losses (Decreased expression).
			AR	Xq11-12	Amplification, mutations (increased expression, altered function)
MSR1	*8P22*	Base substitutions leading to Arg293X, Pro36Ala, Ser41Tyr, Val113Ala, Asp174Tyr, Gly369Ser, and His441Arg alleles.			
AR	Xq11-12	Polymorphic polyglutamine (CAG) and Polyglycine (GGC) repeats.			
CYP17	10q24.3	Base substitution in transcriptional promoter (T → C transition leading to new Sp1 recognition site).			
SRD5A2	2p23	Base substitutions leading to Val89Leu and Ala49Thr alleles.			

cancer. Some somatic gene alterations associated with prostate cancer are given in Table 3.1. The alterations associated with *GSTP1* have been used as DNA-based detection of prostate cancer. The possible role of these alterations in the molecular pathogenesis of prostate cancer is schematically represented in Figure 3.1. In a recent study, *GSTP1* methylation was quantified in non-neoplastic prostatic tissue and organ-confined prostate adenocarcinoma (Jerónimo et al., 2001). The authors used

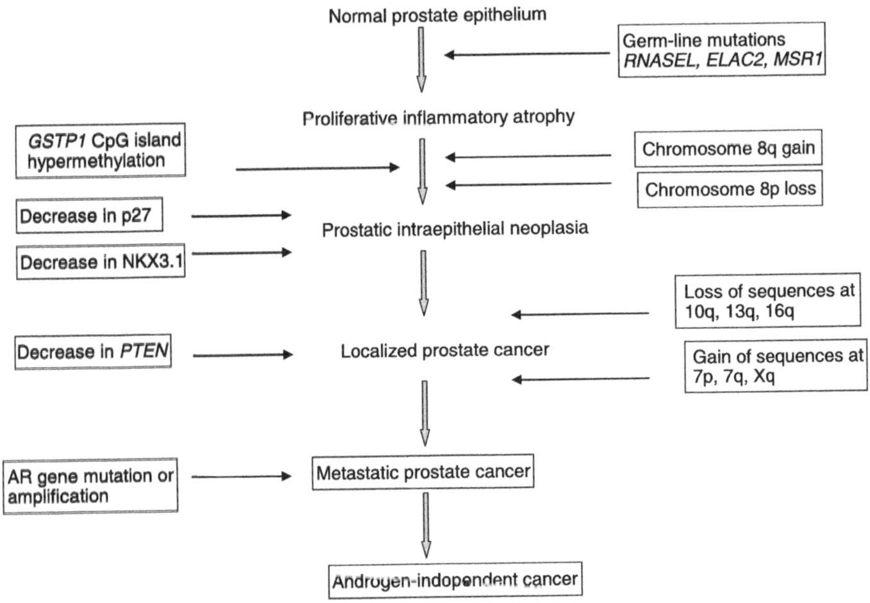

Figure 3.1. A schematic representation of the molecularpathogenesis of prostate cancer. Adapted from Nelson et al. (2003) with permission.

fluorogenic real-time methylation-specific polymerase chain reaction assay to assess cytidine methylation in the *GSTP1* promoter in prostate tissue samples derived from 69 patients with early stage prostatic adenocarcinoma. The authors concluded that the distinctly different levels of *GSTP1* methylation in non-neoplastic tissues and in those with evidence of prostate cancer suggest that quantification of *GST1* methylation could be more useful than measurements of serum PSA levels in distinguishing men with a very low risk of harboring prostate cancer from those who carry a clinically silent prostate adenocarcinoma. The same authors added that the absence of a correlation between PSA levels and *GST1* methylation in prostate cancer patients in their study may enforce the idea that such methylation can be used as an independent marker.

A simple survey of current literature on PSA performance indicates that such performance can be influenced by other factors, which are not included in the sections above. One of these factors is diet and lifestyle. In a massive study, the risk of prostate cancer among 44,788 pairs of twins in Sweden, Denmark, and Finland (Lichtenstein et al., 2000), 42% of the cases of prostate cancer were attributed to inheritance, with the remainder most likely attributable to environmental factors. This deduction is supported by epidemiologic evidence. For example, the incidence of prostate cancer and associated mortality are higher in the United States and Western Europe, with the highest rates among black men, whereas lower rates are more of the characteristic of Asians (Hsing, 2000). The risk of prostate cancer among Asians increases when they emigrate to North America, which may implicate their environment and lifestyle. In a recent clinical trial, men given tomato-sauce-based dishes for three weeks before radical prostatectomy had increased lycopene levels in the blood and the prostate, decreased oxidative genomic damage in leukocytes and prostate cells, and a reduction in PSA level (Chen et al., 2001). Other antioxidants, such as vitamin E and selenium have also been associated with reduced risk of prostate cancer (Clark et al., 1998; Heinonen et al., 1998). A large clinical trial of supplementation with vitamin E and selenium had been initiated 4 years ago (Hoque et al., 2001) aimed at reducing the risk from prostate cancer. In a recent review (Nelson et al., 2003), the authors discussed the mechanism of disease of prostate cancer and the factors that may influence the development and progression of the disease including diet, lifestyle, and genetic factors. The authors commented that genes, diet, and lifestyle-related factors contribute to the development of prostate cancer. This comment was supported by two considerations: First, they argued that two inherited susceptibility genes, *RNASEL* and *MSR1*, may have roles in responses to infections, raising the possibility that prostate infection or inflammation initiates prostatic carcinogenesis. *RNASEL* gene encodes a widely expressed latent endoribonuclease that participates in an interferon-inducible RNA-decay pathway that is thought to degrade viral and cellular RNA (Zhou et al., 1993; Silverman et al., 1988). *RNASEL* has been linked to the hereditary prostate cancer *(hpc1)* gene (Carpten et al., 2002). *MSR1* is another gene, which emerged as a candidate prostate-cancer-susceptibility gene (Xu et al., 2002). It encodes subunits of a macrophage-scavenger receptor that is capable of binding a variety of ligands, including bacterial lipopolysaccharide and lipoteichoic acid, and oxidized high- and low-density lipoproteins in serum (Platt and Gorgon, 2001). Germ-line *MSR1* mutations have been linked to prostate cancer in some families with hereditary prostate cancer (Dejager et al., 1993; Xu et al., 2002).

The possible relationship between diet and prostate cancer has been investigated in a recent study (Marks et al., 2004). This study focused on the effects of animal fats and soy on prostatic tissues among native Japanese and second generation or third generation Japanese Americans. The study was conceived to explore elements within Western diet, which may result in carcinogenic effects in prostatic tissues. This study has concluded that prostate-cancer-specimens from the two groups were histologically similar, but tissue biomarker expression, especially of lipoxygenase and the caspase family, suggested differing mechanisms of carcinogenesis. Furthermore, differences in nuclear morphometry suggested the additional possibility of gene-nutrient interactions.

3.6.2. Phosphorylated Akt

Akt, which is also known as protein kinase B(PKB), consists of a family of highly conserved serine/threonine kinases including Akt1, Akt2, and Akt3 (Alessi and Cohen, 1998; Zinda et al., 2001). Akt can be expressed in both non-neoplastic and tumor tissues of a variety of origins including prostate (Zinda et al., 2001). Although the regulatory pathways in prostate cancer progression remain poorly understood, there is an increasing evidence that Akt plays some role in promoting cell survival in prostate cancer. Recently, there have been a number of studies that investigated the possible correlation between phosphorylated Akt and the progression of prostate cancer. Malik et al. (2002) used immunohistochemical staining to assess the extent of Akt activation (phosphorylation) in 74 cases of resected prostate cancer. This study was based on the following consideration: Although the early stage of prostate cancer is marked by excessive proliferation, in advanced stages of the disease, a decreased apoptotic death rate (increased cell survival) also contributes to net tumor growth. Akt-regulated cell survival pathways are one of the suspected causes of the above consideration. The authors concluded that an immunohistochemical examination of paraffin-embedded human prostate cancer showed that 92% of the poorly differentiated adenocarcinomas of the prostate stained strongly for phosphor-Akt in a membrane location. In all other grades of prostate cancer including proliferative intraepithelial neoplasia (PIN), only 10% stained for phosphor-Akt. This study was the first of its type in which phosphor-Akt was detected using a phosphor-specific antibody in paraffin-embedded human prostate cancer. It is of interest to note that in an earlier study using reverse-phase protein microarrays (Paweletz et al., 2001) demonstrated that prostate cancer progression was associated with an increased phosphorylation of Akt together with a suppression of apoptotic pathways. In a second and more recent study (Kreisberg et al., 2004), an attempt was made to use the extent of phosphorylation of Akt(Ser^{473}) to predict poor clinical outcome in prostate cancer. This study was designed to establish whether increased phosphorylation of Akt and/or decreased phosphorylation of ERK (extracellular signal regulated kinase) could be used to predict poor clinical outcome in prostate cancer. It is also worth noting that within the same study, other commonly used indicators for predicting disease occurrence, including the cell proliferation antigen Ki67 and Gleason grading, were also compared as biological markers for clinical outcome. Prostate cancers were obtained after radical prostatectomy from a mixed population in the age range 42–81 years. PSA data available from patient follow-up were evaluated to assess good and poor outcome after radical prostatectomy.

3.7. CONCLUDING REMARKS

Based on early and recent literature regarding the performance of PSA as a marker for prostate cancer, the following observations can be made:

- At present there is no other marker, which can perform better than the PSA. This marker has its shortcomings but remains the most widely used serum marker for both the detection and monitoring of prostate cancer. PSA as a prostate cancer marker can be criticized in many different ways, yet a closer look at the performance of this marker over the last 20 years renders some of these criticisms rather unjustified. To support this statement, here are some considerations: Limiting ourselves to serum markers, the PSA enjoys sensitivity, which over 20 years and after thousands of scientific works has never been indicated as a motive to look for another marker. At the usually indicated level of 4.0 ng/mL and the proposed lower level of 2.5 ng/mL, this marker is more than capable of doing the job. What I am trying to say is that the sensitivity of this marker should not be listed as one of the reasons behind current efforts to find an alternative marker(s). On the contrary, the major drawback, which has been pointed out by many experts in the field, is that such levels and even higher do not correlate with the stage of the disease, particularly advanced stages, which become life threatening.

- PSA level in serum can be influenced by a host of parameters including age, ethnicity, genetic polymorphism, lifestyle, year-to-year fluctuations, and so forth. Screening trials, which compensate for all these parameters, are almost impossible. Furthermore, randomized trials that test a variety of thresholds are nonexistent. The absence of data from such trials renders the suggestions of lower/higher thresholds difficult to sustain. The recent publication of data regarding the detection of prostate carcinoma in men with what is termed "normal" PSA level makes this a challenging time for clinicians and patients alike. At present the question of PSA specificity is the focus of various research activities. These activities can be divided into two main streams: The first is mainly concerned with the improvement of PSA, whereas the other aims at finding alternative markers with better specificity. There are three areas that have demonstrated an initial potential for finding markers to enhance PSA specificity or even finding new alternatives. On the proteomic side, Chapter 2 has given a fairly detailed description of various technologies and their initial contribution to this search. It also discussed possible future developments, which will no doubt enhance the contribution of a number of emerging technologies. On the genetic side, there have been reasonable indications that both somatic and susceptibility gene alterations may provide the missing answers to a number of questions associated with prostate cancer. For example, *RNASEL* and *MSR1* may have roles in responses to infections, raising the possibility that prostate infection or inflammation initiates prostatic carcinogenesis. A new prostate cancer-precursor lesion, proliferative inflammatory atrophy, may be another link between prostatic inflammation and prostate cancer. Loss of the *GSTP1* caretaker function, as cells of proliferative inflammatory atrophy give rise to cells of prostatic intraepithelial neoplasia and to prostate-cancer cells, increases the prostate's vulnerability to genomic damage caused by inflammatory oxidants and dietary carcinogens. Somatic targets of genomic damage include *NKX3.1,* a candidate gatekeeper gene, as well as *PTEN* and *AR,* genes that may modulate the progression of prostate cancer. Inherited polymorphic variants of genes mediating androgen action, *AR, CYP17,* and *SRD5A2,* also influence the development and progression of prostate cancer. Although mutations in several genes are associated with prostate adenocarcinoma, some mutations, such as those that activate the ras oncogene or inactivate the tumor suppressor *p53*, occur in only a small number of prostate cancer cases (Isaacs and Isaacs 1997), whereas other mutations, such as those that inactivate the tumor suppressor gene *PTEN*, are found mainly in advanced stages (Cairns et al., 1997). This implies that the identification of genetic or epigenetic alterations that occur more frequently and in earlier stages of the disease as well as in premalignant lesions, such as prostate intraepithelial neoplasia (PIN), may allow DNA-based detection of prostate cancer. The application of this approach has been described in a preliminary study involving a limited number of patients with organ-confined prostate adenocarcinoma (Jerónimo et al., 2001). This study distinctly reported different levels of methylamine of *GSTP1* in non-neoplastic tissues and those with evidence of prostate cancer. Gene-expression profiles that molecularly distinguish prostatic neoplasmas have been investigated by using microarrays of complementary DNA (Dhanasekaran et al., 2001); the authors integrated data generated by *c*DNA with those generated by high density tissue microarrays and clinical/pathological tests. The authors reported an association between various stages of prostate cancer and two genes *hepsin*, a transmembrane serine protease, and *pim-1*, a serine/thrionine kinase.

The P13K-AKt signaling pathway regulates many normal cellular processes including cell proliferation, survival growth, and motility (Luo et al., 2003). The role of this pathway in oncogenesis has been extensively investigated, and altered expression or mutation of many components of this pathway has been implicated in various forms of human cancer including prostate (Vivanco and Sawyers, 2002). A number of investigations targeting the P13K-AKt pathway have generated preliminary yet encouraging data. These studies indicated that signaling pathways encompassing the Akt family of protein kinases is associated with prostate cancer progression (Ayala et al., 2004). This hypothesis is partially supported by the role played by the Akt kinases as intermediaries in signaling the growth and antiapoptotic regulators and the frequent carcinogenic mutations targeting components of the signaling pathways (Vivanco and Sawyers, 2002; Fresno Vara et al., 2004). Future studies, which may expand and add to the above findings, will no doubt contribute to a better understanding of the mechanism of the disease, and possibly contribute to a new generation of biomarkers.

3.8. EXISTING BIOMARKERS FOR OVARIAN CANCER

Despite progress in cancer therapy, ovarian cancer mortality has remained virtually unchanged over the last two decades (Jemal et al., 2003). The steep survival gradient in relation to the stage at which the disease is diagnosed has provided the rationale for more efforts to detect this form of cancer in its early stages. Although ovarian cancer is an important cause of mortality, its low prevalence, the lack of a clearly defined precursor lesion, and the high cost together with possible complications associated with surgical confirmatory procedures have placed stringent requirements on any test intended for general population screening. A direct result of these stringent requirements is that none of the existing serum markers including carcinoma-associated glycoprotein antigen (CA-125), or macrophage colony-stimulating factor, can be used individually for screening (Bast et al., 1998). The above "guidelines" have their good reasons, including: first, at present there is no single test with the desired sensitivity and specificity to allow the detection of ovarian cancer in its early stages. Unfortunately, most serum markers are identified on the basis of samples derived from patients suffering from an advanced stage of the disease. This means that existing markers with high sensitivity for clinically diagnosed cancer do not have sufficient sensitivity for the preclinical tests. Second, the absence of a specific and sufficiently sensitive test for the early detection of this disease together with the sporadic nature of ovarian cancer render the design of screening strategies and the choice of target population extremely difficult. These two factors may explain why more than 70% of patients have stage III or IV disease at the time of diagnosis. Despite the introduction of new and improved chemotherapeutic agents for the treatment of this malignancy, the fact that most patients are diagnosed with advanced-stage disease translates into poor 5-year survival rates of 20–30% (Lippincott, 1997; Dupont et al., 2004). On the contrary, 90% of women diagnosed with disease confined to the ovary survive more than 5 years. Furthermore, when ovarian cancer is diagnosed in stage I, more than 90% of patients can be cured with conventional surgery and chemotherapy. At present, however, only 25% of ovarian cancer is detected in stage I, which underlines the pressing need for an improved sensitivity and specificity to allow such early detection. Currently, there are a number of markers, which are considered of value for the detection and monitoring of ovarian cancer. Table 3.2 lists some of these markers, which have been given in a number of recent articles (Bast, 2003; Jacobs and Menon, 2004). CA125 remains the most thoroughly investigated biomarker in ovarian cancer; however, a number of shortcomings associated with this marker have encouraged its use with other markers or in combination with other approaches such as transvaginal sonography and transabdominal ultrasonography. Research efforts to improve the performance of this and other markers are discussed below. However, before discussing the current role of CA 125, both on its own and/or in combination with other markers, it is relevant to consider some of the biological aspects of ovarian cancer. This choice is motivated by a number of considerations: First, there is no substantial evidence to date to suggest that screening reduces mortality, either in general population or in women at increased genetic risk. Having said that, it is fair to point out that there is a paucity of randomized controlled trials of women at risk. One of the rare examples along these lines is a study by Jacobs et al. (1999), and even this study was not adequately designed to detect a reduction in mortality. Second, it is becoming more apparent that the hereditary proportion of ovarian cancer is among the highest of all common forms of cancer in adults (Risch et al., 2001). There is also substatantial evidence that germline mutations in *BRCA1* and *BRCA2* can result in cancer predisposition in the majority of families with the breast ovarian cancer syndrome (Norad et al., 1995; Frank et al., 2002; Norad and Foulkes, 2004). However, it is not yet clear as to what proportions of ovarian cancer in randomized general populations are due to mutations in these genes. The above statement and other related aspects are discussed in the sections below.

TABLE 3.2. Serum, Plasma, and Urine Markers for Epithelial Ovarian Cancer. Adapted from Bast, Jr (2003 with permission).

Marker	Marker
Carcinoembryonic antigen	Mesothelm*
Placental alkaline phosphatase	Beta-chain–human chorionic gonadotropin
CA-15-3*	Inhibin*
CA-19-9*	Urinary gonadotropin peptide*
CA-50*	Ceruloplasmin
CA-54-61	C-reactive protein*
CA-72-4 (TAG72)	CYFRA21-1
CA-195	Immunosuppressive acidic protein*
Cancer-associated serum antigen*	NB/70K*
Human milk fat globule protein	Tissue peptide antigen*
Human milk fat globule-2	Lipid-associated sialic acid*
Mucinlike carcinoma antigen	Lysophosphatidic acid*
Ovarian serum antigen*	
OVX1	
Sialyl TN	
Galactosyltransferase	
Cathepsin L	
Matrix metalloproteinase 2*	
Prostasin	
Kallekrein-6 and -10	
Tetranectin	
Alpha-1-antitrypsin	
HE4	
Tumor-associated trypsin inhibitor*	
p110 epidermal growth factor receptor	
ErbB-2 (*HER-2-neu*)	
Tumor necrosis factor receptor	
Soluble Fas ligand	
Interleukin-2 receptor	
Interleukin-6	
Interleukin-8	
Interleukin-10	
Macrophage colony-stimulating factor*	

*Reported to complement CA-125.

3.8.1. Genetic Disorder and Increased Risk of Ovarian Cancer

Cancer-causing genetic alterations fall broadly into two functional classes: Those that activate cellular genes, known as oncogenes, and those that inactivate cellular genes, known as tumor suppressor genes (Sansal and Sellers, 2004). Most breast and ovarian cancers are sporadic (i.e., not inherited); however, there is an accumulating body of evidence to suggest that a strong history of breast cancer, ovarian cancer, or both may be related to the presence of an inherited mutation in one of the two genes, known as *BRCA1* and *BRCA2*; the first is located on chromosome 17q, whereas the second is located on chromosome 13q, and their gene products are involved in DNA repair (Risch et al., 2001; Venkitaraman, 2002; King et al., 2003; Cannistra, 2004; Norad and Foulkes, 2004). Although clinical evaluation of the *BRCA1* and *BRCA2* genes had begun almost 10 years ago, there is still no clear consensus on what specific personal and family history should prompt a consideration of hereditary cancer risk assessment. It is also interesting to note that *BRCA1* and *BRCA2* are frequently referred to as breast cancer genes; however, some recent findings indicate that the mutations in these two genes are more prevalent among women with a diagnosis of ovarian cancer. In a population series of women with ovarian cancer the authors estimated the prevalence of mutations associated with that diagnosis to be approximately 12% (Risch et al., 2001) compared to approximately 5% for breast cancer (Ford et al., 1998). Furthermore, certain ethnic groups, such as Ashkenazi Jews, have an increased probability of harboring germ-line (inherited) *BRCA1* or *BRCA2* mutations (Struewing et al., 1997).

The role of *BRCA1* and *BRCA2* in the development of noninherited cases of ovarian cancer is still difficult to pin down. On the one hand these genes virtually never undergo somatic mutation of the sort frequently encountered in the *p53* gene found in many solid tumors (King, 2004). On the other hand, the expression of proteins of both genes is reduced in most sporadic breast and ovarian cancer. One of the mechanism(s) behind such loss of expression is tentatively attributed to somatic deletion of one complete copy of either *BRCA1* or *BRCA2*, as revealed by genomic analysis of sporadic forms of both types of cancer. The well-established link between germ-line mutations in the BRCA1 and BRCA2 genes and inherited susceptibility to various types of human cancer is in part responsible for unprecedented scientific interest in the biological functions of the proteins that are encoded by these genes (Scully and Livingston, 2000; Jasin, 2002; Venkitaraman, 2002). The functions, type of mutations, and interactions with other proteins in particular EMSY (Hughs-Davies et al., 2003; King, 2004) and RAD51 (Davies et al., 2001) are briefly discussed below.

3.8.2. Association of *BRCA1* and *BRCA2* with Cancer-susceptibility

In humans these genes encode large, nuclear localized proteins of 1,863 and 3,418 amino acids, respectively. One of the unusual features of *BRCA2* is the presence of a cluster of eight repeat sequences (called BRC repeats), which are highly conserved between mammalian species (Bork et al., 1996; Bignell et al., 1997). It has been shown that the more conserved repeats (*BRC1*–4, *BRC7*, and *BRC8*) are involved in the interaction between *BRCA2* and *RAD51* (Mizuta et al., 1997; Wong et al., 1997). The main features of BRCA proteins are given in Figure 3.2. The sequence of these proteins provides few clues about their biological roles, because of the absence of similarity to other proteins with known functions. Although many of the proteins encoded by both the genes have been discovered during the last decade, many functions of such proteins are still to be established. So far we know that *BRCA2* is involved in homologous recombination, but little else is known about its

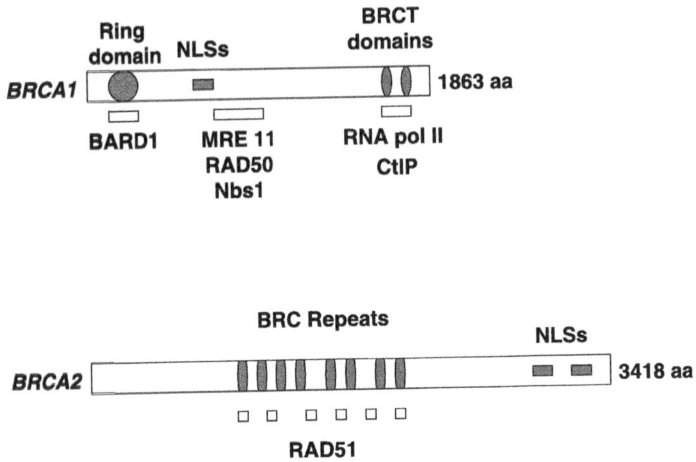

Figure 3.2. Features of the human BRCA proteins: BRCA1 contains an N-terminal ring domain, nuclear localization signals (NLSs), and two C-terminal BRCT domains of ~110 residues. BRCA2 contains eight repeats of the ~40 residue BRC motifs. Six of the eight motifs in human BRCA2 can bind directly to RAD51 when expressed in vitro. Adapted from Venkitaraman (2002) with permission.

other functions. On the contrary, several functions have been attributed to *BRCA1*, which underline its role in carcinogenesis. These functions include DNA repair, cell cycle checkpoint control, protein ubiquitylation, and chromatin remodeling (Scully and Livingston, 2000; Venkitamaran; 2002, 2004). A better understanding of some of these functions has been furnished by studying the interactions between these genes and other proteins in particular RAD51 and EMSY. RAD51 belongs to an evolutionary conserved family of ATPases that are represented in prokaryotes, eukaryotes, and archaea. This protein is known to coat single-stranded DNA substrates to form a helical nucleoprotein filament, which then invades and pairs with homologous sequences in duplex DNA, initiating the strand exchange reactions needed for homologous recombination (Venkitaraman, 2004).

Since *BRCA2* was cloned in 1995 (Wooster et al., 1995), its involvement in the cellular response to DNA damage has become more and more apparent (Zhang et al., 1998; Scully and Livingston, 2000). However, the biological nature of this response together with a number of intrinsic biochemical activities of BRCA2 protein remains unclear. The RAD51 protein possesses various biochemical activities required for homologous recombination and DNA repair, including the ability to promote DNA strand exchange between homologous DNA molecules (Benson et al., 1994; Gupta et al., 1997). A prerequisite for such activities is the binding of RAD51 with DNA to form highly ordered nucleoprotein filaments in which the DNA is encased within a protein sheath (Benson et al., 1994; Mcllwraith et al., 2000). The interaction between RAD51 and the various repeat sequences of *BRCA2* has been investigated by Davies et al. (2001), a work in which the authors have addressed the question as to whether the genetic instability associated with *BRCA2* disruption can be in some way a reflection of defects in the activities of its partner protein, the RAD51 recombinase. This and other works (Yu et al., 2003; Venkitaraman, 2004) gave valuable information on DNA-repair reactions as mediated by RAD51 and *BRCA2*. A schematic model for such mediation is given in Figure 3.3.

A number of recent works (Garcia-Higuera et al., 2001; Howlett et al., 2002; Venkitaraman, 2003) have indicated a growing network of cancer susceptibility genes. Of particular interest to the present discussion are the mutations in *BRCA2*, which have been linked to the Fanconi anemia (FA) disorder. FA is an autosomal recessive disease, which renders patients more susceptible to several types of cancers including acute myeloid leukemias and squamous cell carcinomas of the head and neck (Huibregtse et al., 1985; Alter, 1996). At least eight subtypes of Fanconi's anemia (designated A, B, C, D1, D2, E, F, and G), each with a distinct

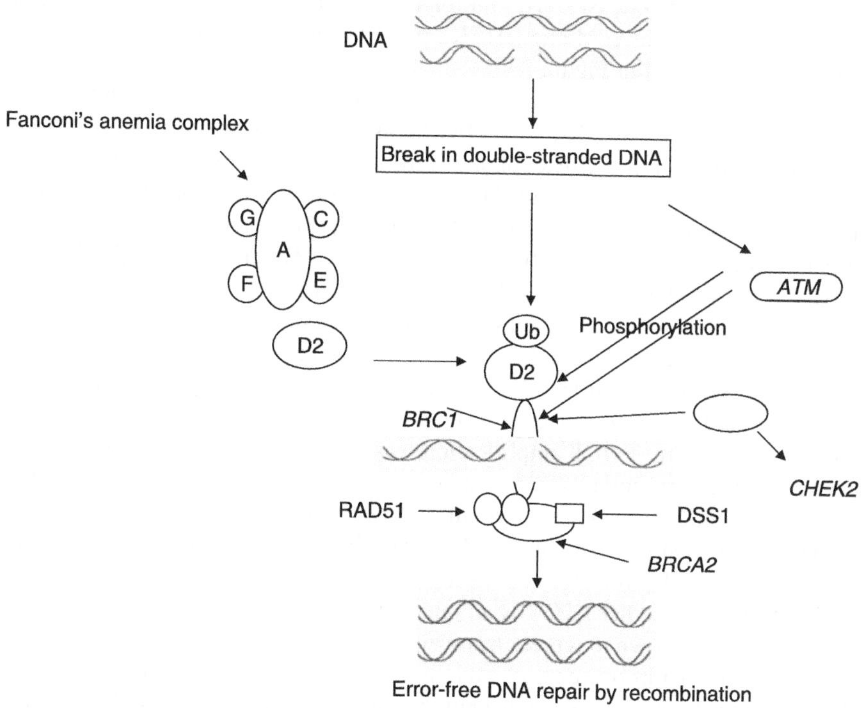

Figure 3.3. A schematic representation of some genes (*ATM, CHEK2, BRCA1, BRCA2*) whose inactivation predisposes people to breast and other cancers participate in the error-free repair of breaks in double-stranded DNA by homologous recombination. The process starts when ATM and CHEK2 protein kinases signal the presence of double stranded breaks by phosphorylating proteins such as BRCA1, inducing their migration to sites where DNA is repaired. BRCA2 carries the DNA-recombination enzyme RAD51 to the same sites. It is guided there by the DNA-binding structures formed between its carboxy terminal and Dss1. The concerted activity of these proteins culminates in error-free DNA repair by recombination. Adapted from Venkitaraman (2003) with permission.

genetic cause, have been recognized (Strathdee et al., 1992; Joenje et al., 1997, 2000; de Winter et al., 1998, 2000). Connection between the FA proteins and *BRCA1*, and 2 has gained momentum because of the spontaneous chromosomal aberrations, that occur in *BRCA*-deficient cells, and are remarkably similar to those encounterd in cell lines taken from patients with FA (Venkitaraman, 2002). The connection between *BRCA* genes and FA disorder has been supported by the work of Garcia-Higuera et al., (2001), who demonstrated that *FANCD2*, the gene mutated in patients with the subtype D2, links Fanconi's anemia proteins to *BRCA1* (Garcia-Higuera et al., 2001). In such work, it was demonstrated that in cells exposed to ionizing radiation, the Fanconi's anemia D2 protein forms a conjugate with a single ubiquitin peptide moity, causing it to co-localize with and bind to *BRCA1*, at sites in the nucleues already occupied by RAD51. In turn, the ubiquitation of Faconi's anemia D2 depends on the proteins of other subtypes (A, C, E, G, and F), which cooperate to move D2 to sites where DNA damage is processed by RAD51, BRCA1, and related proteins. The involvement of these proteins together with ATM and CHEK2 protein kinases in DNA repairs is schematically represented in Figure 3.4.

Another *BRCA2* interacting protein EMSY has been also implicated in DNA-repair functions. EMSY was discovered by Hughs-Davies and co-workers (2003) and was named EMSY after the sister of one of the authors who is a breast cancer nurse. One of the most relevant roles attributed to this protein is its capability to bind to the part of *BRCA2* that is responsible for transcriptional activation; an excess of EMSY silences this important function of *BRCA2* (Milner et al., 1997). Accordingly, it was hypothesized that if the overexpression of EMSY suppress a critical function of this gene, then it is reasonable to expect such overexpression in breast and in ovarian cancers. The authors (Hughes-Davies et al., 2003) have shown that the *EMCY* gene was amplified almost exclusively in sporadic breast cancer (13%) and higher grade ovarian cancer (17%) but not in borderline or low-grade ovarian cancer. Furthermore, *EMCY* amplification was found to be associated with worse survival, particularly in node-negative breast cancer. The discovery of EMSY has generated considerable interest particularly regarding its overexpression in some sporadic breast and ovarian cancers. However, biochemical and biological implications of *BRCA2*-EMSY interactions are still to be fully defined. For instance, it is not yet known whether *EMSY* is an oncogene for breast and ovarian cancer, although EMSY is known to repress transcriptional activation of *BRCA2*, it is not yet established whether its overexpression is sufficient to drive tumorigenesis. This and other questions have been raised in a number of recent articles (Haber, 2003; King, 2004). Another question that has been raised in the same articles and yet remain to be addressed is the influence of such overexpression on DNA repair, chromatin remolding, or transcriptional regulation.

3.8.3. *p53* Mutations in *BRCA1*-linked and Sporadic Ovarian Cancer

As already mentioned, a small proportion of cancers in adults are attributable to the effects of mutations in known cancer susceptibility genes. The heredity proportion

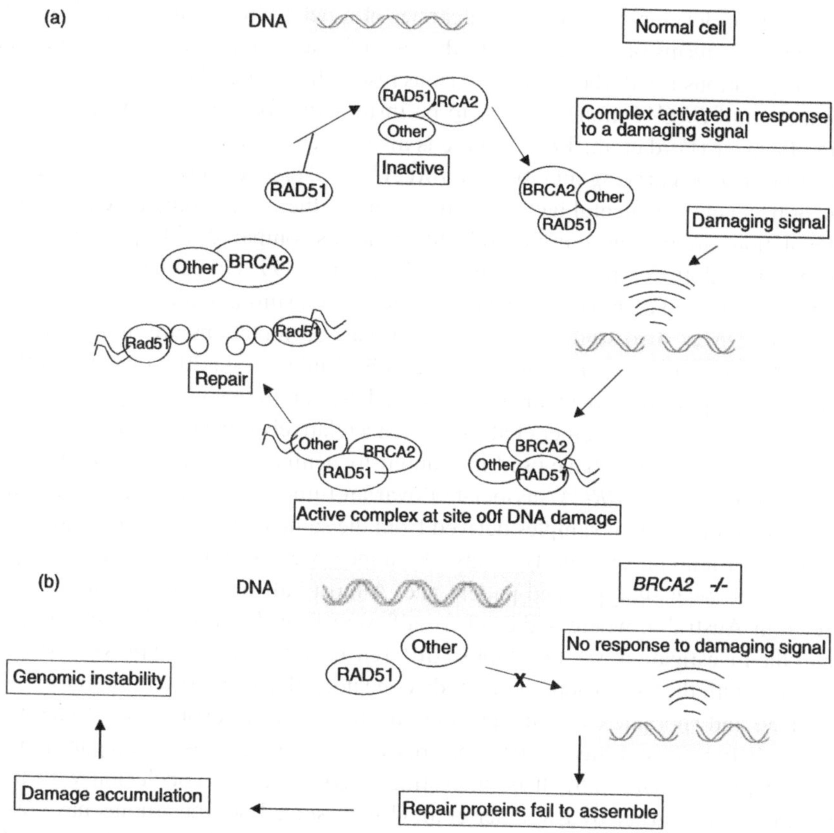

Figure 3.4. Schematic model for the role of *BRCA2 and RAD51* in DNA repair: (a) In a normal cell, RAD51 and BRCA2 interact to form a complex with each other and with other proteins (other). The complex may include proteins such as RAD52, RAD54, XRCC3, the complex is activated, possibly by posttranslational modification of BRCA2 or RAD51, and is recruited to the sites of DNA repair. There, RAD51 protein forms nucleoprotein filaments that, in conjunction with other repair proteins, effect double-strand break repair using the sister chromatid as a template. (b) In *BRCA2* mutant cells typified by the BRCA2 truncation cell line CAPAN-1, complex formation between RAD51 and BRCA2 is disrupted, and much of the RAD51 resides in the cytoplasm along with the truncated BRCA2. The RAD51 that remains in the nucleus lacks BRCA2 control and may bind nonproductively at undamaged regions of DNA. In these cells, the introduction of double-strand breaks fails to stimulate the recruitment of BRCA2, RAD51, and other repair proteins, leading to inefficient homologous recombination and genomic instability. Adapted from Davies et al. (2001) with permission.

varies according to the type of cancer, and the relative importance of genetic and environmental factors varies between populations. Having said that the heredity proportion of ovarian cancer is among the highest of all common forms of cancer in adults (Risch et al., 2001). The term penetrance is commonly used to describe the probability of developing disease in a carrier of a deleterious mutation and is usually defined in terms of a given age. In the case of ovarian cancer, the penetrance of *BRCA1* mutations is still the focus of various research efforts, which suggest that the dysfunction of *BRCA1* alone appears to be insufficient for the development of this form of cancer (Ford et al., 1998). There is also increasing evidence that the risk of cancer in subjects carrying mutations in *BRCA1* and *2* can be influenced by a second gene or by an environmental factor. In breast cancer, there are strong indications that p53 mutations are increased in *BRCA1*-linked tumors compared with sporadic breast cancer (Eiriksdottir et al., 1998; Osin and Lakhani, 1999; Greenblatt et al., 2001). On the contrary, no convincing data exist to support a similar deduction in the case of ovarian cancer. Two studies have reported that *p53* mutations are more frequent in the *BRCA1*-related group (Ramus et al., 1999; Buller et al., 2001) whereas other studies have reported no detectable difference between the two investigated groups (Auranen et al., 1997; Zweemer et al., 1999). In a relatively recent study (Aghmesheh et al., 2004), the authors have used immunohistochemistry to examine samples derived from forty-eight *BRCA1*-associated ovarian tumor, these samples were compared with another set of samples derived from the same number of individuals with sporadic ovarian cancer. The two sets of samples were matched for tumor stage, grade, histological subtype, and patient age. Almost half of the individuals chosen were from Australia, whereas the other half were from Norway. The main objective of the investigation was to establish differences in the protein expression of p53 associated with the two groups. This study concluded that over 70% of both *BRCA1*-associated and sporadic ovarian cancers have shown an overexpression of p53 protein, but there is no significant difference between the two groups with regard to p53 protein expression ($p = 0.5$). It is interesting to compare this conclusion with that reported 5 years earlier by Zweemer et al. (1999), since both the studies have used the same detection technique and involved a comparable number of individuals. In the latter investigation, the authors have assessed the accumulation of *p53* protein in ovarian cancers which occurred in *BRCA1* or *BRCA2* germline mutation carriers and compared the results with a panel of ovarian cancers from patients who gave negative tests for germline mutations in both *BRCA1* and *BRCA2*. Each group consisted of about 40 individuals and the accumulation of p53 protein was assessed using immunohistochemistry. This study concluded that similar rate of p53 protein accumulation was observed for *BRCA1* and *2* related ovarian cancer and negative controls. The conclusions of both the studies (Zweemer et al. 1999; Aghmesheh et al., 2004) are in disagreement with the findings of a study by Buller et al. (2001) who reported more frequent *p53* mutation in *BRCA1*-linked ovarian tumors compared with sporadic ovarian cancers. The difference between the findings of the above investigations can be tentatively attributed to a number of reasons: First, mutations in *p53* are more frequent in stage III/IV compared with stage I/II ovarian cancers (Berchuck and Carney, 1997); a similar situation can also manifest in different grades or in histological subtypes (Ramus et al., 1999). It is, therefore, imperative to match *BRCA1*-linked and sporadic ovarian cancers to enhance the chances of getting representative p53 protein accumulation. In the study by Aghmesheh et al. (2004), the *BRCA1*-linked and sporadic samples were matched for tumor stage, grade, histological subtype, and patient age, whereas in the study by Buller et al. (2001), the *BRCA1*-linked samples had a higher stage and grade compared with their sporadic counterparts. Second, in the latter study the sporadic cohort was tested for *BRCA1* mutation, whereas such a test was not conducted by Aghmesheh et al. (2004) due to ethical issues. Third, the limited number of published works dealing with p53 mutation and its correlation with sporadic and *BRCA1*-linked ovarian cancers and the limited number of samples examined in such studies render the assessment of either conclusion rather difficult. In other words, further studies involving higher number of subjects together with strict sample matching is still required before a common consensus regarding this argument can be developed.

3.8.4. Carcinoma-associated Glycoprotein Antigen (CA-125)

The cancer antigen 125 (CA-125) is a high molecular weight glycoprotein, which was first identified by Bast et al. (1981, 1983). The first method developed to measure CA-125 was a radioimmunometric assay that used the murine monoclonal antibody OC-125 as both capture and indicator antibodies (Bast et al., 1983). OC-125 was obtained after immunization with OVCA-433 cell line, which was derived from a patient with papillary cystadenocarcinoma of the ovary (Bast et al., 1981). Although CA-125 antigen was identified almost 25 years ago, little is known about its biochemistry. Most studies have concluded that this antigen is a high molecular mass glycoprotein, although estimation of its size ranges from 200 to 2000 k Da. Most studies have shown that CA-125 is a mucin-type molecule, whereas other studies have claimed that it is a typical glycoprotein with asparagines-linked sugar chains (Zurawski et al., 1988). In another study, the same antigen has been described as glycosylphosphoinositol-linked glycoprotein (Nagata et al., 1991a). In one of these studies, Yin and Lloyd (2001) have used a rabbit antiserum to purified CA-125 to clone a long partial *c*DNA sequence corresponding to a new mucin species (designated MUC16), which was retained as a good candidate for being the peptide core of the CA-125 antigen. The last 20 years have demonstrated that the absence of a general consensus regarding the biochemistry of this antigen did not interfere with its rapid diffusion as a frontline marker for ovarian cancer. Serum concentration of CA-125 has been shown to be elevated in various forms of cancer including, ovarian, pancreatic, breast, colon, lung, and endometrial carcinoma (Bast et al., 1998). It is commonly agreed that CA-125 lacks both sensitivity and specificity as a marker for early stage disease (stage I/II). On the contrary, the same marker has become an integral part of the management of epithelial ovarian cancer. There is increasing evidence to suggest that almost 80% of patients with epithelial ovarian cancer have elevations in CA-125, which encouraged various research groups to emphasize the role of this marker for disease monitoring, assessing response to therapy, and for the detection of relapse (Rustin et al., 1992; Eisenhauer et al., 1994). Most existing studies regarding the value of CA-125 as a marker for ovarian cancer seem to agree on two main points: First, specificity of CA-125 is inadequate for screening stages I and II of ovarian cancer, particularly in premenopausal population in which endometriosis, adenomyosis, and retrograde menstruation can result in false elevation of the level of this antigen. Specificity can, however, be improved by combining CA-125 with various forms of sonography in two-stage strategy and by sequential monitoring of the antigen values over time (Bast, 2003). Second, when ovarian cancer is diagnosed in stage I, more than 90% of patients can be cured with conventional surgery and chemotherapy. At present, however, only 25% of ovarian cancers are detected in this early stage. To date, the largest trial to use CA-125 was conducted by Jacobs et al. (1999) in the United Kingdom. Postmenopausal women aged 45 years or older were randomized to a control group ($n = 10{,}977$) or screened group ($n = 10{,}958$). Women randomized for screening were offered three annual screens that involved the determination of CA-125 level in serums, pelvic ultrasonography if such levels were 30 U/mL or more, and referral for gynecological opinion if ovarian volume was 8.8 mL or more on ultrasonography. All women were followed up to see whether they developed invasive epithelial cancer of the ovary. The main findings of this study were as follows: Among 10,985 women screened, 29 operations were performed to detect six cancers, providing a positive predictive value of 21%. During the 7-year follow-up, 10 more cancers were diagnosed in the screened group. During the same interval, 21 ovarian cancers were diagnosed in the control group. Median survival in the screened group (72.9 months) was significantly higher than that in the control group (41.8 months). Currently, a massive trial is under way in the United Kingdom involving 200,000 postmenopausal women randomly assigned to three groups. A control group (100,000 patients) will be followed up with conventional pelvic examination, a second group (50,000 patients) will undergo annual transvaginal sonography (TVS), and a third group (50,000 patients) will have their CA-15 levels measured at least once an year. On the basis of risk of cancer (ROC) algorithm, patients in the third group will be referred for TVS, surgery, or both. Women will be screened for 3 years and followed up for 7 years. Such a massive trial may provide more reliable information on the feasibility of screening for ovarian cancer and the effect of early detection on survival (Bast, 2003).

The initial clinical investigations of newly introduced tumor markers are normally designed to evaluate the potential of such markers as screening, diagnostic, staging, or prognostic tools. The case of CA-125 is no exception, where numerous clinical studies have indicated a relatively poor performance of this antigen for screening early stages of ovarian cancer. Such findings have encouraged the search for different ways to exploit the potential of this molecule. There is accumulating evidence to suggest that the two main application areas are as follows: First, the use of CA-125 to assess response to treatment (including new drugs) and to monitor progression of the disease, second, its use in combination with other markers to enhance its specificity, particularly in the early stages of the disease.

3.8.5. Potential Uses of CA-125 in Prognosis and Patient Management

Since its development, measurement of serum levels of CA-125 has become an integral part of the management of epithelial ovarian cancer. While the levels of this antigen at the time of diagnosis are of limited prognostic significance, the measurements of the same antigen are now performed almost routinely during the course of various treatments. Over 70% of women diagnosed with ovarian cancer are given chemotherapy, yet they die from their disease. We know that the two main goals of clinical trials of new anticancer drugs are to establish the response rates to the treatment and to assess the progression-free survival. Given that a substantial proportion of ovarian cancer is poorly visualized by scans, the evaluation of response is rather poor. On the contrary, there is strong evidence that serum levels of CA-125 become elevated in more than 90% of patients with advanced epithelial ovarian cancer (Bast et al., 1983; Tuxen et al., 1995). This antigen is used routinely in clinical practice to determine whether patients are responding to therapy or have progression of their ovarian cancer (Rustin, 2003). In such trials, a serial decrease usually leads to a continuation in the prescribed therapy, whereas a serial increase is likely to lead to a change in the patient's management. For example, decreasing levels of CA-125 after cytoreductive surgery and during initial chemotherapy courses have been used as an indicator of clinical outcome (Van Dalen et al., 2000). Such a decision, however, has to be based on standard criteria otherwise there is the risk of starting a second-line therapy before a progression date has been reliably confirmed.

Over 25 years ago, the World Health Organization (WHO) published classifications and criteria of tumor response designed to standardize measurements for solid tumors (WHO handbook, 1979; Miller et al., 1981). Despite their wide use, these criteria had a number of shortcomings, which attracted a number of modifications and clarifications by various organizations. To address the shortcomings of the WHO criteria, the Response Evaluation Criteria in Solid Tumors (RECIST) group has recently proposed guidelines to evaluate the response to treatment of solid tumors (Therasse et al., 2000). Although the RECIST criteria have addressed many shortcomings of the WHO criteria, it will have less influence on the evaluation of the ovarian cancer (Rustin et al., 2004). This limitation is due to the fact that the newly introduced criteria define progression of the disease on the basis of a measurable disease; such characteristics would exclude almost half of ovarian cancer patients. The same authors argued that there is currently a wealth of data to support various roles of CA-125 in clinical trials. Therefore, CA-125 should be given a role in the design of clinical trials, from prognosis to follow-up.

Other attempts to enhance the capability of CA-125 to predict early-stage ovarian cancer have used other promising markers, some of which are considered below.

3.9. OSTEOPONTIN

Osteopontin (OPN) is commonly described as an acidic member of the small integrin-binding ligand N-linked glycoprotein family of extracellular matrix proteins/cytokines that undergoes extensive posttranslational modifications, including phosphorylation on serine and threonine residues (Prince et al., 1987), glycosylation with primarily O-linked glycosaccharides (Sorensen et al., 1995; Neame and Butler, 1996), sulfation (Nagata et al., 1989), and proteolic cleavage (Fresno et al., 1981; Agnihotri et al., 2001), yielding a range of molecular masses between 25 and 75 kDa (Johnson et al., 2003). This range of variants is in part responsible for the versatility of this protein, which has multiple functions derived from its role as a mediator of cell–cell and cell–extracellular matrix communication that encompass both normal and tumorigenic. Originally isolated from bone, OPN has been found in epithelial cells and in secretions of gastrointestinal tract, kidneys, thyroid, breast, uterus, and testes (Young et al., 1990; Johnson et al., 1999; Luedtke et al., 2002). OPN protein contains several cell adhesive domains, including an arginine–glycine–aspartate sequence, which mediates cell substrate attachment and chemotaxis (Denhardt and Guo, 1993). A number of studies using pathological specimens have shown that OPN expression occurs in various forms of cancer including breast, colon, stomach, ovary, lung, thyroid, kidney, prostate, and pancreas (Brown et al., 1994; Coppola et al., 2004). The correlation between the expression of OPN and various forms of cancer has been enforced by a number of recent studies. In one of these studies, Fisher et al. (2004) have used a cancer profiling array containing normalized cDNA from tumor and corresponding normal tissues from 241 individual patients to screen the expression of various members of the small integrin binding family including OPN. This approach was developed to enable quantification of expression of a single gene across multiple tissue types and tumor stages. According to the authors, such approach yields a high representation of mRNA transcripts, reduce biased amplification, and has a good linearity of the signal. Regarding the expression of OPN, this study reported that its expression was elevated two- to four-fold in five types of cancer including ovary, breast, uterus, colon, and lung. In an earlier study by Wong et al. (2001), the authors have used cDNA microarrays to investigate the up- or downregulation of genes in ovarian cancer cells or tissues. This study was followed by another investigation to establish the clinical relevance of the expression of OPN (Kim et al., 2002). To verify their earlier data regarding OPN expression, the authors used immunohistochemistry, ELISA, and real time-PCR. Based on these investigations, it was concluded that the acquired data provided evidence for an association between the levels of OPN and ovarian cancer. In my opinion, this conclusion together with the experimental approach deserves further consideration. First, any attempt to evaluate new markers for the early detection of ovarian cancer is a valuable contribution to the continuous search aimed at substituting or complementing the role of CA-125. Kim et al. (2002) have used a multidisciplinary approach, which encompasses high sensitivity methods. This sensitivity, however, was not accompanied by the desired specificity, a parameter considered central in this kind of analysis. The lack of specificity is associated with some intrinsic characteristics of osteopontin, which can be partially attributed to two elements: First, Osteopontin is a multifunctional protein, which appears to regulate aspects of inflammation and tissue repair. In particular, it is associated with responses characterized by the presence of macrophages and T cells. A systematic assessment of OPN protein levels in a large series of human tumors and normal tissue has recently been reported (Coppola et al., 2004). The authors used immunohistochemistry to compare OPN expression in a group of 350 human cancers from a large variety of anatomical sites and in 113 corresponding normal tissues. This study reported that high cytoplasmic OPN staining was observed in a wide range of carcinomas including, gastric, colorectal, translational cell, renal pelvis, pancreatic, renal cell, lung, head and neck, and ovarian. Second, OPN can be expressed by a variety of tumor cells; it is also widely expressed by macrophages which infiltrate tumor tissue (Brown et al., 1994; Chambers, 1995; O'Regan and Berman, 2000). Although OPN produced by tumor cells and macrophages tend to fulfill different roles (Crawford et al., 1998), the measurements by Kim et al. (2002) were not designed to examine the role of this molecule, but only its expression. The studies cited above and others in the literature suggest that assays of total serum osteopontin alone will lack sufficient specificity to be clinically applicable as a marker for ovarian cancer. It is worth reminding ourselves that PSA as a marker for prostate cancer is still considered a low-specificity marker. One of the approaches to enhance such specificity has been the use of its isoforms (see Section 2.5.2.). A parallel can be drawn between the case of PSA and that of OPN. The latter is known to have at least three transcriptional variants as well as numerous posttranslationally modified forms (Prince et al., 1987; Singh et al., 1990; Nagata et al., 1991b; O'Regan and Berman, 2000). The presence of these isoforms and the knowledge that tumor-derived and macrophage-derived OPN have different roles may present future opportunities to design assays based on specific isoforms of OPN.

3.9.1. Human Kallikrein 10

Human tissue kallikrein are secreted serine proteases encoded by 15 structurally similar, hormonally regulated genes that colocalize to chromosome 19q13.4 in a 300-kB region (Diamandis et al., 2000; Yousef and Diamandis, 2001). An international group of investigators has established human kallikrein gene nomenclature (Diamandis et al., 2000). This gene family now contains 15 genes designated *KLK1-KLK15*; their encoded proteins are designated hK1-hK15. All of the kallikreins studied to date have been found to be differentially expressed at the mRNA and/or protein level in endocrine-related malignancies including cancers of the ovary, breast, prostate, and testis. The potential role of the various members of this family in carcinogenesis will be discussed in Chapter 4 and therefore this section will be limited to hK10 and its potential as a biomarker for ovarian cancer. Recently, Diamandis's group (Luo et al., 2001) has developed and tested an immunofluorometric assay to detect human kallikrein in biological fluids and in tissues. These authors reported that the assay was specific for hK10 and had no detectable cross-reactivity with other homologous kallikrein proteins, such as PSA( ) hK3, hK2 and, hK6; the detection limit was 0.05 μg/L. In a subsequent study, this assay was used to quantify hK10 in sera from 97 controls, 141 patients with benign gynecologic diseases, and 146 patients with ovarian cancer (Luo et al., 2003). These measurements were based on the assumption that the secretion of this protein is likely to be altered in ovarian tissue (Liu et al., 1996). The results by Luo et al. (2001, 2003) gave preliminary indications on the potential of hK10 as a marker for ovarian cancer. For example, both studies have indicated weak correlation between hk10 and CA-125, which suggests that both molecules can be combined to increase the diagnostic sensitivity of each of the biomarkers alone. The potential of hK10 and other members of killikreins protein family as potential markers for ovarian cancer and other diseases is discussed in more detail in chapter 4.

3.9.2. Prostasin

Prostasin is a serine protease discovered in ejaculated human semen over 10 years ago (Yu et al., 1994). Using SDS polyacrylamide gel electrophoresis, the same authors have shown that it has a molecular mass around 40 kDa. The predicted mature prostasin peptide sequence has a potential carboxyl terminal hydrophobic membrane anchorage domain followed by a short cytoplasmic tail. The translated amino acid residue sequence has a similarity with testisin, plasma kallikrein, hepsin, and plasminogen (Yu et al., 1995; Hooper et al., 1999; Nelson et al., 1999). Serine proteases are known to play an important role in a diverse range of physiological processes and are implicated in various pathological conditions such as cardiovascular disorders and cancers (Rawlings and Barrett, 1994). Prostasin is present at high levels in normal human semen and in the prostate gland. It is believed that prostasin is synthesized in prostatic epithelial cells, secreted into the ducts, and excreted into the seminal fluid. It is also expressed at much lower levels in a variety of human tissues, including kidnies, liver, pancreas, salivary glands, lungs, bronchi, and colon. Curiously, prostasin has not been detected in the ovaries (Yu et al., 1995).

As pointed out earlier, microarray technology permits the simultaneous comparison of the expression of thousands of genes in samples to allow the identification of those that are differentially expressed. The same technology has the capability to identify overexpressed complementary DNA (cDNA) corresponding to secretory proteins that might serve as serum markers for cancer. Mok et al. (2001) have described the use of microarray technology on RNA pooled from ovarian cancer and from normal human ovarian surface epithelial (HOSE) cell lines to assess the expression of a gene that produces prostasin. This is one of the first and possibly the only study to demonstrate the overexpression of prostasin gene in ovarian cancer. As a preliminary investigation, it also has its limitations, including: First, that this study demonstrates the potential of the technology, yet the sample size was relatively small to demonstrate the value of this serine proteinase as a screening marker. Second, the same study does not provide any evidence that prostasin has a higher sensitivity than CA-125, which despite its limitations remains the best characterized marker for advanced epithelial ovarian cancers. Having said that, the lack of correlation between CA-125 and prostasin may suggest that a combination of the two entities may result in an increase in sensitivity. Third, prostasin is found in at least eight normal human tissues (Yu et al., 1994, 1995). Whether the prostasin level will be elevated in serum from patients harboring other forms of cancer remains to be established. Fourth, prostasin, along with other putative biomarkers, must undergo rigorous testing, which satisfy existing criteria for the development and evaluation of potential biomarkers. One of these criteria has been established by the National Cancer Institute's Early Detection Research Network (EDRN) (Pepe et al., 2001). The five phases within this criteria are as follows: (a) a preclinical exploratory phase to identify promising directions; (b) a clinical assay and validation phase to evaluate the ability of the assay to detect established disease; (c) a retrospective/longitudinal phase to determine the putative biomarker's ability to detect preclinical disease and to define a "screen positive" rule; (d) a prospective screening phase to identify the extent and characteristics of disease detected by the test and the false-positive rate; and (e) a definitive trial (prospective randomized trial) to determine the impact of screening on reducing the burden of disease in general population.

3.10. COMBINATION OF CA-125 WITH OTHER POTENTIAL BIOMARKERS

During the last two decades, a large number of serum tumor markers have been evaluated for their ability to detect early stage epithelial ovarian cancer. In a relatively recent review (Bast, Jr. et al. 2002, Bast, Jr. 2003), 29 different serum tumor markers have been evaluated in combination with CA-125, and some of these were reported to increase sensitivity and specificity. To appreciate the value of combining CA-125 with other potential serum biomarkers, a number of specific examples are considered below.

Among the early attempts to evaluate CA-125 in combination with other potential biomarkers were conducted by Bast's group (Xu et al., 1993, 1991). Lewis X mucin determinant (OVX1) and a cytokine machrophage colony-stimulating factor (M-CSF) were evaluated for their ability to detect stage I ovarian cancer and to complement CA-125. Among 89 serum samples obtained from patients with stage I ovarian cancer before surgery, CA-125 was greater than 35 U/mL in 69%. The authors reported that a combination of CA-125, OVX1, and M-CSF detected 84% of early stage/ovarian cancer, while the specificity went down from 99 to 84%. In a relatively recent study, Skates et al. (2004) have assessed the preoperative sensitivity and specificity for early-stage ovary cancer when combining CA-125II, CA 15-3, CA72-4, and machrophage colony-stimulating factor (M-CSF). CA15-3 is a marker developed for breast cancer, CA72-4 was developed for pancreatic cancer, and M-CSF is a cytokine that stimulates proliferation and differentiation of monocytes, but that can also act as an autocrine or paracrine growth factor for some epithelial cancers. These markers were measured in preoperative serum and serum from healthy controls in a training set from early- and late-stage patients from a hospital in the United Kingdom and a validation set from only early-stage patients from two hospitals, one in the USA and the other in Holland. Three theoretical models were fitted to a training data set of preoperative serum measurements. These models were as follows: (a) logistic regression, which estimates the probability of having ovarian cancer (Cox and Snell, 1989); (b) classification trees model, which splits the data into two on the basis of a cut-off point for a marker, beginning with the marker for which a single cut-off point yields the highest classification accuracy. These two parts of the data form the basis for a tree, and each part is split independently of the other by identifying the marker for which a subsequent split accurately distinguishes patients from controls. The tree continues to be generated until no significant improvement in accuracy occurs (Breiman, 1984; S-Plus 2000 guide, 1999); and (c) mixture discriminant analysis approximates the density for patient cases and density for controls using a mixture of multivariate normal distributions (Hastie and Tibshirani, 1996). The patient cases are distinguished from the controls through ratio of the density for patient cases with the density for control using Bayes theorem (Ripley, 1996) which gives the predicted probability of having ovarian cancer. Based on this analysis, the authors reported that the combination of CA-125II, CA 72-4, and M-CSF increased the preoperative early-stage sensitivity from 45% with CA-125II alone to 70%, while maintaining 98% first-line specificity. This conclusion together with the overall design of this study deserves a number of considerations: First, this substantial improvement in sensitivity, while maintaining 98% first-line specificity has to be considered together with the size of total samples investigated (347 samples). Second, having independent training and validation data sets from separate institutions to estimate model parameters and then independently evaluating the model's performance helps to render the sensitivity estimate less biased and therefore more applicable to other sets of subjects outside this study. However, the same authors pointed out that the preoperative sensitivity estimate of 70% is likely to be an overestimate in the early-stage screening. This is mainly due to the design of this study, which measured the markers in preoperative serum of the patients. These patients are often symptomatic by the time of diagnosis, and therefore an upwardly biased estimate in the measured sensitivity for early-stage screening cannot be ruled out. In fact, only a prospective clinical trial can evaluate the sensitivity for early-stage preclinical disease, while simultaneously estimating the positive predictive value.

Another recent study by Dupont et al. (2004) assessed the capability of serum YLK-40 for the detection and prognoses of early-stage ovarian cancer. This capability was also compared with that of CA 15-3 and CA-125. YKL-40 is a glycoprotein in the chitinase protein family. The gene for this protein is located on chromosome 1q32 and is a mammalian member of the 18-glycosyl-hydrolase family, a family that includes bacterial and fungal chitinases (Renkema et al., 1995, 1998). This study involved a relatively small number of samples including 946 healthy subjects, 61 high-risk individuals, 33 patients with benign gynecologic processes, and 50 preoperative patients subsequently diagnosed with predominantly early-stage ovarian cancer. Based on these analyses, it was reported that patients with an elevation of more than 80 ng/mL YKL-40 values were significantly more likely to have recurrence of disease compared with patients with lesser values. Furthermore, 64% of patients with more than 80 ng/mL YKL-40 values died of disease, whereas none of the patients with lower preoperative YKL-40 values died. The same authors reported that preoperative levels of CA-125 and CA15-3 in these patients did not correlate with a poor outcome (no data were presented within the same article). Setting aside the limited number of samples used, which is always an important element to consider, the above study has a limitation related to what we already know about YKL-40. This glycoprotein has been described as a serum marker for a list of diseases, including rheumatoid arthritis (Matsumoto and Tsurumoto, 2001), severe osteoarthritis (Register et al., 2001), hepatic fibrosis, primary colorectal cancer (Cintin et al., 1999), glioblastoma (Tanwar et al., 2002), metastic breast cancer (Johansen et al., 1995), and recurrent ovarian cancer (Dehn et al., 2003). This means that the specificity of this potential marker for the detection of early-stage ovarian cancer will be reduced in patients suffering from any of these pathologies. Furthermore, it is reasonable to ask whether such specificity would surpass the well-tested CA-125.

Other studies, which evaluated promising ovarian cancer biomarkers, included: expression of trypsinogen-1, 2, and tumor-associated trypsin inhibitor in ovarian cancer (Paju et al., 2004); comparison of CA-125 and tissue polypeptide specific antigen (TPS) levels after chemotherapy courses in ovarian cancer patients (Van Dalen et al., 2000); haptaglobin-α-subunit as a potential biomarker for ovarian cancer (Ye et al., 2003), and HE4 (WFDC2) protein in ovarian carcinoma (Hellström et al., 2003).

3.11. PROFILING PROTEINS AND GENE EXPRESSION IN OVARIAN CANCER

The last decade has witnessed intense research activities focused on finding more specific biomarkers for the early detection of ovarian cancer. Such efforts involved various technical approaches aimed at profiling genes and proteins in tissue and serum samples. Over the past few years, several gene expression-profiling methods have been successfully used in cancer research including ovarian cancer. Differential screening of cDNA libraries has been commonly used to detect differences in gene expression between different cell types (Perou et al., 1999; Ross et al., 2000). With the advent of rapid DNA-sequencing techniques, investigations have shifted toward the sequencing of randomly picked cDNA clones from libraries and the comparison of the resulting spectra of the sequenced clones. Serial analysis of gene expression (SAGE) is another powerful approach for the identification of differentially expressed genes, providing comprehensive and quantitative gene expression profiles in the form of short tag sequences. Each tag represents a unique transcript, and the relative frequencies of tags in the SAGE library are equal to the relative proportions of the transcripts they represent (Velculescu et al., 1995, 2000; Coyne et al., 2004). It is relevant to point out that existing SAGE libraries have two main drawbacks: First, there is some measure of ambiguity in assigning the 10-nucleotide tag sequences to particular genes. Although the SAGE database uses a sophisticated algorithm to make such identification as error free as possible, incorrect identifications may still arise. The 22-nucleotide "long SAGE" protocol (Saha et al., 2002) will largely remove this source of ambiguity, but most of the existing SAGE libraries are still based on the 10-nucleotide tags, and "short SAGE" tags will remain the major source for SAGE analysis for the immediate future. Second, genes that are expressed at low levels will give rise to only a small number of tags. A tag that is represented by only a few examples in a tissue might, however, itself arise as an artefact from erroneous sequencing of a tag present at high levels. DNA sequencing errors can be high; hence a 10-nucleotide tag present at 1000 copies might appear with dozens of "neighbours" that differ from it by a single nucleotide in 10. Any such artifactual tag would appear as a low copy tag and may be incorrectly assigned. Thus, low copy number tags are inherently suspected, so that a gene that is indeed represented by a low number of tags cannot be unambiguously reported (Stein et al. 2004).

High-density comparative hybridization is another approach, which demonstrated its potential in the identification of transcripts (Schena et al., 1998; Shummer et al.,

1999). Despite the well-established power of mRNA-based techniques, a number of studies, which compared the abundance of mRNA transcripts and the corresponding proteins, have shown poor correlation. Anderson and Seilhamer (1997) were the first to report a correlation coefficient less than 0.5 when mRNA and protein abundance of some gene products were compared in human liver. This result was further supported when expression of only one protein (GST-π) in 60 different human cell lines was studied by LC and quantitative northern blot analysis methods for protein and mRNA measurement, respectively (Anderson et al., 1998). This poor correlation between mRNA and protein abundance, albeit in few studies, has underlined the need for both types of data to allow a clear picture of the gene expression profile of a particular cell type. Another good reason for obtaining information on both mRNA and protein abundance is that posttranslational events such as protein modifications will escape detection at the mRNA level. Chapter 2 gave a list of proteomic approaches, which are currently applied in the search for potential cancer biomarkers including ovarian cancer. Within the same text a number of representative examples were given and therefore I shall use the remaining part of this section to cite some examples on the use of SAGE to assess genes expression in ovarian cancer.

Hough et al. (2000) have used SAGE to generate global gene expression profiles from various ovarian cell lines and tissues, including primary cancers, ovarian surface epithelia cells, and cystadenoma cells. These profiles were used to compare overall patterns of gene expression and to identify differentially expressed genes. Although gene expression alterations that arise during malignant transformation can be identified in a number of ways, the authors used the unbiased, comprehensive method SAGE to create global gene expression profiles from 10 different ovarian sources. This study involved the sequencing of 385,000 tags, yielding >56,000 genes expressed in 10 different libraries derived from ovarian tissues. Using the expression profiles described above and stringent selection criteria, the authors have identified a number of genes highly differentially expressed between nontransformed ovarian epithelia and ovarian carcinomas. Four of the upregulated genes were further confirmed by performing immunohistochemical analysis. Many of the genes upregulated in ovarian cancer were surface or secreted proteins such as claudin-3 and -4, HE4, mucin-1, epithelial cellular adhesion molecule, and mesothelin. Interestingly, both apolipoprotein E (ApoE) and ApoJ, two proteins involved in lipid homeostasis, were among the genes highly upregulated in ovarian cancer. Selected serial analysis of gene expression results were further validated through immunohistochemical analysis of ApoJ, claudin- 3, claudin-4, and epithelial cellular adhesion molecule in archival material.

Cell lines are widely used in testing new anticancer agents, despite a commonly recognized observation that cell lines are more sensitive to cytotoxic drugs than are their corresponding solid tumors. Drug sensitivity and resistance of cell lines and solid tumors were assessed by SAGE database to identify genes, which might be responsible for the observed differences (Stein et al., 2004). The authors used SAGE libraries available for both solid tumors and cell lines from ovarian, breast, colon, pancreatic, and prostate cancers. The same study used the SAGE database to identify genetic differences between tumor types that convey a broad range of survival to the patients that bear them as distant metastases. SAGE gene expression data were correlated with 5-years survivals documented in SEER (Surveillance, Epidemiology and End Results) database for patients diagnosed with distant or metastatic cancers. Based on these analyses, a number of conclusions were drawn, which can be briefly summarized as follows: First, compared with cell lines, solid tumors (a) overexpressed genes associated with cell–cell communication and with the extra cellular matrix; (b) overexpressed genes involved in the immune response; and (c) underexpressed genes involved in protein synthesis. Second, the 5-year survival of a patient with a given tumor is correlated with (a) and (c) of the above factors. Using the SEER 5-year survival data as a surrogate for chemosensitivity (and recognizing that factors other than chemosensitivity also influence the survival data), it was observed that improved survival was associated with (a) relative overexpression of genes coding for protein synthesis components; and (b) underexpression of cell adhesion and cytoskeletal genes.

Peters et al. (2005) have generated SAGE libraries from three serous adenocarcinomas of the ovary, and used statistical tools to compare these libraries with SAGE data derived from two pools of normal human ovarian surface epithelium (HOSE). The innovative element in this study is that the SAGE libraries were not derived from cultured cell lines. The authors reported several known and novel genes with elevated expression in ovarian cancer, including *CLDN3, WFDC2, FOLR1, COL18A1, CCND1,* and *FLJ12988*. The same study reported marked differences in gene expression patterns in primary HOSE tissue compared with cultured HOSE.

3.12. GENERAL OBSERVATIONS

The bulk of existing literature dealing with the diagnosis and prognosis of ovarian cancer seems to underline a number of elements:

- Currently, the only clinically accepted serum marker for ovarian cancer is CA-125. This marker has proved to be of a better value for clinical prognosis and for monitoring response to treatment but not for the early detection (phases I and II) of the disease. The utility of CA-125 as an early screening tool has been hampered by the fact that its level can be elevated in various benign diseases, including endometriosis, ovarian cysts, uterine fibroids, and chronic liver disease. Alternative markers and/or markers to be combined with CA-125 are urgently needed. Efforts to address this need are currently proceeding on three main fronts: first, searching for alternative biomarkers through the use of a wide range of proteomic and gene expression-profiling methods. SELDI, 2-DE, 2-D-LC coupled to MS, SAGE, and cDNA are some of these technologies, which have shown some promise in this search. Second, enhancing the screening capability of CA-125 to detect early-stage epithelial ovarian cancer through the use of multiple markers. New technologies, such as gene-expression analysis and serum proteomics, have already identified numerous potential markers for detecting and classifying cancers. These and other approaches could eventually improve the future availability of multiple markers for the early detection of various forms of cancer, allowing more complete coverage than ever before of the spectrum of diseases that we call cancer. Over the last two decades, a large number of serum tumor markers have been evaluated for their ability to detect early-stage epithelial ovarian cancer (see Table 3.2).

- At present, screening for ovarian cancer has to be regarded as a work in progress. Such work is made difficult by the fact that randomized trials may require 10 years or more to complete and still may provide false-negative outcomes. Another element that complicates screening for ovarian cancer is the absence of sufficient information on how often stage III and IV disease evolves from clinically detectable stage I disease and what the coefficient of variation will be around this average (Bast, 2003). Furthermore, the carcinogenesis rarely involves a single transforming step from normal cells to aggressive cancer cells. This characteristic complicates the question of screening. A method that identifies only fully transformed neoplastic cells with metastic potential might indeed result in earlier detection than would be the case if patients were assessed only by clinical examination (Pollak and Foulkes, 2003). Having said that, a significant proportion of such lesions might have already resulted in micrometastases when detected, thereby limiting the benefits of early surgical removal and the entire screening exercise. These and other challenges for existing screening methods have been recently discussed by Pollak and Foulkes (2003). The WHO has listed a number of conditions for early detection to be an appropriate disease control approach (Winawer et al., 1995). First, the disease must be common and associated with serious morbidity and mortality. Second, screening tests must be able to accurately detect early-stage disease. Third, treatment after detection through screening must have been shown to improve prognosis relative to usual diagnosis. Fourth, evidence must exist that the potential benefits outweigh the potential harms and costs of screening. Although screening tests are in use for a range of cancers, almost none of these tests satisfy the criteria set by the WHO. Luckily, such skeptical statements are contradicted by the success story of cervical cancer, which demonstrates the power of early detection and subsequent treatment in reducing the burden of cancer (Etzioni et al., 2003). At the beginning of the last century, mortality due to invasive cervical cancer was among the highest for women. 50 years later, pathologists have demonstrated that the natural history of this type of cancer progressed through stages of increasingly severe cervical intraepithelial neoplasia, and that these stages could be identified histologically using exfoliated cells. The elucidation of the natural history of cervical neoplasia led to the development of the "PAP SMEAR," and subsequent introduction of programmes and policies in developed countries to implement widespread early detection of preneoplastic cervical lesions (Boyes, 1981). Since 1950, there has been ~70% decline in the incidence of, and mortality due to, invasive cervical cancer in the USA (Christopherson et al., 1970; Schiffman et al., 1996). In developing countries where "PAP SMEAR" screening is not widespread, cervical cancer remains a major problem (Parkin et al., 1993).

- Although there is currently no screening strategy for ovarian cancer with proven effectiveness, the outlook for patients with epithelial ovarian cancer has clearly improved over the last decade, largely as a result of taxane- and platinum-based first line chemotherapy, as well as an increase in options for the management of recurrent disease (Cannistra, 2004). Several molecular targets for drug development have been identified, including pathways mediated by *p53* (Strobel et al., 1996), lysophosphatidic acid (Fang et al., 2002), and the epidermal growth factor receptor (Alper et al., 2001).

Most ovarian and breast cancers are sporadic (i.e., not inherited), but some are the result of inherited predisposition, principally due to mutations in the tumor suppressor genes *BRCA1* and *BRCA2*. Women born with mutations in these two genes are at significantly higher risk of developing breast and ovarian cancer than are women in the general population. This observation remains rather controversial because it is not yet clear as to what proportion of ovarian cancer in unselected general populations is due to mutations in these genes. Future studies and screening trials, which may provide more accurate knowledge of cases carrying mutations in these genes, may facilitate future introduction of genetic screening as well as more informed counseling to women with either ovarian cancer or family histories of cancer.

REFERENCES

Abdalla, I., Ray, P., Ray, V., et al. (1998) Urology **51**, 300.

Abdalla, I., Ray, P., Ray, V., et al. (1999) Am. J. Clin. Oncol. **22**, 537.

Adams, J. (1853) Lancet. **1**, 393.

Aghmesheh, M., Nesland, J. M., Kaern, J., et al. (2004) Gyn. Oncol. **95**, 430.

Agnihotri, R., Crawford, H. C., Haro, H., et al. (2001) J. Biol. Chem. **276**, 28261.

Albin, R. J., Soanes, W. A., Bronson, B., et al. (1970a) J. Reprod. Fertil. **22**, 573.

Albin, R. J., Soanes, W. A., Bronson, B., et al. (1970b) J. Immunol. **104**, 1329.

Alessi, D. R., Cohen, P. (1998) Current Opin. Genet. Dev. **8**, 55.

Allard, W. J., Cheli, C. D., Morris, D. L., et al. (1999) Int. J. Biol. Markers. **14**, 73.

Alper, O., Bergmann-Leitner, E. S., Bennett, J. A., et al. (2001) J. Natl. Cancer Iust. **93**, 1375.

Alter, B. P. (1996) Am. J. Hematol. **53**, 99.

Anderson, L., Seilhamer, J. (1997) Electrophoresis **18**, 533.

Anderson, K., Andrews, R., Yin, L., et al. (1998) Hum. Exp. Toxicol. **17**, 131.

Auranen, A., Grenman, S., Klemi, P. J. (1997) Cancer **79**, 2147.

Ayala, G., Thompson, T., Yang, G., et al. (2004) Clin. Cancer Res. **10**, 6572.

Barak, M., Mecz, Y., Lurie, A., et al. (1989) J. Lab. Clin. Med. **113**, 598.

Bast, R. C., Jr, Freeney, M., Lazarus, H., et al. (1981) J. Clin. Invst. **681**, 1331.

Bast, R. C., Jr, Klug, T. L., St John, E., et al. (1983) N. Engl. J. Med. **309**, 883.

Bast, R. C., Jr, Xu, F. J., Yu, Y. H., et al. (1998) Int. J. Biol. Markers. **13**, 179.

Bast, Jr, R. C. (2003) J. Clin. Oncol. **21**, 200s.

Bast, Jr, R. C., Urban N., Shrider, V., et al. (2002) in *Ovarian Cancer*, Stack, M. S., Fishman, D. A. (Eds), Boston, MA, Kluwer, pp 61–79.

Baylin, S. B., Herman, J. G. (2000) Trends Genet. **16**, 168.

Beahrs, O. H., Henson, D., Hutter, R. V. P., et al. (1992) *Manual for Staging of Cancer* (4th edition), Philadelphia, J. B. Lippincot.

Benson, M. C., Whang, I. S., Olsson, C. A., et al. (1992) Urology **147**, 817.

Benson, F. E., Stasiak, A., West, S. C. (1994) EMBO J. **13**, 5764.

Bignell, G., Micklem, G., Stratton, M. R., et al. (1997) Hum. Mol. Genet. **6**, 53.

Bork, P., Blomberg, N., Nilges, M. (1996) Nat. Gnet. **13**, 22.

Boyes, D. A. (1981) Cancer **48**, 613.

Brawer, M. K., Aramburu, E. A., Chen, G. L., et al. (1993) J. Urol. **150**, 369.

Bratt, O., Kristoffersson, U., Lundgren, R., et al. (1999) Eur. J. Cancer **35**, 272.

Brawn, P. M., Speights, V. O., Kuhl, D, et al. (1991). Cancer **68**, 1592.

Breiman, L. (1984) *Classification and Regression Trees*, Belmont, CA, Wadsworth Int. Group.

Brown, L. F., Papadopoulos-Sergiou, A., Berse, B., et al. (1994) Am. J. Pathol. **145**, 610.

Buting, P. S., DeBoer, G., Choo, R., et al. (2002) Clin. Biochem. **35**, 471.

Buller, R. F., Lallas, T. A., Shahin, M. S., et al. (2001) Clin. Cancer Res. **7**, 831.

Burchuck, A., Carney, M. (1997) Biochem. Pharmacol. **54**, 541.

Cairns, P., Okami, K., Halachmi, S., et al. (1997) Cancer Res. **57**, 4997.

Canto, M. T., Chu, K. C. (2000) Cancer **88**, 2642.

Cannistra, S. A. (2004) N. Engl. J. Med. **351**, 2519.

Carpten, J., Nupponen, N., Isaacs, S., et al. (2002) Nat. Genet. **30**, 181.

Carter, H. B., Pearson, D. J., Metter, E. J., et al. (1992) JAMA **267**, 2215.

Carter, H. B., Pearson, J. D., Waclawiw, Z., et al. (1995) Urology **45**, 591.

Carter, H. B., Epstein, J. I., Partin, A. W. (1999) Urology 53, **126**.

Carter, B., Bova, G., Beaty, T., et al. (1993) J. Urol. **150**, 797.

Carter, H. B., Piantadosi, S., Isaacs, J. T. (1990) J. Urol. **143**, 742.

Carter, H. B. (2004) N. Engl. J. Med. **350**, 2292.

Carter, H. B., Epestein, J. I., Chan, D. W., et al. (1997) JAMA **277**, 1456.

Catalona, W. J., Richie, J. P., Ahmann, F. R., et al. (1994) J. Urol. **151**, 1283.

Catalona, W. J., Partin, A. W., Slawin, K. M., et al. (1998) JAMA **279**, 1542.

Catalona, W. J., Bartsch, G., Rittenhouse, H. G., et al. (2003) J. Urol. **170**, 2181.

Chen, l., Stacewicz, -Sapuntzakis, M., Duncan, C., et al. (2001) J. Natl. Cancer Res. **93**, 1872.

Chambers, A. F. (1995) Ann. NY Acad. Sci. **760**, 101.

Christopherson, W. M., Parker, J. E., Mendez, W. M. (1970) Cancer **26**, 808.

Cintin, C., Johansen, J. S., Christensen, I. J., et al. (1999) Br. J. Cancer **79**, 1494.

Clark, L. C., Dalkin, B., Krongrad, A., et al. (1998). Br. J. Urol. **81**, 730.

Coppola, D., Szabo, M., Boulware, D., et al. (2004) J. Clin. Cancer Res. **10**, 184.

Cooner, W. H., Mosely, B. R., Rutherford, C. L., Jr, et al. (1990) J. Urol. **143**, 1146.

Cox, D. R., Snell, E. J. (1989) *Analysis of Binary Data* (2nd edition), Chapman and Hall, New York.

Coyne, K. J., Burkholder, J. M., Feldman, R. A., et al. (2004) Appl. Environ. Microbiol. **70**, 5298.

Crawford, H. C., Matrisian, L. M., Liaw, L. (1998) Cancer Res. **8**, 5206.

Davies, A. A., Masson, J-Y., Mcllwraith, M. J., et al. (2001) Mol. Cell. **7**, 273.

D'Amico, A. V., Chin, M-H., Kimberly, A. R., et al. (2004). N. Engl. J. Med. **351**, 125.

de Winter, J. P., Waisfisz, Q., Rooimans, M. A., et al. (1998) Nat. Genet. **20**, 281.

de Winter, J. P., Waisfisz, Q., Rooimans, M. A., et al. (2000) Hum. Mol. Genet. **9**, 2665.

Dehn, H., Hogdall, E. V., Johansen, J. S., et al. (2003) Acta Obstet. Gynecol. Scand. **82**, 287.

Dejager, S., Mietus-Snyder, M., Friera, A., et al. (1993) J. Clin. Invest. **92**, 894.

Denhardt, D. T., Guo, X. (1993) FASEB J. **7**, 1475.

Dhanasekaran, S. M., Barrette, T. R., Ghosh, D., et al. (2001) Nature **412**, 822.

Diamandis, E. P. (2004) Mol. Cell. Proteomics **3**, 367.

Diamandis, E. P., Yousif, G. M., Clements, J., et al. (2000) Clin. Chem. **46**, 1855.

Dmochowski, A., Assenhajm, D. (1935) Naturwissen-schaften. **23**, 501.

Dupont, J., Tanwar, M. K., Thaler, H. T., et al. (2004) J. Clin. Oncol. **22**, 3330.

Eastham, J. A., May, R. A., Whatley, T., et al. (1998) J. Natl. Cancer Inst. **90**, 756.

Eastham, J. A., Riedel, E., Scardino, P. T., et al. (2003) JAMA **289**, 2695.

Egawa, S., Arai, Y., Tobisu, K., et al.(2000) Prostatic Dis. **3**, 269.

Eiriksdottir, G., Barkardottir, R. B., Agnarsson, B. A., et al. (1998) Oncogene **16**, 21.

Eisenhauer, E. A., ten Bokkel Huinink, W. M., Swenerton, K. D. (1994) J. Clin. Oncol. **12**, 2654.

España, F., Martinez, M., Royo, M., et al. (2002) BJU Int. **90**, 672.

Etzioni, R., Urban, N., Ramsey, S., et al. (2003) Nat. Rev. **3**, 1.

Fang, X., Schummer, M., Mao, M., et al. (2002) Biochem. Biophys. Acta **1582**, 257.

Fearon, E. R., Vogelstein, B. (1990) Cell **61**, 759.

Fisher, L. W., Jain, A., Tayback, M., et al. (2004) Clin. Cancer Res. **10**, 8501.

Ford, D., Easton, D. F., Stratton, M., et al. (1998) Am. J. Hum. Genet. **62**, 676.

Fowler, J. E., Bigler, S. A., Nirmal, K., et al. (2002) Cancer **94**, 1661.

Fowler, J. E., Bigler, S. A., Kilambi, N. K., et al. (1999) Urology **53**, 1175.

Frank, T. S., Deffenbaugh, A. M., Reid, J. E., et al. (2002) J. Clin. Oncol. **20**, 1480.

Freeland, S. J., Dorey, F., Aronson, W. J. (2001) Urology **57**, 476.

Fresno Vara, J. A'., Casado, E., de Castro, J., et al. (2004) Cancer Treat Rev. **30**, 193.

Fresno, M., McVay-Boudreau, L., Nabel, G., et al. (1981) J. Expt. Med. **153**, 1260.

Gann, P. H., Hennekens, C. H., Stampfer, M. J. (1995) JAMA **273**, 289.

Garcia-Higuera, I., Taniguchi, T., Ganesan, S., et al. (2001) Mol. Cell. **7**, 249.

Goluboff, E. T., Heitjan, D. E., DeVries, G. M., et al. (1997) J. Urol. **158**, 1876.

Greenblatt, M. S., Chappuis, P. O., Bond, J. P., et al. (2001) Cancer Res. **61**, 4092.

Greenlee, R. T., Hill-Harmon, M. B., Murry, T., et al. (2001) CA. Cancer J. Clin. **51**, 15.

Gronberg, H., Damber, L., Damber, J. E. (1994) J. Urol. **152**, 1484.

Guinan, P., Bhatti, R., Ray, P. (1987) J. Urol. **137**, 686.

Gupta, R. C., Bazemore, L. R., Golub, E. I., et al. (1997) Proc. Natl. Acad Sci. USA **94**, 463.

Gustafsson, O., Mansour, E., Norming, U., et al. (1998) Scan. J. Urol. Nephrol. **36**, 373.

Gutman, E. B., Sproul, E. E., Gutman, A. B., et al.(1936) Am. J. Cancer **28**, 485.

Hara, M., Koyanagi, Y., Inoue, T., et al. (1971) Nippon Hoigaku Zasshi **25**, 322.

Haber, D. A. (2003) Cell **115**, 507.

Hastie, T., Tibshirani, R. (1996) J. R. Stat. Soc. B Mthodol. **58**, 155.

Hays, R. B., Liff, J. M., Pottern, L. M., et al. (1995) Int. J. Cancer **60**, 361.

Hemminki, K., Czene, K. (2002) Cancer **95**, 1346.

Henderson, R. J., Easstham, J. A., Culkin, D. J., et al. (1997) J. Natl. Cancer Inst. **89**, 134.

Heinonen, O. P., Albanes, D., Virtamo, J., et al. (1998) J. Natl. Cancer Inst. **90**, 440.

Hellström, I., Raycraft, J., Hayden-Ledbetter, M., et al. (2003) Cancer Res. **63**, 3695.

Hoque, A., Albanes, D., Lippman, S. M., et al. (2001) Cancer Causes Control **12**, 627.

Hooper, J. D., Nicol, D. L., Dickinson, J. L., et al. (1999) Cancer Res. **59**, 3199.

Horninger, W., Cheli, C. D., Babian, R. J., et al. (2002) Urology. **60**, 31.

Hough, C. D., Sherman-Baust, C. A., Pizer, E. S., et al. (2000) Cancer Res. **60**, 6281.

How Lett, N. G., Taniguchi, T., Olson, S., et al. (2002) Science **297**, 606.

Hsing, A. W. (2000) Int. J. Cancer. **85**, 60.

Hughes-Davies, L., Raus, M., Fuks, F., et al. (2003) Cell **115**, 523.

Huibregtse, J. M., Scffner, M., Beaudenon, S., et al. (1985) Proc. Natl. Acad. Sci. USA **92**, 2563.

Isaacs, W. B., Isaacs, J. T. (1997) Molecular Genetics of Prostate Cancer Progression. in *Principles and Practice of Genitourinary Oncology* (Ist edition), Raghavan, D., Scher, H. I., Leibel, S. A., Lang, P. H. (Eds), Lippincott Haven, Philadelphia, p. 403.

Jacobs, I. J., Skates, S. J, MacDonald, N., et al. (1999) Lancet **353**, 1207.

Jacobs, I. J., Menon, U. (2004) Mol. Cell. Proteomics **3**, 355.

Jasin, M. (2002) Oncogene **21**, 8981.

Jemal, A., Murry, T., Samuels, A., et al. (2003) CA. Cancer J. Clin. **53**, 5.

Jerónimo, C., Usadel, H., Henrique, R., et al. (2001) J. Natl. Cancer Inst. **93**, 1747.

Joenje, H., Oostra, A. B., Wijker, M., et al. (1997) Am. J. Hum. Genet. **61**, 940.

Joenje, H., Levitus, M., Waistisz, Q., et al. (2002) Am. J. Hum. Genet. **67**, 759.

Johansen, J. S., Cintin, C., Jorgensen, M., et al. (1995) Eur. J. Cancer **31**, 1437.

Johnson, G. A., Burghardt, R. C., Bazer, F. W., et al. (2003) Biol Reprod. **69**, 1458.

Johnson, J. A., Spencer, T. A., Brghardt, R. C., et al. (1999) Biol. Reprod. **61**, 884.

Jones, P. A., Baylin, S. B. (2002) Nat. Rev. **3**, 415.

Kolonel, L. (1997) In Fortner, J. and Sharp P. (Eds) *Accomplishments in Cancer Research*, Lippincot-Raven, Philadelphia, p. 221.

Khan, M. A., Han, M., Partin, A. W., et al. (2003) Urology **62**, 86.

Kim, J-H., Skates, S. J., Uede, T., et al. (2002) JAMA **287**, 1671.

King, M-K., Marks, J. H., Mandell, J. B. (2003) Science **302**, 643.

King, M-C. (2004) N. Engl. J. Med. **350**, 1252.

Kreisberg, J. I., Malik, S. N., Prihoda, T. J., et al. (2004) Cancer Res. **64**, 5232.

Krumholtz, J. S., Carvalhal, G. F., Ramos, C. G., et al. (2002) Urology **60**, 469.

Kutscher, W., Worner, A. (1936) Physiol. Chem. **239**, 109.

Labrie, F., Candas, B., Dupont, A., et al. (1999) Prostate **38**, 83.

Lam, J. S., Cheung, Y. K., Benson, M. C., et al. (2003) J. Urology **170**, 451.

Lee, W-H., Mrton, R. A., Epstein, J. I., et al. (1994) Proc. Natl. Acad. Sci. USA **91**, 11733.

Li, T. S., Beling, C. G. (1973) Fertil. Steril. **24**, 134.

Lichtenstein, P., Holm, N. V., Verkasalo, P. K., et al. (2000) N. Engl. J. Med. **343**, 78.

Lilja, H., Christensson, A., Dahlen, U., et al. (1991) Clin. Chem. **37**, 1618.

Lilja, H., Haese, A., Bjork, T., et al. (1999) J. Urology **162**, 2029.

Lilja, H. (1985) J. Clin. Invest. **76**, 1899

Lippincot, J. B. (1997) in *Principles and Practices of Gynocologic Oncology* (2nd edition) Hoskins, W. J., Perez, C., Young, R. C., (Eds), Lippincot-Raven Philadelphia.

Liu, X. L., Wazer, D. E., Watanbe, K., et al. (1996) Cancer Res. **56**, 3371.

Luedtke, C. C., McKee, M. D., Cyr, D. G., et al. (2002) Biol. Reprod. **66**, 1437.

Luo, L-Y., Grass, L., Howarth, D. J. C., et al. (2001) Clin Chem. **47**, 237.

Luo, L-Y., Katsaros, D., Scorilas, A., et al. (2003) Cancer Res. **63**, 807.

Luo, JI, Manning, B. D., Cantley, L. C. (2003) Cell **4**, 257.

Malik, S. N., Brattain, M., Ghosh, P. M., et al. (2002) Clin. Cancer Res. **8**, 1168.

Marks, L. S., Kojima, M., Demarzo, A., et al. (2004) Urology **64**, 765.

Martinez, M., España, F., Royo, M., et al. (2002) Clin. Chem. **48**, 1251.

Matsumoto, T., Tsurumoto, T. (2001) Clin. Exp. Rheumatol. **19**, 655.

Mettlin, C. J., Murphy, G. P., McGionnis, l. s., et al. (1995) Cancer **76**, 1104.

McCammon, K. A., Schellhammer, P. F., Wright, G. L., et al. (1996) J. Urol. **155**, 426.

McNeal, J. E., Bostwick, J. D., Kindrachuk, R. A., et al. (1986) Lancet **1**, 60.

Mcllwraith, M. J., Van Dyck, E., Masson, J. Y., et al. (2000) J. Mol. Biol. **304**, 151.

Mikolajczyk, S. D., Grauer, L. S., Millar, L. S., et al. (1997) Urology **50**, 710.

Mikolajczyk, S. D., Millar, L. S., Wang, T. J., et al. (2000a) Cancer Res. **60**, 756.

Mikolajczyk, S. D. (2000b) Urology **55**, 41.

Mikolajczyk, S. D., Marker, K. M., Millar, L. S., et al. (2001) Cancer Res. **61**, 6958.

Mikolajczyk, S. D., Marks, L. S., Partin, A. W., et al. (2002) Urology **59**, 797.

Miller, A. B., Hoogstraten, B., Staquet, M., et al. (1981) Cancer (Phila). **47**, 207.

Milner, J., Ponder, B., Hughes-Davies, L., et al. (1997) Nature. **386**, 772.

Mizuta, R., LaSalle, J. M., Cheng, H. L., et al. (1997) Proc. Natl. Acad. Sci. USA **94**, 6927.

Morgan, T. O., Jacobsen, S. J., McCarthy, W. F., et al. (1996) N. Engl. J. Med. **335**, 304.

Moul, J. W., Sesterrhenn, I. A., Connelly, R. R., et al. (1995) JAMA **274**, 1277.

Mok, S. C., Chao, J., Skates, S., et al. (2001) J. Natl. Cancer Inst. **93**, 1458.

Nagata, T., Todescan, R., Golberg, H. A., et al. (1989) Biochem. Biophys. Res. Commun. **165**, 234.

Nagata, T., Bellows, C. G., Kasugai, S., et al. (1991b) Biochem J. **274**, 513.

Nagata, A., Hirota, N., Sakai, T., et al. (1991a) Tumor Biol. **12**, 279.

Neame, P. J., Butler, W. T. (1996) Connect Tissue Res. **35**, 145.

Nelson, P. S., Gan, L., Ferguson, C., et al. (1999) Proc. Natl. Acad. Sci. USA. **96**, 3114.

Nelson, W. G., De Marzo, A., Isaacs, W. B. (2003) N. Engl. J. Med. **349**, 366.

Nixon, R. G., Wener, M. H., Smith, K. M., et al. (1997) J. Urol. **157**, 2183.

Norad, S., Ford, D., Devilee, P., et al. (1995) Am. J. Hum. Genet. **57**, 957.

Norad, S. A., Foulkes, W. D. (2004) Nat. Rev. **4**, 665.

Oesterling, J. E., Jacobsen, S. J., Chute, C. G., et al. (1993) JAMA **270**, 860.

Okihara, K., Fritsche, H. A., Ayala, A., et al. (2001) J. Urol. **165**, 1930.

O'Regan, A., Berman, J. S.(2000) Int. J. Exp. Pathol. **81**, 373.

Ornstein, D. K., Smith, D. S., Rao, G. C., et al. (1997) J. Urol. **157**, 2179.

Osin, P. P., Lakhani, S. R. (1999) Breast Cancer Res. **1**, 36.

Paju, A., Vartiainen, J., Haglund, C., et al. (2004) Clin. Cancer Res. **10**, 4761.

Parkin, D. M., Pisani, P., Feday, J. (1993) Int. J. Cancer **54**, 594.

Parkin, D., Whelan, S., Ferlay, J., et al. (1997) Cancer Incidence in Five Continents, Vol. VII. Int. Agency for Research on Cancer, Lyon, France.

Partin, A. W., Criley, S. R., Subong, E. N., et al. (1996) J. Urol. **155**, 1336.

Partin, A. W., Brawer, M. K., Bartsch, G., et al. (2003) J. Urol. **170**, 1787.

Parsons, J. K., Partin, A. W. (2004) Urology. **815**, 818.

Paweletz, C. P., Charboneau, L., Bichsel, V. E., et al. (2001) Oncogene **20**, 1981.

Pepe, M. S., Etzioni, R., Feng, Z., et al. (2001) J. Natl. Inst. Cancer **93**, 1054.

Perou, C. M., Jeffrey, S. S., Van Derijn, M., et al. (1999) Proc. Natl. Acad. Sci. USA **96**, 9212.

Peter, J., Unverzagt, C., Krogh, T. N., et al. (2001) Cancer Res. **61**, 957.

Peters, D. G., Kudla, D. M., DeLoia, J. A., et al. (2005) Cancer Epidemiol. **14**, 1717.

Platt, N. and Gorgon, S. (2001) J. Clin. Invest. **108**, 649.

Pollak, M. N., Foulkes, W. D. (2003) Nat. Rev. 3, 297.

Prince, C. W., Oosawa, T., Butler, W. T., et al. (1987) J. Biol. Chem. **262**, 2900.

Punglia, R. S., D'Amico, A. V., Catalona, W. J., et al. (2003) N. Engl. J. Med. **349**, 335.

Ramus, S. J., Bobrow, L. G., Pharoah, P. D. P., et al. (1999) J. Clin. Pathol. **52**, 372.

Rawlings, N. D., Barrett, A. J. (1994) Methods Enzymol. **244**, 19.

Register, T. C., Carlson, C. S., Adams, M. R. (2001) Clin. Chem. **47**, 2159.

Reissigl, A., Pointner, J., Horninger, W., et al. (1995) Urology **46**, 662.

Renkema, G. H., Boot, R. G., Muijsers, A. O., et al. (1995) J. Biol. Chem. **270**, 2198.

Renkema, G. H., Boot, R. G., Au, F. L., et al. (1998) Eur. J. Biochem. **251**, 504.

Ripley, P. D. (1996) Pattern Recognition and neural Networks, Cambridge University Press, NY.

Risch, H. A., McLaughlin, J. R., Cole, D. E. C., et al. (2001) Am. J. Hum. Genet. **68**, 700.

Ross, D. T., Scherf, U., Eisen, M. B., et al. (2000) Nat. Genet. **24**, 227.

Rustin, G. J., Bast, R. C. Jr., Kelloff, G. J., et al. (2004) Clin Cancer Res. **10**, 3919.

Rustin, G. J, Nelstrop, A., Stiwell, J., et al. (1992) Eur. J. Cancer. **28**, 79.

Rustin, G. J. (2003) J Clin Oncol. **21**, 187s.

Saha, S., Sparks, A. B., Rago, C., et al. (2002) Nat. Biotechnol. **20**, 508.

Sansal, S., Sellers, W. (2004) J. Clin. Oncol. **22**, 2954.

Schena, M., Heller, R. A., Theriault, T. P., et al. (1998) Trends Biotechnol. **16**, 301.

Schaid, D. J. (2004) Human Mol. Genet. **13**, R103.

Schiffman, M. H., Brinton, L. A., Devesa, S. S., et al. (1996) In Cancer Epidemiology and Prevention 2nd edn. Schottenfeld and Fraumeni (eds). Oxford University Press, NY. pp 1090–1116.

Scully, R., Livingston, D. M. (2000) Nature. **408**, 429.

Sensabough, H. J., Crim, D. J. J. (1978) Forensic Sci. **23**, 106.
Shummer, M., Ng, W. V., Bumgarner, R. E., et al. (1999) Gene **238**, 375.
Silverman, R. H., Jung, D. D., Nolan-Sorden, N. L., et al. (1988) J. Biol. Chem. **263**, 7336.
Singh, R. P., Patarca, R., Schwartz, J., et al. (1990) J. Expt. Med. **171**, 1931.
Skates, S. J., Horick, N., Yu, Y., et al. (2004) J. Clin. Oncol. **22**, 4059.
Smith, D. S., Catalona, W. J., Herschman, J. D. (1996) JAMA **276**, 1309
Sokoll, L. J., Chan, D. W., Mikolajczyk, S. D., et al. (2003) Urology. **61**, 274.
Sokoll, L. J., Mangold, L. A., Partin, AS. W., et al. (2002) Urology. **60**, 18.
Sorensen, E. S., Hojrup, P., Petersen, T. E. (1995) Protein Sci. **4**, 2040.
S-Plus 2000 (1999) Guide to Statistics, Seattle, WA, Mathsoft Inc.
Stamey, T. A., Caldwell, M., McNeal, J. E., et al. (2004) J. Urol. **172**, 1279.
Stamey, T. A., Kabalin, J. N., McNeal, J. E., et al. (1989) J. Urol. **141**, 1076.
Stamey, T. A., Yang, N., Hay, A. R., et al. (1987) N. Engl. J. Med. **317**, 909.
Stamey, T. A., Kabalin, J. N., McNeal, J. E., et al. (2001) Clin. Chem. **47**, 631.
Stamey, T. A., Johnstone, I. M., McNeal, J. E., et al. (2002) J. Urol. **167**, 103.
Stein, W. D., Litman, T., Fojo, T., et al. (2004) Cancer Res. **64**, 2805.
Strathdee, C. A., Gavish, H., Shannon, W. R., et al. (1992) Nature **356**, 763.
Strobel, T., Swanson, L., Korsmeyer, S., et al. (1996) Proc. Natl. Acad. Sci. USA **93**, 14094.
Struewing, J. P., Hartge, P., Wacholder, S., et al. (1997) N. Engl. J. Med. **336**, 1401.
Sullivan, T. J., Gutman, E. B., Gutman, A. B. (1942) J. Urol. **48**, 426.
Tanwar, M. K., Gilbert, M., Holland, E. C. (2002) Cancer Res. **62**, 4364.
Therasse, P., Rrbuck, S. G., Eisenhauer, E. A., et al. (2000) J. Natl. Cancer Inst. **92**, 205.
Thompson, I. M., Donna, M. D., Pauler, K., et al. (2004) N. Engl. J. Med. **350**, 2239.
Tuxen, M. K., Soletormos, G., Dombernowsky, P. (1995) Cancer Treat Rev. **21**, 215.
Van Dalen, A., Favier, J., Burges, A., et al. (2000) Gynecol. Oncol. **79**, 444.
Van Iersel, M. P., Witjes, W. P., de la Rosette, J. J., et al. (1995) Br. J. Urol. **76**, 47.
Velculescu, V. E., Zhang, L., Vogelstein, B., et al. (1995) Science **270**, 484.
Velculescu, V. E., Vogelstein, B., Kinzler, K. W. (2000) Trends Genet. **16**, 423.
Veltri, R. W., Miller, M. C., O'Dowd, G. J., et al. (2002) Urology. **60**, 47.
Velculesco, V. E., et al. (1995) Science **270**, 484.
Venkitaraman, A. R. (2003) N. Engl. J. Med. **348**, 1917.
Venkitaraman, A. R. (2004) Natl. Rev. Cancer. **4**, 266.
Venkitaraman, A. R. (2002) Cell. **108**, 171.
Vijayakumar, S., Karison, T., Weishselbaum, R. R., et al. (1992) Cancer Epidemiol. Biomarkers Prev. **1**, 541
Vis, A. N., Kranse, R., Roobol, M., et al. (2002) BJU Int. **89**, 384.
Vivanco, I., Sawyers, C. L., Nature Rev. **2**, 489 (2002).
Wang, M. C., Valenzuela, L. A., Murphy, G. P., et al. (1979) Investig. Urol. **17**, 159.
Welch, H. G., Schwarz, L. M., Woloshin, S. (2005) J. Natl. Cancer Inst. **97**, 1132.
Whittemore, A. S., Wu, A. H., Kolonel, L. N., et al. (1995) Am. J. Epidemiol. **141**, 732.
WHO Handbook for Reporting Results of Cancer Treatment, No. 48, Geneva, 1979.
Wingo, P. A., Bolden, S., Tong, T., et al. (1996) CA. Cancer J. Clin. **46**, 113.

Winawer, S. J., et al. (1995) Bull. World Health Organization. **73**, 7.
Wong, A. K., Pero, R., Ormonde, P. A., et al. (1997) J. Biol. Chem. **272**, 31941.
Wong, K. K., Cheng, R. S., Mok, S. C. (2001) Biotechniques. **30**, 670.
Wooster, R., Bignell, G., Lancaster, J., et al. (1995) Nature. **378**, 789.
Xu, F-J, Ramakrishnan, S., Daly, L., et al. (1991) Am. J. Obstet. Gynocol. **165**, 1356.
Xu, F-J, Yu, Y-A., Daly, L., et al. (1993) J. Clin. Oncol. **11**, 1506.
Xu, J., Zheng, S. L., Komiya, A., et al. (2002) Nat. Genet. **32**, 321.
Ye, B., Cramer, D. W., Skates, S. J., et al. (2003) Clin. Cancer Res. **9**, 2904.
Yin, B. W. T., Lloyd, K. O. (2001) J Biol Chem. **276**, 27371.
Young, M. F., Kerr, J. M., Termine, J. D., et al. (1990) Genomics. **7**, 491.
Yousef, G. M., Diamandis, E. P. (2001) Endocr. Rev. **22**, 184.
Yu, D. S., Sonoda, E., Takeda, S., et al. (2003) Mol. Cell **12**, 1029.
Yu, J. X., Chao, L., Chao, J. (1994) J. Biol. Chem. **269**, 18843.
Yu, J. X., Chao, L., Chao, J. (1995) J. Biol. Chem. **270**, 13483.
Zhang, H., Somasundaram, K., Peng, Y., et al. (1998) Oncogene **16**, 1713.
Zhou, A., Hassel, B. A., Silverman, R. H. (1993) Cell **72**, 753.
Zinda, M. J., Johnson, M. A., Paul, J. D., et al (2001) Clin. Cancer Res. **7**, 2475.
Zurawski, v. R. Jr., Davis, H. M., Finkler, N. J., et al. (1988) Cancer Rev. **11–12**, 10.
Zweemer, R. P., Shaw, P. A., Verbeijen, R. M. H., et al. (1999) J. Clin. Path. **52**, 372.

4

POTENTIAL CANCER BIOMARKERS

4.1. INTRODUCTION

Our increasing understanding of the various forms of cancer including genetic, molecular, and cellular mechanisms is now providing clear objectives for their early detection, prevention, and therapy. Over the past 30 years, much has been learned about the complex interplay of genetic and environmental factors that cause cancer. An extraordinary research achievement in humans as well as in model organisms has, to a large extent, defined the type of changes necessary for the transformation of a normal cell to one that is cancerous (Hanahan and Weinberg, 2000; van Dyke and Jacks, 2002). Such changes include insensitivity to antigrowth signals, evasion of immuno surveillance, evasion of apoptosis, unlimited replicative potential, sustained angiogenesis, tissue invasion, and metastasis. Although spectacular advances in molecular medicine, genomics, and proteomics have been made, current efforts to combat cancer remain extremely disappointing. One main reason for the lack of the desired success is that in many cases, cancer is diagnosed and treated too late, when the cancer cells have already invaded adjacent tissues and established new colonies. The capability for invasion and metastasis enables cancer cells to escape the primary tumor mass and colonize new terrain in the body where, at least initially, nutrients and space are not limited (Hanahan and Weinberg, 2000). These distant settlements of tumor cells are the cause of 90% of human cancer deaths (Sporn, 1996). It is commonly agreed that one of the most promising strategies to combat this devastating disease is to discover biomarkers, which can detect the disease in

Cancer Biomarkers: Analytical Techniques for Discovery, by Mahmoud H. Hamdan

its early stages or even predict the occurance of the disease before its manifestation. Over the last 15 years, protein- and DNA-based analyses have identified a substantial number of proteins, genes, and other processes, which can be linked to various forms of cancer. Furthermore, the recently acquired knowledge of how chromatin organization moderates transcription has highlighted the importance of epigenetic mechanisms in the initiation and progression of cancer. These epigenetic changes, in particular, aberrant promoter hypermethylation that is associated with inappropriate gene silencing virtually affect every step in tumor progression (Jones and Baylin, 2002). In the present chapter, an attempt is made to have a closer look at the biological functions and biochemical properties of some protein families, which have been described as potential biomarkers for various forms of cancer. Some gene alterations and associated mechanisms, in particular, DNA methylation that may impact on the development of certain forms of cancer are also addressed.

4.2. HUMAN TISSUE KALLIKREINS

Human tissue kallikreins are secreted serine proteases encoded by 15 structurally similar, steroid hormone-regulated genes that co-localize to chromosome 19q13.4 in a 300-kb region (Diamandis and Yousef 2000; Yousef et al., 2001a). Tissue kallikreins form a subgroup of secreted serine proteases within the S1 family of clan SA. Before describing the emerging role of human tissue kallikreins in cancer, it is helpful to consider some brief comments regarding the serine proteases to which tissue kallikereins belong. Proteases are commonly defined as enzymes that catalyze peptide bond hydrolysis and are known to perform a wide range of essential functions in almost all living organisms (Schultz and Liebman, 1997). They were initially recognized as gastric juice proteolytic enzymes that were involved in the nonspecific degradation of dietary proteins. However, recent advances have provided a clearer picture of the various functions in living organisms. It is now known that besides mediating nonspecific protein hydrolysis, proteases also act as processing enzymes that perform highly selective, limited, and efficient cleavage of specific substrates, which initiates irreversible decisions at the post-translational level that impact on a wide range of biologic processes (Horl, 1989; Henderson et al., 1992). Proteolytic processing events are fundamental in various biologic roles, including ovulation, fertilization, embryonic development, bone formation, control of homeostatic tissue remodeling, neuronal outgrowth, antigen presentation, cell-cycle regulation, immune and inflammatory cell migration and activation, wound healing, angiogenesis, and apoptosis (Barrett et al., 1998). Accordingly, alterations in the structure and expression patterns of proteases underlie many human pathological processes, including cancer, arthritis, osteoporosis, neurodegenerative disorders, and cardiovascular diseases (Esler, 2001; Krane, 2003). On the basis of their catalytic mechanisms, they can be classified into five distinct classes: aspartic, metallo, cysteine, serine, and threonine proteases. The first two classes use an activated water molecule as a nucleophile to attack the peptide bond of the substrate, whereas in the remaining classes, the nucleophile is a catalytic amino acid residue (Cysteine,

Serine, or Threonine) that is located in the active site from which the class names derive. According to a widely used and probably the most comprehensive database of proteases (MER-OPS; http://www.merops.co.uk), enzymes of each catalytic type are classified into distinct "clans," and each clan is subdivided into "families." The first classification is based on similarities in their three-dimensional folding, whereas classification to families is based on homology in amino acid sequence (Rawlings and Barrett, 1993; Rawlings et al., 2002).

Tissue kallikreins form a subgroup within a family (S1) of clan (SA) of secreted serine proteases. Serine proteases including human tissue kallikreins have been shown to be associated with many diseases including various forms of cancer, arthritis, and emphysema (Henderson et al., 1992; Froelich et al., 1993; Yousef and Diamandis, 2001; Diamandis et al., 2002). The remaining part of this section will be mainly concerned with the function of tissue kallikreins, its association to disease, and its emerging role as potential biomarkers for various forms of cancer.

4.2.1. Background and Nomenclature

The term "Kallikrein" was first introduced in 1930s by Werle and colleagues (Kraut et al., 1930; Werle, 1934). Currently, these enzymes are divided into plasma and tissue kallikreins (Fiedler, 1979; Movat, 1979). These two categories have significant differences in their molecular masses, substrate specificity, immunologic characteristics, gene structure, and type of kinin released. Plasma kallikrein or Fletcher factor is encoded by a single gene, which is located on human chromosome 4q35 (Asakai et al., 1987; Yu et al., 1998). Plasma kallikrein is exclusively expressed by lever cells and has a number of functions, including participation in the process of blood clotting, fibrinolysis, and in the regulation of vascular tone and inflammatory reactions through the release of bradykinin (Bhoola et al., 1992 a,b). Plasma kallikrein will not be discussed further in this section, which will be fully concerned with tissue kallikreins.

The human tissue kallikrein gene family was discovered in the 1980s, and it was concluded that the entire family was composed of only three genes, namely the pancreatic/renal kallikrein, the human glandular kallikrein, and prostate specific antigen. Research efforts during the 1990s by independent groups have resulted in the cloning of new serine protease genes that showed significant homologies with the three genes discovered earlier. Furthermore, the molecular description of the human kallikrein gene locus (Yousef and Diamandis, 2000a; Yuosef et al., 2000a) allowed the construction of a physical map containing 15 genes that share significant structural similarities. The identification of these genes and the fact that they were cloned by different investigators who used various empirical names highlighted the need for a revised nomenclature to describe the entire list of these genes and their encoded proteins. A combined effort by The Human Genome Organization (HUGO) and an international working party (Berg et al., 1992) has resulted in a nomenclature for the newer human kallikreins, consistent with that already defined for the three early genes (*KLK1-3*). Table 4.1 gives a summary of the new and the previous nomenclature. According to this nomenclature, the proteins are designated hK(n),

TABLE 4.1. Old and new nomenclature for human kallikreins. On the basis of data from listed references.

Official gene/protein symbol	Other names/symbols	Gene bank Accession No.	References
KLK1/hK1	Tissue, Pancreatic, renal, urinary kallikrein, hPRK	M25629, M33105	Evans et al. (1988); Fukushima et al. (1985)
KLK2/hK2	Human grandular kallikrein 1, hGK-1	M18157	Schedlich et al. (1987)
KLK3/hK3	Prostate-specific antigen, PSA,APS	X14810, M24543, M27274	Lundwall, (1987); Riegman et al. (1988, 1989)
KLK4/hK4	Prostase, *KLK-L1*, EMSP1, PRSS17,ARM1	AF113141	Nelson et al. (1999); Yousef et al. (1999c); Stephenson et al. (1999); Hu et al. (2000)
KLK5/hK5	*KLK-L2*, HSCTE	AF135028	Yousef and diamandis (1999); Brattsand and Egelrud (1999)
KLK6/hK6	hK6/Zyme, protease M, neurosin	D78203 (mRNA), AF149289(full gene)	Anisowicz et al. (1996); Little et al. (1997); Yamashiro et al. (1997); Yousef et al. (1999b)
KLK7/hK7	HSCCE,PRSS6	L33404(mRNA), AF166330(full gene)	Hansson et al. (1994); Yousef et al. (2000c)
KLK8/hK8	Neuropsin, ovasin, TADG-14, PRSS19	AB009849	Yoshida et al. (1998); UnderwooD(1999)
KLK9/hk9	*KLK-L3* protein	AF135026	Yousef and Diamandis (2000a)
KLK10/hK10	*NES1*, PRSSL1	NM_002776 (mRNA), AF055481(full gene)	Liu et al. (1996); Luo et al. (1998)
KLK11/hK11	TLSP/hippostasin, PRSS20	(AB012917 (mRNA), AF164623 (full gene)	Yoshida et al. (1998); Yousef et al. (2000c)
KLK12/hK12	*KLK-L5*	AF135025	Yousef et al. (2000d)
KLK13/hK13	*KLK-L4*	AF135024	Yousef et al. (2000e)
KLK14/hk14	*KLK-L6*	AF161221	Yousef et al. (2001b)
KLK15/hK15	Prostinogen, HSRNASPH	AF242195	Takayama et al. (2001); Yousef et al. (2001c)

whereas the genes are designated *KLK(n)*, where *n* is given the value 1, 2, 3,...15. If new kallikrein genes/proteins are identified in the same locus, they will be sequentially numbered starting with *KLK16* and encoded as hK16.

4.2.2. Gene Locus and Gene Organization of Human Kallikreins

The first report on the expanded human kallikrein multigene locus was reported by Yousef et al. (1999a). An updated version and further characterization of the same locus was given a year later (Yousef et al., 2000a; Harvey et al., 2000; Gan et al., 2000). The human kallikrein gene locus on chromosome 19q13.4 is formed of 15 tandemly localized kallikrein genes with no intervention from other genes and is the largest cluster within the human genome. Centromeric to the *KLK1* gene lies a nonkallikrein gene, testicular acid phosphate (*ACPT*) (Yousef et al., 2001a). Telomeric to the last kellikrein gene, *KLK14*, lies another nonkallikrein gene, *Siglec 9*, a member of the Siglec multigene family (Foussias et al., 2000). Diamandis's group (Yousef et al. 2000a) has constructed the first detailed map of the human kallikrein gene locus with single-base-pair accuracy and has defined the direction of transcription of all genes (see Fig. 4.1). The three early discovered genes, *KLK1–3*, cluster together within a 60-kb region, whereas the newly discovered gene, *KLK15*, maps between *KLK1* and *KLK3*. The remaining genes are aligned within this locus as shown in Figure 4.1. The direction of transcription is from telomere to centromere with the exception of *KLK3* and *KLK2*. The genomic lengths of all these genes are relatively small, ranging from 4 to 10 kb.

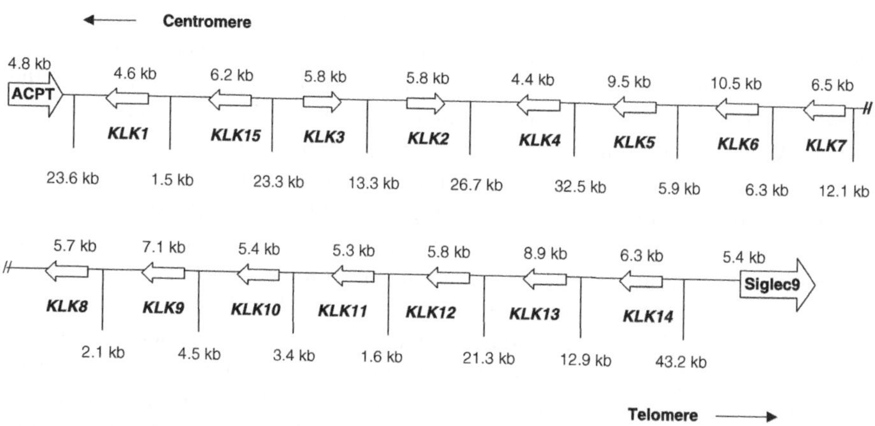

Figure 4.1. Schematic representation of the human kallikreins gene locus on chromosome 19q13.4. Arrow heads show the direction of transcription. Gene lengths are indicated above each gene, and the lengths between genes are given in approximate kilobase below the arrows. The genes *Siglec 9* and *ACPT* are not members of the kallekreins family. *NES1*, normal epithelial cell specific 1 gene; TLSP, trypsin-like serine portrease. Adapted from Diamandis and Yousif (2002) with permission.

According to Diamandis and Yousef (2002) all 15 genes consist of five coding exons with a very similar organization. The first coding exon having a short 5′-untranslated region, the second exon harboring the amino acid histidine of the catalytic triad toward the end of the exon, the third exon harboring the aspartic acid of the catalytic triad around the middle, and the fifth exon harboring the serine of the catalytic triad at the beginning of the exon. Beyond the stop codon, there is a 3′-unsaturated region of variable length. The same authors underlined two common features of such organization: First, the intron phases (the location of the intron within the codon) of the coding exons are conserved in all 15 genes, and the pattern of such phases is always "I-II-I-0." Phase I indicates that the intron occurs after the first nucleotide of the codon, phase II indicates that the intron occurs after the second nucleotide, and phase 0 refers to intron between codons. Second, the position of the residues of the catalytic triad of serine protease is conserved, with the histidine always occurring near the end of the second coding exon, the aspartate in the middle of the third coding exon, and the serine residue at the beginning of the fifth coding exon.

4.2.3. Tissue Expression and Regulation

Various research groups have used northern blot, reverse-transcription-PCR, and ELISA to demonstrate the expression of kallikreins at the mRNA and the protein levels in a wide range of tissues (Yousef et al. 1999a; Yousef and Diamandis, 2001). This expression was found to be relatively high in few major tissues, whereas lower levels of expression were detected in many others. Some of these studies have demonstrated that kallikreins are often co-expressed within the same tissues (Howarth et al., 1997; Clements et al., 2001; Kapadia et al., 2003; Kishi et al., 2003; Komatsu et al., 2003; Borgońo et al., 2004). A representative example is the concurrent and almost exclusive expression of *KLK2, 3, 4, 11*, and *15* in the prostate, at the mRNA level. Almost every kallikrein is expressed in the salivary gland; *KLK5, 6, 10*, and *13* in the breast; *KLK1* and *KLK6–13* in pancreas; and KLK6–9, and 14 in the central nervous system. A substantial group of these genes resides in the skin (*KLK1, 4-11, 13*, and *14*). Tissue-specific patterns of expression have been documented for a number of alternative mRNA transcripts of kallikrein genes. *KLK2* and *KLK3* splice variants, both with a partially retained intron, were found exclusively expressed in prostatic epithelium (David et al., 2002). Splice variants of *KLK4, 8*, and *13* gene transcripts were found to be the predominant mRNA species in the skin (Komatsu et al., 2003). One variant of the *KLK8* gene is predominantly expressed in the pancreas, whereas another variant is preferentially expressed in adult brain and hippocampus (Mitsui et al., 1999). The *KLK11* gene has two tissue-specific mRNA isoforms, known as the brain type and prostate type; the former is expressed in both organs, whereas the latter is expressed exclusively in the prostate (Mitsui et al., 2000). More details on the tissue distribution of Kallikrein genes have been given in an extensive review by Borgońo et al. (2004).

The regulation of gene expression by steroid hormones, mediated on binding to their cognate receptors, plays an important role in the normal development and function of many organs as well as in the pathogenesis of endocrine-related cancers (Klein-Hitpass et al., 1998; Henderson and Feigelson, 2000). Numerous in vitro and in vivo studies have demonstrated that all human Kallikrein genes are under steroid hormone regulation in endocrine-related tissues and cell lines (Riegman et al., 1991; Yousef et al., 2000b; Myers et al., 2001; Yousef et al., 2003a). The up-regulation of *KLK2* and 3 transcriptions in response to androgens and progestins in prostate and breast cell lines is the most recognized example of such a regulation (Riegman et al., 1991; Young et al., 1995; Cleutjens et al., 1997). On the contrary, *KLK1*, *6*, and *10* are known to be more responsive to estrogens (Clements et al., 1994; Yousef et al., 1999b; Luo et al., 2000). There are a number of observations regarding the regulation of kallikreins, which are still difficult to be fully explained: First, there are certain differential patterns associated with a number of these genes, for example, *KLK4* is upregulated by androgens in prostate and breast cancer cell lines (Nelson et al., 1999; Yousef et al., 1999c) and by estrogens in endometrial cancer cell lines (Myers et al., 2001). On the contrary, *KLK12* is upregulated by androgens and progestins in prostate cancer cell lines and by estrogens and progestins in breast cancer cell lines (Yousef et al., 2000b). Second, several recent studies have hypothesized some form of cross talk between steroid hormone signaling with other signal transduction pathways in the regulation of kallikrein gene transcription. One of these studies by Sadar (1999) suggested that such a cross talk between androgen receptor (AR) and protein kinase A signal induction pathways could contribute to the androgen-independent induction of *KLK3* gene expression. Third, epigenetic control of gene expression such as DNA methylation may also be implicated in the regulation of some kallikrein gene transcription, particularly during carcinogenesis. This suggestion has been supported by two recent works. In the first work, methylation-specific PCR and sequence analysis of sodium bisulfite-treated genomic DNA were used to demonstrate a strong correlation between hypermethylation and the loss of *KLK10* mRNA expression in a panel of breast cancer cell lines and in primary tumors. The same study also demonstrated that treatment of *KLK10*–nonexpressing cells with a demethylating agent led to reexpression of this gene (Li et al., 2001). In a more recent work it was also demonstrated that *KLK10* was downregulated by hypermethylation in acute lymphoblastic leukemia (Roman-Gomez et al., 2004).

4.2.4. Physiologic Roles

The completion of various genome projects has provided a reasonable overview of the composition and organization of the proteolytic machinery that is used by different organisms. However, neither all proteases have been discovered yet, nor have their in vivo substrates and functions been identified. Indeed, many human protease genes that have been predicted from genome computer searches remain to be cloned and the proteolytic activity of their gene products confirmed (López-Otin et al., 2002). Therefore, one of the future challenges is to determine the structures and to assign physiological and pathological functions to the proteins that are encoded by these genes.

Serine proteases were among the first enzymes to be studied extensively (Neurath, 1985). The study included their structural characteristics, catalytic mechanism, and roles in normal physiology and pathology in a number of diseases including cancer and neurodegenerative disorders (Turgeon and Houenou, 1997; Barrett et al., 1998). The number of human serine proteases for which a specific physiologic function has not yet been elucidated is very large. This holds true for the 15 kallikreins encoded by genes clustered in a small part of chromosome 19 (19q13.3-q13.4). With sensitive RNA techniques, expression of the majority of kallikreins can be detected in most tissues, including skin. Although some human kallikreins have been connected to physiologic processes and pathologic conditions, none has been assigned to cleave a specific substrate. Kallikreins are often co-expressed in the skin, breast, prostate, pancreas, and brain, primarily by secretory epithelial cells, from which they enter bodily fluids such as sweat, milk, saliva, seminal plasma, and cerebrospinal fluid. As such, they are implicated in a wide range of normal and pathological processes, where they act independently and/or as part of one or more proteolytic cascades. On the basis of existing knowledge, these cascades are assumed to contribute to a range of processes in specific tissues, including extracellular matrix remodeling and the regulation of cellular proliferation, differentiation, and apoptosis in the normal and neoplastic prostate tissues (Borgoño and Diamandis, 2004). However, it cannot be excluded that such cascades can also operate in other organs, such as the pancreas and pituitary, and participate in hormone processing or in signal-transduction pathways involving cell-surface receptors such as protease-activated receptors. Proteolytic cascade of kallikreins in stratum corneum has been demonstrated in a number of studies (Ekholm et al., 2000; Brattsand et al., 2005). In simple terms, the stratum corneum can be described as a stockade made up of corneocytes held together by corneodesmosomes and surrounded by lipids. Its continuous production as a result of epidermal differentiation is balanced by desquamation. Regulated transition from a state of strong intercellular cohesion in deeper layers to a state where individual cells are being shed at the skin surface implies a well-coordinated proteolytic degradation of adhesive structures. The presence of hK5 and hK7 in stratum corneum extracts in active as well as in precursor forms raised the possibility that zymogen activation may be an important part of the regulation mechanisms in stratum corneum dependent on proteolysis, including desquamation. In other words, enzymes present in the tissue may form a proteolytic cascade in which already activated enzymes serve as activators of precursors of other enzymes, resulting in amplification effects as well as a multitude of targets for regulatory influences (Brattsand et al., 2005). In the same study, the authors advanced a speculative hypothesis on how hK5 and hK7 may be involved in the desquamation. It was suggested that precursors of both the enzymes could be secreted to the extracellular space at the transition between the granular and confined layers, where at neutral pH of the deepest layers of the stratum corneum, hK5 undergoes autoactivation or activation by other proteases such as hK14. The same authors argued that active hK5 starts activating precursor hK7 at a rate at which, owing to the neutral pH, the reaction is maintained throughout the stratum corneum. Owing to the slow rate of hK7 activation, a significant amount of precursor hK7 will remain in layers close to the skin surface (Ekholm et al., 2000). As a result of this process, there will be a concentration gradient of active hK7 between deep and superficial layers of this region, and hence also an increase in the rate of degradation of the intercellular cohesive structures as the corneocytes move toward the skin surface. Expression and localization of KLK mRNA in normal human skin has been examined by RT-PCR and in situ hybridization (Komatsu et al., 2003). The authors reported abundant expression of mRNA for *KLK1* and *KLK11*; moderate expression of *KLK4-7* and *13*; and low expression of *KL8*. Furthermore, for *KLK4*, *8*, and *13*, mRNA splice variants were identified to be their major mRNA species. Interestingly, the expression of the two proteins, hK5 and hK7, and the two genes that encode them, *KLK5* and *KLK7*, has been examined in epithelial ovarian cancer (Dong et al., 2003). Semiquantitative RT-PCR, various forms of western blot, and immunohistochemistry were used to analyze primary cultured cells together with normal, benign, and malignant ovarian tissues. On the basis of these analyses, the authors reported concordant higher expression of both *KLK5* and *KLK7* and their respective proteins in ovarian carcinomas compared with normal ovaries and benign adenomas. In addition to the high expression of *KLK5*/hK5 and *KLK7*/hK7 in ovarian carcinomas, the authors also reported two novel *KLK5* splice variants with an additional upstream exon found in the ovarian cancer cell line (OVCAR-3), whereas only one transcript was found in the normal ovarian epithelial (NOE) and in ovarian cancer cell line (PEO1). Two *KLK7* mRNA transcripts from NOE and PEO1 were also identified. These transcripts have a similar additional upstream 188-bp exon, whereas the short form which was predominantly expressed by NOE cells has no exon 2 and has only 177 bp in the 3′-UTR. The presence of different mRNA transcripts of *KLK5* and *KLK7* in NOE cells compared with ovarian cancer cells was interpreted as an indication that variant *KLK* transcripts may have a role to play in ovarian tumorigenesis with the possibility of their future use as biomarkers for this form of cancer (Dong et al., 2003).

The three kallikreins, hK1, 2, and 3 were the first to be discovered, and till date they are the only members of the kallikrein family that can be commercially purchased. Several biological roles have been assigned to these proteins. The primary activity of hK1 involves the cleavage of low molecular weight kininogen to release lysyl-bradykinin (kallidin), which in turn binds to its receptors, bradykinin B1 and B2, in target tissues and mediates varied processes such as regulation of blood pressure, smooth muscle contraction, neutrophil chemotaxis and pain induction, vascular permeability, vascular cell growth, electrolyte balance, and inflammatory cascades (Clements, 1997; Bhoola et al., 2001). A role for the hK1-kinin system in the establishment and maintenance of placental blood flow through vasodilation, platelet antiaggregation, cell proliferation, and trophoblast invasion during different stages of pregnancy has also been suggested (Valdes et al., 2001a,b). It has been shown that both hK2 and hK3 have relatively lower kininogenase activity than Hk1 (Deperthes et al., 1997). Although both hK1 and hK2 have trypsin-like enzymatic activities, hK3 has chymotrypsin-like substrate specificity. A number of independent studies have reported the activation of the pro-form PSA(hK3) through a process similar to the autoactivation of hK2 (Lovgren et al., 1997; Kumar et al., 1997; Takayama et al., 1997). However, this deduction was disputed by a latter study that showed that hK2 was unable to cleave the fluorogenic pro-hK3 peptide substrate "APLILSR-AMC" (Denmeade et al., 2001).

Other kallekreins have been assigned potential roles in the central nervous system. In a recent review (Yousef et al., 2003b), the authors explored the putative functions of hK6 and hK8 through the extrapolation of their actions in rodent orthologs. Such extrapolation was based on high amino acid sequence similarity of hK6 and hK8 in humans and their rodent orthologs (~70%). The same authors argued that such similarity makes it conceivable that the proteins exhibit similar activities. For example, the rat ortholog of hK6, called myelencephalon-specific protease, may play a role in the regulation of central nervous system demyelinating disease (Scarisbrick et al., 2002), including the development of multiple sclerosis lesions (Scarisbrick et al., 2000), whereas the mouse ortholog may function in myelination and myelin turnover (Yamanaka et al., 1999). Human kallikrein 6 has been implicated in the development of Alzheimer's disease partly because of its ability to cleave amyloid precursor protein, in vitro, and possibly generate β-amyloid peptides (Little et al., 1997; Magklara et al., 2003), which are known to aggregate and form one of the major pathological lesions characteristic of this disease. Several reports indicate that mouse hK8/neuropsin might be involved in synaptogenesis, neural development (Suzuki et al., 1995), and seizures in kindled brain (Okabe et al., 1996).

4.2.5. Kallikreins as Potential Cancer Biomarkers

Up to the mid-1990s, there were only three known kallikrein genes (*KLK1-3*) colocalized to the 19q13.4 chromosomal region. For the last 20 years, one of the proteins, hK3 (PSA), encoded by *KLK3* has been in use as a biomarker for the detection and monitoring of prostate cancer (see Chapter 3). The second half of 1990s saw the expansion of the kallikrein family to 15 genes and the complete description of the human kallikrein locus. With the full identification and characterization of the 15 genes, a number of reports have implicated a number of these genes in different forms of cancer (see Table 4.2). One of the examples, which stands out in the same Table, is the concurrent upregulation of twelve *KLK* genes in ovarian cancer. The mechanisms responsible for the differential expression of kallikrein genes in cancer are still to be elucidated. Transcriptional alterations due to genetic polymorphisms, epigenetic modifications or alterations in transacting transcription factors have been indicated as the most likely determinants of aberrant *KLK* expression in tumors (Borgoño and Diamandis, 2004). This hypothesis is supported by a number of recent reports. For example, the downregulation of *KLK10* in breast, prostate, ovarian cancers, and in acute lymphoblastic leukemia have been attributed to the hypermethylation of coding exon 3 (Li et al., 2001; Roman-Gomez et al., 2004). Epigenetic events including DNA methylation are among the causes responsible for gene silencing in carcinogenesis (Jones and Baylin, 2002; Strathdee and Brown, 2002). Furthermore, steroid hormones are generally implicated in the etiology of hormone-related malignancies such as ovarian, breast, prostate, and testicular cancers (Henderson and Feigelson, 2000). Given the accumulating evidence that kallikrein gene expression is regulated by steroid hormones, it is reasonable to assume that kallikreins are part of steroid hormone-driven cascade pathway that is activated during the promotion and progression of cancer (Borgoño et al., 2004). A range of genetic polymorphisms, primarily single nucleotide polymorphisms (SNPs), exists within *KLK* coding and promoter enhancer regions. These polymorphisms affect transcriptional regulation and/or confer increased susceptibility

TABLE 4.2. Some representative studies to assess *KLKs* and hKs in various forms of cancer using different samples, including tissues, cell lines, and biofluids.

Type of cancer/ sample	Technique	Summary/reference
Ovarian, Breast/ Tissues, and biologic fluids.	Immunofluorometric ELISA, Immunohistochemistry.	hK14 was quantified in human tissue extracts and biologic fluids. The authors reported that hK14 levels were elevated in a proportion of patients with ovarian(65%) and breast (40%) cancers. Immunohistochemical analyses indicated strong cytoplasmic staining of hK14 by the epithelial cells of normal and malignant skin, ovary, breast, and testis. Borgoño et al. (2003) Cancer Res. **63**, 9032.
Ovarian/tissues and cell lines.	SAGE/Expressed sequence tag databases of the Cancer Genome Anatomy project.	Bioinformatic tools were used to examine *KLK* gene expression in normal and cancerous ovarian tissues and cell lines. The authors reported up-regulation of seven genes (*KLK5-8, 10, 11, 14*). The overexpression of six of these genes was experimentally verified at the protein level in cancer tissues. Yousef et al. (2003c) Cancer Res. **63**, 2223.
Ovarian/benign, malignant ovarian tissues, primary cultured cells, ovarian cancer cell lines.	Semiquantitative RT-PCR, Immunohistochemistry, Southern, Northern, and Western blot analyses.	Higher expression both of *KLK5*/hK5 and *KLK7*/hK7 in ovarian carcinoms, especially late-stage serious carcinomas, compared with normal ovaries and benign adenomas. The same study found that one novel *KLK5* transcript with a short 5′-untranslated region and a novel *KLK7* transcript with long 3′-untranslated region were highly expressed in some cancer cell lines but were expressed at very low levels in normal epithelial cells. Western blot and immunohistochemical analyses showed that both enzymes are secreted from ovarian carcinoma cells. Dong et al. (2003) Clin. Cancer Res. **9**, 1710.

(*continued*)

TABLE 4.2. (***Continued***)

Type of cancer/ sample	Technique	Summary/reference
Prostate/human prostate cell lines.	RT-PCR, Immunofluorescence, Western blot, Tissue microarray.	The authors provided a detailed mapping of the KLK4 mRNA 5′ end. This conclusion was supported by predominantly nuclear localization of the hK4 protein in the cell. It was also shown that in addition to androgens, hK4 expression is also regulated by estradiol and progesterone in prostate cancer cells. *KLK4* was found predominantly expressed in the basal cells of the normal prostate gland and overexpressed in prostate cancer. Xi et al. (2004) Cancer Res. **64**, 2365.
Breast, Ovarian/ Tissue extracts, biologic fluids (normal, malignant)	ELISA/Time-resolved fluorometric detection.	This study describes the development of the first immunofluorometric assay for hK5 and the use of this assay to investigate hK5 expression in various tissues and biologic fluids. The authors reported higher expression of this protein in a proportion of patients with ovarian (69%) and breast (49%) cancers. High levels were also detected in ascites fluid from metastatic ovarian cancer patients. Yousef et al. (2003d) Cancer Res. **63**, 3958.
Breast/cell lines	Methylation-specific PCR, sequence analysis of sodium bisulfite-treated DNA.	The authors reported strong correlation between exon 3 hypermethylation and loss of *KLK10* mRNA expression in a panel of breast cancer cell lines and in primary tumors. The same study demonstrated that treatment of *KLK10*-nonexpressing cells with a demethylated reagent led to reexpression of this gene. Li et al. (2001) Cancer Res. **61**, 8014.

or resistance to cancer. The PSA gene *KLK3* contains a 6-kb promoter in the 5′ region that contributes to tissue and hormone specificity of PSA expression. This promoter contains androgen responsive elements (ARE) that regulate promoter activity by binding to ARs. AREI and AREII are located in the proximal region of the PSA promoter and are centered at −170 base pairs (bp) and −394 bp, respectively, with respect to the transcription start site (Cleutjens et al., 1996; 1997). Specific genetic polymorphism in AREI was identified by Rao and Cramer (1999) and subsequently was found to be associated with PSA level (Xue etal., 2001). These authors have examined the association between PSA levels in healthy White men (in a multiethnic cohort) and polymorphisms in the *PSA* and *AR* genes. The authors concluded that a specific genetic polymorphism in *AR* gene was associated with serum PSA levels. Furthermore, this SNP was found to be G to A change at position −158 bp with respect to the start of transcription; the two alleles were found at approximately equal frequencies among the tested whites. This polymorphism has been associated with an increased risk for the development of prostate cancer (Xue et al., 2000; Medieros et al., 2002). In other words, the −158 G/A polymorphism directly contributes to differences in PSA gene promoter activity. Such deductions however, were contradicted in other works, where no relationship between this polymorphism and the PSA level was established (Xu et al., 2002; Rao et al., 2003). This disagreement was further supported through in vitro assessment of the activity of PSA gene promoter constructs that differed only by the −158 G/A polymorphism (Rao et al., 2003). The disagreement between these studies regarding the association between −158 G/A polymorphism and PSA level has been tentatively attributed to the following: first, linkage disequilibrium (the dependence of an allele at one locus on alleles at another locus) of the −158 G/A polymorphism with other polymorphisms in the PSA gene and its promoter (Cramer et al., 2003). Second, Additional confounders such as ethnicity, diet, and lifestyle might also contribute to these inconsistent results (Wang et al., 2003).

There is increasing evidence that many members of the human kallikrein gene family are differentially regulated in ovarian cancer and have a potential as diagnostic and/or prognostic markers. Yousef et al. (2003c) have used a serial analysis of gene expression and expressed sequence tag databases of the Cancer Genome Anatomy Project (CGAP) to perform in silico analyses of the expression pattern of the 15 human *KLK* genes in normal and cancerous ovarian tissues and cell lines. Probing two normal and 10 ovarian cancer serial analyses of gene expression libraries with gene-specific tags for each *KLK* revealed that seven genes, *KLK5-8*, *10*, *11*, and *KLK14* were upregulated in ovarian cancer. The overexpression of six KLK proteins in cancer versus normal or benign tissues was experimentally verified using immunofluorometric assays. It is worth noting here that in silico analyses of kallikrein mRNA expression levels in normal and in cancerous breast tissues and cell lines have also indicated the downregulation of *KLK5*, *6*, *8*, and *10* in breast cancer. Other studies have also shown that *KLK5* and *14* mRNA levels were reduced in breast cancer, whereas elevated serum levels of hK5 and hK14 proteins were observed in a subgroup of breast cancer patients (Borgoño et al., 2003; Yousef et al., 2003d).

To underline the potential of kallekreins as a future source for cancer biomarkers, I have listed in Table 4.2. a number of recent works describing the expression of these proteins and their encoding genes in various types of cancer. The list is by no means exhaustive, but it does give us some useful indications ON current and future challenges, which have to be resolved before some of these proteins/genes can make it as biomarkers for at least some type of cancers.

4.2.6. Concluding Remarks

Currently we know that the hK family comprises 15 serine protease genes on the 19q13.4. Numerous reports have shown an association between dysregulated kallekrein expression and various types of cancer and their potential use as diagnostic/prognostic biomarkers for cancer. These reports also indicate that before translating such a potential into clinical use, a number of challenges have to be confronted. The following considerations may highlight some of these challenges:

- hK3 is a well-recognized member of this family, which is accepted as a biomarker for the diagnosis, monitoring, and population screening for prostate cancer. This biomarker has been in use for more than 25 years, yet there are still strong points of contention regarding its specificity, particularly in the advanced stages of the disease. It has to be pointed out that the very restricted tissue expression of hK3 in the prostate was one of the reasons behind its acceptance as a biomarker. The lessons drawn from a long experience with hK3 should not be ignored in the evaluation of the potential of other more recently identified members of kallikreins. For example, a number of studies have associated the overexpression of multiple kallikreins and their encoding genes with various types of cancer. The most impressive example is the parallel overexpression of seven kallikrein genes in ovarian cancer (Yousef et al., 2003c). This multiple overexpression may indicate some form of *KLK* involvement in ovarian cancer. Such possible involvement can be supported by a number of observations: First, being serine proteases, such overexpressed genes could be implicated in tumor progression through extracellular matrix degradation. Other studies have shown that serine protease expression can be associated with unfavorable clinical prognosis of various forms of cancer (Duffy, 2002). Second, being under steroid hormone regulation (Yousef and Diamandis, 2001), *KLKs* may represent downstream targets through which hormones affect the initiation or progression of ovarian cancer. These observations and the idea that such parallel expression may indicate an involvement in a cascade enzymatic pathway still lack direct evidence.
- On the basis of tissue expression patterns and putative substrates, tissue kallikreins seem to be implicated in diverse physiologic processes, ranging from the regulation of cell growth to tissue remodeling, where they may act individually or in cascade pathways. However, before this group of genes and associated proteins can deliver on their promise as a potential source for cancer biomarkers, a number of questions have to be addressed. These questions have been identified by Diamandis and collaborators who made a substantial contribution to the area of kallekreins and their possible roles in various forms of cancer (Borgoño et al., 2004). These authors pointed out that future activities must aim at the identification of physiologic substrates and delineation of the functional intersections between kallekreins and other proteolic systems, including those involved in cell-signaling. The completion of the kallekrien transcriptome, including variant mRNA transcripts and proteins, is another aspect that needs to be encompassed in future investigations.

4.3. PROTEIN FAMILY 14-3-3

The term 14-3-3 denotes a large family of ~25–30 kDa acidic proteins that exist primarily as homo- and heterodimers within all eukaryotic cells (Aitken, 1996; Fu et al., 2000; Tzivion, et al., 2001). 14-3-3 was initially described as an abundant, acidic brain protein by Moore and Perez (1967) and named after its fraction number on two-dimensional DEAE (diethylaminoethyl)-cellulose chromatography and migration position in starch-gel electrophoresis. Molecular cloning and biochemical characterization of 14-3-3 proteins have subsequently revealed seven homologous isoforms in mammalian cells, which were designated with the Greek letters β, γ, ε, η, σ, τ (sometimes referred to as θ), and ζ (Fu et al., 2000; Ichimura et al., 1988). Most of these isoforms are expressed in all human tissues although the σ form expression is restricted to epithelial cells (Leffers et al., 1993). These homologous isoforms are not the result of alternative splicing but rather the products of different genes from separate chromosomes, implicating the importance of these isoforms in cellular physiology. The crystal structures of 14-3-3 proteins (Xiao et al., 1995; Liu et al., 1995) showed that they exist as dimers, with residues 5-21 in one monomer forming contacts with residues Ser58-Glu89 in the opposing monomer. The same studies showed that these proteins are highly helical and form a large negatively charged channel, the interior of which contains amino acids that are almost invariant throughout the family. This channel would recognize common features of target proteins, so the specificity of interaction of 14-3-3 isoforms with diverse target proteins may involve the outer surface of the protein. The role of these proteins had been unclear till the mid-1990s when they gained acceptance as a novel type of chaperone proteins that modulate interactions between components of signal–transduction pathways (Aitken, 1996). Over the last 10 years, numerous cellular proteins have been shown to interact with 14-3-3 isoforms, and many of these interactions result in altered function of the proteins interacting with the various isoforms, which may account for the diverse roles of this family of proteins in cellular regulation. The interacting proteins include protein kinases such as Raf-1, apoptosis signal-regulating kinasel (ASK1), Mitogen-activated protein kinase kinase 1(MEKK1), kinase suppressor of Ras (KSR), and phosphatidylinositol 3-kinase (Fu et al., 2000; Subramanian et al., 2001) and apoptosis regulators such as the Bcl-2 family member BAD (Zha et al., 1999) and forkhead transcription factor (FKHLR1) (Brunet et al., 1999). Other more recent studies have also shown that several proteins including 14-3–3 family prevent apoptosis through the sequestration of *Bax* (Samuel et al., 2001; Nomura et al., 2003). Apoptosis is essential for normal development and maintenance of tissue homeostasis. Furthermore, insufficient induction of apoptosis is now recognized as a hallmark of cancer (Johnstone et al., 2002). It has been suggested that one of the functions of 14-3-3 is to support cell survival (Masters et al., 2002). These authors proposed that such survival is partially promoted by antagonizing the activity of associated proapoptotic proteins, including BAD and apoptosis signal-regulating kinase 1 (ASK1). This hypothesis was supported by performing a number of measurements to assess the effect of 14-3-3 antagonists in lung, prostate, and cervix cancer cell lines. Before considering various analyses that identified some members of the 14-3-3 family as potential cancer biomarkers, it is helpful to consider some of the biological functions attributed to these proteins.

4.3.1. Functions Attributed to the 14-3-3 Proteins

Processes that are relevant to cancer biology and that are regulated by 14-3-3 protein interactions include cell-cycle progression, apoptosis, and mitogenic signaling. The correct execution of the cell cycle is essential for the maintenance of genomic integrity, and therefore for tumor suppression. Checkpoint mechanisms guarantee that the next cell-cycle phase is only entered after error-free completion of the previous stage; for example, cell-cycle progression is inhibited after DNA damage to allow DNA repair or to ensure a permanent arrest of cells that have suffered severe damage. There is substantial evidence that various members of the 14-3-3 family are involved in this regulation of the cell-cycle machinery at several key points. The relevance of 14-3-3 to cancer biology is supported by their mediation in negative cell-cycle regulation, which is integrated into tumor-suppressive pathways, such as the ATM-*p53* pathway (Shiloh, 2003; Vogelstein et al., 2000). Existing literature attributes a number of specific roles to 14-3-3 proteins, including regulation of the following: enzyme activities, subcellular localization, and protein–protein interactions. The same literature underlines the direct correlation between the binding capability of 14-3-3 proteins and the roles attributed to these proteins. Some of these roles and the basis for their binding to other proteins are discussed below.

4.3.2. Binding of 14-3-3 Proteins to Different Partners

The biological action of 14-3-3 proteins is thought to depend on their ability to bind to a variety of proteins. Currently there are at least 100 cellular proteins, which can bind to the 14-3-3 proteins (van Hemert et al., 2001; Aitken et al., 2002). In many, but not all cases, 14-3-3 proteins bind to the phosphorylated form of these proteins. This binding capability made 14-3-3 proteins important players in various cellular processes like signal transduction, cell-cycle regulation, apoptosis, stress response, cytoskeleton organization, and malignant transformation. In the mid-1990s, Muslin et al. (1996) have shown that 14-3-3 isoforms interact with a novel conserved phosphorylated motif that has been found in a number of signal transduction proteins: RS*X*p**S***X*P, where p**S** represents phosphoserine and *X* represents any amino acid. Substitution of amino acids surrounding the central phosphoserine (position 0) demonstrated that Arg at position −4 or −3, Ser at position −2, and Pro at position +2 were critical for high-affinity association. Within this motif, phosphorylation of the serine at position −2 alone does not support 14-3-3 binding, and when the Ser at position −2 is phosphorylated in addition to the Ser at position 0, phosphopeptide binding is completely abrogated. Works by Yaffe et al. (1997) and Rittinger et al. (1999) using oriented phosphopeptide libraries provided evidence for two distinct 14-3-3 binding motifs: RS*X*p**S***X*P (mode 1) and R*XXX*p**S***X*P (mode 2). The binding of 14-3-3 to the mode 1 motif is favored by aromatic or positively charged amino acid at position −1, whereas 14-3-3 binding to the mode 2 motif exhibits a preference for aromatic residues at position −2, positive residues at position −1, and Leu, Glu, Ala, or Met at position +1. These motifs define the requirements for an efficient binding of a single phosphoserine motif to a 14-3-3 partner, and many of such partners identified so far do contain either a mode 1 or mode 2 motif. Nevertheless, several well-characterized proteins that interact with 14-3-3 in a phosphorylation-dependent manner do not contain either of these motifs, including insulin-like growth factor 1 (IGF-1), insulin receptor substrate-1 (IRS-1), and the tyrosine kinase Weel (Tzivion et al., 2002). Some 14-3-3 partners have been shown to bind monomeric and dimeric forms of 14-3-3 with similar efficiency (Luo et al.,1995; Tzivion et al., 1998; Ichimura et al., 1997); however, the overall target protein binding profiles of monomers and dimers were found to differ substantially (Cahill et al., 2001). For instance, c-Raf-1 that contains a high-affinity binding site, which matches both modes of binding, can bind monomeric 14-3-3 in a stable manner, whereas proteins that contain only low-affinity binding sites tend to bind monomeric 14-3-3 weakly, whereas a more stable reaction is achieved with the dimers of 14-3-3. More stable binding with the dimers is also encountered with Weel, keratin, K18, and insulin-like growth factor 1(IGF-1). These proteins, which lack a high-affinity binding motif, seem to require the dimer that binds at two sites resulting in a more stable association.

The binding of 14-3-3 proteins to various partners can have a substantial impact on the regulation of the function(s) of such partners. Such impact can take different forms, including the alteration of the ability of the partner protein to interact with other partners. A representative example of such effect is the binding of 14-3-3 to BAD, which is discussed in the following section. Another effect that can be provoked by the binding to 14-3-3 is the modification of the cytoplasmic/nuclear partition of the protein partner by increasing nuclear export rate, decreasing nuclear import, or both (Muslin and Xing, 2000). Among the Proteins that can be subjected to such regulation are the cyclin-dependent kinase (Cdc25), insulin-regulated forkhead domain transcription factors (FKHRL1), and histone deacetylase. There is still an evolving debate regarding the binding specificity of the various isoforms of 14-3-3 proteins. Given that the residues lining the phosphopeptide-binding groove of the various isoforms are highly conserved, a considerable overlap of specificity would be anticipated. Nevertheless, differences in the abilities of 14-3-3 isoforms to bind synthetic peptides and proteins are well documented (Yaffe et al., 1997; Vincenz and Dixit, 1996; Van Der Hoeven et al., 2000; Tzivion et al. (2000)). Furthermore, there are a number of examples illustrating some form of isoform-specific biologic responses. A known example is the overexpression of the σ isoform in colorectal carcinoma, which has been implicated in eliciting G2 arrest in colorectal carcinoma, a response that is not observed on comparable overexpression of the β isoform (Hermeking et al., 1997). Whether such difference between the two isoforms is due to difference in their binding specificity or due to different subcellular localization, or simply due to unknown property of one of the isoforms is still unclear. However, the well-documented selective increase in the expression of the σ isoform in response to DNA damage suggests a physiologic significance to such difference in expression between the two isoforms (Hermeking et al., 1997; Suzuki et al., 2000).

4.3.3. The Role of 14-3-3 Proteins in Apoptosis

Apoptosis is a physiological process of cell death that plays a critical role in normal development as well as in the pathophysiology of a variety of diseases (Jacobson et al., 1997; Ruden and Thompson 1997). The regulation of programmed cell death, or apoptosis, is essential for the successful development of multiple tissues and for the maintenance of normal tissue homeostasis. 14-3-3 proteins promote the cytoplasmic localization of many binding partners, including the proapoptotic protein BAD and the cell-cycle regulatory phosphatase Cdc25C.The following considerations underline the role of 14-3-3 proteins in the regulation of signal transduction, apoptotic and checkpoint control. First, the first indication that 14-3-3 is involved in regulation of cell death came when it was found to be associated with BAD in response to the survival factor interleukin-3 (Zha et al., 1999), resulting in the inhibition of BAD proapoptotic activity. Greenberg and colleagues later discovered that 14-3-3-BAD association is regulated by Akt/PKB through phosphorylation of BAD, providing a direct link between a survival signaling kinase and a death promoter (Datta et al., 1997). BAD contains three known phosphorylation sites, Ser-112, Ser-136 and Ser-155, and a host of kinases, including PKA, p21- activated kinase (PAK) and RSK, have been shown to phosphorylate one or more of them (Harada et al., 1999; Schumann et al., 2000; Shimamura et al., 2000). Although all the three phosphorylation sites have been reported to be important for 14-3-3 binding, Masters et al. (2002) argued that 14-3-3-BAD interaction is strictly dependent on phosphorylated Ser-136. Second, the Bcl-2 homology 3 (BH3) domain of prodeath Bcl-2 family members mediates their interaction with prosurvival Bcl-2 family members and promotes apoptosis. Proteins of the Bcl-2 family are components of a survival factor-regulated checkpoint in the cellular death machinery (Reed, 1998). Bcl-2 family members promote either survival (e.g., BCL-2, BCL-XL, and the *C. elegans* Bcl-2 homolog CED-9) or apoptosis (e.g., BAK, BAD, and the *C. elegans* BCL-2 homolog EGL-1). Although the precise mechanism by which the Bcl-2 proteins function remains unclear, their ability to homo- and heterodimerize has given rise to the hypothesis that the balance between prosurvival and prodeath Bcl-2 family members determines whether a cell lives or dies (Oltvai and Korsmeyer, 1994). When cells are challenged with apoptotic stimuli, death-promoting Bcl-2 family members become activated, and the equilibrium within the Bcl-2 family protein network shifts toward apoptosis. Apoptosis then occurs through a cascade of events that culminates in the release of cytochrome c from the mitochondria and subsequent death protease activation (Reed, 1998). Datta et al. (2000) described a mechanism by which survival signals inactivate the proapoptotic Bcl-2 family member BAD. This model for the inactivation of BAD is schematically represented in Figure 4.2. This model describes how 14-3-3 proteins cooperate with survival kinases to inactivate BAD by BH3 domain phosphorylation. Basically, it shows that the survival signals block apoptosis by phosphorylating the proapoptotic Bcl-2 family member BAD at Ser-155, a site within its BH3 domain. In the absence of survival stimuli, BAD is dephosphorylated and tightly bound to prosurvival Bcl-2 family members such as BCL-X_L. Upon exposure to survival factors, Ser-136 phosphorylation is induced, leading to the recruitment

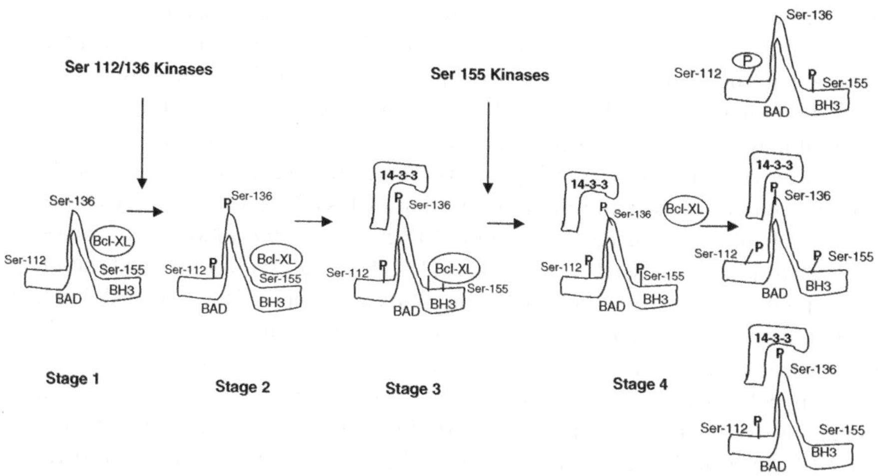

Figure 4.2. A model for the inactivation of BAD. This model suggests that survival signals activate kinases that induce phosphorylation(indicated by P) of BAD at Ser-112 and Ser-136 (stages 1 and 2). This is followed by the binding of 14-3-3 proteins to the BAD/Bcl-XL complexes in a Ser-136 dependent manner (stage 3). At this stage, Ser-112 phosphorylation is assumed to enhance the binding of 14-3-3 to BAD. This binding in turns weakens the effective interaction between BAD and BCL-XL and increase the access of Ser-155 kinases to BAD ser-155. Phosphorylation of Ser-155 permenantly disrubts the BCL-XL/BAD complex and inactivates BAD. Adapted from Datta et al. (2000) with permission.

of 14-3-3 proteins to BAD. Although the binding between these proteins and BAD/ Bcl-X_L complex was described to be weak, it is sufficient to increase the accessibility of Ser155 to survival kinases. This phosphorylation was found to disrupt the interaction between the hydrophobic face of the BAD BH3 domain and the hydrophobic groove in Bcl-X_L, resulting in BAD dissociation and its subsequent translocation from the outer mitochondrial membrane to the cytoplasm, where it forms a stable complex with 14-3-3 proteins. The above model and other works (Reed, 1998) along these lines indicate that the net outcome of 14-3-3 binding to BAD is an inhibition of apoptosis and the promotion of cell survival.

Other examples on the role of 14-3-3 proteins in apoptosis include their participation in the inhibition of a forkhead transcription factor (FKHRL1), suppression of apoptosis signal-regulating kinase 1 (SRK1), and the association in isoform-specific manner with the zinc finger protein, A20.

There is a substantial evidence that growth factors promote cell survival at least in part by activating the phosphatidylinositol 3-kinase (PI3K) and its downstream target, the serine/threonine kinase (AKT). One function of AKT is to phosphorylate and inhibit proapototic components of the intrinsic cell death machinery present within the cytoplasm (Brunet et al., 1999; Datta et al., 1998; Segal et al., 1997). Brunet et al. (1999) demonstrated that the PI3K/AKT pathway regulates transcription by phosphorylating FKHLR1, a member of the forkhead transcription factor family.

This leads to its interaction with 14-3-3 proteins and subsequent sequestration in the cytoplasm, away from its transcriptional targets. Under conditions of growth factor deprivation, the PI3K/AKT pathway is inactivated, FKHLR1 is dephosphorylated at its AKT sites with the direct result of FKLHR1 accumulation within the nucleus where it may activate death genes, including the *Fas ligand* gene, and thereby participate actively in the process of apoptosis. There are several mechanisms that have been put forward to explain the translocation of FKHLR1 to the nucleus. One of these mechanisms has been proposed by Brunet et al. (1999) following the analysis of other forkhead transcription factors (FH). This analysis has revealed the presence of a domain enriched in basic amino acids at the end of the FH DNA-binding domain that may form part of a nuclear localization signal (NLS). As in FKHLR1 this basic domain overlaps with the second site of AKT phosphorylation (Ser253), it was hypothesized that phosphorylation of this site could add a negative charge to the positively charged basic region, resulting in the disruption of the function of the NLS. In growth factor-deprived cells, when AKT is inactive and FKHLR1 is dephosphorylated, the NLS may function effectively and thus promote the translocation of FKHLR1 to the nucleus. An alternative to such hypothesis is that the phosphorylation of FKHRL1 could influence its interaction with other proteins such as 14-3-3, resulting in the sequestration of phosphorylated FKHLR1 in the cytoplasm or in localizing the dephosphorylated species within the nucleus.

Another protein that associates with some isoforms of 14-3-3 and can influence apoptosis is A20. This is a zinc finger protein that inhibits the tumor nercosis factor-α (TNF-α). This is catabolic proinflammatory cytokine that is capable of inducing apoptotic death in a number of tumor cell lines (Dixit et al., 1990). A20 associates with several isoforms of 14-3-3 proteins, which function as an adapter to mediate the interaction between A20 and the c-Raf kinase. Furthermore, the association with 14-3-3 proteins was found to enhance the solubilization of A20 within the cytoplasmic compartment.

Apototis signal-regulating kinasel (ASK1) is considered a central component of a signaling pathway induced by many death stimuli, including tumor necrosis factor-α, Fas, and a number of anticancer drugs, including cisplatin and paclitaxel (van Hemert et al., 2001). It is argued that the regulation of ASK1 by both pro- and antiapoptotic signals may provide a critical point of control for cell death and cell survival (Zhang et al., 1999). The same authors reported that the overexpression of 14-3-3 proteins in HeLa cells blocked ASK1-induced apoptosis whereas disruption of the ASK1/14-3-3 interaction dramatically accelerated ASK1-induced cell death.

4.3.4. The Role of 14-3-3 Proteins in Cell-cycle Regulation

There are a number of examples that show that the association of 14-3-3 with a protein ligand contributes to inhibition of cell-cycle progression. In other words, the 14-3-3 proteins seem to act as natural breaks for cell-cycle progression. This statement can be partially supported by the following considerations: First, progress through the cell cycle can be regulated at the G0/G1, G1/S, and G2/M transitions. At each of these stages, the activity of a cyclin-dependent kinase (CDK) is required to proceed through the transition (Nigg, 1995). CDK activity is, in turn, dependent upon the presence of the appropriate cyclin and activating phosphorylation, and the absence of CDK inhibitors as well as inhibitory phosphorylation. 14-3-3 proteins have been shown to have a crucial role during undisturbed cell divisions. The cyclin-dependent kinase (Cdc25C) and Cdc25B proteins are dual-specificity phosphotases that remove inhibitory phosphate groups from Thr14 and Thr15, and thereby activate the cyclin-dependent kinase Cdc2, the main protein involved in driving cells through mitosis (Smits et al., 2001). The mechanism of Cdc25C activation during the G2/M transition is still under investigation; however, Bulavin et al. (2003) have provided strong evidence that Cdc2, once activated by Cdc25C at the G2/M boundary, phosphorylates Cdc25 at Ser214. This in turn prevents Ser216 phosphorylation, and thereby the inactivation of Cdc25C by 14-3-3-mediated inhibition during mitosis. Other proteins that impact the activity of Cdc2 are the Wee1 tyrosine kinase and the checkpoint kinase 1 (Chk1). The latter kinase can phosphorylate Cdc25c on Ser216, thereby creating a binding site for 14-3-3 proteins (Davenzac et al., 2000; Jiang et al., 2003). It has been shown that Rad 24 and Rad 25 (*Schizosaccharomyces pombe* 14-3-3 proteins) associate preferentially with the phosphorylated form of Chk1, which is generated in response to DNA damage. The connection between 14-3-3 proteins and DNA damage checkpoints was first discovered in a study by Ford et al. (1994), who showed that two 14-3-3 genes, *rad24* and *rad 25*, are required for G2/M checkpoint in the fission yeast *Schizosaccharomyces pombe*. Furthermore, *rad24*-null mutants, and to a lesser extent *rad25*-null mutants, enter mitosis prematurely, which supports the hypothesis that 14-3-3 proteins have a role in determining the timing of mitosis during undisturbed proliferation. Besides regulating kinases and phosphatases following activation of DNA-damage checkpoints, 14-3-3 proteins also regulate the activity of transcription factors that induce negative regulators of the cell-cycle machinery.

4.3.5. The Potential of Some 14-3-3 Proteins as Cancer Biomarkers

Of all the *14-3-3* genes, *14-3-3* σ has been most directly linked to cancer. It is thought to function as a tumor suppressor by inhibiting cell-cycle progression and by causing cells to leave the stem-cell compartment and undergo differentiation (Hermeking, 2003). Inactivation of this isoform occurs at various levels, and the high frequency of its inactivation indicates that it has an important role in tumor formation. Before discussing in more details some of the works that support the potential role of 14-3-3σ in some forms of cancer, it is worth pointing out two differences between this isoform and the rest of 14-3-3 isoforms: First, most of these isoforms are expressed in all human tissues although the σ form expression is restricted to epithelial cells. Second, the structural basis for 14-3-3σ functional specificity has recently been investigated by Wilker et al. (2005). These authors reported that the X-ray crystal structure of 14-3-3σ bound to an optimal phosphopeptide ligand. This structure revealed two features that the authors believe are uniquely associated with the σ isoform: (a) 14-3-3σ preferentially forms homodimers, and the unique stabilizing ring–ring and salt bridge interactions reinforce this interaction. On the contrary, heterodimers of this isoform are destabilized by electrostatic interactions. (b) The phosphopeptide ligand interacts with 14-3-3σ in a manner conserved throughout all isoforms, but the structure suggests that there is a second binding site involved in 14-3-3σ-specific ligand discrimination. According to the same authors, this site contains residues that are specific to this isoform.

Despite the many processes regulated by 14-3-3 proteins that could, if perturbed, contribute to cancer development, so far, it is mainly the σ isoform of the human 14-3-3 proteins that has been directly implicated in the etiology of cancer. Over the last 5 years, various techniques have been used to generate data implicating the involvement of this isoform in various forms of cancer and at the same time underlying its potential as a biomarker for some of these forms of cancer. Some representative examples underlying this potential are considered below.

4.3.5.1. Down-regulation of 14-3-3σ in Various Types of Cancer. Over the last 10 years, there have been various lines of evidence to suggest that functional inactivation of 14-3-3σ (also called stratifin) may be linked to carcinogenesis, a hypothesis that is supported by the frequent observation of 14-3-3σ downregulation in various human malignancies, including cancer of the breast, stomach, lung, liver, prostate, and vulva (Iwata et al., 2000; Lodygin et al., 2004; Osada et al., 2002; Suzuki et al., 2000; Vercoutter-Edourt et al., 2001). It is relevant to point out that up- rather than downregulation of this protein has been also reported in other forms of cancer including lung, head and neck, gastric, and pancreatic (Nakanishi et al., 1997; Villaret et al., 2000; Shimomura et al., 2002; Friess et al., 2003; Iacobuzio-Donahue et al., 2003). Both proteomic and genomic-based approaches have generated sufficient data to support the role of 14-3-3σ in various types of cancer. The same data provided strong indications on the potential of this isoform as a biomarker for some malignancies, including breast cancer. Before describing some of these analyses and the type of technology used to generate such data, it would be helpful to underline two elements that are commonly cited as the basis for the link between 14-3-3σ and carcinogenesis. First, *14-3-3σ* has been identified as a p53-inducible gene that is responsive to DNA-damaging agents. It apparently plays a crucial role in the G2 checkpoint by sequestrating Cdc2/cyclinB1 complex in the cytoplasm. This isoform is induced in a *p53*-dependent manner and prevents the Cdc2/cyclinB1 complex from entering the nucleus. This event provides an opportunity for DNA repair of damage before further cell-cycle progression (see also sections 4.2.3, 4.2.4). Second, although inactivation of p53 is one of the mechanisms that lead to decreased 14-3-3σ expression in tumor cells, such an expression can also be downregulated by aberrant CpG methylation.

4.3.5.2. Down-regulation of 14-3-3σ in Breast Cancer. Breast cancer is one of the most common malignancies accounting for ~18% of all cancers in women (Parkin, 2004). At present, routine mammography is the most widely used tool for the early detection of breast cancer. To be detected, however, a tumor should be at least a few millimeters in size, a requirement that can negatively influence the odds of survival and cure. This dilemma is further complicated by the fact that till date few serum tumor markers are available for detection of breast cancer, and the most commonly used markers such as serum levels of CA 15-3, mucinous-like cancer antigen, carcinoembryonic antigen, tissue polypeptide, Neu oncoprotein, and tissue polypeptide-specific antigen are of limited value for detecting this form of cancer in its early stages (Eskelinen et al., 1997). As it has been pointed out that the 14-3-3 proteins have the ability to bind to a variety of functionally diverse signaling proteins, including kinases, phosphotases, and transmembrane receptors. Such binding capability allows them to play important roles in a wide range of regulatory processes, including signal transduction, cell-cycle control, and apoptosis. To appreciate some of the arguments that support the promise of 14-3-3σ as a biomarker for breast cancer, it would be helpful to consider some experimental data and associated interpretations, which have been generated over the last few years.

Ferguson et al. (2000) used serial analysis of gene expression together with northern blot analysis to assess the expression of 14-3-3σ in breast carcinoma cells. These analyses were conducted on four different breast cancer cell lines and two human mammary epithetlial cell lines. The main conclusions of this study were as follows: (a) σ mRNA was undetectable in 45 of 48 primary breast carcinomas. (b) Genetic alterations at σ such as loss of heterozygosity were rare, and no mutations were detected. (c) hypermethylation of CpG islands in the σ gene was detected in 75 of 82 (91%) breast tumors and was found to be associated with lack of gene expression. Before commenting on these deductions, it is useful to compare them with other set of deductions based on protein profiling study by Celis' group (Moreira et al., 2005). In this relatively recent study, the authors have assessed the expression of 14-3-3σ in 65 breast carcinomas analyzed by Western blotting, 2-D PAGE, and immunochemistry using blot specific antibodies against this isoform. These authors reported three main findings: First, in most cases, matched malignant, nonmalignant, and nodal metastatic tissue displayed only minor differences in the expression of this protein. Second, the immunoreactivity of 14-3-3σ was restricted to epithelial cells and was significantly stronger in the myoepithelial cells that line the mammary ducts and lobules. Third, the lack of expression of 14-3-3σ in the three breast carcinomas was not associated with high levels of expression of the dominant negative transcriptional regulator ΔNp63 or with increased expression of estrogen-responsive finger protein, an ubiquitin-protein ligase (E3) that targets 14-3-3σ for proteolysis. A comparison of these findings with those reported by Ferguson et al. (2000) reveals a number of differences, which merit some considerations: First, the disagreement between the two studies regarding the expression levels of 14-3-3σ can be tentatively attributed to a fairly well-established line of evidence that advocates that mRNA quantities based on gene expression do not always correlate to protein quantities (Anderson and Seihamer, 1997; Gygi et al., 1999). Second, the absence of association between lack of expression of 14-3-3σ and the increased expression of estrogen-responsive finger protein is difficult to interpret, particularly when other works dealing with the same effect are considered. In a recent article, Suzuki et al. (2005) have used immunohistochemistry and laser capture microdisection/real time polymerase chain reaction (PCR) to examine the expression of estrogen-responsive finger protein (Efp) in breast carcinoma tissues. These authors reported the detection of 14-3-3σ in 38.4% of the examined carcinomas and that they were inversely associated with Efp immunoreactivity. This result seems to support the downregulation of 14-3-3σ by methylation of the gene and/or proteolysis by Efp in breast carcinoma tissues. This finding is inconsistent with that reported by Moriera et al. (2005). Considering that immunohistochemistry was used by both groups for detection and the results by both groups were published within a period of 5 months, such inconsistency merits further assessment. In fact, Suzuki et al. (2005) suggested further examination, including validation of the immunohistochemical results by other laboratories, to clarify the expression of 14-3-3σ in breast carcinoma tissues. It is also relevant to mention an earlier study by Urano et al. (2002), which showed that 14-3-3σ was a primary target for proteolysis by Efp and the regulation of the first protein was due to posttranslational modifications mediated by Efp.

The potential of 14-3-3σ as a cancer biomarker is not limited to breast cancer. Many studies have reported its up- or downregulation in various forms of cancer, some of these works are summarized in Table 4.3.

4.3.5.3. Perspectives. Over the last 10 years, numerous reports have described the interaction of 14-3-3 proteins with over 100 proteins, both in vivo and in vitro. The combination of new proteomic techniques and available genomic data should allow the identification of additional 14-3-3-binding/regulated proteins. These future investigations are likely to provide a deeper insight into the role of 14-3-3 in cellular regulation, which over the last few years has focused on changes in target protein phosphorylation as the initiating regulatory event, viewing the 14-3-3 proteins as the passive element in the interaction. Other potential modes of regulation, including isoform-specific expression, subcellular localization, and differential target binding specificity are still to be elucidated. We already know that this family of proteins participate in a wide variety of signal transduction processes, including Ras-Raf mediated activation of the mitogen-activated protein kinase pathway, regulation of apoptosis, adhesion-dependent integrin signaling, and cell-cycle control in response to genotoxic stress (Aitken, 1996; Tzivion et al., 2001; van-Hermet et al., 2001; Wilker and Yaffe, 2004; Fu et al., 2000). Most, if not all, of these processes appear to involve multiple 14-3-3 isoforms. Although some isoform-specific differences in signaling clearly exist, the molecular basis behind such differences remains unclear. Given the strong sequence conservation and the broad heterodimerization observed among the individual members of the group, rendering a rationalization of an explanation for the differences in signaling between the isoforms, becomes a very challenging task. However, it is encouraging to note that some results toward such a clarification have already appeared in the literature. A structural basis for 14-3-3σ functional specificity has been reported (Wilker et al., 2005). These researchers have demonstrated that this isoform preferentially forms homodimers and has a second ligand binding site, which has residues that are specific to 14-3-3σ. Mutation of these 14-3-3σ specific residues to the corresponding sequence in other 14-3-3 proteins causes this isoform to bind to Cdc25C, a molecule that normally binds to other isoforms but not to 14-3-3σ. Future structural analysis might also facilitate the design of specific inhibitors or stabilizers of 14-3-3-ligand interactions that have therapeutic potential.

TABLE 4.3. Assessment of 14-3-σ expression in various types of cancer using different techniques.

Cancer type	Technique	Summery/references
Breast carcinoma	Two-dimensional-PAGE/ MALDI-MS	The purpose of the study is to assess the role of 14-3-3σ in breast epithelial cell neoplasia. 2-D PAGE combined with MALDI-MS was used to quantify in two different prototypic breast cancer cell lines and in primary breast carcinomas as compared with normal breast epithelial cells. The authors reported strong down-regulation of this protein in MCF-7, MDA-MB-231 cell lines and in primary breast carcinomas compared with normal breast epithelial cells. Vercoutter-Edouart et al. (2001) Cancer Res. **61**, 76.
	Serial analysis of gene expression (SAGE)/Northern blot.	Using SAGE analysis, the authors reported 7-fold down-regulation of σ gene in breast carcinoma cells compared with normal breast epithelia. This finding was verified using Northern blot. Furthermore, genetic alterations at σ were rare and no mutations were detected. Ferguson et al. (2000) PNAS **97**, 6049.
	Immunohistochemical analysis	This study has analyzed the immunohistochemical distribution of 14-3-3σ in normal breast tissue and in a large series of benign and malignant breast lesions on whole tissue sections and by tissue microarray. The authors reported that this protein was consistently expressed in the citoplasmic compartment and occasionally in the nuclei of myoepithelial cells, whereas luminal epithelial, stromal, endothelial, pericytic, lipomatous, and neural cells showed no staining. On the basis of survival data for 452 patients with invasive breast carcinoma and the distribution of 14-3-3σ, the authors concluded that the sublocalization of this protein was a statistically significant prognostic factor for the whole series of invasive carcinomas, as well as for those positive for estrogen(ER) or progesterone receptors (PR). Simpson et al. (2004) J. Pathology **202**, 274
	Methylation-specific polymerase chain reaction (PCR)	This study investigated the status of *14-3-3*σ methylation during the progression from normal breast epithelium to invasive breast cancer. Laser capture microdissection was used to reduce contribution of contaminating nonepithelial cells. The authors reported that hypermethylation of this gene was a consistent alteration in invasive breast cancer, which is acquired in the late preinvasive phase of tumor progression. The authors pointed out that when hypermethylation occurs, it appears to involve breast epithelium beyond the margins of the primary tumor. Unless a great care is taken to avoid contamination by nonepithelial cells, the potential of such an alteration as a potential tumor marker may be compromised.Umbricht et al. (2001) Oncogene **20**, 3348.

(continued)

TABLE 4.3. ***(Continued)***

Cancer type	Technique	Summery/references
Bladder	Two-dimensional PAGE, Western blot, Immunohistochemistry.	The expression of 14-3-3σ protein in normal, urothelium, and bladder transitional cell carcinomas (TCCs) was examined using 2-D PAGE in combination with Western blotting and immunochemistry. This study showed that the expression of this protein was down-regulated in Invasive TCCs, particularly in lesions that were undergoing epithelial to mesenchymal conversion. This down regulation was confirmed by indirect immunofluorescence using a peptide-based rabbit polyclonal antibody specific to 14-3-3σ. The same study showed that the expression of this protein was up-regulated in pure squamous cell carcinomas. Moreira et al. (2004) Mol. Cell. Proteomics **3.4**, 410.
Lung	One-dimensional PAGE/ ESI-MS-MS; ELISA.	This study was conducted in two phases: In the first phase protein profile in the conditioned medium of lung cancer primary cell or organ cultures and in the adjacent normal bronchus was examined by one-dimensional PAGE in combination with nano ESI tandem mass spectrometry, which allowed the identification of 299 proteins. Thirteen of these proteins (considered interesting) were analyzed in 628 blood samples using ELISA. The authors reported that the plasma levels of 14-3-3σ, β, and η were significantly lower in lung cancer patients than those observed in control subjects. Xiao et al. (2005) Mol. Cell Proteomics **4.10**, 1480.
	Methylation-specific polymerase chain reaction (PCR); Northern and Western blot.	The DNA methylation status of *14-3-3σ* gene in a number of both small and nonsmall lung cancer (SCLC/NSCLC) cell lines was analyzed by methylation specific-PCR, and the correlation between *14-3-3σ* expression and DNA methylation was confirmed by Northern and Western blot analysis. This study concluded that SCLC cell lines showed 69% hypermethylation and subsequent silencing of *14-3-3σ*, whereas NSCLC cancer cell lines showed rare hypermethylation (6%). This was one of the first studies to show that the inactivation of *14-3-3σ*

		gene in lung cancer is strongly related to DNA hypermethylation in SCLC but not in NSCLC. This observation was interpreted as an indication that the involvement of this gene in lung tumorigenesis is histological type-specific. Osada et al. (2002) Oncogene **21**, 2418
Ovarian	Methylation-specific PCR; Reverse transcription PCR; Immunohistochemistry	The objective is to evaluate methylation status and expression of *14-3-3*σ gene. The listed techniques were used to examine ovarian cancer cell lines, ovarian surface epithelial cell lines, normal, benign, border line, and ovarian cancer tissues. The authors concluded that the inactivation of *14-3-3*σ gene in the investigated samples was mainly due to aberrant DNA methylation and that it may play an important role in the pathogenesis of epithelial ovarian cancer. Akahira et al. (2004) Clin. Cancer Res. **10**, 2687.
Prostate	Immunohistochemistry, Methylation-specific PCR; Western blot, Immunohistochemistry	Radical prostatectomy specimens (n=111) containing prostatic adenocarcinoma with adjacent high-grade prostatic intraepithelium neoplasia (PIN) and normal prostate epithelium were examined by immunohistochemical analysis. The authors reported high-level ubiquitous expression of 14-3-3σ in normal prostate epethilium and a significant decrease of the same protein in PIN and prostatic adenocarcinoma. It was also pointed out that the suppression of this protein occurred during the development of NIP from normal epithelium. Cheng et al. (2004) Clin. Cancer Res. **10**, 3064.

Extensive research efforts over the last decade have provided us with valuable information regarding the structure, biochemistry, and various physiological roles of this family of proteins. However, translation of such knowledge into clinical applications, particularly their use as cancer biomarkers, still requires further efforts. The case of 14-3-3σ can be considered a representative example. Various lines of evidence have linked the functional inactivation of *14-3-3σ* gene to various forms of cancer, including breast, colon, lung, liver and other types of cancer, yet there is still an apparent disagreement between some of the experimental data regarding the expression of this gene in breast cancer. The loss of this gene during neoplastc transformation of breast epithelial cells has been reported by different research groups (Umbrecht et al., 2001; Ferguson et al., 2000). These findings have been challenged by a recent study (Moriera et al., 2005), whereas the same findings have been supported by another recent study (Suzuki et al., 2005). Disagreement between experimental data generated by different research groups is not new in research; however, when such a disagreement can impact the destiny of a potential cancer biomarker then further efforts are needed to clarify the source of disagreement. I am more than convinced that further studies will be able to provide more direct evidence on the role of this gene in breast epithelial cell carcinogenesis breast cancer. Such future investigations will undoubtedly involve larger clinical studies with substantially more data sets containing matched malignant and nonmalignant specimens.

4.4. HEAT SHOCK PROTEINS (HSPs)

Following exposure to environmental or physiological insults, the cells in most tissues dramatically increase the production of a small group of proteins that are collectively known as "heat shock or stress proteins" (HSPs). Research over the last 30 years has shown that in the course of their evolution, cells have developed complex and varied mechanisms to respond to the many physiological and environmental insults they encounter. Analysis of these responses has led to the discovery of highly conserved proteins, whose synthesis is transiently induced in response to low levels of stress, in a process referred to as the stress response (Parsell and Lindquist, 1993). HSPs encompass several groups of proteins and may be divided into five major families on the basis of their size, structure, and function (Lindquist and Craig, 1988): the HSP110, HSP90, HSP70, HSP60, and small HSP families. HSPs were originally named because of their rapid induction in response to elevated temperatures (Tissieres et al., 1974); however, since then it has been shown that a wide variety of different physical, chemical, and biological stimuli are also capable of inducing HSPs, including oxidative stress, heavy metals, hypoxia, acidosis, and metabolic poisons (Lindquist and Craig, 1988). Under normal conditions, HSPs play various roles in cell function, including modulating protein activity by changing protein conformation, promoting multiprotein complex assembly/disassembly, regulating protein degradation within the proteasome pathway, facilitating protein translocation across organellar membranes, and ensuring proper folding of nascent polypeptide chains during protein translation. When cells are stressed, a common response is to undergo cell death by one of the two pathways, either necrosis or apoptosis. Necrosis is the result of acute cellular dysfunction, which can be provoked by an extreme trauma or injury to the cell. It is a passive and disruptive process in which the cell rapidly loses control of ion flux. This results in an uptake of water, giving rise to cell and organelle swelling, resulting in cytolysis. The release of cell contents into the surrounding tissue perpetuates a local inflammatory response (Haslett et al., 1992). By contrast, apoptosis is a genetically regulated process that, unlike necrosis, occurs through the activation of a cell-intrinsic death cascade. This cascade can be modulated by many exogenous signals. The dying cell undergoes a relatively ordered form of cell death characterized morphologically by cell shrinkage, membrane blebbing, chromatin condensation, internucleosomal DNA fragmentation, and the formation of apoptotic bodies (Arends and Wyllie, 1991). Injured/stressed cells may undergo necrosis or apoptosis, depending on the level of the stress. Under extreme conditions, when the stress level eliminates the capacity for regulated activation of the apoptotic cascade, the cells undergo necrosis. At lower levels, injured cells activate their own apoptotic programs. However, if the level of stress is low enough, cells attempt to survive and activate a stress response system. A simple representation of these two pathways is given in Figure 4.3. This response involves a shut down of all cellular protein synthesis apart from a rapid induction of heat shock proteins, which results in a transient state of thermotolerance (Hahn and Li, 1990). Once the stress source

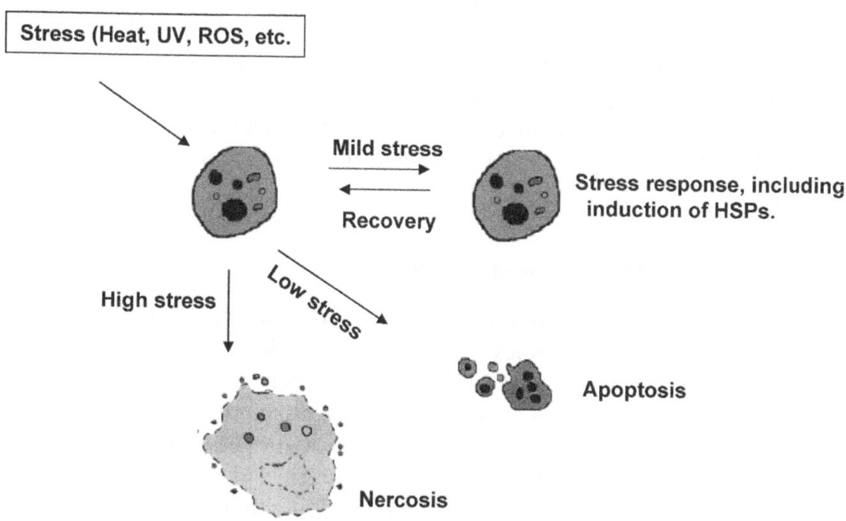

Figure 4.3. A simplified model of cell response to stress and other damaging stimuli. Mild stress triggers the rapid induction of HSPs, resulting in thermotolerence. On the contrary, increasing the levels of stress result in the apoptotic response and under extreme conditions nercosis occurs. Heat shock proteins (HSPs); reactive oxygen species (ROS). Adapted from Creagh et al. (2000) with permission.

is removed, these cells function normally and the levels of HSPs drop back to basal levels with time. During the intervening period, while HSPs levels are elevated, cells are refractory to the toxicity of various agents (Lindquist and Craig, 1988). Many tumor cells, however, appear to have constitutively elevated levels of HSPs that may serve to protect them from otherwise harmful conditions/agents.

The increased expression of HSPs that is observed in many tumor types, has been interpreted as an effort by the malignant cells to maintain homeostasis in a hostile environment. However, there is further evidence that in addition to facilitating the survival of tumor cells within their stressful microenvironment, these proteins also allow tumor cells to tolerate alterations from within (Whitesell and Lindquist, 2005). A substantial number of publications have indicated that the proliferation and cell death are both tightly regulated processes that are controlled by signal transduction pathways that transmit information from the environment to effector molecules. In other words, cells under stress must react rapidly to prevent the inappropriate transmission of these death signals or to restrain passage through checkpoints leading to cell division. Inappropriate activation of signaling pathways could occur during acute or chronic stress as a result of protein misfolding or disruption of regulatory complexes. The action of HSPs through their properties in protein homeostasis is thought to restore balance. For example, when normal growth conditions are restored following a severe exposure to elevated temperatures, heat-damaged proteins are sequestrated through interactions with chaperones and are then either refolded to the native state or transferred to the degradative machinery (Mosser and Morimoto, 2004). The volume of existing literature dealing with the physiological roles of these chaperones under normal and stressed conditions renders a fair description of such roles, which is almost impossible in a single section. To bypass this daunting task and at the same time give the reader a reasonable account of some of the roles attributed to these proteins, I have chosen three representative members with high and low molecular masses, namely HSP90, HSP70, and HSP27. When necessary, other members of this family will be cited.

4.4.1. Structure and Functions of HSP90

Members of the HSP90 family are ubiquitous and abundant protein chaperones that have several physiologic roles. They play a key role in the cellular stress response by interacting with a host of proteins after their native conformation has been altered by various stresses, ensuring adequate protein folding and preventing nonspecific aggregation (Smith et al., 1998). Other studies have suggested that this protein may also play a role in buffering against the effect of mutations, possibly by correcting the inappropriate folding of mutant proteins (Rutherford and Lindquist, 1998). Two distinct genes encode inducible and constitutively-expressed isoforms, namely HSP90α and HSP90β, respectively (Hickey et al., 1999). The functional differences between these isoforms are still poorly understood. These isoforms are found in the cytosol, where they are required for the stability and functional maturation of certain signaling proteins such as steroid receptors, the Raf serine kinases, cyclin-dependent kinase 4 (cdk4), and some receptor tyrosine kinases (Czar et al., 1997; Webb et al., 2000; Xu et al., 2001; Whitesell and Cook, 1996). Homologues of HSP90 are also found in the endoplasmic reticulum, namely glucose related protein 94 (GRP94) (Argon and Simen, 1999) and the mitochondria (tumor nercosis factor receptor associated protein1, TRAP1) (Felts et al., 2000). The HSP90 family members contain an ATP-binding pocket. ATP binding and hydrolysis are required for the final stage of refolding and release of the native protein from the chaperone complex (Obermann, 1998). Detailed description of the structural and biochemical features of HSP90 is outside the scope of this work and can be found in some excellent reviews (Pearl and Prodromou, 2001; Prodromou and Pearl, 2003). However, a brief overview of such features is necessary to appreciate some of the functions attributed to this family of proteins. HSP90 reside primarily in the cytoplasm, where it exists predominantly as a homodimer. Each homodimer is made up of monomers that consist of three main domains that have important functional interactions (Whitesell and Lindquist, 2005). Because of the intrinsic conformational flexibility of the intact protein, atomic resolution crystal structures have only been solved for its individual structural domains. The N-terminal domain consisting of residues 1-220 (yeast) is highly conserved in a sequence among the HSP90 family, the yeast HSP90 sharing 62% sequence identity with human HSP90α, β and 38% identity with the E. *coli* HtpG protein. This domain binds adenine nucleotides and is essential for the ATP-dependent function of the chaperone in vivo; its high-resolution crystal structure has been determined for human HSP90α isoform (Stebbins et al., 1997). In the eukaryotic HSP90s, the N-terminal nucleotide-binding domain is connected to the rest of the protein by a highly charged and proteolytically sensitive segment that is variable both in length and in composition between different species and between different isoforms in the same specie. This segment seems to function as a simple covalent tether, connecting the N-terminal with the remainder of the protein. At the opposite end of the protein, proteolysis and yeast two-hybrid screens (Young et al., 1998) have identified a C-terminal domain of ~12 kDa to which two essential functions have been ascribed: First, the C-terminal region provides a strong inherent dimerization interface, which is essential for some functions. For example, It has been hypothesized that the removal of this domain drastically impairs the ATPase activity of HSP90, emphasizing the role of highly cooperative intermolecular and intramolecular interactions in regulating the use of ATP by chaperones (Soti et al., 2002). Second, this domain carries a conserved motif (EEVD) that is responsible for recruiting various tetratricopeptide-containing repeats (TRP). This conserved motif occurs at the extreme C-terminus of HSP90 and binds specifically in the groove of the TRP domain (Carrello et al., 1999; Scheufler et al., 2000). Between the N-terminal nucleotide-binding domain and the C-terminal dimerization/TPR-binding domain is the central region of ~45 kDa. Although this middle domain comprises the main part of the protein, its role in HSP90 chaperone function is still unclear. It has been suggested (Pearl and Prodromou, 2001) that it may provide the binding site for client proteins associated with HSP90; however, there are no experimental data to sustain such hypothesis.

4.4.2. Association of HSP90 with Cancer

Within the last decade, HSP90 has emerged as being of prime importance to the survival of cancer cells. Some studies have shown that this protein is constitutively expressed at 2- to 10- fold higher levels in tumor cells as compared to their normal counterparts, suggesting that it may be critical for tumor cell growth and/or survival (Isaacs et al., 2003). Many roles of HSP90 in cancer biology can be traced back to its interaction with a wide range of proteins. A recent study using global proteomic and genomic methods in yeast to map HSP90 interactions has identified an extended network consisting of 198 putative physical interactions and 451 putative genetic and chemical-genetic interactions (Zhao et al. (2005). Posttranslational interactions with its client proteins allows HSP90 to link the cell to its environment and couple the stress response to integrated changes in signal-transduction pathways (Morimoto, 2002) and transcriptional responses (Freeman and Yamamoto, 2002; Morimoto, 2002).

There is substantial evidence that HSP90 and its co-Chaperones modulate tumor cell apoptosis. Much of these activities seem to be mediated through effects on serine/thrionine kinase, AKT (Basso et al., 2002), tumor-nercosis factor (TNF)-receptors (Vanden Berghe et al., 2003), and nuclear factor-κB (NF-κB) function (Chen et al., 2002). A brief description of such mediation is given below.

- Inhibition of HSP90 function has been shown to cause selective degradation of important signaling proteins involved in cell proliferation, cell cycle regulation, and apoptosis (Maloney and Workman, 2002). The deregulation of some signaling networks associated with HSP90 inhibition is depicted in Figure 4.4. It is now well accepted that many signaling cascades exist, which provide the interface between external cellular stimuli and the control of numerous biological processes within the cell, including cell-cycle control and apoptosis. These pathways are commonly deregulated in cancer cells, and kinases/phosphatases within these pathways have become important anticancer drug targets. Inhibition of HSP90 is particularly attractive in the search for anticancer drugs, which can cause simultaneous degradation of several key oncogenic HSP90 client proteins operating within these pathways.
- A corollary to the above point is the role of HSP90 in the regulation of some cell-cycle mechanisms. It is known that the cell cycle consists of four phases: G1, S, G2, and M. During the cell cycle, a number of ckeckpoints operate to prevent the production of cells with damaged DNA. The G1/S transition requires the phosphorylation of the retinoblastoma protein by CDK4/cyclin D, whereas the G2/M transition requires the dephosphorylation of CDC2/cyclin B by the phosphatase CDC25C. It has been demonstrated that HSP90 inhibition results in the depletion of several kinases that are important for the regulation of both these checkpoints, including CDK4, which operates during the G1/S transition.
- HSP90 has a fairly complex role in facilitating neoplastic transformation. Its dynamic and low-affinity interactions with its client proteins such as hormone receptors, transcription factors, and kinases maintain them in a latent but readily activated state (Smith et al., 1995). However, oncogenic mutation of such clients

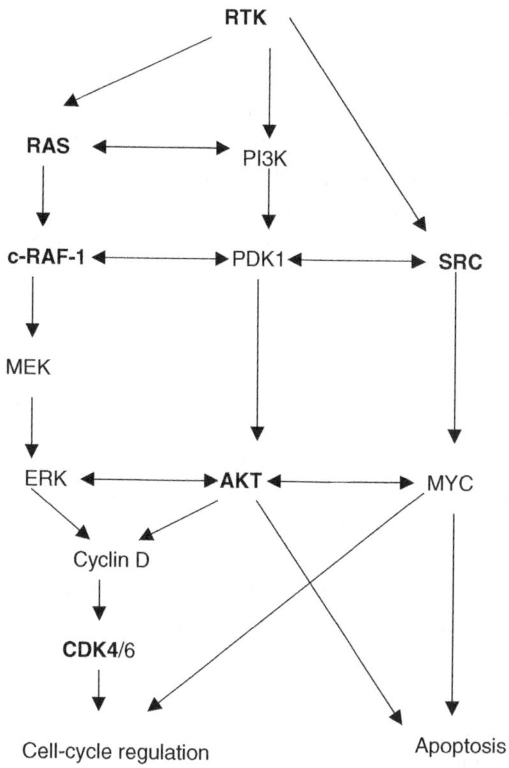

Figure 4.4. Some signaling pathways (indicated in bold) which may be regulated by the heat shock protein, HSP90. Abbreviations are: CDC (cyclin dependent kinase); ERK (extracellular signal-regulated kinase); MAPK (mitogen activated protein kinase); PI3K (phosphatidyl inositol 3-kinase); PDK (phosphoinositidependent kinase); RTK (receptor tyrosine kinase). Adapted from Maloney and Workman (2002) with permission.

places higher requirements for HSP90 function, possibly because of an enhanced conformational instability of the mutant. The literature (Whitesell and Lindquist, 2005) cites two representative examples of such an effect. The first example, which illustrates altered chaperone use by mutant versus wild-type protein, is provided by the tumor-suppressor protein p53, which is encoded by *TP53*. Most mutations in this gene result in the expression of a protein with an altered conformation and impaired cycle–cycle checkpoint activity. *TP53* mutations are known to result in the expression of a protein with an altered conformation and impaired cell-cycle checkpoint activity. Wild-type p53 is a short-lived protein that undergoes transient interactions with elements of the HSP90 machinery that maintain it in an activation-competent state and regulate its degradation through the ubiquitin-proteasome system (Muller et al., 2004; Walerych et al., 2004). Presumably because of their aberrant conformations, most p53 mutants display extended interactions with the chaperone machinery that prevent their normal ubiquitylation and subsequent degradation. As a result, increased cellular levels of dysfunctional protein (a pathological hallmark of mutation) accumulate within the tumor cell (Whitesell et al., 1998; Blagosklonny et al., 1996). Mutant p53 proteins that are bound to HSPs do not function as tumor suppressors. They might also interfere with the function of normal p53 by forming heterodimers (dominant-negative effect) or inappropriately transactivating other target genes (positive tumor-promoting effect). The second, and possibly the earliest example of how oncogenetic mutation of client proteins places higher requirements on HSP function is provided by the SRC tyrosine kinase. Most oncogenic SRC mutations involve truncation of the C-terminus of the protein, resulting in the deletion of its crucial regulatory domain. This domain normally undergoes an intramolecular interaction with an SH2 domain in the protein that stabilizes its structure and represses its kinase activity. Truncation leads to a constitutively active but conformationally unstable kinase (Falsone et al., 2004). Normal c-SRC requires only limited assistance from the HSP90 machinery for its maturation and function within cells (Xu et al., 1999). By contrast, v-SRC mutants display unusually stable physical association with HSP90, which was noted soon after their discovery as the first oncogenes (Oppermann et al., 1981; Brugge et al., 1983). Using genetic and pharmacological approaches, this aberrant chaperone interaction was eventually shown to be essential for both acquisition and maintenance of the increased kinase activity that underlies the transforming activity of v-SRC (Xu and Lindquist, 1993; Whitesell et al., 1994).

4.4.3. HSP90 as a Therapeutic Target

Owing to its role in regulating a number of signaling pathways that are important in driving the phenotype of tumor, the inhibition of HSP90 has attracted a considerable attention for its development as an anticancer drug. The rationale behind such research efforts is based on the knowledge that natural products and synthetic small molecules that bind to the ATP-binding pocket in the N-terminal domain of HSP90 inhibit its function and cause the degradation of its client proteins. The diversity of its client proteins and their involvement in multiple signaling pathways renders HSP90 an attractive therapeutic target. A number of existing anticancer drugs may target only one signaling pathway or act at just one molecular target within a given pathway. In other words, such drugs are likely to miss substantial and important crosstalk between these signaling networks. In such scenario, the tumor cells may have the opportunity to adapt and utilize an alternative signaling pathway and hence circumvent the desired effect of the drug, resulting in drug resistance and the persistence of the malignant phenotype (Maloney and Workman, 2002). There is an increasing evidence suggesting that the inhibition of HSP90 can cause simultaneous, combinatorial blockade of multiple cancer-associated pathways This blockade is commonly attributed to the degradation of many oncogenic client proteins, including ErbB2, cRaf-1, AKT/PKB, CDK4, mutant p53, and polo-1 kinase. Currently, there are two lines of thinking on how to best treat multigene malignancies, including various types of cancer. The first approach advocates the development of a cocktail of specific inhibitors for each individual malignancy or at least the main combinatorial genotypes. The second approach, on the contrary, is to develop drugs that act on specific targets, which then modulate multiple downstream genes and pathways. In both approaches, however, the risk of toxic side effects is increased, whereas the therapeutic selectivity is reduced. Currently, there is substantial evidence to suggest that the development of HSP90 inhibitors fall within the second approach. This statement is supported by a number of potential mechanisms by which such inhibitors may achieve therapeutic selectivity, including the following: (a) HSP90 inhibitors cause simultaneous combinatorial depletion of oncogenic client proteins, including kinases, hormone receptors, and transcription factors (Isaacs et al., 2003; Maloney and Workman, 2002). (b) Cancer cells are known to produce increased levels of HSP90 and become more dependent upon it for the correct folding and function of the large amounts of mutated and overexpressed client proteins (Weinstein, 2002; Workman, 2002). (c) Recent evidence indicates that tumor HSP90 is present entirely in multichaperone complexes with high ATPase activity, whereas HSP90 in normal cells is in a latent, uncomplexed state (Kamal et al., 2003). This latter study showed that HSP90 extracted from cancer cells was able to bind 17-allylaminogeldanamycin (17-AAG) (see Fig. 4.5.

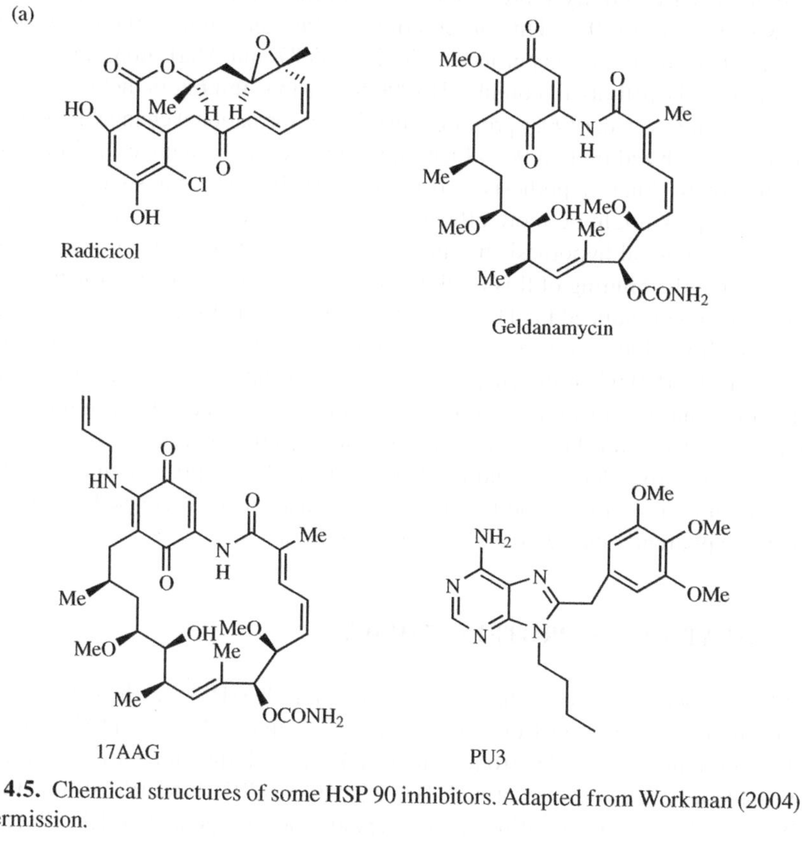

Figure 4.5. Chemical structures of some HSP 90 inhibitors. Adapted from Workman (2004) with permission.

for structure) approximately 100 times more tightly than HSP90 extracted from normal cells. This remarkable difference was interpreted by the authors as a strong indication that HSP90 in tumor cells exists in a functionally distinct molecular form, which is more biochemically active as well as more readily inhibited by the drug. These conclusions and their interpretation by Kamal et al. (2003) attracted a number of comments, which are worth considering. First, previous studies have shown that the ATPase activity of HSP90 can be enhanced by the addition of co-chaperones, particularly the recently discovered Aha1 protein (Panaretou et al., 2002), or a known client protein such as the hormone-binding domain of the glucocorticoid receptor (McLaughlin et al., 2002). In the work of Kamal et al. (2003), purified HSP90 was reconstituted with four accessory proteins p23, Hop, HSP70, and HSP40. Subsequent measurements by this group showed that the apparent affinity for the drug was increased from an IC_{50} value of 600 nM for native HSP90 on its own to 12 nM for the reconstituted complex. It has been pointed out by Workman (2004) that the list of co-chaperones studied did not include two particularly important players, namely Cdc37 and Aha1 proteins. The first protein is known to form a complex with Akt and HSP90, which can be destabilized by the inhibitors of the latter (Basso et al., 2002). The Aha1, on the contrary, has been shown to increase HSP90 ATPase activity more efficiently than the mixture of co-chaperones used by Kamal et al. (2003). This observation implies that future studies, which may use a richer mix of co-chaperones including Cdc37 and Aha1, may furnish further insight into the affinity of complex HSP90 for various inhibitors including 17-AAG. Second, Chiosis et al. (2003) proposed that the conformational change in the 17-AAG might be catalyzed by HSP90, which might, in a sense, recognize the inhibitor as a client protein. Such hypothesis also implies that in this catalysis the super chaperone complex might be more efficient than the native HSP90. An alternative and more conventional hypothesis has been proposed by Workman (2004). The author argued that the binding of this inhibitor to HSP90 is an example of the well-known phenomenon of induced fit. This hypothesis envisage that such induced fit might be more readily achieved between the 17-AAG ligand and the deep nucleotide-binding pocket in HSP90 when the chaperone is part of a super-chaperone complex rather than in its native, uncomplexed form. It has to be said that in the absence of sufficient experimental data both scenarios remain rather speculative. Experimental data derived from detailed binding studies will certainly provide much needed information on the kinetic and thermodynamic basis of the increased affinity in the altered, apparently cancer-specific state of HSP90.

4.5. HEAT SHOCK PROTEIN 27 (HSP27)

HSP27 belongs to the small heat shock protein family (sHSPs), which are detectable in virtually all organisms from prokaryotes to mammals (Arrigo and Mehlen, 1994). These proteins vary in size from 15 to 30 kDa, and till date nine different members of this family have been identified: HSP27, p20, HSPB3, MKBP/HSPB2, HSPpB8, HSPB9, cvHSP, α-A crystallin, and α-B crystalline (Arrigo and Welch, 1987; Iwaka et al., 1997; Krief et al., 1999; Ingolia and Craig, 1982; Kappe et al., 2001). Although members of this family share low amino acid homology, they are grouped together on the basis of similar structural and functional properties, with all sHSPs having a conserved core region that was first identified within the crystallin proteins of the vertebrate eye (De Jong et al., 1993). This domain, termed the crystallin box, comprises 80–100 amino acids in the C-terminus of the protein and has an IgG-like fold, which is followed by a short, more freely flexing C-terminal extension. By contrast, the N-terminus of sHsps is much more variable both in sequence and in length (De Jong et al., 1998) and contains the WDPF motif, which is involved in oligomerization of the protein (Bova et al., 2000). Within unstressed cells, Hsp27 levels are generally low, and it exists predominantly as a large oligomeric unit of up to 800 kDa, usually comprised of six tetrameric complexes of the protein. The size of this oligomeric unit is dependent on a number of physical and chemical parameters, which include temperature, pH, ionic strength, and the degree of phosphorylation of the individual monomers. During the stress response, an increase in the level of Hsp27 expression is preceded by a phosphorylation-induced reorganization of the multimeric status of the protein. Phosphorylation occurs on three different serine residues, Ser-15, Ser-78 and Ser-82, resulting in the redistribution of the large oligomer into smaller tetrameric units (Lavoie et al., 1995; Zantema et al., 1992).

Higher levels of HSP27 expression, compared to those in normal cells have been reported in various types of cancer, including breast (Love and King, 1994; Oesterreich et al., 1993), prostate (Cornford et al., 2000), gastric (Ehrenfried et al., 1995), and ovarian cancer (Langdon et al., 1995). Whether such overexpression can be correlated to the clinical prognosis and progression of the tumor is still to be demonstrated. Discussing some of the roles attributed to this protein and examples on its overexpression in some types of cancer may provide useful indications on the future of this protein as a potential tool for monitoring the progression of certain types of cancer.

4.5.1. The Role of HSP27 in Apoptosis

Dysregulation of apoptosis is well known to play a role in many diseases, including degenerative disorders and many types of cancer. In this regard, the intercellular levels of molecules such as HSP27 that are involved in the regulation of apoptosis are crucial for maintaining the balance between cell death and cell survival within an organism. The involvement of HSP27 in apoptosis has been demonstrated by different research groups. Before discussing specific works dealing with this argument, the following general considerations are relevant: First, characterization of signal pathways that regulate apoptosis have identified phosphoinositide 3-kinase (PI-3K) as a transducer of survival signals (Lee et al., 1993). It is also known that serine/thrionine kinase, Akt, is present in the cytosol of unstimulated cells and is a major target of PI-3K (phosphatidylinositol 3-kinase). Activation of the latter generates 3'-phosphorylated phosphoinositides that in turn induce the translocation of Akt to the plasma membrane (Marte and Downward, 1997; Franke et al., 1997; Stokoe et al., 1997; Downward, 1998). Second, A number of Akt substrates, including BAD, caspase 9, apoptosis signal-regulating kinase1(ASK1), and the forkhead transcription factors (FKHR, FKHRL1) play an important role in cell survival. It has been shown that HSP27 binds to and inactivates a number of proapoptotic molecules, including caspase 3, 9, and cytochrome c (Garrido et al., 1999; Pandey et al., 2000; Paul et al., 2002).

Recently, a model for Akt activation has been proposed by Rane et al. (2001; 2003). On the basis of a number of measurements, it was suggested that inactive Akt exits in a signalling complex with p38 MAPK (mitogen-activated protein kinase), MAPKAPK-2 (MAPK-activated protein kinase-2), and HSP27. This model suggests that through the interaction with both Akt and MAPKAPK-2, HSP27 may act as a scaffolding protein. PI-3K-generated phosphoinositides induce the translocation of the Akt complex to the plasma membrane and activate PDK1 (phosphoinositides-dependent kinase1) and p38 MAPK (Rane et al., 2001). Binding of Akt to phosphoinositides is assumed to result in conformational changes that facilitate the phosphorylation of Akt in ser-473. This docking site binds active PDK1 and the phosphorylation of Akt at Thr-308. Phosphorylation of HSP27 by MAPKAPK-2 or Akt leads to the dissociation of HSP27 from the complex and may promote independent survival signals.

Several studies have assessed the correlation between HSP27 and the regulation of various apoptotic pathways. A mechanism by which apoptosis is initiated is associated with changes in the intercellular redox balance and production of reactive oxygen species. This causes certain changes in the mitochondria and the release of various apoptotic factors. In recent years, evidence has accumulated to show that HSP27 can inhibit apoptosis through a direct inhibition of caspase activation (Garrido et al., 1999; Samali et al., 2001). Caspases are aspartate-specific cysteine proteases present in cells as catalytically inactive zymogens composed of three subunits: a prodomain and two catalytic subdomains (Martin and Green, 1995; Alnemri et al., 1996). A number of works demonstrated a significant presence of HSP27 in the mitochondrial fraction of thermotolerant Jurkat cells (Samali et al., 2001). Such presence is reminiscent of another antiapoptotic protein, Bcl-2, which inhibits apoptosis by blocking the release of cytochrome-c from the intermembrane space of mitochondria (Kluck et al., 1997; Yang et al., 1997). Such a parallel implies that Hsp27 may protect against apoptotic stimuli by blocking the release of cytochrome-c. This hypothesis, however, does not draw a full agreement between the various research groups. For example, the work reported by Samali et al. (2001) supports such hypothesis, whereas Bruey et al. (2000) disagreed that the release of cytochrome-c could be blocked by Hsp27.

4.5.2. Expression of HSP27 in Cancer

Although heat-shock proteins are only induced transiently after periods of cell stress, they are often constitutively overexpressed in tumor cells. Elevated expression of Hsp90, Hsp70, and Hsp27 either individually or in combination has been widely reported in breast, uterine, renal, osteosarcoma, and endometrial cancer, and various leukemias (Jaattela, 1999; Helmbrecht et al., 2000; Jolly and Morimoto, 2000). The overexpression of these proteins in biopsy samples has been suggested to be of prognostic value in breast, renal, and endometrial cancer, and in some cases such overexpression has been suggested to be an indicator of poor therapeutic outcome. For example, gene-expression profiling of lung cancer tumor samples placed HSP70 among a group of tumor-specific genes whose expression was indicative of poor patient outcome (Beer et al., 2002). Proteome profiling has also revealed that several classes of heat-shock proteins are overexpressed and detected on the cell surface of tumor cells (Shin et al., 2003). Targeting heat-shock protein expression or function has been suggested as an effective anticancer strategy based on the speculation that higher levels of chaperones are protective against cell death and increase survival against modalities used in chemotherapy.

HSP27, in contrast to the other major chaperones, is ATP independent yet can efficiently associate with unfolded proteins and maintain them in a folding-competent state. The chaperone activity of this protein is regulated by heat-induced changes in phosphorylation and oligomerization (Haslbeck and Buchner, 2002). HSP27 is found in the cytoplasm as large aggregates that have native molecular mass in the range of 100–800 kDa. Following cell stress stimulation, HSP27 can be rapidly phosphorylated by mitogen activated protein kinase activated protein (MAPKAP) kinase 2 (Oesterreich et al., 1990).

Till date, the precise molecular mechanism(s) responsible for the overexpression of heat-shock proteins, including HSP27, in cancer cells are not known but may be tumor specific. As it has been pointed out within this text, uncontrolled proliferation and unregulated signaling apoptosis are mechanisms strongly associated with neoplastic transformation and tumor progression. Apoptosis has emerged as a major mechanism by which anticancer agents eliminate the preneoplastic and cancer cells (Samali et al., 1999). Furthermore, the overexpression of HSP27 was found to confer resistance against actin fragmentation, mediating an adaptive response to oxidative stress, including carcinogens, anticancer drugs, and other xenobiotics (Huot et al., 1996). Therefore, it is not surprising that a number of cancers show increased levels of HSP27 associated with different expressions of phosphorylated isoforms compared with normal cells (Lavoie et al., 1993).

In a recent study (Tremolada et al., 2005), it was demonstrated that the phosphorylation of HSP27 occurs differently in human renal cell carcinoma (RCC) compared with homologous normal kidney tissue. Tumor and normal samples, obtained after surgery resection from patients with RCC, were examined by 2-D PAGE in combination with tandem mass spectrometry and immunostaining. A similar proteomic approach has also demonstrated the overexpression of HSP27 in human gastric carcinomas (Nishigaki et al. (2005).

MCF-7 cell lines are extensively used as a cell model to investigate human breast tumors and the cellular mechanism of antitumor drugs. Two-D PAGE in combination with N-terminal amino acid sequencing has been used to investigate protein profiling in MCF-7 cell lines following their treatment with doxorubicin, an anthracyclin drug widely used in clinical chemotherapy (Chen et al., 2002). One of the conclusions reported by the authors is that the exposure of these cell lines for 2 days to 0.1 μM of the drug resulted in a marked decrease in the levels of three isoforms of HSP27, whereas the levels of other stress-associated proteins including HSP60 were not significantly altered.

Multidrug resistance (MDR) is another area where some members of the HSP family seem to have a role to play. MDR may arise from alterations at any step in the cell-killing pathway. Alterations have been described in drug transport, drug metabolism, and cellular repair mechanisms and in the ability of the cells to undergo apoptosis (McKenna and Padua, 1997). In addition to being target for several pharmacological drugs, HSP90 chaperone complexes influence the sensitivity of cells to many drugs, and high Hsp90 expression is often associated with multidrug resistance. Liu et al. (1999) reported that upon lowering the HSP90β concentration the sensitivity of cancer cell lines to chemotherapeutic drugs. Many of the cytotoxic drugs target topoisomerase enzymes. Sinha et al. (2003) have used 2-DE in conjunction with MALDI mass spectrometry to investigate chemoresistance development in melanoma cell lines. To probe molecular factors potentially associated with the drug-resistant phenotype of malignant melanoma, the authors used a panel of human melanoma cell variants exhibiting low and high levels of resistance to four commonly used anticancer drugs in melanoma treatment. This study revealed that in the neutral and weak acidic range (pH 4-8), a total number of 14 proteins showed alterations in their expression, whereas 20 proteins were differentially expressed in the basic region (pH 8-11). The authors reported that in various chemoresistant variants, there was an increased expression of the small stress protein HSP27, whereas in most cell lines, upregulation of HSX70 and HSP60 isoforms were registered. It is interesting to note that existing literature (Sarto et al., 2000) shows that upon cell stimulation, HSP27 becomes a frequent target of phosphorylation. On the contrary, HSP70 has long been recognized as one of the primary heat-shock proteins in mammalian cells. Both HSP70 and HSP27 have been frequently associated with the inhibition of apoptosis induced by different chemotherapeutics, particularly those that target topoisomerase II enzymes, such as anthracyclins and etoposide. Indeed Creagh et al. (2000) have shown that the production of reactive oxygen species by these drugs plays an important role in their induction of apoptosis, whereas treatment with antioxidants increases cellular resistance to these agents.

4.6. HEAT SHOCK PROTEIN 70 (HSP70)

The HSP70 family of stress proteins constitutes the most conserved and best studied class of HSPs. Human cells contain several HSP70 family members including stress inducible HSP70, constitutively expressed HSC70, mitochondrial HSP75, and GRP78, localized in the endoplasmic reticulum (Jaattela, 1999a). 70-kDa heat-shock proteins assist a wide range of folding and unfolding processes, including the folding and assembly of newly synthesized proteins and aggregated proteins, membrane translocation of organellar and secretory proteins, and control of the activity of regulatory proteins (Bukau et al., 2000; Hartl and Hayer-hartl, 2002; Pratt and Toft, 2003; Wegele et al., 2003). All these activities appear to be based on the property of HSP70 to interact with hydrophobic peptide segments of proteins in an adenosine triphosphate (ATP)-controlled fashion. The broad spectrum of cellular functions of HSP70 proteins is achieved through (i) the amplification and diversification of

HSP70 genes in evolution, which has generated specialized HSP70 chaperones, (ii) co-chaperones that are selectively recruited by chaperones to fulfil specific cellular functions, and (iii) cooperation of HSP70s with other chaperone systems to broaden their activity spectrum. HSP70 proteins with their co-chaperones and cooperating chaperones thus constitute a complex network of folding machines. The HSP70 protein is known to have a high affinity for extended hydrophobic peptides, and its diverse functions depend upon ATP-regulated binding and release of exposed hydrophobic surfaces on proteins. Under normal conditions, HSP70 proteins function as ATP-dependent molecular chaperones by assisting the folding of newly synthesized polypeptides, the assembly of multiprotein complexes, and the transport of proteins across cellular membranes (Shi and Thomas, 1992; Murakami et al., 1988). Under stressful conditions, the synthesis of inducible HSP70 enhances the ability of cells to cope with increased concentrations of unfolded or denatured proteins (Nollen et al., 1999). Although hardly or never expressed in normal tissues and basal conditions, HSP70 is constitutively expressed in human tumor samples from various origins, and its expression has been known to increase after chemotherapy (Parcellier et al., 2003). HSP70 up-regulation, as a consequence of either oncogenic transformation or cellular stress, may inhibit apoptosis induced by a wide range of cellular insults, as suggested by transfection experiments in vitro (Ravagnan et al., 2001). The overexpression of this protein in cancer cells also increases their tumorigenicity in rodent models (Jaattela, 1995), and high HSP70 expression in human breast cancer, glioblastoma, endometrial, or renal tumors has been associated with metastasis, poor prognosis, and resistance to chemotherapy or radiation therapy (Vargas-Roig et al., 1998; Nanbu et al., 1998). The downregulation of HSP70 has been associated with the death of tumor cells or their sensitization to cytotoxic drug-induced apoptosis in vitro and decreasing their tumorigenicity in vivo (Gurbuxani et al., 2001). For all these reasons, HSP70 seems to be an interesting molecular target for sensitizing tumor cell to cancer therapy. To gain some insight into the potential of HSP70 as a biomarker for some types of cancer, some of its functions and its overexpression in cancer are briefly considered in the following sections.

4.6.1. Structure and Mechanism of Action

HSP70 have a modular structure consisting of three interdependent domains, which are a highly conserved 44-kDa N-terminal ATPase domain, an 18-kDa peptide binding domain, and a 10-kDa C-terminal helical lid domain (Chappel et al., 1987; Flaherty et al., 1990). The lid domain gates the polypeptide binding pocket in an ATP-dependent fashion such that when ATP is bound in the N-terminal domain, the peptide binding channel is exposed and HSP70s exhibit fast on and off rates for substrate binding. Transient interactions with peptide can stimulate ATP hydrolysis, triggering a conformational change in the chaperone (McCarthy et al., 1995). This increases the affinity of HSP70s for peptides by closing the lid domain on the peptide binding pocket and trapping bound substrates (Misselwitz et al., 1998). The C-terminal domain of many eukaryotic cytosolic HSP70s contains a conserved octapeptide that mediates interactions with a variety of co-chaperones, including

HSP40s and others that contain tetratricopeptide repeat (TPR) protein interaction motifs (Scheufler et al. 2000). TPR co-chaperones have been shown to stimulate ATPase of yeast cytosolic HSP70 (Hainzl et al. 2004). The ATPase cycle of HSP70 consists of an alternation between the ATP state with low affinity and fast exchange rates for substrates, and the ADP state with high affinity and low exchange rates for substrates. The molecular mechanism of the ATPase and substrate binding/release cycles has been analyzed in detail only for a few HSP70 homologs including DnaK and HscA (HSC66) of *E. coli*, DnaK of *Thermus thermophilus*, Ssa1 of *S. cerevisiae*, HSC70 of bovine, and BiP of hamster and *S. cerevisiae* (Harrison et al., 1997; Sondermann et al., 2001; Zhang and Zuiderweg, 2004). The ATPase domain consists of two large globular subdomains, each further divided into two small subdomains. The subdomains are separated by a deep cleft at the bottom of which the nucleotide binds in complex with one Mg^{2+} and two K^+ ions contacting all four subdomains (Flaherty et al., 1990). The X-ray structures of the bovine HSC70 ATPase domain complexed with several adenosine nucleotides (ADP, AMPPNP, ATP to mutant proteins) revealed that the adenosine nucleotide is positioned in the active site by interactions with two *b*- and *g*-phosphate-binding loops and a hydrophobic adenosine binding pocket (Flaherty et al., 1990). In addition, nuclear magnetic resonance (NMR) investigations have demonstrated a high flexibility of the ATPase domain with a shearing and tilting motion of the different subdomains toward each other, leading to an opening and closing of the nucleotide binding cleft (Zhang and Zuiderweg, 2004). A continuous opening and closing of the nucleotide binding cleft was proposed earlier to depend on the nucleotide present in the nucleotide binding site with the opening frequency being largest in the nucleotide-free state and decreasing in the order: nucleotide-free > ADP > ADP$^+$ inorganic phosphate >ATP (Gässler et al., 2001). These observations explain the different dissociation rates for the different nucleotides and the influence of inorganic phosphate on these dissociation rates.

4.6.2. Anti-apoptotic Role of HSP70

There are two principal functions attributed to HSP70: de novo folding of nascent polypeptides and interaction with signal transduction proteins. The first function, which deals with the structural and biochemical aspects of this molecule, is outside the scope of this text, which will mainly deal with the interaction of HSP70 with some signal transduction proteins and its impact on the antiapoptotic role of this protein. HSP70 is known to interact with key regulators of many signal transduction pathways controlling cell homeostasis, proliferation, differentiation, and cell death. The interaction of HSP70 with these regulatory proteins continues in activation cycles that also involve HSP90 and a number of co-chaperones. The regulatory proteins, called clients, are thereby kept in an inactive state from which they are rapidly activated by the appropriate signals. HSP70 and HSP90, thus, repress regulators in the absence of the upstream signal and guarantee full activation after the signal transduction pathway is switched on (Pratt and Toft, 2003). HSP70 can be titrated away from these clients by other misfolded proteins that may arise from internal or external stresses. Consequently, through HSP70 disturbances of the cellular system induced by environmental, developmental, or pathological processes act on these signal transduction pathways. In this way, stress response and apoptosis are linked to each other. HSP70 inhibits apoptosis acting on the caspase dependent pathway at several steps both upstream and downstream of caspase activation and on the caspase independent pathway. For example, upregulation of HSP70 leads to increased resistance against apoptosis-inducing agents such as tumor necrosis factor-α (TNF-α), staurosporin, and doxorubicin, whereas downregulation of HSP70 levels leads to increased sensitivity toward these agents (Jaattela et al., 1998; Jaattela 1999b). This observation relates to many pathological processes including oncogenesis. In many tumor cells, increased HSP70 levels are observed and correlate with increased malignancy and resistance to therapy, whereas downregulation of the same protein in cancer cells induce differentiation and cell death (Nylandsted et al., 2000).

As it has been pointed out in other parts of this text, regulated cell death is essential for the proper functioning of all multicellular organisms. Cells that have been damaged by stress have the option of actively engaging the intrinsic apoptotic pathway leading to their self-destruction (Adams, 2003; Danial and Korsmeyer, 2004). Dismantling of the dying cells is carried out by the caspase family of cysteine proteases, which normally lie dormant in healthy cells (Riedl et al., 2004). In normal cells, caspases are constitutively expressed as inactive single polypeptide chains, known as procaspases, and their activation requires specific proteolytic cleavage (Cohen, 1997; Thomberry and Lazebnik, 1998). Active caspases can typically amplify apoptotic events by their ability to cleave their own precursor forms as well as those of other caspases (Stennicke et al., 1998). In addition, mitochondria have a crucial function in initiating the cascade of caspase activation in response to different apoptotic signals (Green and Reed, 1998). For example, disruption of the outer mitochondrial membrane by apoptotic stimuli results in the release of cytochrome c into the cytoplasm (Goldstein et al., 2000). Cytochrome c binds to the cytosolic apoptotic-protease-activating factor 1 (Apaf-1), thereby promoting Apaf-1-mediated activation of caspase-9 in an ATP/dATP-dependent manner (Saleh et al., 1999). In other words, the activation of the capses requires the formation of the apoptosome, an oligomeric complex containing an active initiator caspase (caspase-9), which processes and activates effector caspases (Cory et al., 2003; Kuwana et al., 2005). Formation of the apoptosome occurs through a cytochrome c-mediated conformational change in the cytosolic adaptor molecule apoptotic protease activation factor-1 (Apaf-1) (Green and Kroemer, 2004) that facilitates its oligomerization and procaspase-9 recruitment (Li et al., 1997). Ultimately, the process of cell death is controlled by the regulated release of cytochrome c and other proapoptogenic factors such as the apoptosis inducing factor from mitochondria (Green and Kroemer, 2004). This step is under the control of the Bcl-2 family of proapoptotic and antiapoptotic proteins (Cory et al., 2003). Each of the antiapoptotic members of this family contains three or four conserved Bcl-2 homology (BH) domains. Their ability to suppress apoptosis is prevented by interactions with proapoptotic BH3-only proteins. It is suggested that discrete stress signals are conveyed to the antiapoptotic Bcl-2 family members by diverse BH3-only proteins. Inhibition of the prosurvival function of these proteins is essential for the activation of the proapoptotic BH1–3 family members Bax and Bak, although, some BH3-only members can directly activate Bax (Kuwana et al., 2005).

Saleh et al. (2000) suggested that HSP70 associates directly and specifically with Apaf-1 and inhibits activation of procaspase-9 and subsequently of caspase-3. According to these authors, this inhibitory effect is in part due to the ability of HSP70 to modulate Apaf-1 oligomerization and thereby affects its interaction with procaspase-9. Immunoprecipitation and affinity-binding experiments showed that HSP70 binds to the caspase recruitment domain (CARD) sequence of Apaf-1 and is able to compete with procaspase-9 for binding to this domain. Although the CARD of Apaf-1 has not been shown to have a function in Apaf-1 oligomerization, it seems that binding of HSP70 to this domain inhibits Apaf-1 oligomerization. Binding of HSP70 to the Apaf-1 CARD could induce a conformational change in the adjacent ATP-binding pocket of Apaf-1, resulting in inhibition of Apaf-1 oligomerization. Another possibility is that association of HSP70 with the CARD of Apaf-1 blocks oligomerization of Apaf-1 by some form of steric-hindrance. The finding by the same study that the association of HSP70 with Apaf-1 requires ATP suggests a critical role of the latter in the association mechanism. This hypothesis was supported by investigating the association between Apaf-1 and HSP70 in the presence of ADP or nonhydrolyzable ATP analog, γ-S-ATP. These measurements showed that in the first case, the association was substantially reduced, whereas in the second case, the association was nonexistent. These results are in reasonable agreement with previous findings, which indicated that the accessibility of the C-terminal substrate binding domain of HSP70 is conformation-dependent and is influenced by ATP binding to the N-terminal region of the protein (Schmid et al., 1994). This is also in line with the findings that deletion of the substrate binding domain of HSP70 prevents association with Apaf-1 and blocks the ability of HSP70 to protect cells against etoposide- or heat-induced cell death (Li et al., 1992).

The role of HSP70 in the prevention of heat-induced apoptosis has been examined by Stankiewicz et al. (2005). The authors attempted to investigate events occurring upstream of cytochrome-c release, in particular, the proapoptotic Bcl-2 family member Bax. These authors reported that heat shock resulted in a number of events including a conformational change in Bax, its translocation to mitochondria, stable membrane association, and oligomerization. According to this study, all these events were inhibited in cells that had elevated levels of HSP70. However, HSP70 did not protect cells expressing a mutant form of Bax that has constitutive membrane insertion capability or cells treated with a small molecule activator of apoptosome formation. Based on these results, the authors concluded that HSP70 blocks heat-induced apoptosis primarily by inhibiting Bax activation and thereby preventing the release of apoptotic factors from the mitochondrial intermembrane space.

4.6.3. Overexpression of HSP70 in Cancer

Although HSP70 protein is mainly induced transiently after periods of cell stress, it is often constitutively overexpressed in tumor cells. Elevated expression of HSP70 either individually or in combination with other members including HSP90 and HSP27 has been widely reported in breast, uterine, renal, and endometrial cancer, and various leukemias (Jaattela, 1999b; Helmbrecht et al., 2000; Jolly and Morimoto, 2000). The overabundance of these proteins in biopsy samples has been suggested to be of prognostic value in breast, renal, and endometrial cancer, and in some cases, overexpression has been suggested to be an indicator of poor therapeutic outcome. For example, gene-expression profiling of lung cancer tumor samples placed HSP70 among a group of tumor-specific genes whose expression was indicative of poor patient outcome (Beer et al., 2002). Proteome profiling has also revealed that this protein is overexpressed and detected on the cell surface of tumor cells (Shin et al., 2003). Targeting heat-shock protein expression or function has been suggested as an effective anticancer strategy based on the speculation that higher levels of chaperones are protective against cell death and increase survival against modalities used in chemotherapy. The molecular mechanisms responsible for the overexpression of HSP70 in cancer cells are still unknown. Whether such expression is tumor specific is still to be elucidated. Takashima et al. (2003) used two-dimensional gel electrophoresis in combination with MALDI-MS to profile the expression of HSP70 family members in heptacellular carcinoma (HCC) associated with hepatitis C virus. The authors reported increased levels of nine proteins in cancerous tissues compared to their levels in corresponding noncancerous liver tissues. 78-kDa glucose regulated protein (GRP78), (GRP75), HSP71-kDa, and 70-kDa protein 1(HSP70.1) were among the proteins that have experienced upregulation in the cancerous tissues.

BAG-1 (Bcl-2 athanogene-1) is a multifunctional protein that associates with various cellular targets including the antiapoptotic Bcl-2 protein, steroid hormone receptors (including the oestrogen receptor, ER), the RAF-1 serine/threonine kinase, and receptors for the hepatocyte growth factor/scatter factor and platelet-derived growth factor (Takayama et al., 1995). BAG-1 binds directly to the highly related 70-kDa heat-shock chaperone proteins, HSC70 and HSP70, which may mediate some of these diverse interactions (Zeiner et al., 1997). The expression of BAG-1 and HSP70 in 160 cases of invasive breast cancer was examined by immunohistochemistry (Townsend et al., 2002). These authors reported that BAG-1 was expressed in 92% of cases; most tumors exhibited cytoplasmic BAG-1, whereas a smaller proportion also had nuclear immunostaining. There was a significant inverse correlation between histological grade and nuclear BAG-1 expression, with higher-grade tumors tending to have reduced nuclear BAG-1 expression, but there was no association with cytoplasmic BAG-1. There was also no significant correlation between nuclear or cytoplasmic BAG-1 expression and oestrogen receptor positivity. HSC70 was also detected in the majority (97%) of cases, although, such expression was not correlated with BAG-1 levels, oestrogen receptor status, or tumor grade.

Several studies have examined the correlation between BAG-1 expression and the outcome in human breast cancer, which generated rather conflicting results. A study by Tang et al. (1999) examined BAG-1 expression and localization in 140 cases of breast cancer. The authors reported such expression in 77% of tumors, of which 18% had nuclear and 57% cytoplasmic staining with 1% having both cytoplasmic and nuclear staining. BAG-1 expression, particularly nuclear staining, was correlated significantly with differentiation but not with histological type or clinical stage. The same expression was significantly associated with shorter disease-free (DFS) and overall survival (OS) in multivariate, but not univariate analysis, with a trend toward shorter DFS and OS in patients whose tumor had high nuclear BAG-1 expression. In a second study by Krajewski et al. (1999), the majority (93%) of invasive cancers were positive for BAG-1 expression, largely exhibiting cytoplasmic immunoreactivity. High levels of nuclear staining correlated significantly with improved OS. Subsequently, for 122 patients with early-stage breast cancer all treated by lumpectomy and radiotherapy, the same group described increased BAG-1 staining in 65% of invasive cancers compared with normal breast tissue (Turner et al., 2001). In contrast to their earlier report, univariate and multivariate analysis demonstrated a significant correlation between cytoplasmic, but not nuclear, overexpression and improved OS.

Considering existing data on the overexpression of HSP70 and its client proteins under cancerous conditions, it is still premature to evaluate the future role of this protein as a prognostic/screening biomarker. On the contrary, future studies involving higher number of samples and more powerful analytical techniques may provide further information on the mechanism(s) responsible for the specificity, localization, and significance of such over-expression.

4.7. GENERAL REMARKS

It is commonly argued that an underlying basis of oncogenesis is the acquisition and accumulation of mutations that provide the transformed cells with the combined characteristics of deregulated cell proliferation and suppressed cell death. Currently, there is sufficient evidence that heat-shock proteins are capable of performing dual roles as regulators of protein conformation and stress sensors, which render them important players in both cell proliferation and apoptosis. Based on the material presented here and references within, a number of observations can be made.

- There are numerous reports demonstrating the overexpression of HSPs in cancer compared to normal tissues. Most of these reports concentrate on HSP27 and HSP70; however, a number of reports have suggested that HSP90 expression and localization is deregulated in human tumors. In my opinion, such suggestions are not supported sufficiently by experimental data, which by definition should include large scale screening and clinical trials. Against this background, it is reasonable to ask whether the overexpression of one or more HSPs have the potential to deliver cancer biomakers. A partial answer to this question can be obtained by considering specific cases associated with the deregulation of HSP27 and HSP70. Increased levels of this protein, relative to its level in nontransformed cells have been detected in a number of cancers, including breast cancer, leukemia (Ciocca et al., 1993; Fuller et al., 1994), and prostate cancer (Cornford et al., 2000). Furthermore, the analysis of the pattern of HSP27 phosphorylation in tumor cells seems to be characterstic and different compared to the phosphorylation pattern in normal cells (Ciocca et al.,

1993; Tremolada et al., 2005). These results are encouraging; however, more comprehensive analyses of tissue and other samples are required to support the suggestions that diversity of HSP27 phosphorylation isoforms could represent a potential tumor marker. A number of studies in vitro and in whole cells have implicated HSP70 in various processes. These include the regulation of apoptotic signaling through the JNK-SAPK (JUN N-terminal kinase-stress-activated protein kinase) pathway, and caspase activation through effects on the assembly of the multiprotein apoptosome complex and its involvement in events downstream of caspase activation, such as binding and inhibition of AIF (apotosis-inducing factor) (Mosser and Morimoto, 2004). Elevated expression of members of the HSP70 family has been reported in high-grade malignant tumors (Ralhan and Kaur, 1995; Santarosa et al., 1997). In breast tumors, elevated expression of this protein was found to be associated with short-term disease-free survival, metastasis, and poor prognosis among patients treated with combined chemotherapy, radiation therapy, and hyperthermia (Liu et al., 1996). Interestingly, inhibiting the expression of certain HSP70 isoforms has been shown to selectively cause cell death in breast cancer cell lines, whereas, nontumorigenic breast epithelial cells are not affected (Rohde et al., 2005). This seems to suggest a role of HSP70 in helping organisms balance their response to stressors and damage is subverted in tumor cells, allowing them to survive when they should otherwise die. This antiapoptotic function of HSP70 has made it an interesting target for anticancer therapy. It has to be said, however, that so far no small molecules that can inhibit such function have been identified.

- Because of its involvement in regulating a number of signaling pathways that are known to be important in driving the phenotype of tumor, HSP90 is currently assessed as a new target for anticancer. The rationale behind such development derives from studies showing that the inhibition of HSP90 function may selectively cause the degradation of important signaling proteins involved in cell proliferation, cell cycle regulation, and apoptosis. Various approaches to develop better HSP90 inhibitors are being pursued by academic laboratories and the pharmaceutical industry. Although the proof-of-principle has been established by several phase I trials, there are a number of aspects that have to be further investigated before the search for HSP90 inhibitors realizes its objective. First, the ability of HSP90 inhibitors to affect multiple oncogenic pathways simultaneously is a unique and therapeutically attractive feature of these compounds. However, there is the possibility that inhibiting HSP90 buffering activity at certain stages of malignant progression, although deleterious to most cancer cells, might also reveal mutations that enhance the survival and malignant progression of some cells within the same environment (Whitesell and Lindquist, 2005). Second, predicting the patients likely to benefit from HSP90 inhibitors is an important element to be considered in developing such inhibitors. Such benefit could be dictated by the constellation of molecular genetic defects that are present in a particular tumor (Panaretou et al., 2002; Maloney et al., 2003). In other words, variations in response to

HSP90 inhibitors caused by differences in the genetic make-up of a particular cancer will involve not only differences in the client proteins involved but also the biological outcome of depletion, e.g., in terms of cell cycle arrest, apoptosis, and the disruption in signaling pathways. Although such qualitative differences in the biological outcome may contribute to differences in the quantitative level of response, it is nevertheless likely that therapeutic activity will be seen across most genetic backgrounds, and thus future HSP90 inhibitors should have a broad spectrum anticancer activity.

4.8. CALCIUM BINDING PROTEINS

The S100 protein family is the largest subgroup within the superfamily of proteins carrying Ca^{2+}-binding EF-hand motif. The first family members were discovered in the brain by Moore (1965) and given the name S100 because of their 100% solubility in saturated ammonium sulphate. The 20th member of this subgroup has only been recently identified in humans (Marenholtz et al., 2004). S100 proteins are small (10–12 kDa), acidic proteins composed of two distinct EF hands flanked by hydrophobic regions at either terminus and separated by a central hinge region. The carboxy-terminal EF hand is usually referred to as the canonical Ca^{2+}-binding loop and encompasses 12 amino acids, whereas the amino-terminal loop is responsible for the binding of Ca^{2+} to S100 proteins, which leads to a conformational change that exposes hydrophobic regions in the molecules and allows for target proteins interaction. The physical and structural properties of S100 proteins suggest that they are trigger or activator proteins by contrast with other Ca^{2+}-binding proteins that act mainly as buffers. A model of the molecular mechanism behind the function of S100 proteins has been adopted from calmodulin; in that the binding of Ca^{2+} to S100 proteins leads to a conformational change that exposes hydrophobic regions in the molecules and allows for target protein interaction. Many S100 proteins remarkably show cell and tissue-specific expression patterns. Furthermore, some members of this family form homo- and heterodimers (Fritz and Heizmann, 2004) and even oligomers (Moroz et al., 2002; Novitskaya et al., 2000), which contribute to the diversification of their functions. Despite their conserved functional domain, S100 proteins have been assigned a wide range of tissue-specific and extracellular functions. Accordingly, various diseases, such as cardiomyopathies, neurodegenerative, inflammatory disorders, and cancer, have been associated with altered levels of this subgroup.

4.8.1. Structure and Chromosomal Location of S100

An S100 protein is generally characterized by the presence of two Ca^{2+} binding motifs of the EF-hand type interconnected by an intermediate region, commonly referred to as the hinge region. In each Ca^{2+}-binding motif of the EF-hand type, a Ca^{2+}-binding loop is flanked by α-helices, resulting in a helix-loop-helix arrangement (see Fig. 4.6). The highest sequence identity among S100 members is found in the Ca^{2+}-binding sites. Instead, the hinge region and the C-terminal

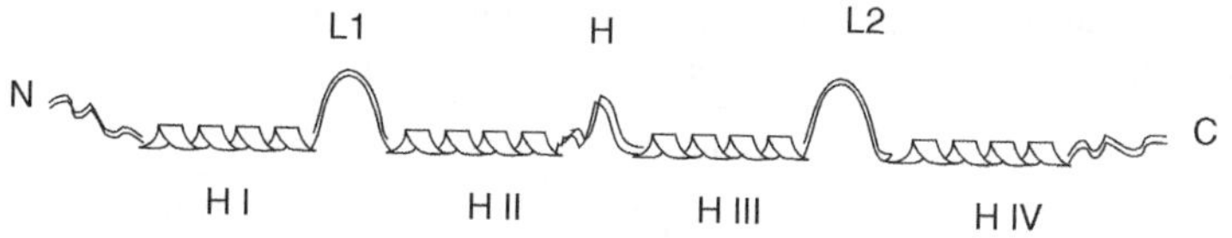

Figure 4.6. Schematic representation of the secondary structure of an S100 protein. Each Ca^{2+}-binding loop (L1 and L2, in the N- and C-terminal half, respectively) is flanked by α-helices (H I- H IV). A linker region(hige-region H) connects helix II-III. Helix IV is followe by a c-terminal extension. Adapted from Donato (2001) with permission.

extension display the least amount of sequence identity suggesting the possibility that these two regions might have a role in the specification of the biological activity of the individual members (Hilt and Kligman, 1991). The two Ca^{2+} binding sites in an S100 protein are attributed different binding affinities, a higher affinity in the case of the C-terminal site and a much lower affinity in the case of the N-terminal site (Donato, 1986). Many S100 proteins bind Zn^{2+} besides Ca^{2+}; however, most of the reported Zn^{2+} dissociation constants are in the micromolar range versus nanomolar concentrations of free Zn^{2+} in the cytoplasm. This low intercellular concentration renders the binding of Zn^{2+} to most S100 proteins rather unlikely as it has been shown for S100B and S100A6 (Fritz et al., 2002). S100B and S100A5 have been shown to bind to Cu^{2+} (Schäfer et al., 2000). Within cells, most S100 proteins exist as homodimers, in which the two monomers are related by a twofold axis of rotation and are held together by noncovalent bonds. Such type of dimerization has been demonstrated for a number of S100 proteins through the use of various techniques including nuclear magnetic resonance and X-ray crystallography (Drohat et al., 1998; Sastry et al., 1998; Moroz et al., 2000). Dimerization of S100 proteins seems to have an important role in their biological activities. Available structural information suggests that upon Ca^{2+}-binding, helix III becomes more perpendicular to helix IV as a consequence of an unusual change of the interhelical angle between these two helices (Smith and Shaw, 1997; Matsumura et al., 1998). Because of this change, the hinge region swings out and a cleft forms in each monomer, which is defined by residues in the hinge region, helices III, IV, and the C-terminal extension, and is buried in apo S100 monomer. Residues defining this cleft are believed to be important for the Ca^{2+}-dependent recognition of S100 target proteins. As an exception, S100A10 does not undergo Ca^{2+}-dependent conformational changes as it is in a permanent 'Ca^{2+}-on' state (Gerke and Weber, 1985; Réty et al., 1999). Experimental evidence confirms that the hinge region and the C-terminal extension play a critical role in the interaction of S100A1, S100B, S100A10, and S100A11 with several target proteins (Rustandi et al., 1998; Garbuglia et al., 1999). Thus, upon Ca^{2+} binding, each S100 monomer opens up to accommodate a target protein (with the exception of S100A10 that is normally in an open-up state), and the S100 dimer can bind target proteins on opposite sides. By this mechanism, an S100 dimer functionally crosslinks two homologous or heterologous target proteins.

According to the established nomenclature (Schäfer et al., 1995), *S100* genes located within the cluster on chromosome 1q21 are designated by consecutive Arabic

TABLE 4.4. Nomenclature, chromosomal localization, and reference sequences of the S100 family. Based on data in Marenholz et al. (2004) with permission.

Gene Symbol	Chromosomal location	NCBI mRNA/protein ref..seqs..
S100A1	1q21	NM_011309 NP_035439
S100A2	1q21	NM_005978 NP_005969
S100A3	1q21	NM_002960 NP_002951
S100A4 (metastasin, calvasculin)	1q21	NM_002961/NM_019554
S100A5	1q21	NM_002962 NP_002953
S100A6 (calcyclin)	1q21	NM_014624 NP_055439
S100A7 (psoriasin)	1q21	NM_002963 NP_002954
S100A7L2 (S100A7-like 2)	1q21	XM_060509 (predicted)
S100A7L3 (*S100A7*-like 3)	1q21	————
S100A7L4 (*S100A7*-like 4)	1q21	————
S100A8 (calgranulin A)	1q21	NM_002964 NP_002955
S100A9 (calgranulin B)	1q21	NM_002965 NP_002956
S100A10 (annexin II ligand;calpactin I, light polypeptide)	1q21	NM_002966 NP_002957
S100A11 (calgizzarin)	1q21	NM_005620 NP_005611
S100A11P (S100A11 pseudogene)	7q22-q31	NM_021039
S100A12 (calgranulin C)	1q21	NM_005621 NP_005612
S100A13	1q21	NM_005979 NP_005970
S100A14	1q21	NM_020672 NP_065723
S100A15	1q21	NM_176823 NP_789793
S100A16	1q21	NM_080388 NP_525127
S100B	21q22	NM_006272 NP_006263
S100P	4p16	NM_005980 NP_005971
S100Z (symbol not approved)	5q14	NM_130772 NP_570128

numbers placed behind the stem symbol S100A. On the contrary, *S100* genes from other chromosomal regions should carry the stem symbol, S100 followed by a single letter. Table 4.4 gives the approved gene symbol, chromosomal location, and reference sequences of the S100 family. Functions attributed to individual members of the S100 family and their potential role as prognostic biomarkers for some forms of cancer are considered below.

4.8.2. S100A4 Protein

S100A4 consists of 100 amino acids excluding the initiating N-terminal methionine. The molecular mass of the natural protein, determined by electrospray mass spectrometry, 11646 Da, differs from that of the recombinant protein (11594 Da), due to the N-terminal acetylation of the former (Pedrocchi et al., 1994). This protein has been purified from natural sources (Barraclough et al., 1990) and also in recombinant form (Gibbs et al., 1994). The *S100A4* gene was cloned independently by several groups under various names, including metastasin (Mts1), fibroblast-specific protein (FSP1), 18A2, P9Ka, CALP, pEl98, calvasculin (Barraclough et al., 1982; Engelkamp et al., 1992; Ebralidze et al., 1989). The majority of the human, mouse, and rat *S100* genes, including *S100A4*, are located as a gene cluster on chromosomes 1q21, 3f3, and 2q34, respectively (Ravasi et al., 2004).

S100A4 protein, as a typical member of the S100 family, exerts dual, intracellular, and extracellular functions. It was shown that this protein binds to several interacellular target proteins and modulates their function. Before citing some specific examples on the possible role of S100A4 in certain forms of cancer, it would be helpful to consider some of the functions, which are commonly attributed to this protein. (a) S100A4 binds to p53 tumor suppressor protein and inhibits its phosphorylation by protein kinase C (PKC1) that modulates the expression of p53-regulated genes, such as *p21* and *bax* (Grigorian et al., 2001). It has also been suggested by the same authors that S100A4 cooperates with wild-type p53 to stimulate apoptosis and selection of more aggressive cancer cell populations. The binding of S100A4 to liprin β-1 and heavy chain IIA nonmuscle myosin results in the inhibition of their phosphorylation by PKC and casein kinase II (CKII) in vitro and subsequent disruption of myosin self-assembly (Kriajevska et al., 2000;2002). (b) Cytoskeletal dysregulation induced by S100A4 has been linked to a redistribution of the membrane-associated adhesive glycoprotein CD44, thus, creating patchy and strongly adhesive CD44 expression patterns on the cell surface. It has been hypothesized that such effect may enable neoplastic cells to acquire an invasive behavior (Lakshmi et al., 1997). Along this line of thought, studies were designed to investigate the interaction between S100A4 and cadherins, a family of transmembrane glycoproteins that mediate Ca^{2+} dependent cell-cell adhesion and suppress invasion (Perl et al., 1998). The expression of E-cadherin and S100A4 was monitored in two mouse tumor cell families and found to be inversely regulated. In addition, transfection experiments showed a reciprocal downregulation of both molecules and suggested that the invasiveness of tumors expressing S100A4 may be at least partially induced by the abrogation of E-cadherin expression (Keirsebilck et al., 1998). Similar mechanisms have been postulated in humans on the basis of immunohistochemical analysis of both proteins in a series of nonsmall cell lung cancers; an inverse correlation of E-cadherin and S100A4 expression was demonstrated (Kimura et al., 2000). (c) An association between S100A4 and cell proliferation has been postulated after the initial cloning experiments that isolated the *S100A4* gene from growth-stimulated cells (Linzer and Nathans, 1983). Recently, it became evident that, as demonstrated for other S100 proteins such as S100B (Lin et al., 2001), a possible mechanism of action may imply binding of S100A4 to the tumor-suppressor protein p53. Using a dexamethasone-inducible clone of B16 murine melanoma transfected with MMTV-*S100A4* (*mts1*), it was shown that S100A4 expression is associated with elevated levels of wild-type p53 (Parker et al., 1994). It can be argued that these results may be biased by the formation of p53 and glucocorticoid receptor complexes, resulting in cytoplasmatic sequestration of both. Nevertheless, it was suggested that a complex of S100A4 with p53 and the sequestration of p53 may result in a stimulation of the cells to enter the

S phase by abrogating the control functions of p53 at the G1-S checkpoint (Parker et al., 1994; Sherbet and Lakshmi, 1998).

4.8.3. Association of S100A4 with Cancer

Studies in rodents have provided evidence supporting the direct involvement of S100A4 in tumor progression and metastasis. The role of S100A4 in cancer has been examined most widely in breast cancer models, which have demonstrated that overexpression of S100A4 in nonmetastatic mammary tumor cells confers a metastatic phenotype (Davies et al., 1993; Grigorian et al., 1996). It was also demonstrated that transgenic mice that overexpress S100A4 in the mammary epithelium are phenotypically indistinguishable from wild-type mice (Ambartsumian et al., 1996), demonstrating that S100A4 itself is not tumorigenic; however, transgenic mouse models of breast cancer have shown that S100A4 expression correlates with metastasis. The association between S100A4 expression and metastasis observed in animal studies has led to a number of studies examining the utility of S100A4 expression as a prognostic marker in human cancers. In a retrospective study of 349 invasive human breast cancer specimens, S100A4 expression and other variables were evaluated for their prognostic significance over a period of 14–20 years (Platt Higgins et al., 2000; Rudland et al., 2000). Analysis of patients with carcinomas that stain positively for S100A4 expression demonstrated that S100A4 expression is highly correlated with patient death. In addition to breast cancer, the overexpression of S100A4 has been evaluated as a prognostic marker in a number of human cancers including esophageal-squamous cancers (Ninomiya et al., 2001), nonsmall lung cancers (Kimura et al., 2000), primary gastric cancers (Yonemura et al., 2000), malignant melanomas (Andersen et al., 2005), prostate cancers (Saleem et al., 2005), bladder cancers (Davies et al., 2002), and pancreatic carcinomas (Lacobuzio-Donahue et al., 2003).

To gain more insight into the potential role of S100A4 and other members of S100 family as prognostic/screening biomarkers for some types of cancer, a number of specific investigations are considered below.

4.8.4. Overexpression of S100A4 in Pancreatic Ductal Adenocarcinoma

Using the National Centre for Biotechnology Serial Analysis of Gene Expression Database, Rosty et al. (2002) investigated the expression of S100A4 in normal and in pancreatic carcinoma cell lines. Reverse transcriptase-polymerase chain reaction and immunohistochemistry were used for confirming such expression. The latter analyses revealed the expression of S100A4 in 93% of the 61 invasive pancreatic carcinoma compared to almost undetected expression of the same protein in 69 low-grade pancreatic intraepithelial neoplasia lesions. The same study reported a significant statistical correlation between the expression of S100A4 and hypermethylation of the first intron of the *S100A4* gene. Additional evidence of the role of hypermethylation inducing S100A4 gene expression was provided by its reexpression after treatment of the methylated pancreatic cancer cell line, Hs766T with 5-aza-2-deoxycytidine. It is of interest to note that the specific expression of S100A4 in cancer cells was questioned in a latter study by Logsdon et al. (2003). These authors have used gene expression profiling to investigate pancreatic adenocarcinoma, pancreatic cancer cell lines, chronic pencreatitis, and normal pancreas. These authors reported that S100A6, S100A11, and S100P were found to be highly and specifically expressed in pancreatic adenocarcinoma. The same study reported significantly higher levels of S100A4 in adenocarcinoma compared with normal pancreas, but contrary to the findings of Rosty et al. (2002), there was no statistically significant difference between the levels in adenocarcinoma compared with chronic pancreatitis.

In a more recent investigation, Shen et al. (2004) have used two-dimensional gel electrophoresis in combination with mass spectrometry to investigate pancreatic tissues. The investigated samples included pancreatic adenocarcinoma, normal adjacent tissues, pancreatitis, and normal pancreatic tissues. Some of the differentially expressed proteins were confirmed by Western blot analyses and/or immunohistochemistry. This investigation identified 40 proteins differentially expressed in pancreatic adenocarcinoma compared to pancreatitis and normal pancreas. S100A8 was found to be 10-fold higher in cancerous tissues compared to their normal counterparts. The overexpression of this protein has been reported in other forms of cancer (Luo et al., 2004). In addition and perhaps more interestingly, enhanced expression of this protein has been linked to drug resistance in breast cancer cells (Sommer et al., 2003).

Complementary DNA (cDNA) microarrays were used by Han et al. (2002) to compare gene expression in pancreatic cancer cell lines and normal pancreas. The authors reported that out of 5289 different genes interrogated by the arrays, 30 of them showed significant upregulation in pancreatic cancer cells. Of particular interest to the present discussion is the upregulation of *S100A11* gene. Using tissue microarrays, Cross et al. (2005) assessed the expression of S100A6, A8, A9, and A11 in normal human tissues and in common cancers. These authors reported that the staining pattern associated with the expression of S100A11 was strictly nuclear in normal tissues and cytoplasmic to more weakly nuclear in cancer tissues. This observation is in reasonable agreement with the results of Sakaguchi et al. (2003), who investigated the role of S100A11 as a mediator of Ca^{2+}-induced growth inhibition in human epidermal keratinocytes. The same study has also shown that increase in extracellular calcium caused phosphorylation of S100A11 at Thr^{10} and Ser^{94} with subsequent binding to nucleolin and translocation to the nucleus. Within the nucleus, S100A11 liberated Spl/3 from nucleolin, leading to increased transcription in $p21^{CIP1/WAF1}$ and $p16^{INK4a}$, which are known negative regulators of cell growth. The data by Cross et al. (2005) imply a movement of S100A11 from the nucleus into the cytoplasm in tumors, an effect, which may result in a decreased nuclear concentration of $p21^{CIP1/WAF1}$ and $p16^{INK4a}$, and thus increased cell proliferation.

4.8.5. S100A4 in Human Breast Cancer

The role of S100A4 in cancer has been examined most widely in breast cancer models, which have demonstrated that the expression of this protein in nonmetastatic mammary tumor cells confers a metastatic phenotype (Davies et al., 1993; Grigorian et al., 1996). As it has been mentioned elsewhere in this text, the process of metastasis is ultimately responsible for the death of most patients suffering from the common carcinomas. Nonetheless, not all patients suffering from primary breast cancer necessarily develop metastases and die of the disease. Among early breast cancer patients who have small invasive carcinomas with negative lymph nodes, approximately 25–30% will develop distant metastasis within 10 years after surgery; these carcinomas are often fatal (Fisher et al., 1983; Clark and McGuire, 1988). Existing guidelines for adjuvant systemic therapy of breast cancer are based on the assessment of tumor size, histological grade, hormone receptor status, age, and menopausal status. Following these guidelines, more than 75% of patients with node-negative breast cancer will receive adjuvant systemic therapy, even though only about one-third will develop distant metastasis (McGuire and Clark, 1992). However, this favorable subgroup may not benefit from adjuvant systemic therapy and may even suffer from its side effects. To date, available prognostic markers do not allow us to identify the minority of early-stage patients at greatest risk for distant metastasis (Mansour et al., 1994; Schwirzke et al., 1999). van't Veer et al. (2002) used DNA microarrays analysis, which identified a gene profile that included 70 genes (excluding *S100A4*) from node-negative breast cancer. This profile was used to classify breast cancer into good-prognosis and poor-prognosis subgroups. This study involved 295 patients with node-negative or node-positive stage I and II breast cancer and found that the gene profile predicted distant metastasis as well as survival. The authors reported that the overall 10-year survival rate was 94.5% in the good-prognosis group and 54.6% in the poor-prognosis group. However, because the poor prognosis group had a 54.6% long-term survival rate, a substantial proportion of patients would receive unnecessary systemic adjuvant therapy. In other words, the results of this study seem to be more beneficial to the good-prognosis group rather than the poor-prognosis group.

The association between immunostaining of S100A4 and prognosis in breast cancer was assessed by Rudland and co-workers (Platt-Higgins et al., 2000; Rudland et al., 2000) and Pedersen et al. (2002). The first research group reported that S100A4 expression was an important predictor of survival in a panel of 349 stages I and II breast cancers. This deduction was not supported by Pedersen et al. (2002) who studied 66 patients with stages I–IV. These authors reported that no association was found between S100A4 expression and clinical outcomes. This disagreement may simply be due to too many different parameters between the two studies, including the number of cases (349 cases versus 66 cases), the stages of cancer investigated (stages I–II versus stages I–IV), and the follow-up period (19 years versus 6.6 years). The combination of S100A4 with other biomarkers in the prediction of metastasis in early stage breast cancer has been evaluated by Lee et al. (2004). Six biomarkers including S100A4, Met, bcl-2, p53, survivin, and HER-2/neu were used to examine a homogeneous cohort of 92 T1-2N0M0 breast carcinoma patients with a long-term follow-up. The choice of this combination was based on a number of considerations. The *Met* proto-oncogene encodes a trans-membrane tyrosine kinase receptor (Met), which modulates cell motility, adhesion, and invasion (Rosen et al., 1994). Earlier studies have shown that high level of Met expression could be used as a prognostic factor in breast cancer and it could be associated with metastatic disease in patients with node-negative breast carcinoma (Camp et al., 1999). Mutant *p53* with its resulting overexpression of p53 protein has been associated with distant metastasis in node-negative breast cancer patients (Silvestrini et al., 1996). Survivin, a novel apoptosis inhibitor, is selectively distributed in common human cancers but not in normal adjacent tissues. It has been suggested that apoptosis inhibition by survivin could be another prognostic parameter of worse outcome in breast carcinoma (tanaka et al., 2000). The main deductions of the study by Lee et al. (2004) were: (a) Comparison between metastasis group and disease-free groups revealed no significant differences in tumor size, histological type, histological grade, hormone status, or hormone therapy. (b) S100A4 expression was found to be correlated to a poor prognosis for T1-2N0M0 breast cancer, while the combination of this protein with Met expression gave the best results regarding risk discrimination. There were no significant differences in Bcl-2, HER-2/neu, or survivin expressions between the two investigated groups.

4.8.6. General Considerations

By considering what is already known about the biochemistry of S100 proteins family and their frequent overexpression in various types of cancer, the following general observations can be made.

- Different forms of cancer exhibit substantial changes in the expression of various members of S100 family. Such changes have been tentatively attributed to possible rearrangements and deletions in chromosomal region 1q21, which are frequently observed in tumor cells. This hypothesis, however, still represents a partial explanation to a widely different mechanisms and functions, which are still to be elucidated. The expression patterns of these proteins in various types of cancer may prove to be a valuable prognostic tool, yet such tool needs to be further strengthened through additional structural, biochemical, and functional information. S100A4 can be taken as a representative example of such situation. The information gathered throughout the past few years have demonstrated that this protein is involved in the regulation of cancer and metastasis. In addition, clinical studies are providing some support to the prognostic significance of this protein in some forms of human cancers. It is likely that additional knowledge on the prognostic role of this protein will be further improved with the development and commercial availability of new and more specific antibodies. However, in parallel with this future development, we still need to understand how the conformational changes in this protein and its interaction with other members of S100 may influence its cellular functions. Further investigations of the interactions of S100A4 with proteins involved in the control of the cell cycle, either in the wild-type or in mutated form, and its interactions with adhesion molecules are likely to shed more light on the functions of this protein in various types of human cancer.

• Many existing studies dealing with the expression of S100 proteins in cancer seem to agree that such expression tend to involve more than one member and more than one type of human cancers. Such trend raises the question of specificity, which is one of the two pillars of biomarkers discovery. Considering existing experience with other biomarkers, it is reasonable to envisage three different approaches, which may result in better specificity. (i) Combination of SA100 expression with other potential markers. Preliminary data in this direction have been already reported by various research groups. For example, *E-cadherin* has an important role in the homophilic cell-cell adhesion and is called an invasion suppressor gene. The adhesive function of this molecule is dependent on the intracellular molecules such as catenin and actin. In the progression of carcinogenesis, irreversible inactivation of E-cadherin at the genomic level gene or through methylation of its promoter is frequently found (Vleminckx et al., 1991; Perl et al., 1998). As a result, the mutual adhesive ability of cancer cells is weakened and cell dispersion occurs. In a study by Yonemura et al. (2000), the expression of *E-cadherin* and S100A4 in gastric cancer cell lines, primary gastric cancers, and their normal counterparts were analyzed by reverse transcription-PCR, Western blot, and immunohistochemical methods. The authors concluded that the combined analysis of *E-cadherin* and S100A4 may be a good prognostic indicator of patients with gastric cancer and that tumors with overexpression of S100A4 and reduced *E-cadherin* can be classified as highly malignant phenotype. Inverse expression of S100A4 and *E-cadherin* seems to be associated with the diffuse histological type and with the invasive ability of gastric cancer. Another example on the combination of S100A4 with other molecules to enhance the potential role of the first in predicting metastasis in early-stage breast cancer has been given by Lee et al. (2004). (ii) A second approach which may enhance the specificity of potential S100-related markers is the use of complexed/free ratio of a given S100 protein rather than the overexpression of the same protein. This approach has given some promising results in the case of PSA as a biomarker for prostate cancer (see Chapter 3, section 3.5.1.).

• Whether or not the interactions of many S100 proteins with other target proteins are all functionally relevant in vivo still to be resolved. Further structural, biochemical, and clinical studies regarding this aspect will no doubt provide new indications on how to design new assays to explore the role of S100 and their target proteins in the prognosis/screening of cancer. In a relatively recent study, Semov et al. (2005) provided some evidence that angiogenic effects of S100A4 are induced through its interaction with annexin II on the surface of endothelial cells. The authors suggested that the secretion of S100A4 by tumor cells induces translocation of annexin II on the cell surface of endothelial cells, where formation of S100A4/annexin II complex takes place. Endothelial cells constitutively secrete t-PA (tissue-type plasminogen activator), and S100A4/annexin II complex on the surface of endothelial cells locally increases the t-PA-mediated plasmin production from plasminogen. This active form in turns activates pro-MMPs (matrix metalloproteinases). The activated forms of MMPS and plasmin may induce interacellular matrix remodeling facilitating angiogenesis and tumor invasion.

4.9. DNA METHYLATION

Although only four bases, adenine, guanine, cytosine, and thymine, form the primary sequence of DNA, there is a covalent modification of postreplicative DNA (i.e., DNA that has replicated itself in a dividing cell) that produces what can be considered a "fifth base." Cytosine methylation occurs after DNA synthesis by enzymatic transfer of a methyl group from the methyl donor S-adenosylmethionine to the carbon-5 position of cytosine, a reaction catalyzed by a family of dedicated enzymes called DNA methyltransferases (DNMTs). The predominant sequence recognition motif for mammalian DNA methyltransferases is 5′-CpG-3′, although non-CpG methylation in mammals has also been reported (Ramsahoye et al., 2000). CpG-rich clusters (commonly called CpG islands) are often found in the promoter region and first oxons of genes, and are mostly nonmethylated. The overall frequency of CpGs in the genome is substantially less than what would be mathematically predicted (Herman and Baylin, 2003). This under representation is tentatively attributed to progressive depletion of CpG dinucleotides. The mechanism of the depletion is related to the propensity of methylated cytosine to deaminate, thereby forming thymidine. If this mutation is not repaired, a cytosine-to-thymidine change remains. The depletion of CpG dinucleotides in the genome corresponds directly to sites of such nucleotide transitions and this change is the most common type of genetic polymorphism in human population (Rjdeout et al., 1990).

The term epigenetic refers to a heritable change in the pattern of gene expression that is mediated by mechanisms other than alterations in the primary nucleotide sequence of a gene (Russo et al., 1996; Bird, 2002). The classification of DNA methylation as a possible epigenetic mechanism of carcinogenesis was predicted over a decade ago (Holliday, 1993). The distribution of the CpG dinucleotide in mammalian genome and differences in methylation patterns between normal and cancer cells are schematically represented in Figure 4.7.

During the past 10 years, a substantial advance in our understanding of the functional consequences of DNA methylation and its interaction with chromatin structure and the transcriptional machinery were seen. Further insight into what causes DNA methylation patterns to undergo changes in cancer cells has also been acquired (Di Croce et al., 2002; Song et al., 2002). From a clinical perspective, DNA methylation changes in cancer represent an attractive therapeutic target, as epigenetic alterations are, in principle, more readily reversible than genetic events (Karpf and Jones, 2002). However, the great strength of DNA methylation lies in its potential in the area of molecular diagnostics and early detection of various types of cancer.

4.9.1. Detection of DNA Methylation

The wide distribution of hypermethylated genes across the human genome and the discovery of hypermethylated candidate tumor-suppressor genes in regions of high-frequency chromosome deletions have attracted various attempts to screen the cancer-cell genome for such genes. Efforts to use diagnostic markers that are based on DNA methylation have been encouraged by a number of characteristics

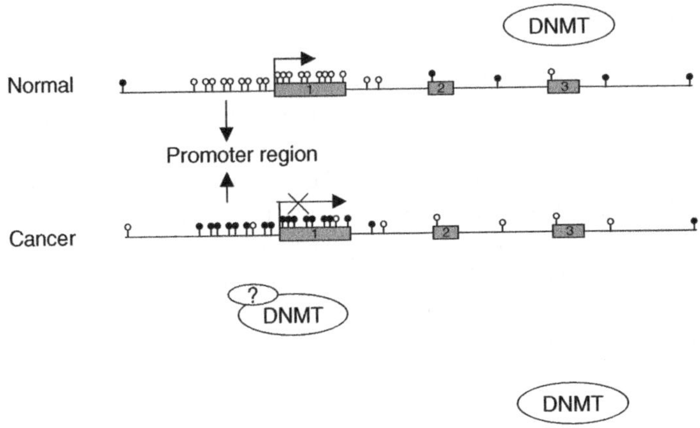

Figure 4.7. Distribution of CpG dinucleotide in human genome and differences in methylation patterns between normal and cancer cells. Exons 1, 2, 3 are represented by boxes. Small regions of DNA, approximately 0.5 to 4.0 kb in size, harbor the expected number of CpG sites and are termed CpG islands. Most of these are associated with promoter regions of approximately half the genes in the genome (numerous circles surrounding and within exon 1 of the sample gene). In normal cells, most CpG sites outside of CpG islands are methylated (black circles), whereas most CpG-island sites in gene promoters are unmethylated (white circles). This methylated state in the bulk of the genome may help suppress unwanted transcription, whereas the unmethylated state of the CpG islands in gene promoters permits active gene transcription (arrow in upper panel). Adapted from Herman and Baylin (2003) with permission.

associated with this epigenetic mechanism. These characteristics can be summarized as follows: (i) DNA can easily be isolated from most bodily fluids as well as from archived fixed tissues. Furthermore, the DNA containing the methylation information is far more stable than the RNA needed for real-time PCR assays. (ii) Once a gene is hypermethylated in a developing tumor that methylation signal generally persists as the tumor continues to progress (Markl et al. (2001). (iii) Although cDNA and protein profiling analyses require fairly strict sample-handling protocols, the methylation status of the DNA is not as vulnerable to factors such as time-lapse before freezing. (iv) The promoter changes potentially provide a positive signal for cancer cells that can be sensitively detected by recently derived polymerase chain reaction-based procedures (Herman et al., 1996a; Sadri et al., 1995). Furthermore, as almost every tumor type appears to have multiple independent promoter hypermethylation events, reasonably sized marker panels might be constructed for each to provide indices for monitoring cancer risk assessment, early cancer diagnosis, and cancer prognosis. Proof of principle for these possibilities has appeared in several studies (Blinsky et al., 1998; Esteller et al., 1999a). Over the last 15 years, a number of techniques have been developed to detect and investigate DNA methylation; some of these techniques are considered below.

4.9.1.1. Restriction Landmark Genomic Screening (RLGS). In this approach (Hatada et al., 1991; Castello et al., 2000), restriction enzymes with different sensitivity to cytosine methylation are used to cut genomic DNAs. One of such enzymes is Not 1, which recognizes large CpG-rich sequences that usually occur in CpG islands. Basically, genomic DNA is radioactively labeled at cleavage sites specific to the restriction enzyme to be used followed by one-dimensional size fractionation. The fractionated DNA is then digested with another more frequently occurring enzyme followed by second dimension separation. The resulting two-dimensional map can exhibit thousands of spots derived from restriction sites. Patterns within the same map reveal missing spots associated with the methylated site(s) of the restriction enzyme, which due to methylation are not cut. The RLGS approach has a number of advantages, including the detection of a large number of CpG islands. This method can be applied to almost all organisms because no DNA probes are needed as is the case with hybridization methods. The same technique has some drawbacks. First, although this technique allows the simultaneous visualization of a large number of CpG islands, these are sometimes not in the promoter region of genes and are, therefore, probably not involved with transcriptional regulation. Second, some enzymes possess a sequence preference in their cleavage pattern, a preference that may compromise spot intensity analysis.

4.9.1.2. Methylation-specific PCR (MSP). MSP was first developed in the mid-1990s (Herman et al., 1996a). Over the last 10 years, it became the most widely used approach for the detection of DNA methylation. This technique uses bisulphite-modified genomic DNA. Bisulphite modification converts cytosine to uracil at an efficiency approaching 99%; however, if the cytosine is methylated, it is not converted. The DNA template can then be amplified for specific genes that could be methylated through the design of methylation-specific primers. Therefore, with the MSP assay, small amounts of DNA template (~50–200 ng) can be used to detect gene-specific promoter hypermethylation in DNA recovered from frozen or paraffin-embedded fixed tumors as well as biological fluids such as sputum and plasma. Because primers are designed in such a way that they recognize only methylated or unmethylated alleles, contaminating normal tissue does not interfere with the ability to detect methylation. Moreover, unlike mutation screens that survey numerous exons to detect gene dysfunction (for example, in *TP53*), the MSP approach uses one primer set to assay a common genomic region where the detection of methylation is associated with the loss of gene function. The MSP method is a sensitive assay (1 in 10^{-4}) for detecting methylation in the presence of contaminating normal tissue or cells. This assay can also be conducted directly on tissue sections (in situ MSP) to identify the clonality of the gene silencing in tumors and premalignant lesions (Nuovo et al., 1999). Palmisano et al. (2000) improved the sensitivity of the MSP procedure to detect one methylated allele in ~50,000 unmethylated alleles by incorporating a nested PCR approach. The nested MSP assay first amplifies the CpG island for the gene being evaluated without preference for methylated or unmethylated alleles. A portion of the stage 1 PCR product is then used in a second stage PCR with primers specific to methylated or unmethylated alleles. This approach improves the sensitivity of MSP for examining biological fluids, such as sputum, which is highly contaminated with normal cells, or formalin-fixed tissues, where the DNA might be highly degraded. High-throughput MSP for population-based screening is feasible using robotics for PCR, multiplex reactions for stage 1 and stage 2 MSP, and non-gel-based detection. The addition of a real-time fluorescence probe to MSP makes the reaction both high throughput and homogenous. If the probe does not contain any CpGs, the reaction is essentially a quantitative version of MSP. However, the fluorescent probe is typically designed to anneal to a site containing one or more CpGs, and this third oligonucleotide may increase the specificity of the assay for completely methylated target strands. This method, called MethyLight, was first described a few years ago (Eads et al., 1999; 2000). In this method, genomic DNA is first chemically modified by sodium bisulfite. This generates methylation-dependent sequence differences at CpG dinucleotides by converting unmethylated cytosine residues to uracil, whereas methylated cytosine residues are retained as cytosine. Fluorescence-based PCR is then performed with primers that either overlap CpG methylation sites or do not overlap any CpG dinucleotides. Sequence discrimination can occur either at the level of the PCR amplification process or at the level of the probe hybridization process, or both. Sequence discrimination at the PCR amplification level requires the primers and probe, or just the primers, to overlap potential methylation sites. The MethyLight assay can also be designed such that sequence discrimination does not occur at the PCR amplification level. If neither the primers nor the probe overlaps sites of CpG dinucleotides, no methylation-dependent sequence discrimination occurs at the PCR amplification or probe hybridization level. This reaction represents amplification of the converted genomic DNA without bias to methylation status, which can serve as a control for the amount of input DNA. When just the probe overlaps methylation sites, sequence discrimination can occur through probe hybridization. The design of separate probes for each sequence variant resulting from different methylation patterns can potentially serve as a quantitative version of the MethyLight technology. The relative sensitivity is similar to the relative sensitivity of gel-based MSP. Because the detection of the amplification occurs in real time, there is no need for a secondary electrophoretic step, and these assays are therefore more suitable for routine clinical use. As there is no post-PCR manipulation of the PCR products, the risk of contamination of subsequent reactions by PCR products is greatly reduced. Theoretically, the MethyLight probe can have any of the standard formats, such as a Taqman probe or a LightCycler hybridization probe pair, and if multiple reporter dyes are used, several assays can be performed in the same way.

It is fair to say that over the last 5 years, MSP has made impressive progress; however, the method still has a number of drawbacks, which limits its use in certain applications. For example, the methylation-specific priming requires all CpGs in the primer binding sites to be fully methylated to allow the visualization on gel of the amplified product, whereas the absence of methylation means that the product cannot be detected on the gel. This means that the method cannot distinguish between partial and complete methylation. Having said that, there are continuous efforts to design primers capable of amplification of different extents of DNA methylation. A second drawback to gel-based MSP assays is that the present format is not highly suitable for a clinical setting. To apply this type of assay for a diagnostic clinical test, the assay must be high throughput and homogeneous (Foy and Parkes, 2001). Ideally, the amplification and detection should be carried out in one tube, both to decrease the amount of time for the test and to reduce the chance of amplicon carryover.Third, the standardization of the MSP assays with respect to primers, amount of DNA template, and PCR conditions should reduce assay variability between laboratories. Such standarization is yet to be implemented.

4.9.1.3. Other Variations

Differential Methylation Hybridization (DMH). This approach can be considered as an array-based technology in which linkers are ligated to digested DNA before it is cut with a methylation-sensitive enzyme. The resulting digests are amplified by PCR and then hybridized to an array of immobilized CpG islands. This approach was first applied to assess the extent of methylation of >276 CpG island loci in breast cancer cell lines compared with their normal counterparts. This first study revealed that 5–14% of these loci was hypermethylated extensively (Huang et al., 1999).

Micraoarray and Gene Reexpression. In this approach, gene reexpression is induced by treating cells with agents that block both promoter hypermethylation and histone deacetylation (Suzuki et al., 2002). Silenced genes that are reexpressed by this treatment are then surveyed by cDNA microarray analysis. The advantage of this approach is that the detection of hypermethylation sites is linked to the transcriptional status of genes, the promoters of which are affected by this change. A disadvantage is that the CpG island that is hypermethylated, and is associated with the gene promoter, is not always easy to identify in genomic databases.

Methylated CpG Island Amplification. In methylated CpG island amplification–representational difference analysis (MCA–RDA), DNA is sequentially restricted with two enzymes, each of which recognises the same CpG-rich sites that occur predominantly in CpG islands (Toyota et al., 1999). The first enzyme is methylation sensitive and the second is not. This produces fragments that differ according to the methylation status of the DNA, which will differ between normal and tumor-derived DNA. After PCR amplification of these fragments, the tumor and normal DNA amplicons are subjected to RDA, which exploits the methylation-sensitive restriction-site differences between the normal and tumor-cell DNA to carry out a comparative hybridization subtraction step. The advantages of this approach are very similar to those of RLGS. However, even though many of the CpG islands are associated with genes, defining the start site of a gene and the exact relationship of the island with the transcriptional regulation of a gene can be laborious.

Methylation-sensitive Arbitrarily Primed PCR. Methylation-sensitive arbitrarily primed PCR (MS-AP-PCR) also uses methylation-sensitive restriction enzymes to cut DNA before it is amplified with random CpG-rich primers that target CpG islands (Gonzalgo et al., 1997). The resulting fragments are displayed on gels and gel-spot patterns between different cell types are compared, which leads to a rapid identification of CpG islands that are differentially methylated in different tissues. The method suffers from the same limitations as those for RLGS and MCA–RDA.

Mass spectrometry, namely, matrix-assisted laser desorption ionization (MALDI) in combination with primer extension for allele discrimination has been shown to be a useful tool for the quantification of allele frequencies in pools of genomic DNA (Werner et al., 2002). In another study, MALDI-based assay for the analysis and quantification of CpG methylation has been described by Tost et al. (2003).

4.10. DNA METHYLATION IN CANCER

At present, the number of cancer-related genes affected by epigenetic inactivation equals or exceeds the number that are inactivated by mutations (Jones and Baylin, 2002; Herman, 1999). In addition, many of the genes modified by promoter hypermethylation have well-recognized tumor-suppressor function. There are a number of examples, including cell cycle-control gene *p16* in many types of cancer, *VHL* gene in renal cancer, and the mismatch-repair gene *MLH1* in colorectal cancer and other neoplasms (Herman and Baylin, 2003). The importance of epigenetic gene silencing in cancer has been underlined by a growing awareness that such changes can result in predisposition to mutational events during tumor progression. This was first demonstrated for the mismatch-repair gene *MLH1*, which is frequently hypermethylated in sporadic tumors that have microsatellite (short DNA sequences of di- or trinucleotide repeats of variable lengths distributed widely throughout the genome) instability (Kane et al., 1997; Herman et al., 1998). These changes in the methylation of the 5′ region of this gene have been observed in the apparently normal colonic epithelium of patients who have colorectal cancer with microsatellite instability (Nakagawa et al., 2001) and in hyperplastic regions, preceding the development of endometrial cancers that develop this type of genetic change (Esteller et al., 1999a,b).

It is commonly agreed that early detection of disease results in an improved clinical outcome for most types of cancer. Therefore, much effort is being put into the discovery and development of early detection strategies. DNA methylation changes have been reported to occur early in carcinogenesis, which render them an attractive indicators in the fight against various forms of cancer (Laird, 1997). The first signs of cancer usually come from one or more of the following sources: presentation of symptoms, direct palpation or visual detection, histopathological analysis of a biopsy specimen, remote imaging, or the detection of cancer biomarkers in a tissue or bodily fluid specimen. For many types of solid malignancies, symptoms often do not arise until after the primary tumor has metastasized. Direct palpation, visual detection, and biopsy analysis are generally limited to accessible sites of the body. For some types of cancer, such as ovarian, pancreatic, and lung cancer, poor accessibility and the late presentation of symptoms thwart the timely detection of malignancy, contributing to high mortality rates. For such diseases, improved remote imaging, such as, spiral computed tomography scanning and the development of cancer biomarkers offer the best hope for early detection (Laird, 2003).

4.10.1. CpG Island Methylation and Gene Silencing

A growing number of cancer genes are being recognized that harbor dense methylation in normally unmethylated promoter CpG islands. The initial reports regarding this methylation and the loss of gene function has generated considerable scepticism that persisted even after the event of methylation was correlated with absent expression of a classic tumor suppressor gene such as *VHL* (Herman et al., 1994). According to Baylin and Herman (2000), it could be argued that methylation is not responsible for initiating gene silencing but is a secondary event that marks the process. The same authors argued that this "chicken or egg" question has yet to be resolved for the timing of methylation in transcriptional repression; it is less relevant for gauging the importance of promoter CpG island hypermethylation in cancer. These authors emphasized that whatever be the initiating event, this methylation both marks and plays a key role in an epigenetically mediated loss-of-gene function that is as critical for, and possibly as frequent in, tumorigenesis as mutations in coding regions. To gain some insight into DNA methylation, it would be useful to consider the following aspects.

4.10.1.1. Proteins that Mediate DNA Methylation. Catalyzing enzymes called DNA methyltransferases (DNMTs) play a central role in the methylation process. There are three known biologically active DNMTs in mammalian cells: DNMT1, DNMT3a, and DNMT3b (Okano et al., 1999). In cancer cells, DNMT1 seems to be responsible for most of the DNA-methylating capacity (Rhee et al., 2000; 2002) and has long been suspected to be the chief factor in maintaining abnormal promoter methylation in neoplastic cells. However, some studies have suggested that interaction between this enzyme and DNMT3b may be vital for the function of DNMT1 in colon cancer cells (Rhee et al., 2000; 2002). Other works have recently revealed that DNMTs may also contribute to transcriptionally repressive chromatin by mechanisms other than methylation. Such chromatin consists of DNA that forms complexes in promoter regions with groups of proteins that act to prevent the transcription of genes. Each of these enzymes interacts directly with histone deacetylases and can recruit them to the site of gene promoters. These deacetylases are known to facilitate the deactivated states of the histone, which as discussed later, and is critical for silencing gene transcription. Binding of DNMTs is by no means limited to histone deacetylases; they can also bind to other proteins with the potential to repress gene transcription. For example, in leukemic cells, abnormal transcription factors arising from translocated genes may recruit complexes of DNMTs with other proteins to gene promoters (De Croce et al., 2002).

4.10.1.2. Nucleosomes. These are structures composed of a core of histone proteins around which DNA winds. These structures are closely linked to the constitution of chromatin. Normally, at sites of active genes with nonmethylated CpG islands in promoter regions, the nucleosomes are widely and irregularly spaced in a manner that favors the access of proteins involved in transcriptional activation (Antequera and Bird, 1993; Bird, 2002; Jones and Baylin, 2002). On the contrary, when these islands in the promoter regions are heavily methylated, nucleosomes are tightly compacted and regularly spaced, a conformation which blocks proteins required for the activation of gene transcription or prevent them from acting as gene activators.

4.10.1.3. Histone Acetylation. Certain amino acids within histone tails can exercise a regulatory effect on nucleosomes. These tails protrude from histones and are kept in an acetylated state in the case of transcribed genes but in a deacetylated state in the case of hypermethylated silenced genes (Bird and Wolffe, 1999; Strahl and Allis, 2000; Jenuwein and Allis, 2001). The histone deacetylases act to maintain the deacetylated state in these histone tails and may be targeted to the appropriate genomic regions by DNMTs and by a group of chromatin proteins known as methyl cytosine–binding proteins (Bird and Wolffe, 1999; Bird, 2002). These methyl cytosine–binding proteins recognize methylated DNA and can themselves repress gene transcription. They also reside in complexes with other proteins including histone deacetylases (Bird and Wolffe, 1999; Bird, 2002) and thus may direct these proteins to regions of gene silencing.

In addition to acetylation, other changes or posttranslational modifications of histones are important in gene silencing. These changes, including acetylation, make up the histone code that governs gene transcription (Strahl and Allis, 2000; Jenuwein and Allis, 2001). In this code, two histone modifications stand out: silenced genes are marked by methylation of lysine 9 on the core histone protein H3, whereas, methylation of lysine 4 on histone H3 is a feature of active genes (Heard et al., 2001; Nakayama et al., 2001; Norma et al., 2001; Xin et al., 2001). These sites of methylation distinguish abnormally silenced genes in one cancer from the same gene when it is expressed in another cancer (Fahrner et al., 2002; Koizume et al., 2002; Nguyen et al., 2002) forming a localized zone of inactive or active chromatin including the CpG island in the promoter (Fahrner et al., 2002). In lower organisms, the methylation of lysine 9 at histone H3 is critical for determining the pattern of DNA methylation (Jackson et al., 2002; Tamaru et al., 2001), possibly by targeting DNMT complexes, a mechanism that has recently been implicated in hypermethylated genes in cancer cells as well (Bachman et al., 2003). Moreover, each of the histone acetylation and methylation changes associated with gene silencing can be converted to the changes appropriate for an active gene during drug-induced demethylation of the CpG island in the promoter and reexpression of the gene (Heard et al., 2001; Koizume et al., 2002; Nguyen et al., 2002).

4.10.2. Methylated Biomarkers in Cancer

Almost 30 years ago, it was shown that cancer patients have increased levels of free DNA in their serum (Leon et al., 1977), which is thought to be released from apoptotic or necrotic tumor cells (Vasioukhin et al., 1994; Jahr et al., 2001). Furthermore, DNA methylation changes have been reported to occur early in carcinogenesis and are therefore potentially good indicators of existing disease, and even of risk assessment for the future development of the disease (Laird, 1997). These reports

TABLE 4.5. Potential DNA methylation markers for lung cancer: (1-7) markers detected in plasma or serum, (8-12) markers detected in remote media (DNA is not obtained directly from the tumor).

	DNA source	Gene	Sensitivity Clin. Anal.	Specificity	Reference
1	Serum(NSCLC)	*CDKN2A(INK4A)*	14%	N/A	Esteller, M. et al. (1999b)
2	Serum(NSCLC)	*DAPK1*	18%	N/A	Esteller, M. et al. (1999b)
3	Serum(NSCLC)	*GSTP1*	5%	N/A	Esteller, M. et al. (1999b)
4	Plasma	*CDKN2A(INKA4)*	18%	N/A	Esteller, M. et al. (1999b)
5	Plasma/Serum	*APC*	47%	100%	Usadel, H. et al. (2002)
6	Plasma	*CDKN2A(INKA4)*	73%	N/A	An, Q. et al. (2002)
7	Plasma (NSCLC)	*CDKN2A(INKA4)*	34%	100%	Bearzatto, A. et al. (2002)
8	Sputum	*CDKN2A(INK4A)*	43%	81%	Belinsky et al. (1998)
9	Broncheoalveolar	*CDKN2A(INK4A)*	24%	N/A	Ahrendt et al. (1999).
10	Brushings	*CDKN2A(INK4A)*	16%	92%	Kersting et al. (2000)
11	Bronchial epithelium	*CDKN2A(INK4A)*	44%	56%	Belinsky et al. (2002)
12	Bronchial wastings	*CDKN2A(INK4A)*	N/A	100%	Kurakawa et al. (2001)

have formed the basis for an ever-expanding area of investigation in the search for DNA methylated markers in blood serum and plasma. Table 4.5 gives a list of DNA methylation biomarkers reported by various research groups (Laird, 2003). Some of these examples are discussed below. Most of these examples have used plasma or serum as a DNA source and used MSP-PCR and/or fluorescence-based variants for detection.

4.10.3. Hypermethylation as a Biomarker in Lung Cancer

Lung cancer is the leading cause of cancer mortality in both sexes in the United States (Jemal et al., 2002). Furthermore, an estimated 1.5 million deaths caused by this disease are predicted in the second decade of this century (Parkin et al., 2001). The high mortality rate (>85%) associated with this type of cancer is in part due to its late detection as well as the absence of optimal therapeutic answers for patients who present with advanced disease. The refractiveness of most lung cancers to both conventional and gene-targeted therapy also contribute to the disappointingly low survival rate. This sad yet realistic projection is one of the reasons behind tremendous research efforts to find biomarkers that can detect this form of cancer in its early onset. The benefit of early detection has already demonstrated its benefits. Patients who present with stage I tumors ($\leq$ 1 cm) and are treated with surgical resection have shown a rate of recurrence less than 50% within 5 years (Harpole et al., 1995). Lung cancer is divided into four major histological subtypes as proposed by the World Health Organization (WHO, 1982): Adenocarcinoma (AD), Squamous cell carcinoma (SCC), large cell lung carcinoma (LC), and small cell lung carcinoma (SCLC)/ non-SCLC (NSCLC). To underline the urgent need for biomarkers capable of detecting this type of cancer in its early stages, it is worth considering some of the screening techniques and their limitations. Currently, there are a number of screening techniques that are employed to reduce the mortality rate in the screened individuals. Spiral computed tomography (CT) is an X-ray imaging procedure where several detectors are arrayed in parallel, enabling an image of the entire thoracic cavity to be acquired in less than 20 seconds. This technique was first introduced in the late 1980s by Japanese scientists (Sone et al., 1998). The CT approach could detect 1–5 mm nodules compared with 1 cm detection capability of chest radiography (Davis, 1991). Despite its impressive sensitivity, CT remains largely a peripheral detection technique for lung cancer. Therefore, SCC and SCLC that arise within the central airways will often go undetected by this screening technique. Screening for these histological subtypes of lung cancer is commonly conducted by fluorescence branchoscopy. This technique, however, has limited use for detecting premalignant lesions with surface diameter of a few millimetres, which are difficult to visualize against the normal epithelial surface. Another screening technique designated laser-light-induced fluorescence endoscopy (LIFE) bronchoscopy has been developed by Palcic et al. (1991). LIFE bronchoscopy is based on the principle that dysplastic and malignant tissues have reduced autofluorescent signals compared with normal tissue (Hung et al., 1991). Most studies have reported higher diagnostic specificity of LIFE bronchoscopy for the detection of premalignant and carcinoma in situ lesions (Lam et al., 1998; Kennedy et al., 2000). In these studies, lesions with moderate dysplasia or worse were scored as positive. However, this technology also results in more false positives than white-light bronchoscopy because of sampling of bronchial surfaces that seem to show reduced autofluorescence. There are several obstacles confronting the use of LIFE bronchoscopy as a screening tool. The first is the invasiveness of the procedure and the second is the lack of clinical data substantiating the risk of lung cancer associated with detecting moderate dysplasia in the bronchial epithelium (Belinsky, 2004). A recent review by Tsou et al. (2002) described a long list of genes that are inactivated by promoter hypermethylation in lung cancer (see Table 4.5). It is not the intention of the author to go through this list but to highlight a number of elements related to the potential of gene-promoter hypermethylation in finding biomarkers, which can contribute to the early detection of lung cancer. In recent years, such potential has been facilitated by two concrete elements: First, it has been demonstrated that an important tumor suppressor gene, *CDKNA2*, which encodes *P16* and *P14* is silenced in many cancers through aberrant promoter hypermethylation

(Merlo et al., 1995). *CDKNA2* is one of the genes that is thought to be prone to early hypermethylation during lung cancer development. A supporting evidence for such hypermethylation came from independent groups (Kersting et al., 2000; Palmisano et al., 2000). Furthermore, *CDKN2A* gene is inactivated at prevalences of up to 67% in adenocarcinoma and 70% in squamous-cell carcinoma of the lung, respectively (Zöchbaur-Müller et al., 2001; Belinsky et al., 1998). Disruption of the cell cycle in lung cancer also involves inactivating other genes through methylation. One of these genes is the cell differentiation and embryonic-development gene paired box gene 5 (*PAX5*), which was identified by using a genome-wide screening approach for detecting methylated promoter regions (Stapleton et al., 1993). This locus is frequently associated with chromosomal translocation and contains two distinct promoters that result in two alternative 5′ exons (α and β) (Busslinger et al., 1996). It has been shown that both genes are methylated in ~65% of adenocarcinoma and squamous-cell carcinomas (Palmisano et al., 2003). The *PAXβ* gene codes for the transcription factor B-cell-specific activating protein that, in turn, directly regulates *CD19*, a gene shown to negatively control cell growth (Kozmik et al., 1992). Other genes found to be hypermethylated in over 30% of lung tumors are *APC, CHD13, RARB,* and *RASSF1A* (Tsou, 2002). Each of these genes has been demonstrated to be transcriptionally silenced in cell lines/tissues showing methylation. Reexpression of these genes was seen in lung cancer cell lines following treatment with the methylation inhibitor 5-aza-2′ deoxycytidine, further supporting the notion that methylation caused their inactivation (Brabender et al., 2001; Zhu et al., 2001). The second advance, which brought gene-promoter hypermethylation to the arena of biomarkers discovery, is the development and optimization of methylation-specific PCR (MSP) assays that allow rapid detection of methylation in genes through the selective amplification of methylated alleles within a specific gene promoter (Herman et al., 1996a).

Current literature on lung cancer highlights a number of aspects of the disease where gene promoter hypermethylation analyses are expected to make a substantial contribution. These aspects include early detection of the disease, diagnosis and discrimination between its histological subtypes, and its potential in screening high risk groups. A brief discussion of these aspects is given below.

- It is becoming fairly clear that the efficacy of a biomarker assay is mainly determined by its sensitivity and specificity. DNA methylation biomarkers for lung cancer are no exception. In simple terms, the sensitivity is mathematically defined as the ratio of correctly identified positive cases divided by all examined positive cancer cases (true positive+false negatives). Specificity, on the contrary, is mathematically defined as the ratio between correctly identified no-cancer cases (true negatives) and all examined cancer-free cases (true negatives+false positives). The methylation biomarker studies performed so far vary in methylation targets, source of DNA, and type of tumor including lung cancer. Therefore, it is too early to draw definite conclusions regarding the sensitivity and specificity of current assays. Nevertheless, a number of preliminary observations can be made. First, existing studies indicate that targeted luminal sources of DNA tend to give higher clinical sensitivity than DNA from serum and plasma (Laird, 2003). Second, the specificity of plasma or serum detection of the methylated markers seems to be remarkably high, approaching 100% in many studies reported in Table 4.5. The sensitivity, on the contrary, did not exceed 50%. This relatively low sensitivity is expected to be increased through the use of panels of multiple markers rather than a single marker. Further improvement in the sensitivity can also come from improved MSP primers and methylLight probes and from other advances, such as the use of nested PCR primers (Palmisano et al., 2000).

- The ability to discriminate between the histological subtypes of lung cancer has a high clinical relevance because of the important differences in treatment and monitoring the progress of the disease. The potential of DNA methylation analysis in the discrimination between the different histological subtypes of lung cancer has been demonstrated by Virmani et al. (2002). In a study of almost 100 NSCLC and SCLC cell lines, the methylation analysis of mere 23 loci yielded seven genes exhibiting statistically significant differences in methylation levels between the two groups. The methylation analysis combined with a hierarchical clustering algorithm using data from the seven genes allowed distinction between the two subtypes with a specificity and sensitivity of 78%. It can be argued that clinically applicable markers require much better sensitivity and specificity, yet the above study should be considered as a proof of the principle in which a fairly modest panel of markers have been tested. Thus, it can be anticipated that studies involving an expanded collection of informative markers will undoubtedly improve the capability of methylation panels to discriminate between different histological subtypes of lung cancer.

- Screening of high-risk groups (e.g., long-term smokers) has always attracted intense debate regarding its feasibility and desirability (Black, 1999; Marcus, 2001). It is argued that previous trials of chest radiography and sputum cytology did not improve survival. It has also been suggested that screening may lead to overdiagnosis and the administration of aggressive treatments to subjects who might never have required treatment. Against this background, the development of highly specific and sensitive methylation analyses conducted by noninvasive means would certainly facilitate such trials or at least make them more acceptable than current screening techniques. A prerequisite for such a large-scale trial is the identification of gene panels that can be reliably used as cancer-specific methylation signature. One way of developing such panels can be the use of a global approach, in which thousands of CpG islands are tested simultaneously. A proof of the principle of such an approach has been demonstrated by Dai et al. (2001). The authors have screened more than 1000 CpG islands in NSCLC cell lines and identified the gene *BMP3B/GDF10* as a possible new methylation marker for lung cancer. The power of methylation analyses to generate specific signatures is tremendous, considering that the genome contains an estimated 45,000 CpG islands, each of which could or could be methylated (resulting in 2^{45000} possible distinct methylation patterns). Although this complexity may not entirely approach that result which may be obtained using gene-expression profiles, technically methylation analysis has several advantages. The first is that the variables studied (CpG island methylation) are largely negative in normal tissues, so positive methylation will usually be informative. By contrast, in expression studies, genes may be transcribed at different levels in a variety of tissues or under a variety of normal conditions that are not necessarily related to cancer. The second advantage is that methylation analyses require DNA, instead of the more labile RNA, and can be carried out on many different kinds of samples (including paraffin-embedded fixed material) using a high throughput approach (Eads et al., 2000).

4.11. INHIBITION OF DNA METHYLATION

Accumulating experimental data on various epigenetically silenced genes suggest that the reactivation of their expression could represent a profound antitumor effect. A number of studies have demonstrated that the treatment of cells with the demethylating agent 5-aza-5′-deoxycytidine (DAC) leads to reexpression of genes silenced by promoter methylation (Jones and Baylin, 2002). This agent inhibits DNA methylation by reducing the DNMT enzymatic activity via the formation of a stable complex between the enzyme and DAC-substituted DNA (Santi et al., 1983). Structures of citdine and its 5-aza-analogs are given in Figure 4.5.

The role of DNA methylation in the transcriptional regulation of several matrix metalloproteinases (MMPs) genes and the effect of DAC on the invasive behavior of pancreatic cancer cells have been investigated by Sato et al. (2003a). The authors have assessed the invasiveness of this type of cancer following the treatment of five different cell lines with the DNA methylation inhibitor, 5-Aza-2′-deoxycytidine. The authors reported that four of the five examined cell lines have exhibited an increased invasiveness, a result that was attributed to the reactivation of invasion-promoting genes. In another study, Sato et al. (2003b) have used high throughput oligonucleotide microarrays to analyze global changes in gene-expression profiles in pancreatic cancer cell lines following exposure to 5-Aza-2′-deoxycytidine and/or trichostatin A (inhibitor of histone deacetylases). The main conclusions of this study were as follows: Following DAC treatment, more than 475 genes were markedly (>5-fold) induced in pancreatic cancer cell lines but not in nonneoplastic epithelial cell lines. The methylation status of 11 of these genes were examined in a panel of 42 pancreatic cancers, and all 11 genes were aberrantly methylated in pancreatic cancer but rarely in 10 normal pancreatic ductal epithelia. The authors concluded that a substantial number of genes are reactivated by DAC treatment and many of them may represent novel targets for aberrant methylation in pancreatic carcinoma. In another study (Cecconi et al., 2003), the effect of 5-aza-2′-deoxycytidine (DAC) on the proteomic profiling of pancreatic ductal carcinoma cell lines was investigated by 2-DE combined with PDQuest statistical analysis and MALDI-TOF mass spectrometry. Comparing standard 2-D maps of the total protein extract before and after DAC treatment showed that out of 700 spots visualized in the pH range 3–10, a total of 45 polypeptide chains were differentially expressed following DAC treatment. Thirty-two were downregulated and 13 upregulated; most of these proteins were characterized by MALDI-TOF-MS.

These and other studies have amply demonstrated that DAC and other analogs are capable of reactivating various genes, which are silenced by promoter hypermethylation. At the same time, some of these studies have highlighted two difficulties in using such demethylating agents in vivo: First, to accomplish DNA demethylation, the methylating reagent must be incorporated into DNA to inhibit each of the DNMTs through covalent binding (Momparler, 1985; Vasely, 1985). Although this effect on DNMTs is relatively specific, these reagents tend to exert cellular toxic effects particularly when high doses are used. Second, although epigenetic changes are reversible in experimental systems, aberrant promoter methylation and gene silencing return once DAC treatment is stopped (Pfeifer et al., 1990). Attempts to attenuate the effect of the above drawbacks of DAC included the use of a combination of low doses of DAC together with inhibitors of histone deacetylases. Cameron et al. (1999) observed that treatment of colorectal tumor cell lines with trichostatin A, an inhibitor of histone deacetylase, did not transcriptionally reactivate genes silenced by promoter hypermethylation. However, when the same cell lines were treated with a combination of Tricostatin A and low dose of DAC, synergistic reexpression was seen. Another example demonstrating the benefit of combining DAC with an inhibitor of histone deacetylase has been provided by Belinsky et al. (2003). The authors reported that short-term treatment of DNMT wild-type female mice with low doses of DAC decreased the incidence of neoplasms by 30%; when these doses of DAC were combined with the histone deacetylase inhibitor sodium phenylbutyrate, lung tumor development was reduced by 50%; no effect was observed with sodium phenylburate on its own. These and other studies imply that persistent gene activation may require the targeting of multiple components within the machinery of DNA methylation.

4.12. CONCLUDING REMARKS

Recently, the concept that cancer is as much a disease of misdirected epigenetics as it is a disease of genetic mutations has been consolidated. DNA methylation is now the most well-catagorized epigenetic change to occur in tumors. The exploitation of this epigenetic modification can have a number of advantages compared with other approaches. Some of these advantages have been pointed out earlier in the text. Two of these advantages should be underlined: First, there is strong evidence that methylation is an early event in carcinogenesis, a characteristic highly desired in cancer biomarkers. Second, the DNA containing the methylation information is highly stable and can be easily isolated from most bodily fluids as well as from archived fixed tissues. Extensive research efforts over the last few years have provided strong indications of the potential of DNA methylation as a source for cancer biomarkers. Some of these indications can be appreciated by the following considerations:

- Although current technologies for the detection of DNA methylation biomarkers in bodily fluids, such as blood or urine, have fairly low sensitivity, they excel in specificity. This is an asset in population-based screening. Future efforts will undoubtedly witness an enhanced sensitivity in present detection methods through the optimization of MSP primers, MethylLight probes, and nested PCR primers. The specificity of DNA methylation will also gain from the identification of panels of multiple rather than single marker. Such panels are expected to provide a powerful tool for risk assessment, early detection, and molecular diagnostics of various forms of cancer. A future yet reasonable scenario of such applications has been hypothesized by Laird (2003), which is worth summarizing here. Such anticipated scenario envisages an annual examination of a patient, where few mL of blood are drawn for a DNA methylation screen with a panel of 20 of the most common cancer-specific methylation markers. Few positive markers in such screen are then selected for further analysis with more specific, diagnostic methylation panels to determine the most likely tissue of origin. On the basis of this information, the physician uses noninvasive or minimally invasive imaging techniques (e.g., spiral coputed tomography) to confirm the diagnosis.

- Over the past 5 years, it has become apparent that DNA methylation is a common molecular alteration in human cancers. CpG islands in promoter regions of several key genes are unmethylated in normal tissues but are methylated to different extents in several human cancers (see Table 4.6). The list of inactivated tumor suppressor genes as a result of DNA methylation is still expanding and will surely represent a

TABLE 4.6. Promoter hyper methylation of a number of genes implicated in carcinogenesis.

Gene	Function	Tumor type	Reference
VHL	angiogenesis	clear cell (renal)	Herman, 1994
P16^{INK4a}	cyclin-dependent Kinase inhibitor	non-Hodgkin's lymphoma	Herman, 1997
E-cadherin	cell–cell adhesion	breast, gastric, prostate	Graff, 1995
Hmlhl	DNA mismatch repair	colon, endometrial, gastric	Kane, 1997 Herman, 1998 Grady, 2001
BRCA1	DNA damage repair	breast, ovarian	Dobrovic and Simpfendorfer, 1997
P14^{ARF}	*P53 stabilization*	colon, prostate, bladder	Nguyen, 2000 Esteller, 2000
P15	cyclin-dependent Kinase inhibitor	acute myeloid lymphoid leukemia	Herman, 1996b
ER	estrogen receptor	breast, colon, leukemia	Issa, 1994
DAPK	apoptosis	B cell lymphomas, lung	Zochbaur-Muller, 2001
p73	*p53* –like gene	lymphomas	Widschwendter, 2000 Sirchia, 2000
RABβ	retinoic acid receptor β2	breast, lung, bladder	Widschwendter, 2000 Sirchia, 2000
IGF-2	insulin-like growth factor	colon, breast, lung	Issa, 1996

rich source for designing future biomarker panels to be used in screening and prognosis-based studies. The exploitation of this long list of methylated genes can be also extended to enhance the specificity of some existing and well-established serum biomarkers. One example representative of such a situation is the use of *GSTP1* methylation in combination with PSA levels in prostate cancer. *GSTP1* methylation analysis was used to correctly distinguish between cancer and noncancer patients, when results were compared with those that are obtained by biopsy (Jerónimo et al., 2001).

- Another promising use of DNA methylation is in the management of cancer. For example, if a methylation assay results in the identification of a group of people at very high risk of a given type of cancer, the question arises as to how to manage this group of people? A partial answer to this question is currently provided by active research on the use of chemopreventive agents for such management. It is encouraging to note that there are several clinical trials in progress using compounds that cause demethylation of CpG islands or prevent histone deacetylation (Belinsky et al., 2003; Johnstone, 2002; Cheng et al., 2004). The reexpression of genes that control cell cycle, apoptosis, and cell growth/differentiation could lead to the elimination of premalignant clones in the aerodigestive tract.

In conclusion, it is fair to state that DNA methylation has the potential to serve a number of vital objectives in the fight against cancer. These include early detection, chemoprevention, and disease monitoring.

REFERENCES

Adams, J. M. (2003) Genes Dev. **17**, 2481.

Ahrendt, S. A., Chow, J. T., Xu, L.-H., et al. (1999) J. Natl. Cancer Inst. **91**, 332.

Ambartsumian, N. S., Grigorian, M. S., Larsen, I. F., et al. (1996) Oncogene **13**, 1621.

Andersen, K., Nesland, J. M., Holm, R., et al. (2004) Mod. Pathol. **17**, 990.

Aitken, A. (1996) Trends Cell Biol. **6**, 341.

Aitken, A., Baxter, H., Dubois, T., et al. (2002) Biochem. Soc. Trans. **30**, 351

Akahira, J.-I., Sugihashi, Y., Suzuki, T., et al. (2004) Clin. Cancer Res. **10**, 2687.

Alnemri, E. S., Livingston, D. J., Nicholson, D. W., et al. (1996) Cell **87**, 171.

An, Q., Liu, Y., Gao, Y., et al. (2002) Cancer Lett. **188**, 109.

Anderson, L., Seihamer, J. (1997) Electrophoresis **18**, 533.

Anisowicz, A., Sotiropoulou, G., Stenman, G., et al. (1996) Mol. Med. **2**, 624.

Antequera, F., Bird, A. (1993) EXS. **64**, 169.

Arends, M. J., Wyllie, A. H. (1991) Int. Rev. Exp. Pathol. **32**, 223.

Argon, Y., Simen, B. B. (1999) Semin. Cell Dev. Biol. **10**, 495.

Arrigo, A. P., Welch, W. J. (1987) J. Biol. Chem. **262**, 15359.

Arrigo, A. P., Mehlen, P. (1994) In the Biology of Heat Shock Proteins and Mol. Chaperones. Morimoto, R., Tisseres, A., Georgopoulos, C. (eds) Gold Spring Harbor Laboratory Press. NY. pp. 335–373.

Asakai, R., Davie, E. W., Chung, D. W. (1987) Biochemistry **26**, 7221.

Bachman, K. E., Park, B., Rhee, I., et al. (2003) Cancer Cell **3**, 89.

Barraclough, R., Dawson, K. J., Rudland, P. S. (1982) Eur. J. Biochem. **129**, 335.

Barraclough, R., Gibbs, F. E. M., Smith, J. A., et al. (1990) Biochem. Biophys. Res. Commun. **169**, 660.

Barrett, A. J., Rawlings, N. D., Woessner, J. F. (1998) Handbook of Proteolytic Enzymes, Academic Press, San Diego.

Basso, A. D., Solit, D. B., Chiosis, G., et al. (2002) J. Biol. Chem. **277**, 39858.

Baylin, S. B., Herman, J. G. (2000) Trends in Genetics **16**, 168.

Bearzatto, A., Conte, D., Frattini, M., et al. (2002) Clin. Cancer Res. **8**, 3782.

Beer, D. G., Kardia, S. L., Huang, C. C., et al. (2002). Nat. Med., **8**, 816

Belinsky, S. A., Nikula, K. J., Palmisano, W. A., et al. (1998) PNAS. USA. **95**, 11891.

Belinsky, S. A., Palmisano, W. A., Gilliland, F. D., et al. (2002) Cancer Res. **62**, 2370.

Belinsky, S. A., Klinge, D. M., Stidley, C. A., et al. (2003) Cancer Res. **63**, 7089.

Belinsky, S. A. (2004) Nat. Rev. Cancer **4**, 707.

Berg, T., Bradshaw, R. A., Cerretero, O. A., et al. (1992) Recent Progress on Kinins. Agents and Actions Supplement, Fritz, H., Muller-Esterl, W., Jochum, M., Roscher, A., Luppertz, K. (eds) Birkhauser Verlag, Basel, Vol. **38/1**, pp. 19–25.

Bird, A. P., Wolffe, A. P. (1999) Cell **99**, 451.

Bird, A. (2002) Gene Dev. **16**, 6.

Black W. C. (1999). Cancer Screening; Theory and Practice. Kramer, B. S., Gohagan, J. K., Prorok, P. C. (eds). New York: Marcel Dekker, Inc., pp. 327–377.

Blagosklonny, M. V., Toretsky, J., Bohen, S., et al. (1996) PNAS. USA. **93**, 8379.

Bhoola, K. D., Figueroa, C. D., Worthy, K. (1992a) Pharmacol. Rev. **44**, 1.

Bhoola, K. D., Ferguson, C. D., Worthy, K. (1992b). The Kinin System, Farmer, S. G. (ed.) Academic Press, San Diego, Vol. **1**, pp. 1–8.

Bhoola, K. D., Ramsaroop, R., Plendl, J., et al. (2001) Biol. Chem. **382**, 77.

Borgoño, C. A., Grass, L., Soosaipillai, et al. (2003) Cancer Res. **63**, 9032.

Borgoño, C. A., Michael, I. P., Diamandis, E. P. (2004) Mol. Cancer Res. **2**, 257.

Borgoño, C. A., Diamandis, E. P. (2004) Nat. Rev. Cancer **4**, 876.

Bova, M. P., McHaourab, H. S., Han, Y., et al. (2000) J. Biol. Chem. **275**, 1035.

Brabender, J., Usadel, H., Denenberg, K. D., et al. (2001) Oncogene **20**, 3528.

Brattsand, M., Egelrud, T. (1999) J. Biol. Chem. **274**, 30033.

Brattsand, M., Stefansson, K., Lundh, C., et al. (2005) J. Invest. Dermatol. **124**, 198.

Bruey, J. M., Ducasse, C., Bnniaud, P., et al. (2000) Nat. Cell Biol. **2**, 645.

Brugge, J. Yonemoto, W., Darrow, D. (1983) Mol. Cell Biol. **3**, 9.

Brunet, A., Bonni, A., Zigmond, M. J., et al. (1999) Cell **96**, 8511.

Bulavin, D. V., Higashimoto, Y., Demidenko, Z. N., et al. (2003) Nature **5**, 545.

Busslinger, M., Klix, N., Pfeffer, P., et al. (1996) PNAS. USA. **93**, 6129.

Bukau B., Deuerling, E., Pfund C., et al. (2000) Cell **101**, 119.

Cahill, C. M., Tzivion, G., Nasrin, N., et al. (2001) J. Biol. Chem. **276**, 13402.

Cameron, E. E., Bachman, K. E., Myohanen, S., et al. (1999) Nat. Genet. **21**, 103.

Camp, R. L., Rimm, E. B., Rimm, D. L. (1999) Cancer **86**, 2259.

Cross, S. S., Hamdy, F. C., Deloulme, J. C., et al. (2005) Blackwell Syn. **46**, 256.

Carrello, A., Ingley, E., Minchin, R. F., et al. (1999) J. Biol. Chem. **274**, 2682.

Castello, J. F., Frühwald, M. C., Smiraglia, D. J., et al. (2000) Nat. Genetics **25**, 132.

Cecconi, D., Astner, H., Donadelli, M., et al. (2003) Electrophoresis **24**, 4291.

Chappel, T. G., Konforti, B. B., Schmidt, S. L., et al. (1987) J. Biol. Chem. **262**, 746.

Chen, G., Cao, P., Goeddel, D. V. (2002) Mol. Cell **9**, 401.

Chen, S.-T., Pan, T.-L., Huang, C.-M. (2002) Cancer Letters **181**, 95.

Cheng, L., Pan, C.-X., Zhang, J.-T., et al. (2004) Clin. Cancer Res. **10**, 3064.

Ciocca, D. R., Oesterreich, S., Chamness, G. C., et al. (1993) J. Natl. Cancer Res. **85**, 1558.

Chiosis, G., H., Huezo, H., Rosen, N., et al. (2003) Mol. Cancer Ther. **2**, 123.

Clark, G. M., McGuire, W. (1988) Oncol **15**, 20.

Clements, J., Mukhtar, A., Ehrlich, A., et al. (1994) J. Med. Biol. Res. **27**, 1855.

Clements, J. (1997) The Kenin System, Farmer, S. (ed.), Academic press, New York, pp. 71–97.

Clements, J., Hooper J., Dong, Y., et al. (2001) Biol. Chem. **382**, 5.

Cleutjens, K. B., van Ekelen, C. C., van der Korput, H. A., et al. (1996) J. Biol. Chem. **271**, 6379.

Cleutjens, K. B., van der Korput, H. A., van Ekelen, C. C., et al. (1997) Mol. Endocrinol. **11**, 148.

Cohen, G. M. (1997) Biochem. J. **326**, 1.

Cory, S., Huang, D. C., Adams, J. M. (2003) Oncogene **22**, 8590.

Cornford, P. A., Dodson, A. R., Parsons, K. F., et al. (2000) Cancer Res. **60**, 7099.

Cramer, S. D., Chang, B.-L., Rao, A., et al. (2003) J. Natl. Cancer Inst. **95**, 1044.

Creagh, E. M., Sheehan, D., Cotter, T. G. (2000) Leukemia **14**, 1161.

Czar, M. J., Galigniana, M. D., Silverstein, A. M., et al. (1997) Biochemistry **36**, 7776.

Dai, Z., Lakshmana, R. R., et al. (2001) Neoplasia **3**, 314.

Danial, N. N., Korsmeyer, S. J. (2004) Cell **116**, 205.

Datta, S. R., Greenberg, M. E. (1998) Horm. Signaling **1**, 257.

Datta, S. R., Dudek, H., Tao, X., et al. (1997) Cell **91**, 231.

Datta, S. R., Katsov, A., Hu, L., et al. (2000) Mol. Cell **6**, 41.

Davenzac, N., Baldin, V., Gabrielli, B., et al. (2000) Oncogene **19**, 2179.

David, A., Mabjeesh, N., Azar, I., et al. (2002) J. Biol. Chem. **277**, 18084.

Davis, S. (1991) Radiology **180**, 1.

Davies, B. R., Davies, M. P., Gibbs, F. E., et al. (1993) Oncogene **8**, 999.

Davies, B. R., O'Donnell, M., Durkan, G. C., et al. (2002) J. Pathol. **196**, 292.

De Jong, W. W., Leunissen, J. A., Voorter, C. E. (1993) Mol. Biol. Evolu. **10**, 103.

De Jong, W. W., Caspers, G. J., Leunissen, J. A. (1998) Int. J. Biol. Macromol. **22**, 151.

Denmeade, S. R., Lovgren, J., Khan, S. R., et al. (2001) Prostate **48**, 122.

Deperthes, D., Marceau, F., Frenette, G., et al. (1997) Biochem. Biophys. Acta **1343**, 102.

Diamandis, P., Yuosef, G. M. (2002) Clin. Chem. **48**, 1198.

Diamandis, E. P., Yousef, G. M., Luo, L. Y., et al. (2000) Trends Endocrinol. Metab. **11**, 54.

Di Croce, L., Raker, V. A., Corsaro, M., et al. (2002) Science **295**, 1079.

Dixit, V., W., Green, S., Sarma, V., et al. (1990) J. Biol. Chem. **265**, 2973.

Dobrovic, A., Simpfendorfer, D. (1997) Cancer Res. **57**, 3347.

Donato, R. (1986) Biochem. Biophys. Acta **1450**, 191.

Dong, Y., Kaushal, A., Brattsand, M., et al. (2003) Clin. Cancer Res. **9**, 1710.

Downward, J. (1998) Curr. Opin. Cell Biol. **10**, 262.

Drohat, A. C., Baldisseri, D. M., Rustandi, R. R., et al. (1998) Biochemistry **37**, 2729.

Duffy, M. J. (2002) Clin. Chem. **48**, 1194.

Eads, C. A., Danenberg, K. D., Kawakami, K., et al. (1999) Cancer Res. **59**, 2302.

Eads, C. A., Danenberg, K. D., Kawakami, K., et al. (2000) Nucleic Acids Res. **28**, e32.

Ebralidze, A., Tulchinsky, E., Grigorian, M., et al. (1989) Gene Dev. **3**, 1086.

Ehrenfried, J. A., Herron, B. E., Towsend, C. M. Jr., et al. (1995) Surg. Oncol. **4**, 197.

Ekholm, E., Brattsand, M., Egelrud, T. (2000) J. Invest. Dermatol. **114**, 56.

Engelkamp, D., Shäfer, B. W., Erne, P., et al. (1992)biochem. **31**, 10258.

Eskelinen, M., Kataja, V., Hamalainen, E., et al. (1997) Anticancer Res. **17**, 1231.

Esler, W. P., Wolfe, M. S. (2001) Science **293**, 1449.

Esteller, M., Hamilton, S. R., Burger, P. C., et al. (1999a) Cancer Res. **59**, 793.

Esteller, M., Sanchez-Cespedes, M., Rosell, R., et al. (1999b) Cancer Res. **59**, 67.

Esteller, M. (2000) Eur. J. Cancer **36**, 2294.

Evans, B. A., Yun, Z. X., Close, J. A., et al. (1988) Biochemistry **27**, 3124.

Fahrner, J. A., Eguchi, S., Herman, J. G., et al. (2002) Cancer Res. **62**, 7213.

Falsone, F. M., Leptihn, S., Osterauer, A., et al. (2004) J. Mol. Biol. **344**, 281.

Felts, S. J., Owen, B. A. L., Nguyen, P., et al. (2000) J. Biol. Chem. **5**, 3305.

Ferguson, A. T., Evron, E., Umbricht, C. B., et al. (2000) Proc. Natl. Acad. Sci. USA **97**, 6049.

Fiedler, F. (1979) Bardykinin, Kallidin and Kallikrein, Erdos E. G. (ed.) Soringer Verlag, Berlin, pp. 103–161.

Fisher, B., Bauer M., Wickerham, L., et al.(1983) Cancer **52**, 1551.

Flaherty, K. M., Deluca-Flaherty, C., McKay, D. B. (1990) Nature **346**, 623.

Ford, J. C., Al-Khodairy, F., Fotou, E., et al. (1994) Science **265**, 533.

Foussias, G., Yuosef, G. M., Diamandis, E. P. (2000) Genomics **67**, 171.

Foy, C. A., Parkes, H. C. (2001) Clin. Chem. **47**, 990.

Franke, T. F., Kaplan, D. R., Cantley, L. C., et al. (1997) Science **275**, 665.

Freeman, B. C., Yamamoto, K. R. (2002) Science **296**, 2232.

Friess, H., Ding, J., Kleef, J., et al. (2003) Cell. Mol. Life Sci. **60**, 1180.

Fritz, G., Mittl, P. R., Vasak, M., et al. (2002) J. Biol. Chem. **277**, 33092.

Fritz, G., Heizmann, C. W. (2004) 3D structures of the calcium and zinc binding S100 proteins, in Messerschmidt, A. and Cygler, W. (eds) Handbook of Metalloproteins, Wiley, Chichester, pp. 529–540.

Froelich, C. J., Zhang, X., Turbov, J., et al. (1993) J. Immunol. **151**, 7161.

Fu, H., Subramanian, R. R., Masters, S. C. (2000) Annu. Rev. Pharmacol. Toxicol. **40**, 617.

Fukushima, D., Kitamura, N., Nakanishi, S. (1985) Biochemistry **24**, 8037.

Fuller, K. J., Issels, R. D., Slosman, D. O., et al. (1994) Eur. J. Cancer **30A**, 1884.

Gan, L., Lee, I., Smith, R., et al. (2000) Gene **257**, 119.

Garbuglia, M., Verzini, M., Rustandi, R. R., et al. (1999) Biochem. Biophys. Res. Commun. **254**, 36.

Gässler, C. S., Wiederkehr, T., Brehmer, D., et al. (2001) J. Biol. Chem. **276**, 32538.

Garrido, C., Bruey, J. M., Fromentin, A., et al. (1999) FASEB. J. **13**, 2061.

Gerke, V., Weber, K. (1985) EMBO J. **4**, 2917.

Gibbs, F. E. M., Wilkinson, M. C., Rudland, P. S., et al. (1994) J. Biol. Chem. **269**, 18992.

Goldstein, J. C., Waterhouse, N. J., Juin, P., et al. (2000). Nature Cell Biol. **2**, 156.

Gonzalgo, M. L., Liang, G., Spurk 3rd, C. H., et al. (1997) Cancer Res. **57**, 594.

Grady, W. M., Rajput, A., Lutterbaugh, J. D., et al. (2001) Cancer Res. **61**, 900.

Graff, J. R., Herman, J. G., Lapidus, R. G., et al. (1995) Cancer Res. **55**, 5195.

Green, D. R., Reed, J. C. (1998) Science **281**, 1309.

Green, D. R., Kroemer, G. (2004) Science **305**, 626.

Grigorian, M., Ambartsumian, N., Lykkesfeldt, A. E., et al. (1996) Int. J. Cancer **67**, 831.

Grigorian, M., Anderesen, S., Tulchinsky, M., et al. (2001) J. Biol. Chem. **276**, 22699.

Gurbuxani, S., Bruey, J. M., Fromentin, A., et al. (2001) Oncogene **20**, 7478.

Gygi, S. P., Rochon, Y., Franza, B. R., et al. (1999) Mol. Cell. Biol. **19**, 1720.

Hahn, G. M., Li, G. C. (1990) in Stress Proteins in Biology and Medicine, Morimoto, R. I., Tissieres, A., Georgeopoulos, C. (eds) Harbor Press, Cold Spring Harbor, NY, p79.

Hanahan, D., Weinberg, R. A. (2000) Cell. **100**, 57.

Hanash, S. M. (2003). J. Biol. Chem. **278**, 7607.

Hansson, L., Stromqvist, M., Backman, A., et al. (1994) J. Biol. Chem. **269**, 19420.

Hainzl, O., Wegele, H., Richter, K., et al. (2004) J. Biol. Chem. **279**, 23267.

Harada, H., Becknell, B., Wilm, M., et al. (1999) Mol. Cell **3**, 413.

Hartl, F. U., Hayer-Hartl, M. (2002). Science **295**, 1852.

Harrison, C. J., Hayer-Hartl, M., Di Liberto, M., et al. (1997) Science **276**, 431.

Harvey, T. J., Hooper, J. D., Myers, S. A., et al. (2000) J. Biol. Chem. **275**, 37397.

Haslbeck, M., Buchner, J. (2002) Prog. Mol. Subcell Biol. **28**, 37.

Haslett, C. (1992) Clin. Sci. **267**, 1456.

Hatada, I., Hayashizaki, Y., Hirotsune, S., et al. (1991) PNAS. USA. **88**, 9523.

Heard, E. C., Rougeulle, C., Arnaud, D., et al. (2001) Cell **107**, 727.

Helmbrecht, K., Zeise, E., Rensing, L. (2000) Cell Profil. **33**, 341.

Henderson, B. E., Feigelson, H. S. (2000) Carcinogenesis **21**, 427.

Henderson, B. R., Tansey, W. P., Philips, S. M., et al. (1992) Cancer Res. **52**, 2489.

Herman, J. G., Latif, F., Weng, Y., et al. (1994) PNAS. USA. **91**, 9700.

Herman, J. G., Graff, J. R., Nelkin, P. D., et al. (1996a) PNAS. USA.**93**, 9821.

Herman, J. G, Jen, J., Merlo, A., et al. (1996b) Cancer Res. **56**, 722.

Herman, J. G., Civin, C. I., Issa, J.-P. J., et al. (1997) Cancer Res. **57**, 837.

Herman, J. G., Umar, A., Polyak, K., et al. (1998) PNAS. USA. **95**, 6870.

Herman, J. G. (1999) Semin. Cancer Biol. **9**, 359.

Herman, J. G., Baylin, S. B. (2003) N. Engl. J. Med. **349**, 2042.

Hermeking, H., Legauer, C., Polyak, K., et al. (1997) Mol. Cell **1**, 3.

Hermeking, H. (2003) Nat. Rev. Cancer **3**, 931.

Hickey, E., Brandon, S. E., Smale, G., et al. (1999) Mol. Cell Biol. **9**, 2615.

Hilt, D. C., Kligman, D. (1991) In: C. W. Heinzmann (ed.) Novel Calcium Binding proteins. Fundamental and Clinical implications, Springer-Verlag, Berlin, pp. 65–103.

Holliday, R. (1993) EXS. **64**, 452.

Horl, W. H. (1989) Design of Enzyme Inhibitors as Drugs, Sandler, M., Smith, H. J. (eds) Oxford University Press, Oxford, pp. 573–581.

Howarth, D. J, Aronson, L. B., Diamandis, E. P. (1997) Br. J. Cancer **75**, 1646.

Hu, J. C., Zhang, C., Sun, X., et al. (2000) Gene **251**, 1.

Huang, T. H., Perry, M. R., Laux, D. E. (1999) Hum. Mol. Genet. **8**, 459.

Hung, J., Lam, S., Le Riche, J. C., et al. (1991) Surg. Med. **11**, 99.

Huot, J., Houle, F., Spitz, D. R., et al. (1996) Cancer Res. **56**, 273.

Iacobuzio-Donahue, C. A., Ashfaq, R., Maitra, A., et al. (2003) Cancer Res. **63**, 8614.

Ichimura, T., Isobe, T., Okuyama, T., et al. (1988) PNAS. USA. **85**, 7084.

Ichimura, T., Ito, M., Itagaki, C., et al. (1997) FEBS Lett. **413**, 273.

Ingolia, T. D., Craig, E. A. (1982) PNAS. USA. **79**, 2360.

Issa, J.-P. J., Ottaviano, Y. L., Celano, P., et al. (1994) Nature Genet. **7**, 536.

Issa, J.-P. J., Vertino, P. M., Boehm, C. D., et al. (1996) PNAS. USA. **93**, 11757.

Isaacs, J. S., Xu, W., Neckers, L. (2003) Cancer Cell **3**, 213.

Iwaka, A., Nagano, T., Nakagawa, M., et al. (1997) Genomics **45**, 386.

Iwata, N., Ymamoto, H., Sasaki, S., et al. (2000) Oncogene **19**, 5298.

Jacbson, M. D., Weil, M., Raff, M. C. (1997) Cell, **88**, 347.

Jackson, J. P., Lindroth, A. M., Cao, X., et al. (2002) Nature **416**, 556.

Jaattela, M. (1995) Int. J. Cancer **60**, 689.

Jaattela, M., Wissing, D., Kokholm, K., et al. (1998) EMBO J. **17**, 6124.

Jaattela, M. (1999a) Ann. Med. **31**, 261.

Jaattela M. (1999b). Exp. Cell. Res. **248**, 30.

Jahr, S., Hentze, H., Englisch, S., et al. (2001) Cancer Res. **61**, 1659.

Jemal, A., Thomas, A., Murry, T., et al. (2002) CA: Cancer J. Clin. **52**, 23

Jenuwein, T., Allis, C. D. (2001) Science **293**, 1074.

Jerónimo, C., Usadel, H., Henrique, R., et al.(2001) J. Natl. Cancer Inst. **93**, 1747.

Jiang, K., Pereira, E., Maxfield, M., et al. (2003) J. Biol. Chem. **278**, 25207.

Johnstone, R. W., Rueffi, A. A., Lowe, S. W. (2002) Cell **108**, 153.

Jolly, C., Morimoto, R. I. (2000) J. Natl. Cancer Inst., **92**, 1564

Jones, P., Baylin, S. B. (2002) Nat. Rev. Genet. **3**, 415.

Kamal, A., Thao, L., Sensintaffar, J., et al. (2003) Nature **425**, 407.

Kane, M., Loda, M., Gaida, G. M., et al. (1997) Cancer Res. **57**, 808.

Kapadia, C., Chang, A., Sotiropoulou, G., et al. (2003) Clin. Chem. **49**, 77.

Kappe, G., Verschuure, P., Philipsen, R. L., et al. (2001) Biochim. Biophys. Acta **1520**, 1.

Karpf, A. R., Jones, D. A. (2002) Oncogene **21**, 5496.

Keirsebilck, A., Bonne, S., Bruyneel, E., et al. (1998) Cancer Res. **58**, 4587.

Kennedy, T., Hirsch, F. R., Miller, Y. E., et al. (2000) Lung Cancer **29** (Supl. 1) 244.

Kersting, M., Friedl, C., Kraus, A., et al. (2000) J. Clin. Oncol. **18**, 3221.

Kimura, K., Endo, Y., Yonemura, Y., et al. (2000) Int. J. Oncol. **16**, 1125.

Kishi, T., Grass, L., Soosaipillai, A., et al. (2003) Clin. Chem. **49**, 87.

Klein-Hitpass, L., Schwerk, C., Kahmann, S., et al. (1998) J. Mol. Med. **76**, 490.

Kluck, R. M., Bossy-Wetzel, E., Green, D. R., et al. (1997) Science **275**, 1132.

Koizume, S., Tachibana, K., Sekiya, T., et al. (2002) Nucleic Acids Res. **30**, 4770.

Komatsu, N., Takata, M., Otsuki, N., et al. (2003) J. Invest. Dermatol. **121**, 542.

Kozmik, Z., Wang, S., et al. (1992) Mol. Cell Biol. **12**, 2662.

Krajewski, S., Krajewska, M., Turner, B. C., et al. (1999) Endocrine-Related Cancer **6**, 29.

Krane, S. M. (2003) Arthritis Res. Ther. **5**, 2.

Kraut, H., Frey, E. K., Werle, E. (1930) Z. Physiol. Chem. **189**, 97.

Kriajevska, M., Bronstein, L. B., Scott, D. J., et al. (2000) Biochim. Biophys. Acta **1498**, 7, 252.

Kriajevska, M., Fisher-Larsen, M., Moertz, E., et al. (2002) J. Biol. Chem. **277**, 5229.

Krief, S., Faivre, J. F., Robert, P., et al. (1999) J. Biol. Chem. **274**, 36592.

Kumar, A., Mikolajczyk, S. D., Goel, A. S., et al. (1997) Cancer Res. **57**, 3111.

Kurakawa, E., Shimamoto, T., Utsumi, K., et al. (2001) Int. J. Oncol. **19**, 277.

Kuwana, T., Bouchier-Hayes, L., Chipuk, J. E. (2005) Mol. cell **17**, 525.

Lacobuzio-Donahue, C. A., Ashfaq, R., Maitra, A., et al. (2003) Cancer res. **63**, 8614.

Laird, P. W. (1997) Mol. Med. Today **3**, 223.

Laird, P. W. (2003) Nat. Rev. Cancer **3**, 253.

Lakshmi, M. S., Parker, C., Sherbet, G. V. (1997) Anticancer Res. **17**, 3451.

Lam, S., Kennedy, T., Unger, M., et al. (1998) Chest **113**, 696.

Langdon, S. P., Rabiasz, G. J., Hirst, G. L., et al. (1995) Clin. Cancer Res. **1**, 1603.

Lavoie, J. N., Hickey, E., Weber, L. A., et al. (1993) J. Biol. Chem. **268**, 24210.

Lavoie, J. N., Lambert, H., Hickey, E., et al. (1995) Mol. Cell Biol. **15**, 505.

Lee, A., Whyte, M. K., Haslett, C. (1993) J. Leukocyte Biol. **54**, 283.

Lee, W.-Y., Su, W.-C., Lin, P.-W., et al. (2004) Oncology **66**, 429.

Leffers, H. P., Madsen, H. H., Rasmussen, B., et al. (1993) J. Mol. Biol. **231**, 982.

Leon, S. A., Shapiro, P., Sklaroff, D. M., et al. (1977) Cancer Res. **37**, 646.

Li, P., Nijhawan, D., Budihardjo, I., et al. (1997) Cell **91**, 479.

Li, B., Goyal, J., Dhar, S., et al. (2001) Cancer Res. **61**, 8014.

Lin, J., Blake, M., Tang, C., et al. (2001) J. Biol. Chem. **276**, 35037.

Lindquist, S., Craige, E. A. (1988) Annu. Rev. Genet. **22**, 631.

Linzer, D. I., Nathans, D. (1983) pnas usa. **80**, 4271.

Little, S. P., Dixon, E. P., Norris, F., et al. (1997) J. Biol. Chem. **272**, 25135.

Liu, D., Bienkowska, J., Petosa, C., et al. (1995) Nature **376**, 191.

Liu, X. L., Wazer, D. E., Watanbe, K., et al. (1996) Cancer Res. **56**, 3371.

Liu, F. F., Miller, N., Levin, W., et al. (1996) Int. J. Hyperthermia **12**, 197.

Liu, X. L., Xiao, B., Yu, Z. C. et al. (1999) Word J. Gastroenterol. **5**, 199.

Lodygin, D., Diebold, J., Hermeking, H. (2004) Oncogene **23**, 9034.

Logsdon, C. D., Simeone, D. M., Binkley, C., et al. (2003) Cancer Res. **63**, 2649.

López-Otin, C., Overall, C. M. (2002) Nat. Rev. **3**, 509.

Lovgren, J., Rajakoski, K., Karp, M., et al. (1997) Biochem. Biophys, Res. Commun. **238**, 549.

Love, S., King, R. J. (1994) Br. J. Cancer **69**, 743.

Lundwall, A., Lilja, H. (1987) FEBS. **214**, 317.

Luo, A., Kong, J., Hu, G., et al. (2004) Oncogene **23**, 1291.

Luo, L. Y., Grass, L., Diamandis, E. P. (2000) Anticancer Res. **20**, 981.

Luo, L., Herbrick, J. A., Sherer, S. W., et al. (1998) Biochem. Biophys. Res. Commun. **247**, 580.

Luo, Z. J., Zhang, X. F., Rapp, U., et al. (1995) J. Biol. Chem. **270**, 23681.

Magklara, A., Mellati, A. A., Wasney, G. A., et al. (2003) Biochem. Biophys. Res. Commun. **307**, 948.

Maloney, A., Clarke, P. A., Workman, P. (2003) Curr. Cancer Drug Targets **3**, 331.

Mansour, E. G., Ravdin, P. M., Dressler, L. (1994) Cancer **74**, 381.

Marcus, P. M. (2001). J. Clin. Oncol. **19**, 83S.

Marenholtz, I., Heizmann, C. W. (2004) Biochem. Biophys. Res. Commun. **313**, 237.

Markl, I. D. C., Cheng, J., Liang, G., et al. (2001) Cancer Res. **61**, 5875.

Maloney, A., Workman, P. (2002) Expert Opin. Biol. Ther. **2**, 3.

Marte, B. M., Downward, J. (1997) Trends Biochem. Sci. **22**, 355.

Martin, S. J., Green, D. R. (1995) Cell **82**, 349.

Masters, S. C., Subramanian, R. R., Truong, A., et al. (2002) Biochem. Soc. Trans. **30(4)**, 360.

Matsumura, H., Shiba, T., Inoue, T., et al. (1998) Structure **6**, 233.

McCarthy, J. S., Buchberger, A., Reinstein, J., et al. (1995) J. Mol. Biol. **249**, 126.

McGuire, W., Clark, G. M. (1992) N. Engl. J. Med. **326**, 1756.

McKenna, S. L., Padua, R. A. (1997) Br. J. Haematol. **96**, 659.

McLaughlin, S. H., Smith, H. W., Jackson, S. E. (2002) J. Mol. Biol. **315**, 787.

Medieros, R., Morais, A., Vasconcelos, A., et al. (2002) Prostate **53**, 88.

Merlo, A., Herman, J. G., Mao, L., et al. (1995) Nat. Med. **1**, 686.

Misselwitz, B., Staeck, O., Rapoport, T. A. (1998) Mol. Cell **2**, 593.

Mitsui, S., Tsuruoka, N., Yamashiro, K., et al. (1999) Eur. J. Biochem. **260**, 627.

Mitsui, S., Yamada, T., Okui, A., et al. (2000) Biochem. Biophys. Res. Commun. **272**, 205.

Momparler, R. L. (1985) Pharmacol. Ther. (1985) **30**, 287.

Moore, B. (1965) Biochem. Biophys. Res. Commun. **19**, 739.

Moore, B. E., Perez, V. J. (1967) in Physiological and Biochemical Aspects of Nervous Integration, Carlson,F. D., (ed.) pp. 343–359, Prentice-Hall Englewood Cliffs, NJ.

Morimoto, R. I. (2002) Cell **110**, 281.

Moreira, J. M. A., Gromov, P., Celis, J. E. (2004) Mol. Cell. Proteomics **3.4**, 410.

Moreira, J. M. A., Ohlsson, G., Rank, F. E., et al. (2005) Mol. Cell Proteomics. **3.4**, 410.
Moroz, O. V., Antson, A. A., Dodson, G. G., et al. (2000) Acta Crystallogr. D Biolo. Crystallogr. 56, 189.
Moroz, O. V., Antson, A. A., Dodson, E. J., et al. (2002) Acta Crystallogr. D Biol. Crystallogr. **58**, 407.
Mosser, D. D., Morimoto, R. I. (2004) Oncogene **23**, 2907.
Movat, H. Z. (1979) Bardykinin, Kallidin and Kallikrein, Erdos E. G. (ed.) Soringer Verlag, Berlin, pp. 1–89.
Muller, L., Schaupp, A., Walerych, D., et al. (2004) J. Biol. Chem. **279**, 48886.
Murakami, H., Pain, D., Blobel, G. (1988) J. Cell Biol. **107**, 2051.
Muslin, A. J., Xing, H. (2000) Cell Signl. **12**, 703.
Muslin, A. J., Tanner, J. W., Allen, P. M., et al. (1996) Cell **84**, 889.
Myers, S. A., Clements, J. A. (2001) J. Clin. Endocrinol. **86**, 2323.
Nakanishi, K., Hashizume, S., Kato, M., et al. (1997) Hum. Antib. **8**, 189.
Nakagawa, H., Nuovo, G. J., Zervos, E. E., et al. (2001) Cancer res. **61**, 6991.
Nakayama, J., Rice, J. C., Strahl, B. D., et al. (2001) Science **292**, 110.
Nanbu, K., Konishi, I., Mandai, M., et al. (1998) Cancer detect. Prev. **22**, 549.
Nelson, P. S., Gan, L., Ferguson, C., et al. (1999) PNAS. USA. **96**, 3114.
Neurath, H. (1985) Fed. Proc. **44**, 2907.
Nguyen, T. T., Nguyen, C. T., Gonzales, F. A., et al. (2000) Prostate **43**, 233.
Nguyen, C. T., Weisenberger, D. J., Velicescu, M., et al. (2002) Cancer Res. **62**, 6456.
Nigg, E. A. (1995) Bioassays **17**, 471.
Ninomiya, I., Ohta, T., Fushida, S., et al. (2001) Int. J. Oncol. **18**, 715.
Nishigaki, R., Osaki, M., Hiratsuka, M., et al. (2005) Proteomics **5**, 3205.
Nollen, E. A., Brunsting, J. F., Roelofsen, H., et al. (1999) Mol. Cell Biol. **19**, 2069.
Norma, K., Allis, C. D., Grewal, S. I. S. (2001) Science **293**, 1150.
Nomura, M., Shimizu, S., Sugiyama, T., et al. (2003) J. Biol. Chem. **278**, 2058.
Nuovo, G. L., Plaia, T. W., Belinsky, S. A., et al. (1999) PNAS. USA. **96**, 12754.
Novitskaya, V., Grigorian, M., Kriajevska, M., et al. (2000) J. Biol. Chem. **275**, 41278.
Nylandsted, J., Rohde, M., Brand, K., et al. (2000) Proc. Natl. Acad. Sci. USA **97**, 7871.
Obermann, W. M., Sondermann, H., Russo, A. A., et al. (1998) J. Cell. Biol. **143**, 901.
Oesterreich, S., Benndorf, R., Bielka, H. (1990) Biomed. Biochim. Acta **49**, 219.
Oesterreich, S., Weng, C. N., Qiu, M., et al. (1993) Cancer Res. **53**, 4443.
Okabe, A., Momota, Y., Yoshida, S., et al.(1996) Brain Res. **728**, 116.
Okano, M., Bell, D. W., Haber, D. A., et al. (1999) Cell **99**, 247.
Oltvai, Z. N., Korsmeyer, S. J. (1994) Cell **79**, 189.
Oppermann, H. (1981) PNAS. USA. **78**, 1067.
Osada, H., Tatematsu, Y., Yatabe, Y., et al. (2002) Oncogene **21**, 2418.
Palcic, B., Lam, S., Hung, J., et al. (1991) Chest, **99**, 742.
Palmisano, W. A., Divine, K. K., Saccomanno, G., et al. (2000) Cancer Res. **60**, 5954.
Palmisano, W. A., Crume, K. P., Grimes, M. J., et al. (2003) Cancer res. **63**, 4620.
Pandey, P., Farber, R., Nakazawa, A., et al. (2000) Oncogene **19**, 1975.

Panaretou, B., Siligardi, G., Meyer, P., et al. (2002) Mol. Cell **10**, 1307.

Parcellier, A., Gurbuxani, S., Schmitt, E., et al. (2003) Biochem. Biophys. Res. Commun. **304**, 505.

Parkin, D. M., (2004) Review of Health Statistics, national Board of Health, Copenhagen, Denmark.

Parkin, D. M., Bray, F. I., Devesa, S. S., et al. (2001) Eur. J. Cancer **37**, S4.

Parker, C., Lakshmi, M. S., Plura, B., et al. (1994) DNA Cell Biol. **13**, 343.

Parsell D. A., Lindquist. S. (1993) Annu. Rev. Genet. **27**, 437.

Paul, C., Manero, F., Gonin, S., et al. (2002) Mol. Cell Biol. **22**, 816.

Pearl, L. H., Prodromou, C. (2001) Adv. Protein Chem. **59**, 157.

Pedersen, K. B., Nesland, J. M., Fodstad, O., et al. (2002) Br. J. Cancer **18**, 1281.

Prodromou, C., Pearl, L. H. (2003) Curr. Cancer Drug Targets **3**, 301.

Pedrocchi, M., Schäfer, I., Durussel, J., et al. (1994) Biochemistry **33**, 6732.

Perl, A. K., Wilgenbus, P., Dahl, U., et al. (1998) Nature **392**, 190.

Pfeifer, G. P., Steigerwald, R. S., Hansen, R. S., et al. (1990) PNAS. USA. **87**, 8252.

Platt-Higgins, A. M., Renshaw, C. A., West, C. R., et al. (2000) Int. J. Cancer **89**, 198.

Potts, B. C. M., et al. (1995) Nat. Struct. Biol. **2**, 790.

Pratt, W. B., Toft, D. O. (2003) Exp. Biol. Med. **228**, 111.

Ralhan, R., Kaur, J. (1995) Clin. Cancer Res. **1**, 1217.

Ramsahoye, B. H., Biniszkiewicz, D., Lyko, F., et al. (2000) PNAS. USA. **97**, 5237.

Rane, M. J., Coxon, P. Y., Powell, D. W., et al. (2001) J. Biol. Chem. **276**, 3517.

Rane, M. J., Pan, Y., Singh, S., et al. (2003) J. Biol. Chem. **278**, 27828.

Rao, A., Chang, B.-L., Hawkins, G., et al. (2003) Urology **61**, 864.

Rao, A., Cramer, S. (1999) Cancer Res. **40**, 60.

Ravagnan, I., Gurbuxani, S., Susin, S. A., et al. (2001) Nat. Cell Biol. **3**, 839.

Rawlings, N. D., O'Brian, E., Barrett, A. J. (2002) Nucleic Acids Res. **30**, 343.

Rawlings, N. D., Barrett, A. J. (1993) Biochem. J. **290**, 205.

Reed, J. C. (1998) Science **275**, 983.

Réty, S., Sopkova, D., Renouard, M., et al. (1999) Nat. Struct. Boil. **6**, 89.

Rhee, I., Jair, K-W., Yen, R-W., et al. (2000) Nature **404**, 1003.

Rhee, I., Bachman, K. E., Park, B. H., et al. (2002) Nature **416**, 552.

Riegman, P. H., Klassen, P., van der Korput, J. A., et al. (1988) Biochem. Biophys. Res. Commun. **155**, 181.

Riegman, P. H., Vlietstra, R. J., van der Korput, J. A., et al. (1989) Biochem. Biophys. Res. Commun. **159**, 95.

Riegman, P. H., Vliestra, R. J., van der Korput, H. A., et al. (1991) Mol. Cell Endocrinol. **5**, 1921.

Rittinger, K., Budman, J., Xu, J., et al. (1999) Mol. Cell **4**, 153.

Rjdeout, W. M., Coetzee, G. A., Olumi, A. F., et al. (1990) Science **249**, 1288.

Rohde, M., Daugaard, M., Jensen, M. H., et al. (2005) Genes Dev. **19**, 570.

Roman-Gomez, J., Jimenez-Velasco, A., Agirre, X., et al. (2004) Leukemia **18**, 362.

Rosen, E. M., Nigam, S. K., Goldberg, I. D. (1994) J. Cell Biol. **127**, 1783.

Rosty, C., Takashi, U., Argani, P., et al. (2002) Am. J. Pathol. **160**, 45.

Ruden, C. M., Thompson, C. B. (1997) Annu. Rev. Med. **48**, 267.

Rudland, P. S., Platt-Higgins, A., El-Tanani, M., et al. (2000) Cancer Res **15**, 1595.

Rudland, P. S., Platt-Higgins, A., Renshaw, C., et al. (2000) Cancer Res. **60**, 1595.

Russo, V. E. A., Martienssen, R. A., Riggs, A. D. (eds) (1996) Epigenetic Mechanisms of Gene Regulation. Cold Spring Harbor Laboratory Press, Plain View NY.

Rustandi, R. R., Drohat, A. C., Baldisseri, P. T., et al. (1998) Biochemistry **37**, 1951.

Rutherford, S. L., Lindquist, S. (1998) Nature **396**, 336.

Sadar, M. D. (1999) J. Biol. Chem. **274**, 7777.

Sadri, R., et al. (1995) Nucleic Acids Res. **24**, 5058.

Sakaguchi, M., Miyazaki, M., Takaishi, M., et al. (2003) J. Cell Biol. **163**, 825.

Saleem, M., Adhami, V. M., Ahmad, N., et al. (2005) Clin. Cancer Res. **11**, 147.

Saleh, A., Srinivasula, S. M., Acharya, S., et al. (1999) J. Biol. Chem. **274**, 17941.

Saleh, A., Srinivasula, S. M., Acharya, S., et al. (2000) Nature **2**, 476.

Samali, A., Zhivotovsky, B., Jones, D., et al. (1999) Cell Death Differ. **6**, 495.

Samali, A., Robertson, J. D., Peterson, E., et al. (2001) Cell Stress and Chaperones **6**, 49.

Samuel, T., Weber, H. O., Rauch, P., et al. (2001) J. Biol. Chem. **276**, 45201.

Santi, D. V., Garrett, C. E., Barr, P. J. (1983) Cell **83**, 9.

Santarosa, M., Favaro, D., Quaia, M., et al. (1997) Eur. J. Cancer **33**, 873.

Sapelton, R., Weith, A., et al. (1993) Nat. Genet. **3**, 292.

Sarto, C., Binz, P. A., Mocarelli, P. (2000) Electrophoresis. **21**, 1218.

Sastry, M., Ketchen, R. R., Crescenzi, O., et al. (1998) Structure **6**, 223.

Sato, N., Maehara, N., Su, G. H., et al. (2003a) J. Natl. Cancer Inst. **95**, 327.

Sato, N., Fukushima, N., Maitra, A., et al. (2003b) Cancer Res. **63**, 3735.

Scarisbrick, I. A., Blaber, S., Lucchinetti, C. F., et al.(2002) Brain **125**, 1283.

Schäfer, B. W., Wicki, R., Engelkamp D., et al. (1995) Genomics **25**, 638.

Schäfer, B. W., Fritschy, J. M., Murmann, P., et al. (2000) J. Biol. Chem. **275**, 30623.

Schedlich, L. J., Bennets, B. H., Morris, B. J. (1987) DNA **6**, 429.

Scheufler, C., Brinker, A., Bourenkov, G., et al. (2000) Cell **101**, 199.

Shi, Y., Thomas, J. O. (1992) Mol. Cell Biol. **12**, 2186.

Schmid, D., Baici, A., Gehring, H., et al. (1994) Science **263**, 971.

Schumann, A., Mooney, A. F., Sanders, L. C., et al. (2000) Curr. Biol. **10**, 127.

Schwirzke M., Schiemann, S., Gnirke, A. U., et al. (1999) Anticancer Res. **19**, 1801.

Shen, J., Person, M. D., Zhu, J., et al. (2004) Cancer Res. **64**, 9018.

Sherbet, G. V., Lakshmi, M. S. (1998) Anticancer Res. **18**, 2415.

Shin, B. K., Wang, H., Yim, A. M., et al. (2003). J. Biol. Chem. **278**, 7607.

Segal, G. M., Kaplan, D. R., Greenberg, M. E. (1997) Science **275**, 661.

Silvestrini, R., Daidone, M. G., Benini, E., et al. (1996) Clin. Cancer Res. **2**, 2007.

Sirchia, S. M., Ferguson, A. T., Sironi, E., et al. (2000) Oncogene **19**, 1556.

Sone, S., Takashima, S., Li, F., et al. (1998) Lancet **351**, 1242.

Sinha, D. P., Poland, J., Kohl, S., et al. (2003) Electrophoresis. **24**, 2386.

Scarisbrick, I. A., Blaber, S., Lucchinetti, C. F., et al. (2002) Brain **125**, 1283.

Shiloh, Y. (2003) Nat. Rev. Cancer **3**, 155.

Shimomura, K., Sakakura, Fujita, Y., et al. (2002) Nippn GEKA Gakkai Zasshi **103**, 386.

Simpson, P. T., Gale, T., Reis-Filho, J. S., et al. (2004) J. Pathology **202**, 274.

Shultz, R. M., Liebman, M. N. (1997) Textbook of Biochemistry with Clinical correlation, Devlin, T. M. (ed.) Wiley-Liss, New York. pp. 1–16.

Smits, V. A., Medema, R. H. (2001) Biochem. Biophys. Acta **1519**, 1.

Smith, D. F., Whitesell, L., Katsanis, E. (1998) Pharmacol. Rev. **50**, 493.

Smith, S. P., Shaw, G. (1997) J. Biomol. NMR **10**, 77.

Sommer, A., Hoffmann, J., Lichtner, R. B., et al. (2003) J. Steroid Biochem. Mole. Biol. **85**, 33.

Song, J. Z., Stirzaker, C., Harrison, J., et al. (2002) Oncogene **21**, 1048.

Sporn, M. B. (1996) Lancet. **347**, 1377.

Soti, C., Racz, A., Csemely, P. (2002) J. Biol. Chem. **277**, 7066.

Stankiewicz, A. R., Lachapelle, G., Foo, C. P. Z., et al. (2005) J. Biol. Chem. **280**, 38729.

Stapleton, P., Weith, A., Urbanek, P., et al. (1993) Nat. Genet. **3**, 292.

Stebbins, C. E., Russo, A. A., Schneider, C., et al. (1997) Cell **89**, 239.

Stennicke, H. R., et al. (1998) J. Biol. Chem. **273**, 27084.

Stephenson, S. A., Verity, K., Ashworth, L. K., et al. (1999) J. Biol. Chem. **274**, 23210.

Stokoe, D., Stephens, L. R., Copeland, T., et al. (1997) Science **277**, 567.

Strahl, B. D., Allis, C. D. (2000) Nature **403**, 41.

Strathdee, G., Brown, R. (2002) Exp. Rev. Mol. Med. **2002**, 1.

Subramanian, R. R., Masters, S. C., Zhang, H., et al. (2001) Expt. Cell Res. **271**, 142.

Suzuki, H., Gabrielson, E., Chen, W., et al. (2002) Nat. Genetics **31**, 141.

Suzuki, H., Itoh, F., Toyota, M., et al. (2000) Cancer Res. **60**, 4353.

Suzuki, T., Urano, T., Tsukui, T., et al. (2005) Clin. Cancer Res. **11**, 6148.

Suzuki, J., Yoshida, S., Chen, Z. L., et al. (1995) Neurosci. Res. **23**, 345.

Takashima, M., Kuramitsu, Y., Yokoyama, Y., et al. (2003) Proteomics **3**, 2487.

Takayama, T. K., Carter, C. A., Deng, T. (2001) Biochemistry **40**, 1679.

Takayama, T. K., Fujikawa, K., Davie, E. W. (1997) J. Biol. Chem. **272**, 21582.

Takayama, S., Sato, T., Krajewski, S., et al. (1995) Cell **80**, 279.

Tamaru, H., Selker, E. U. (2001) Nature **414**, 277.

Tanaka, K., Iwamoto, S., Gon, G., et al. (2000) Clin. Cancer Res. **6**, 127.

Tang, S. C., Shaheta, N., Chernenko, G., et al. (1999) J. Clin. Oncol. **17**, 1710.

Thornberry, N. A., Lazebnik, Y. (1998) Science **281**, 1312.

Tissieres, A., Mitchell, H. K., Tracy, U. M. (1974) J. Mol. Biology **85**, 389.

Tost, J., Schatz, P., Schuster, M., et al. (2003) Nucleic Acids Res. **31**, e50.

Townsend, P. A., Dublin, E., Hart, I. R. (2002) J. Path. **197**, 51.

Tremolada, L., Magni, F., Valsecchi, C., et al. (2005) Proteomics **5**, 788.

Tsou, J. A., Hagen, J. A., Carpenter, C. L., et al. (2002) Oncogene **21**, 5450.

Toyota, M., Ho, C., Jaur, K-W., et al. (1999) Cancer Res. **59**, 2307.

Turner, B. C., Krajewski, S., Krajewska, M., et al. (2001) J. Clin. Oncol. **19**, 992.
Turgeon, V. L., Houenou, L. J. (1997) Brain Res. Rev. **25**, 85.
Tzivion, G., Luo, Z., Avruch, J. (1998) Nature **394**, 88.
Tzivion, G., Luo, Z. J., Avruch, J. (2000) J. Biol. Chem. **275**, 29772.
Tzivion, G., Shen, Y. H., Zhu, J. (2001) Oncogene, **20**, 6331.
Tzivion, G., Luo, Z. J., Avruch, J. (2002) J. Biol. Chem. **277**, 3061.
Umbricht, C. B., Gabrielson, E., Ferguson, A., et al. (2001) Oncogene **20**, 3348.
Underwood, L. J., Tanimoto, H., Wang, Y., et al. (1999) Cancer Res. **59**, 4435.
Urano, T., Saito, T., Tsukui, T., et al. (2002) Nature **417**, 871.
Usadel, H., Danenberg, K. D., Jerónimo, C., et al. (2002) Cancer Res. **62**, 371.
Valdes, G., Chacon, C., Corthom, J., et al. (2001a) Endocrine **14**, 197.
Valdes, G., Germain, A. M., Corthom, J., et al. (2001b) Endocrine **16**, 207.
Van Hemert, M. J., Steensma, H. Y., van Heusden, G. P. (2001) Bioassays **23**, 936.
van der Hoeven, P. C., Van Der Wal, J. C., Ruurs, P., et al. (2000) Biochem. J. **345**, 297.
van Dyke, T., Jacks, T. (2002) Cell **108**, 135.
Vanden Berghe, T., Kalai, M., van Loo, G., et al. (2003) J. Biol. Chem. **278**, 5622.
van 't Veer, L. J., Dai, H., van de Vijver, M. J., et al. (2002) Nature **415**, 530.
Vargas-Roig, L. M., Gago, F. E., tello, O., et al. (1998) Int. J. Cancer **79**, 468.
Vasioukhin, V., Anker, P., Maurice, P., et al. (1994) Br. J. Haematol. **86**, 774.
Vercoutter-Edouart, A-S., Le Bourhis, X., Louis, H., et al. (2001) Cancer Res. **61**, 76.
Vincenz, C., Dixit, V. M. (1996) J. Biol. Chem. **271**, 20029.
Villaret, D. B., Wang, T., Dillon, D., et al. (2000) Lryngoscope **110**, 374.
Vleminckx, K., Vakaet, L., Mareel, M., et al. (1991) Cell **66**, 107.
Vogelstein, B., Lane, D., Levine, A. (2000) Nature **408**, 307.
Walerych, D., Kudla, G., Gutkowska, M., et al. (2004) J. Biol. Chem. **279**, 48836.
Wang, L-Z., Sato, K., Tsuchiya, N., et al. (2003) Cancer Lett. **2002**, 53.
Webb, C. P., Hose, C. D., Koochekpour, S., et al. (2000) Cancer Res. **60**, 342.
Wegele, H. M., Haslbeck, J., Reinstein, J., Buchner, J. (2003) J. Biol. Chem. **278**, 25970.
Werle, E. (1934) Biochem. Z. **269**, 415.
Werner, M., Sych, M., Herbon, N., et al. (2002) Hum. Mutat. **20**, 57.
Weinstein, I. B. (2002) Science **297**, 63.
Whitesell, L. M. (1994) PNAS USA. **91**, 8324.
Whitesell, L., Cook, P. (1996) Mol. Endocrinol. **10**, 705.
Whitesell, L., Stphin, P. D., et al. (1998) Mol. Cell Biol. **18**, 1517.
Whitesell, L., Lindquist, S. (2005) Nat. Rev. Cancer **5**, 761
Widschwendter, M., Berger, J., Hermann, M., et al. (2000) J. Natl. Cancer Res. **92**, 826.
Wilker, E. W., Grant, R. A., Artim, S. C., et al. (2005) J. Biol. Chem. **280**, 18891.
Wilker, E., Yaffe, M. B. (2004) J. Mol. Cell. Cardiol. **37**, 633.
Workman, P. (2002) Expert Rev. Anticancer Ther. **2**, 611.
Workman, P. (2004) Trends in Mol. Med. **10**, 47.
Xi, Z., Klokk, T. I., Kormaz, K., et al. (2004) Cancer Res. **64**, 2365.

Xiao, B., Smerdon, S. J., Jones, D. H., et al. (1995) Nature **376**, 188.
Xiao, T., Wing, W., Li, L., et al. (2005) Mol. Cell Proteomics **4.10**, 1480.
Xin, Z., Allis, C. D., Wagstaff, J. (2001) AM. J. Hum. Genet. **69**, 1389.
Xu, Y., Lindquist, S. (1993) PNAS. USA. **90**, 7074.
Xu, Y., Singer, M. A., Lindquist, S. (1999) PNAS. USA. **96**, 109.
Xu, W., Mimnaugh, E., Rosser, M. F., et al. (2001) J. Biol. Chem. **276**, 3702.
Xue, W. M., Coetzee, G., Ross, R. K., et al. (2001) Cancer Epidemiol. Biomarkers **10**, 575.
Xue, W. M., Irvine, R. A., Yu, M. C., et al. (2000) Cancer Res. **60**, 839.
Yamashiro, K., Tsuruoka, N., Kodama, S., et al. (1997) Biochim. Biophys. Acta **1350**, 11.
Yaffe, M. B., Rittinger, K., Volinia, S., et al. (1997) Cell **91**, 961.
Yang, J., Liu, X., Bhalla, K., et al. (1997) Science **275**, 1129.
Yamanaka, H., He, X., Matsumoto, K., et al. (1999) Brain Res. Mol. Brain Res. **71**, 217.
Yonemura, Y., Endou, Y., Kimura, K., et al. (2000) Clin. Cancer Res. **6**, 4234.
Yoshida, S., Taniguchi, M., Suemoto, T., et al. (1998) Biochim. Biophys. Acta **1399**, 225.
Young, J. C., Obermann, W. M. J., Hartl, F. U. (1998) J. Biol. Chem. **273**, 18007.
Young, C. Y., Andrews, P. E., Tindall, D. J. (1995) J. Androl. **16**, 97.
Yousef, J. M., Diamandis, E. P. (1999) J. Biol. Chem. **274**, 37511.
Yousef, G. M., Luo, L-Y., Diamandis, E. P. (1999a) Anticancer Res. **19**, 2843.
Yousef, G. M., Luo, L. Y., Sherer, S. W., et al. (1999b) Genomics **62**, 251.
Yousef, G. M., Obiezu, C. V., Luo, L. Y., et al. (1999c) Cancer Res. **59**, 4252.
Yousef, G. M., Diamandis, E. P. (2000a) Genomics **65**, 184.
Yousef, G. M., Chang, A., Scorilas, A., et al. (2000a) Biochem. Biophys. Res. Commun. **276**, 125.
Yousef G. M., Magklara, A., Diamandis, E. P. (2000b) Genomics **69**, 331.
Yousef G. M., Scorilas, A., Magklara, A., et al. (2000c) Gene **254**, 119.
Yousef, G. M., Scorilas, A., Diamandis, E. P. (2000c) Genomics **63**, 88.
Yuosef, G. M., Magklara, A., Diamandis, E. P. (2000d) Genomics **69**, 331.
Yuosef, G. M., Chang, A., Diamandis, E. P. (2000e) J. Biol. Chem. **275**, 11891.
Yousef, G. M., Diamandis, P. (2001) Endocrine Rev. **22(2)**, 184.
Yousef, G. M., Diamandis, M., Jung, K., et al. (2001a) Genomics **74**, 385.
Yousef, G. M., Mgklara, A., Chang, A., et al. (2001b) Cancer Res. **61**, 3425.
Yousef, G. M., Scorilas, A., Yung, K., et al. (2001c) J. Biol. Chem. **276**, 53.
Yousef, G. M., Fracchioli, S., Scorilas, A., et al. (2003a) AM. J. Clin. Pathol. **119**, 346.
Yuosef, G. M., Kishi, T., Diamandis, EP. (2003b) Clin. Chim. Acta **329**, 1.
Yousef, G. M., Polymeris, M-E., Yacoub, G. M., et al. (2003c) Cancer Res. **63**, 2223.
Yousef, G. M., Polymeris, M-E., Grass, L., et al. (2003d) Cancer Res. **63**, 3958.
Yu, H., Bowden, D. W., Spray, P. J., et al. (1998) Hypertension **31**, 906.
Zantema, A., Verlaan-De Vries, M. Maasdam, D., et al. (1992) J. Biol. Chem. **267**, 12936.
Zeiner, M., Gebauer M., Gehring U. (1997) EMBO J. **16**, 5483.
Zha, J., Harada, H., Yang, E., et al. (1999) Cell **87**, 619.
Zhang, L. X., Chen, J., Fu, H. A. (1999) PNAS. USA. **96**, 8511.

Zhang, Y., Zuiderweg, E. R. (2004) Proc. Natl. Acad. Sci. USA **101**, 10272.
Zhao, R., Hsu, Y.-C., Kaplanek, P., et al. (2005) Cell, **120**, 715.
Zhu, W-G., Dai, Z., Ding, H., et al. (2001) Oncogene **20**, 7787.
Zöchbaur-Müller, S., Fong, K. M., Virmani, A. K., et al. (2001) Cancer Res. **61**, 249.

5

PROTEIN NETWORKS AND PROTEIN PHOSPHORYLATION IN CANCER

5.1. INTRODUCTION

Most gene products mediate their function within complex networks of interconnected macromolecules. Studies in model organisms have shown that complex macromolecular networks possess topological and dynamic properties that reflect biological phenomena (Jeong et al., 2001; Han et al., 2004). Thus, an understanding of biological mechanisms and disease processes demand a "systems" approach that goes beyond one-at-a-time studies of single components to more global analyses. For almost a century, reductionism approach in biological research has provided a wealth of knowledge about individual cellular components and their functions. Despite the enormous success of such approach, it is increasingly evident that discrete biological function can rarely be attributed to an individual biological identity or a single pathway. Instead, most biological events are the result of complex interactions between the cell's numerous constituents, such as proteins, DNA, RNA, and small molecules. Deciphering such interactions has become a key challenge for contemporary biology. Although metabolic network analysis dates back to the 1940s, data-driven genome-scale analyses of gene and protein network can be considered newcomers, which over the last 10 years has started to receive a deserved attention. Despite its revolutionary implications, systems biology still lacks a standard definition. It is still a shared belief, which hinges on the availability of vast databases of biological information and the technical know-how to create detailed maps of protein–protein interactions. These so-called intractomes have already been explored in yeast and other model

Cancer Biomarkers: Analytical Techniques for Discovery, by Mahmoud H. Hamdan

organisms. A limited number of investigations have also provided initial human interactome. Despite disagreement on a specific definition, systems biology has begun to piece together the first detailed sketch of how cells process various biochemical signals, essential information that one day could push more molecular-based cancer treatments closer to reality. As it has been pointed out in various parts of this text, cancer is caused by a variety of genetic, epigenetic, and chromosomal alterations. However, the variations in the molecular alterations that can give rise to cancer can be captured by a handful of traits that cancer cells must acquire toward their route to malignant transformation (Hanahan and Weinberger, 2000). The advent of microarrays and other large scale technologies makes it feasible to obtain a global view of molecular events impacting on cancer development.

In the present chapter, an attempt is made to underline the impact of large-scale functional studies on our understanding of cancer, including the search for new generation of biomarkers. To perform such large-scale studies, contemporary biology has amassed a battery of methods to survey the global features of the cells, from DNA, RNA, and proteins to small molecules (Ideker et al., 2001; Kitano, 2002). A number of these technologies and their derivatives have been developed to elucidate functions of genes and proteins. For example, studies examining global gene expression (transcriptomics) have become highly popular in the field of functional genomics (Cho and Campbell, 2000; Lockhart and Winzeler, 2000). On the basis of cDNA or oligonucleotide arrays and other systems such as serial analysis of gene expression (SAGE), these technologies can now provide simultaneous monitoring of mRNA expression rates of large sets of genes. Genome-wide mutagenesis is another approach in functional genomics based on systematic alteration of many or all genes in a genome, one by one, and to observe the resulting phenotype. Several groups have carried out genome-wide mutagenesis studies on cellular and animal models. These include systematic gene disruption in *Saceharomyces cerevisiae* by deletion and transposon-tagging (Ross-Macdonald et al., 1999; Winzeler et al., 1999). The potential of functional protein microarrays has also been demonstrated by different research groups (Zhu et al., 2000; MacBeath and Schreiber, 2000). Some of the technologies cited herein have been discussed in some detail in Chapter 2 of this book. Therefore, the rest of this chapter will be mainly concerned with other technologies which have not been covered earlier in this text, together with two main arguments relevant to cancer research: protein–protein interaction networks (Interactomes) and protein phiosphorylation together with PI3K/Akt signaling pathway.

5.2. PROTEIN INTERACTION NETWORKS

Protein–protein interactions are central to most biological processes; the systematic identification of all protein interactions is considered a key strategy for uncovering the inner workings of a cell. The systematic efforts at creating proteome-scale data sets of protein–protein interactions, which are represented as complex networks or "intractome" maps have been facilitated by the availability of genome-scale sets of cloned open reading frames (ORFs). Protein–protein interaction mapping projects that follow stringent criteria, coupled with experimental validation in orthogonal systems, can provide information to interrogate disease development and mechanisms at a system level. Efforts in this direction were boosted at the end of 1990s by the computational efforts that used the genomic context of genes. On the experimental side, the first large scale protein interaction data sets were presented by Ito et al. (2001) and Uetz et al. (2000) both studies used yeast two-hybrid technology. Two years later, the first large scale protein complex purification data sets were published (Ho et al., 2002; Gavin et al., 2002). Till date, most of the protein interaction networks that have been detected experimentally have relied on one of the two following technologies: the yeast two-hybrid system (Fields and Sternglanz, 1994) and mass spectrometry-based methods (Rigaut et al. 1999; Pandey and Mann, 2000). These two approaches provide what can be considered as complementary data sets. The MS-based approach identifies the constituents of multiprotein complexes but does not reveal the individual binary contacts that make up each complex. Without data on such binary contacts, the likely paths of energy or information flow through the complex and its relationship to other cellular components may not be apparent. Yeast-two hybrid data, on the contrary identifies likely binary interactions that may indicate possible pathways through a complex, but cannot reveal the constituents of multiprotein complexes.

In an interactome, proteins are usally represented as nodes (e.g. circles or boxes) and interactions as edges (lines) connecting them. Such representation organizes information so that attributes of both proteins and possible interactions are easily accessible. Such graphical representation offers a number of advantages: First, analysing networks of protein–protein interactions increases confidence levels of individual associations and allows the identification of otherwise unexpected links between previously unconnected cellular processes. Second, the same analysis can contribute to the uncovering of protein–protein interactions that unexpectedly link diverse cellular processes or that indicates crosstalk between cellular compartments. These advantages have been elegantly illustrated by the work of Schwikowski et al. (2000) who compiled a list of about 2700 published protein interactions from the *S. cerevisiae* literature and found that 1548 yeast proteins could be depicted in a single large network. By classifying these proteins into different types of functional categories, the same authors generated a functional linkage map, which is given in Figure 5.1. This functional linkage map shows that certain functional classes, such as cell-cycle regulation, transcription, and chromatin regulation have interactions with proteins of many other classes, consistent with their central roles in the cell. Other processes such as membrane fusion are more isolated, with proteins interacting mainly within this group or with a related group such as vesicular transport. The functional classification of proteins also allowed Schwikowski et al. (2000) to evaluate the plausibility of the network. For example, they found that over 72% of all interactions between experimentally characterized proteins in this network are between two partners of the same functional class. On the contrary, when the interactions were randomized among the same set of proteins, only 12% of all interactions were found to belong to the same class. Regulatory networks similar

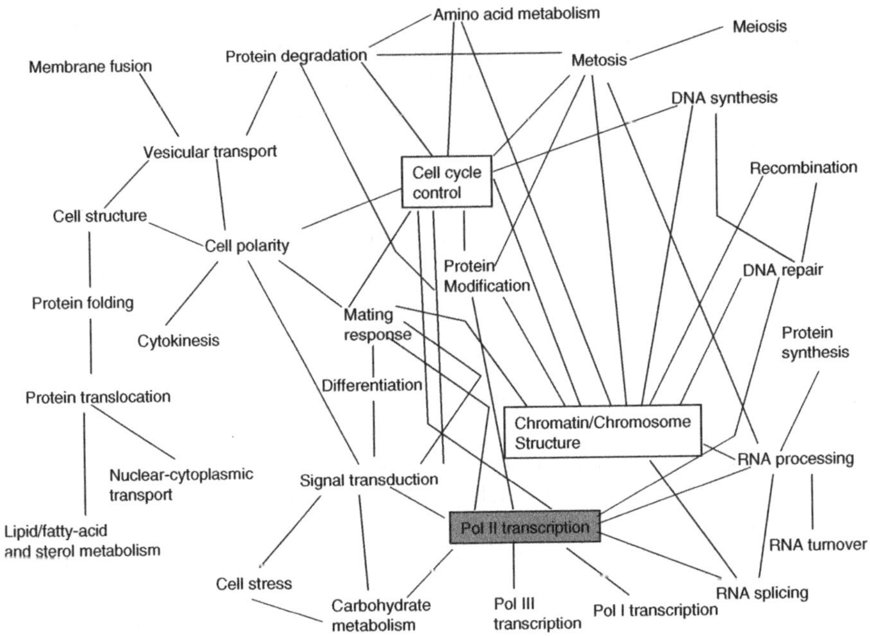

Figure 5.1. Interactions between functional groups of yeast proteins, which are based on yeast protein interaction network of about 1200 interacting proteins. Only proteins with known functions are included. Adapted from Tucker et al. (2001) with permission.

to that for yeast proteins can also be drawn for vertebrates. Figure 5.2 shows an interaction map of human proteins generated by the Myriad Pronet database (www.myriad-pronet.com). The map is based on the signaling pathway of the tumor suppressor gene, *BRCA1* (see also Chapter 3). The original map was simplified by removing proteins, which interact with only one other protein (single interactors) and all single diverging branches (pathways that branch off to a dead end) (Tucker et al. 2001).

Although the impact of protein interaction networks on biomarkers discovery is still difficult to assess, there have been a number of recent studies, which demonstrated the proof of principle. However, before discussing some of these examples, it would be helpful to consider some of the basic aspects of these interaction maps.

5.2.1. Experimental Approaches

A comprehensive interactome can be considered as the complete collection of all physical protein–protein interactions that can take place within a cell. Construction of comprehensive sets of protein–protein interactions requires the creation of genome-scale resource collections of open reading frames cloned to facilitate protein expression, generated iteratively on the basis of improved gene predictions and experimental verification and capturing all expressed isoforms. The realization that

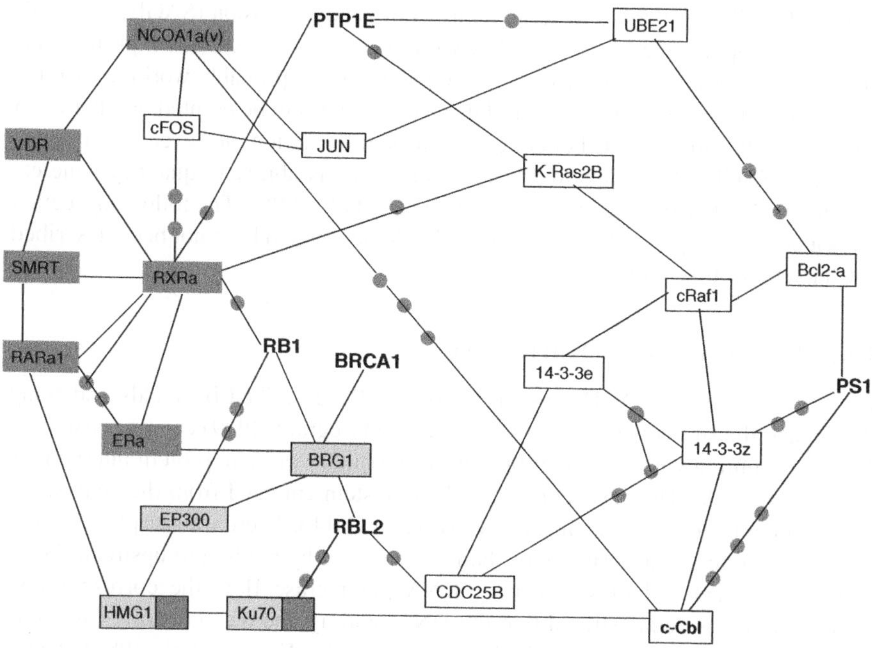

Figure 5.2. Simplified protein-interaction map generated using Myriad Pronet software (www.myriad-pronet.com), showing core proteins involved in the *BRCA1* signaling network. Boxes with white background refer to proteins involved in growth control, dark grey boxes refer to proteins involved in transcription, and boxes in light grey refer to chromatin/chromosome structure proteins. Some proteins have multiple functions indicated in double color boxes. Adapted from Tucker et al. (2001) with permission.

proteins exert virtually all of their activities via interactions with other molecules, be they are proteins, nucleic acids, lipids, carbohydrates or small molecules has driven the development of technologies to examine these interactions. Currently, there are a number of methods, which are used for high-throughput detection of protein interactions. The first global experimental analysis of protein–protein interactions in yeast (*S. cerevisiae*) using the two hybrid system was reported by Ito et al. (2001) and Uetz et al.(2000). This system provides a large-scale experimental approach to determine whether the each pair of proteins from the yeast proteome physically interacts. Another large-scale experimental approach based on proteomics technology, also applied to yeast, is the systematic isolation of previously tagged multiprotein, followed by tandem affinity purification and mass spectrometry detection (Rigaut et al. 1999, Gavin et al., 2002). Another experimental approach that has been applied to explore large-scale protein binary interactions is protein microarrays. The first high-density yeast proteome microarray composed of 5800 fusion proteins was built and used to identify novel colmodulin-binding and phospholipids-binding proteins (Zhu et al., 2001). Other high-throughput technologies used to determine the gene expression profile of thousands of genes are DNA microarrays (DeRisi, 1996; Schena et al.,

1995, 2002; Zhang, 1999) and serial analysis of gene expression (SAGE) (Ye et al., 2000). Both approaches measure mRNA levels, producing expression profiles that indicate those proteins that are coexpressed and that are probably working together in a cellular state or in a specific cellular response. It has to be pointed out that DNA microarrays usually provide better genome wide information; however, an attractive feature of SAGE when compared to microarrays is its ability to quantify gene expression without prior sequence information (Ye et al., 2000). The following section gives a brief description of some of the technologies, which have not been described in Chapter 2 of this book.

5.2.2. Yeast Two Hybrid (Y2H) System

The yeast two hybrid (Y2H) system, as originally described by Fields and Song (1989), and the subsequent derivatives (Vidal and Legrain, 1999) can be considered the most commonly used system for identifying binary protein–protein interactions at proteome scale. The basic concept of Y2H system emerged from the analysis of transcription factors such as the archetypal yeast Gal4p. These transcription factors increase the rate of transcription of their target genes by binding to upstream DNA sequences (UAS) and thus activating RNA polymerase II at the corresponding promoters. It was demonstrated that the DNA binding and the activating functions are located in physically separable domains of Gal4p (Keegan et al., 1986). These two domains are usually referred to as the DNA-binding domain (DB) and the activation domain (AD), respectively. In the most extreme version of such structure-function experiments, a hybrid protein, consisting of the bacterial LexA DB fused to the Gal4p AD, was shown to activate, in yeast cells, the transcription of a bacterial reporter gene containing the LexA operator site in its promoter (Brent and Ptashne, 1985). The Y2H system detects the interaction between two proteins through an assay involving a transcriptional activation of one or several reporter genes. Protein X (generally referred to as the bait) is fused to BD domain, whereas protein Y (generally referred to as the prey) is fused to the AD domain. The transcription of the reporter gene will take place only if X and Y interact together. The basic strategy of Y2H assay is schematically represented in Figure 5.3.

In a pair of landmark papers, Uetz et al. (2000) and Ito et al. (2001) have described how Y2H can be applied for large-scale, high throughput analysis of pair-wise protein interactions. The authors collectively identified over 4000 protein–protein interactions in *S. cerevisiae*. In a latter study Giot et al. (2003) have published a two-hybrid-based protein interaction map of the fly proteome. A total of 10,623 predicted transcripts were isolated and screened against standard and normalized complementary DNA libraries to produce a draft map of 7048 proteins and 20,405 interactions. A computational method of rating the interaction confidence was also applied resulting in a higher confidence map of 4679 proteins and 4780 interactions.

The Y2H system has a number of advantages over alternative assays for gene identification. First, the system is based on a powerful genetic selection scheme performed with a convenient microorganism, thus allowing very high numbers of

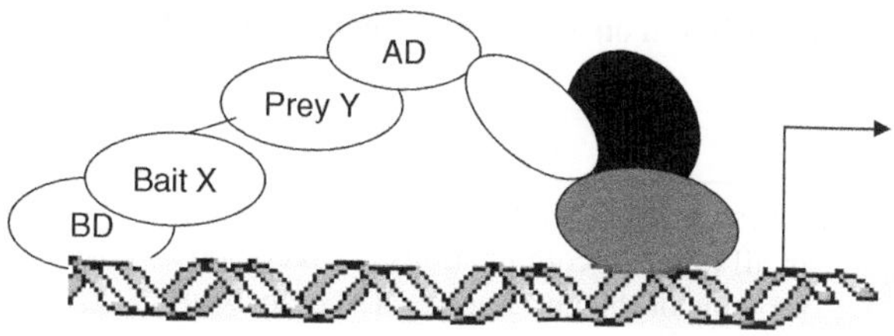

Figure 5.3. Representation of the Yeast two-hybrid (Y2H) assay, which detects the interaction between two proteins through an assay involving transcriptional activation of one or more reporter genes. In this assay, protein X is fused to a protein domain that binds specifically a DNA sequence in the promoter of the reporter gene (the DNA binding domain; BD). Protein Y is fused to a domain that recruits the transcription machinery (the activation domain; AD). Transcription of the reporter gene will occur only if X and Y interact together. Adapted from Lagrain et al. (2001) with permission.

potential coding sequences to be assayed in relatively simple experiments. Second, it relies on an assay performed in vivo and thus it is not limited by the artificial conditions of in vitro assays. Third, the assay is based on a physical binding, and therefore it can handle a wide variety of protein–protein interactions by applying a single protocol. This versatile assay also has a number of limitations. It is commonly recognized that large scale approaches for identifying protein–protein interactions will always experience some degree of false-positives (artificial or spurious interactions) and false-negatives (undetected, genuine associations). An example of the latter class is the failure of Y2H assay to detect interactions associated with post-translational modifications and specific classes of proteins such as integral membrane proteins. Cusick et al. (2005) argued that false-positives can be classified into biological false-positives in which the interaction can be confirmed by multiple, different methods, but the two proteins are never present in the same cell or sub-cellular cell compartment at the same time. This class of false-positives is nearly impossible to detect using interaction assays alone. The second class of false-positives is generally associated with the type of experimental setup used for such measurements.

The Y2H system has been successfully applied in a number of large-scale studies (see Table 5.1), which yielded enough data to allow a reasonable assessment of the impact of false-positives and negatives on the confidence in the generated data. In its general form this assay has a number of limitations: (i) The interaction is forced to occur in the nucleus, which poses problems for certain classes of proteins (e.g. integral membrane proteins) (Cusick et al., 2005). (ii) A major limitation with the conventional two-hybrid strategy is that neither the bait protein nor the potential interacting protein should be able to activate transcription on their own. This can represent a problem with transcriptional activators which naturally contain domains that activate the reporter genes when fused to the BD domain. Furthermore, many proteins other than transcription factors are also found to activate transcription when fused to this

TABLE 5.1. Some studies based on Y2H screening to construct interactomes of different species.

Species	Description	Reference
S. cerevisiae	Y2H system was used to examine interactions in all possible combinations between 6000 proteins of the budding yeast *S. cerevisiae*. This study identified 4549 interactions among 3278 proteins.	Ito et al. (2001)
	Two large-scale Y2H screens were conducted to identify protein–protein interactions between full length open reading frames from *S. cerevisiae* genome sequence. These screens resulted in the detection of 957 putative interactions involving 1004 proteins.	Uetz et al. (2000)
Metazoan C. *elegans*	In this study a large fraction of the *Caenorhabditis elegans* interactome was examined by high throughput Y2H screening. Starting with a subset of metazoan-specific proteins, more than 4000 interactions were identified.Independent coaffinity purification assays were used to experimentally validate the quality of Y2H data set.	Li et al. (2004); Walhout et al. (2000)
D. melanogaster	Y2H was used to generate protein interaction map. A total of 10,623 predicted transcripts were isolated and screened against standard and normalized cDNA libraries to produce a draft map of 7048 proteins and 20,405 interactions. This draft map was refined by a computational method resulting in a higher confidence map of 4679 proteins and 4780 interactions.	Giot et al. (2003)
Homo sapiens	In this work the authors described an initial version of a proteome-scale map of human binary protein–protein interactions. Using a stringent, Y2H system, they have tested pair wise interactions among the products of ~8100 currently available Gateway cloned open reading frames. This investigation resulted in ~2800 interactions with a verification rate of ~78%.	Rual et al. (2005)
	Automated Y2H screening was used to detect interacting pairs of human proteins. A protein matrix consisting of 4456 baits and 5632 preys was screened. Over 3100 interactions among 1705 proteins were detected. Independent pull-down and co-immunoprecipitation assays were used to validate the overall quality of Y2H data set. By applying topological and GO criteria, a scoring system was developed to define 911 high-confidence interactions among 401 proteins.	Stelzl et al. (2005)

TABLE 5.1. ***(Continued)***

Species	Description	Reference
	Y2H screening was used to study Smad signaling system, which is regulated by members of the transforming growth factor β superfamily. This screening identified 755 protein–protein interactions, involving 591 proteins. According to the authors, 179 of these proteins were previously poorly or not annotated.	Colland et al. (2004)
H. pylori	The authors used high throughputY2H strategy to build large-scale protein–protein interaction map of the human gastric pathogen *H. pylori.* Two hundred and sixty one *H. pylori* proteins were screened against a complex library of genome encoded polypeptides (Formont-Racine et al., 1997). Over 1200 interactions were identified, connecting 46.6% of the proteome of the pathogen.	Rain et al. (2001)

domain (Ma and Ptashne, 1987). It has been also argued that transcription factors and other proteins, which represent about 5–10% of randomly selected open reading frames (ORFs), auto-activate transcription of the reporter gene (Cusick et al., 2005). (iii) Another potential limitation of the two hybrid system is the failure of the yeast cells to undergo various post-translational modifications which are required for particular interactions in higher organisms; in particular, glycoproteins or lipoproteins for which the nonpeptidyl part of the molecule is suspected to be involved in the interaction (Vidal and Legrain, 1999). In other words, interactions involving specific post-translational modifications can be missed, unless enzymes responsible for such modifications happen to be present in the yeast nucleus. (iv) The Y2H assay has proved to have the capability to detect protein interactions from almost any organism, although certain type of proteins, such as membrane or extracellular proteins, have demonstrated to be less amenable to this approach. Despite this difficulty, attempts have been made to identify interactions for a set of about 700 *S. cerevisiae* proteins annotated as integral membrane proteins by applying a modified membrane Y2H assay on a large-scale (Johnsson and Varshavsky, 1994; Stagljar et al., 1998). This modified assay relies on the fusion of two membrane proteins to the halves of ubiquitin, an N-terminal and a C-terminal domain, the latter in turn, is fused to a LexA-VP16 transcription factor. Interaction of the membrane proteins reconstitutes a quasi-native ubiquitin, leading to a cleavage by cellular ubiquitin-specific proteases after the ultimate ubiquitin residue. This cleavage releases the transcription factor, which enters the nucleous and activates the expression of reporter genes. In this study, the expression of reporter genes were detected by expression of the *HIS3* gene, with the growth on media lacking histidine. From duplicate set of screens, the authors identified 2,000 putative protein interactions (Miller et al., 2005). However, the same authors noted that, as with the traditional Y2H assay, many of the interactions detected with this modified assay were likely to be false-positives.

5.2.3. Tandem Affinity Purification/Mass Spectrometry (TAP-MS)

Until recently, the Y2H system was the only suitable global approach to investigate protein–protein interactions. The emergence of powerful, sensitive high-throughput mass spectrometry techniques, allowing the detection of peptides in the lower femtomolar range has opened the door for new methods employing biochemical purification of whole cellular protein assemblies, and their subsequent identification by mass spectrometry. In the late 1990s this approach for the investigation of protein–protein interactions has been reported by Rigaut et al. (1999). The authors described a general procedure combining tandem affinity purification (TAP) tag and mass spectrometry. Basically, the method involves the fusion of the TAP tag to the target protein and the introduction of the construct into the host cell or organism, maintaining the expression of the fusion protein at, or close to, its natural level. The fusion protein and associated components are then recovered from the cell extracts by successive affinity selections and use of mass spectrometry to identify the investigated proteins. Following this work, two different groups have used a similar approach to analyze protein–protein interactions (Gavin et al., 2002; Ho et al., 2002). In these studies, bait proteins were created by attaching affinity tags to hundreds of different proteins. They then introduced DNA encoding these bait proteins into yeast cells, allowing the modified proteins to be expressed in the cells and to form physiological complexes with other proteins. The protein complex is then eluted from the affinity resin using the fusing tag peptide itself (affinity elution). The tandem affinity purification procedure, developed by Neubauer et al. (1997), employs a fusion tag consisting of two proteins: immunoglobulin (IgG)-binding domains and a calmodulin binding peptide tag, separated by a spacer containing the specific recognition site for a protease (e. g. tobacco ech virus, TEV). A schematic representation of this procedure is given in Figure 5.4.

5.2.4. Y2H and TAP-MS as Complementary Approaches

It is reasonable to argue that false-positives/negatives are not the only reason for a relatively poor overlapping between data sets generated by different approaches. Biases in interaction coverage have been identified as another possible source of divergence between such studies. Von Mering et al. (2002) identified three areas where the high-throughput interaction data are biased. On the basis of plots between interaction coverage versus mRNA abundance generated by different experimental approaches, the authors identified three areas of bias: First, most protein interactions used to generate the plots in question were heavily biased toward proteins of high abundance. However, the two genetic approaches, Y2H and synthetic lethality appeared to be less biased compared to other approaches. Second, the data sets were found to be biased toward particular cellular localization of interacting proteins. Third, there is a bias in interaction coverage that relates to the degree of evolutionary novelty of proteins. The authors noted that proteins restricted to yeast were less covered than ancient, evolutionary restricted proteins.

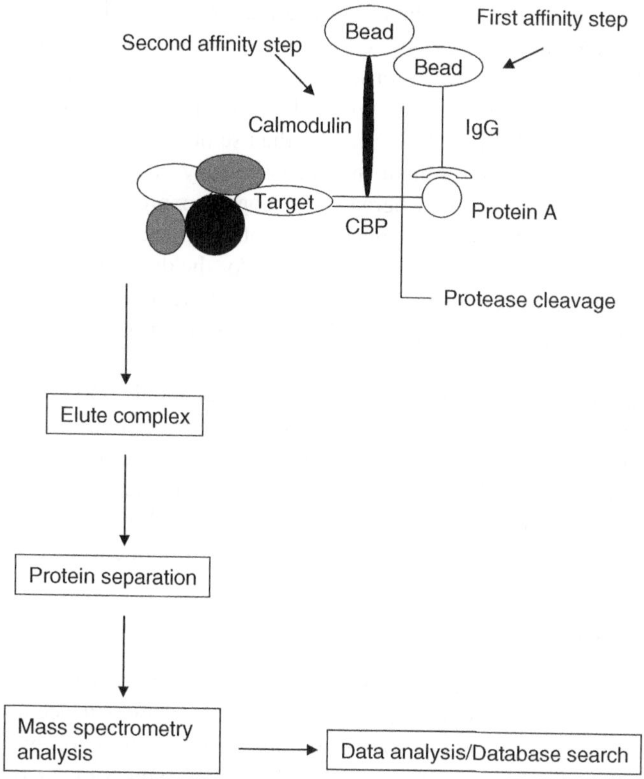

Figure 5.4. The main steps in tandem affinity purification/mass spectrometry approach to isolate and identify targeted protein complexes from cells or tissue extracts. In this two-step affinity purification, a target protein is transgenetically expressed with two high-affinity tags attached to its C- or N-terminus. A specific protease cleavage site is placed between the two high-affinity tags. This allows elution or release of the targeted complex from the first affinity (IgG) step. The complex is then recovered in a second step using a calmodulin matrix and eluted for MS analysis using calcium chelators. CBP: Clamodulin-binding peptide; IgG: Immunoglobulin.

Data on protein–protein interactions provided by the Y2H, protein microarrays, and TAP-MS can be broadly classified as physical and functional interactions. The Y2H and protein microarrays yield information about binary physical interactions, whereas the TAP-MS detect a group of proteins in a stable complex, which are likely to function together. This functional interaction can be compared with genetic interactions, in which the combination of alleles of two different genes has specific phenotypic consequences. This is often taken to suggest that the two genes function in the same or parallel pathways to influence a particular biological process. In other words, a genetic interaction is a measured functional interaction that may or may not correspond to a physical interaction (Uetz and Finley, 2005).

The complementary nature of data sets generated by Y2H and microarrays on one hand and TAP-MS on the other can be underlined by the following considerations: First, as well as the above observations, Y2H and protein microarrays identify specific, binary protein–protein interactions, which are not readily identified within complexes investigated by TAP-MS. Second, the use of ATP-MS to investigate protein expression in their environment in cell/tissue may allow the identification of post-translational modifications, whereas heterologous expression in Y2H generally precludes such modifications. Third, purification of complexes in the TAP-MS procedure may result in the loss of real interactions and/or the detection of spurious ones. Furthermore, the same procedure may fail to detect weak, transient associations, which are more likely to be detected by Y2H (Ito et al., 2002; James et al., 1996). Third, the Y2H approach is amenable to high throughput methods for identifying mutations that disrupt interactions (Endoh et al., 2002; Vidal, 1997) and for interrogating protein domains, whereas TAP-MS is less suitable for both tasks. The notion that these methods are complementary is supported by several independent in silico comparisons of the available global yeast protein–protein interaction maps.

There are other methods, which have been applied for high-throughput detection of protein interactions, some of which are briefly described in the following sections.

5.2.5. DNA Microarrays

DNA microarray-based gene expression profiling relies on nucleic acid hybridization and the use of nucleic acid polymers, immobilized on a solid surface as probes for complementary gene sequences (Southern et al., 1999). Expression profiling techniques have been used to simultaneously monitor the expression of thousands of genes from human tumor samples. They are relatively easy to use and can be applied to large numbers of samples in parallel. Although a number of competing microarray technologies exist, two platforms, mainly complementary DNA (cDNA) and oligonucleotide microarrays are currently used by a majority of investigators.

With cDNA arrays, polymerase chain reaction products of cDNA clone inserts representing genes of interest which are spotted systematically on nitrocellulose filters or glass slides (Schena et al., 1995). Spotted arrays are commonly constructed using cDNA collections (i.e., libraries) that can be focused on genes expressed in a particular context or cell type. The primary benefit of spotted arrays is that they can be made by individual investigators, can be easily customized, and do not require a prior knowledge of cDNA sequence because clones can be used and then sequenced later if of interest. However, we have to bear in mind that on the experimental side managing large clone libraries can be a daunting task for most laboratories, and making high-quality arrays can be difficult.

Oligonucleotide microarrays has a number of differences with their cDNA counterparts. Oligonucleotide probes for different genes can be deposited or synthesized directly on the surface of a silicon wafer in a patterned manner (Lockhart et al., 1996). This approach offers greater specificity than cDNAs because they can be tailored to minimize chances of cross-hybridization, and sequences up to 60 nucleotides have been used effectively (Hughs et al., 2001). Major advantages of this approach include uniformity of probe length and the ability to discern splice variants. Until recently, the design of specific oligonucleotides has been limited by sequence availability, but the completion of the genome sequencing of a number species, including human-made probe design is a much easier task. Another advantage is the ability to recover samples after hybridization to a chip. This allows for a single biologic sample to be sequentially hybridized to multiple arrays, a considerable advantage when dealing with limited biologic material. The hybridization of a test sample to an array can be detected in one of two ways. cDNA microarrays are commonly queried simultaneously with cDNAs derived from experimental and reference RNA samples that have been differentially labeled with two fluorophores to allow for the quantification of differential gene expression, and expression values are reported as ratios between two fluorescent signals. Alternatively, the oligonucleotide system uses a single color fluorescent label, where experimental mRNA is enzymatically amplified, biotinlabeled for detection, hybridized to the wafer, and detected through the binding of a fluorescent compound (streptavidin-phycoerythrin). A comparison between oligonucleotides and cDNA-based microarrays is schematically represented in Figure 5 5.

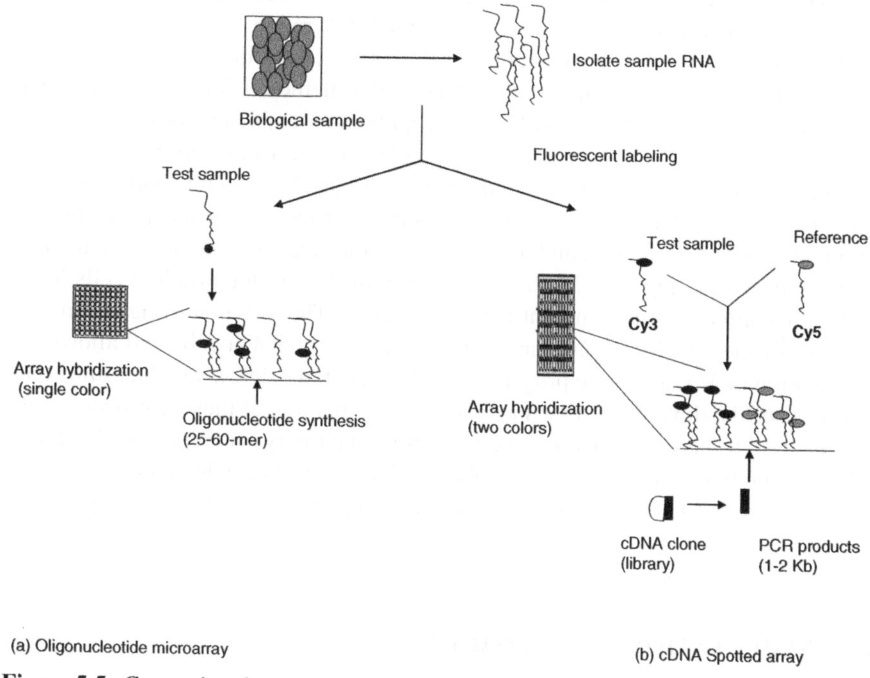

Figure 5.5. Comparison between oligonucleotide (a) and cDNA microarrays (b). In the first mode, direct synthesis or deposition of oligonucleotides onto solid surface and single color readout of gene expression from test sample. In the cDNA approach, deposition of polymerase chain reaction products from cDNA libraries onto a solid surface and simultaneous, two color readout of gene expression in test and reference samples. Adapted from Ramaswamy and Todd (2002) with permission.

5.2.6. Other Approaches

• Correlated mRNA expression (synexpression) in which mRNA levels are systymatically measured under a variety of different cellular conditions, and genes are grouped if they exhibit a similar transcriptional response to these conditions. The investigated gene groups are enriched in genes encoding physically interacting proteins (Ge et al., 2001). Besides being an in vivo method, it has a relatively broad coverage of cellular conditions compared to other methods. Among the drawbacks of this method is its incapability to predict direct physical interactions, and its high sensitivity to the parameters of choice and clustering approach during analysis.

• Synthetic lethal analysis is a method based on strong evidence that redundant functions can often be uncovered by synthetic genetic interactions, usually identified when a specific mutant is screened for second-site mutations that either suppress or enhance the original phenotype. In particular, two genes show a "synthetic lethal" interaction if the combination of two mutations, neither by itself lethal, causes cell death (Novick et al., 1989; Guarente, 1993). Synthetic lethal relationships may occur for genes acting in a single biochemical pathway or for genes within two distinct pathways if one process functionally compensates or buffers the defects in the other (Hartman IV et al., 2001). Synthetic lethal screens have been used to identify genes involved in cell polarity, secretion, and DNA repair (Bender and Pringle, 1991; Chen and Graham, 1998; Mullen et al., 2001). To perform a high-throughput synthetic lethal analysis, Tong et al. (2001) used a method, which they designated synthetic genetic array (SGA) analysis. The authors assembled an ordered array of ~4700 viable gene-deletion mutants and developed a series of pinning procedures in which the mating and meiotic recombination are used to generate haploid double mutants. In this protocol, a query mutation is first introduced into a haploid starting strain, of mating type *MATα*, and then crossed to the array of gene-deletion mutants of the opposite mating type *MATa*. Sporulation of the resultant diploid cells leads to the formation of double mutant meiotic progeny. The *MATα* starting strain carries a reporter, *MFA1pr-HIS3*, that is only expressed in *MAT*a cells and allows for germination of *MAT*a meiotic progeny (Herskowitz et al., 1992), which ensures that carryover of the diploid parental strain and/or conjugation of meiotic progeny does not give rise to false negative interactions. Both the query mutation and the gene-deletion mutations are linked to a number of dominant selectable markers to allow for the selection of double mutants. The main steps in SGA analysis are given in Figure 5.6.

5.3. COMPUTATIONAL APPROACHES

In parallel to the experimental approaches, a series of computational bioinformatics methods have been designed to predict protein–protein interactions. Most of the computational methods described in the literature are based on the assumption that it is possible to predict the interaction of two proteins, when such proteins form an associated pair which interacts at a certain biomolecular and cellular level. Some computational

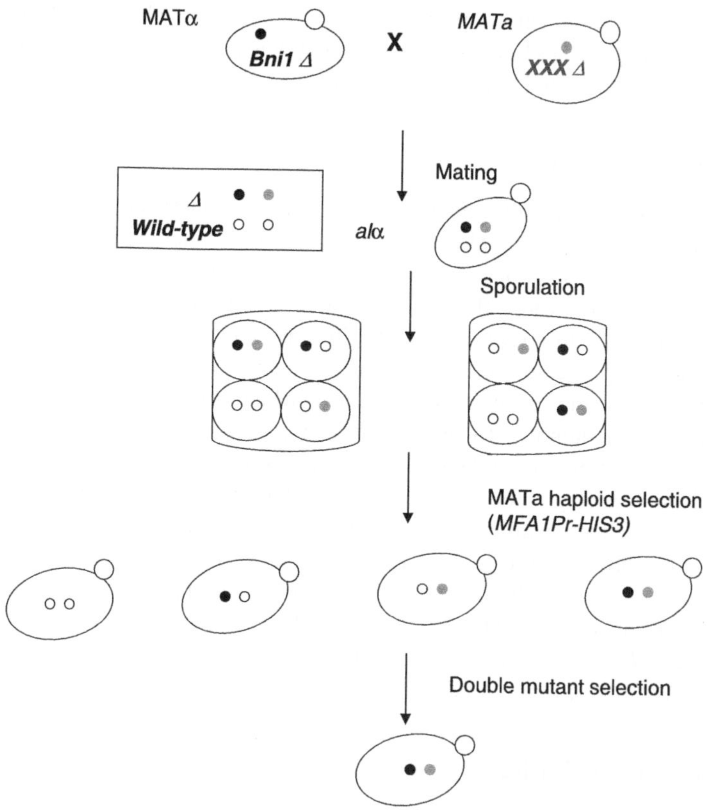

Figure 5.6. Partial representation of synthetic genetic array approach. (a) *MATa* strain carrying a query mutation (*bni1*D) linked to a dominant selectable marker, such as the nourseothricin-resistance marker *natMX* that confers resistance to the antibiotic nourseothricin, and an *MFA1pr-HIS3* reporter is crossed to an ordered array of *MAT***a** viable yeast deletion mutants, each carrying a gene deletion mutation linked to a kanamycinresistance marker (*kanMX*). Growth of resultant heterozygous diploids is selected for on medium containing nourseothricin and kanamycin. (b) The heterozygous diploids are transferred to medium with reduced levels of carbon and nitrogen to induce sporulation and the formation of haploid meiotic spore progeny. (c) Spores are transferred to synthetic medium lacking histidine, which allows for selective germination of *MAT***a** meiotic progeny because these cells express the *MFA1pr-HIS3* reporter specifically. (d) The *MAT***a** meiotic progeny are transferred to medium that contains both nourseothricin and kanamycin, which then selects for growth of double-mutant meiotic progeny. Adapted from Tong et al. (2001) with permission.

methods explore features at the proteome level by comparing the phylogenetic profiles of orthologous proteins in complete proteome (De Las Rivas et al., 2002; Pellegrini et al., 1999); identification of proteins with similar phynogentic trees, using multiple sequence alignments of families of homologous proteins (Pazos and Valencia., 2001; Ramani and marcotte, 2003), and identification of correlated mutations between the multiple sequence alignments of pairs of proteins (Pazos and Valencia, 2002). Various computational approaches have been described by Valencia and Pazos (2002), three of which are briefly described in the following sections.

5.3.1. Phylogentic Profiles

A proof of principle of phylogentic profiles was first demonstrated by Pellegrini et al. (1999). The underlying principle of this method is that functionally linked proteins evolve in a correlated fashion, and, therefore they have homologs in the same subset of organisms. In other words, the phylogenetic profile of a protein describes the presence or absence of homologs in organisms. To represent the subset of organisms that contain a homolog, the authors constructed a phylogenetic profile for each protein. This profile is a string of "*n*" entries, each of which represent one "bit"; where *n* corresponds to the number of genomes (16 genomes were used). The presence of a homolog in a given protein in the *n*th genome is assigned an entry of unity at the *n*th position. If no homolog is found, zero is assigned to the entry. The basic principle of this method is schematically represented in Figure 5.7. This method was tested by examining the phylogenetic profiles for two proteins that are known to participate in structural complexes, the ribosome protein RL7 and flagellar structural protein FlgL. The authors computed phylogenetic profiles for the 4290 proteins encoded by the genome of *Escherichia coli* by aligning each protein sequence with the proteins from 16 other fully sequenced genomes.

5.3.2. Similarity of Phylogenetic Trees (Mirrortree)

In a number of studies dealing with closely related cases, it has been possible to show that interacting proteins coevolve, for example insulin and its receptors (Fryxell, 1996). In such cases, the corresponding polygenetic trees of the interacting proteins show a greater degree of similarity (symmetry) than noninteracting proteins would be expected to show. Goh et al. (2000) used this approach to quantify the similarity of polygenetic trees for the two domains of phosphoglycerate kinase, a similarity based on the linear correlation between the distance matrices used to construct the two trees (see Fig. 5.8). To obtain quantitative information on the interaction between protein "a" and protein "b", the MSAs (multiple sequence alignments) of both proteins are reduced to the set of organisms common to the two proteins. Each of the reduced alignment is then used to construct the corresponding inter-sequence distance matrices, which are then used to construct the corresponding polygenetic trees. The final step in this method is to calculate the linear correlation between these distance matrices. High correlation values are generally interpreted as an indicator of similarity between polygentic trees and hence can be taken as predicted interactions between proteins "a" and "b".

Interactions between proteins can also be deduced from the presence of different genomes in the same protein domains, which either form part of a single polypeptide chain (multidomain protein) or act as independent proteins (single domain). This approach is designated gene-fusion events. Methods based on recursive sequence

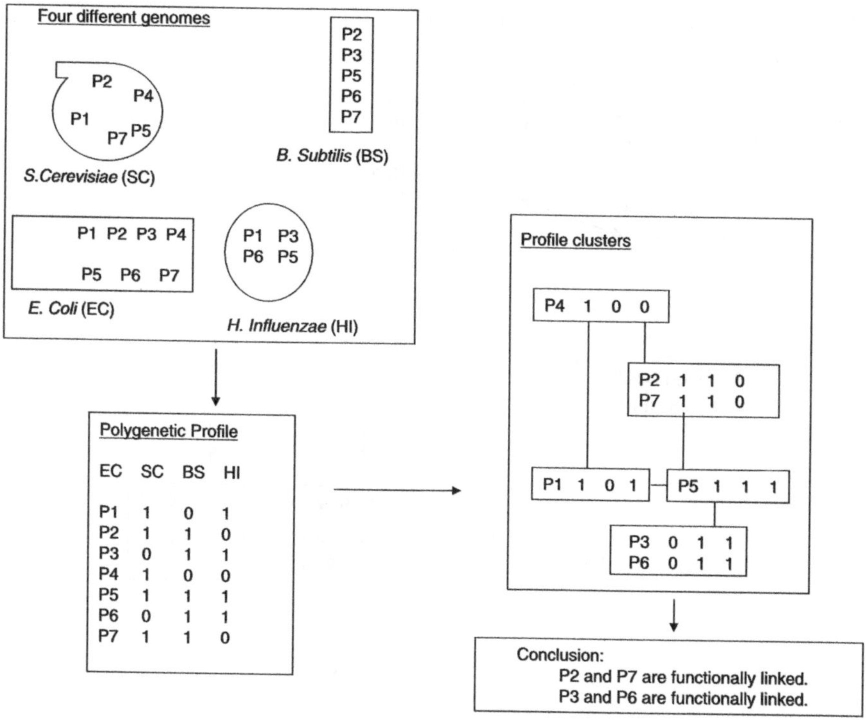

Figure 5.7. Analysis of protein phylogenetic profiles for the hypothetical case of four fully sequenced genomes, focusing on seven proteins (P1-P7). For each *E. coli* protein, a profile is constructed, indicating which genomes code for homologous of the protein. Next the profiles are clustered to determine which proteins share the same profiles. Proteins with identical or similar profiles are boxed to indicate that they are likely to be functionally linked. Boxes connected by lines have phylogenetic profiles that differ by one bit and are termed neighbors. Adapted from Pellegrini et al. (1999) with permission.

searches and multiple sequence alighnments (MSAs) have been combined to detect such domain fusion events (Marcotte et al., 1999; Enright et al., 1999). By definition, this approach is restricted to shared domains in distinct proteins; a phenomenon whose true extent is still to be further clarified, particularly in prokaryotic organisms (Sprinzak and Margalit, 2001).

5.3.3. In Silico Two-hybrid Method

This approach is based on the hypothesis that the coevolution of interacting proteins can be better followed more closely by quantifying the degree of covariation between pairs of amino acid residues from these proteins (correlated mutations). These positions may correspond to compensatory mutations that stabilize the mutations in one protein due to changes in the other. Information about correlated mutations in single proteins has been used in particular to predict proximal pairs of

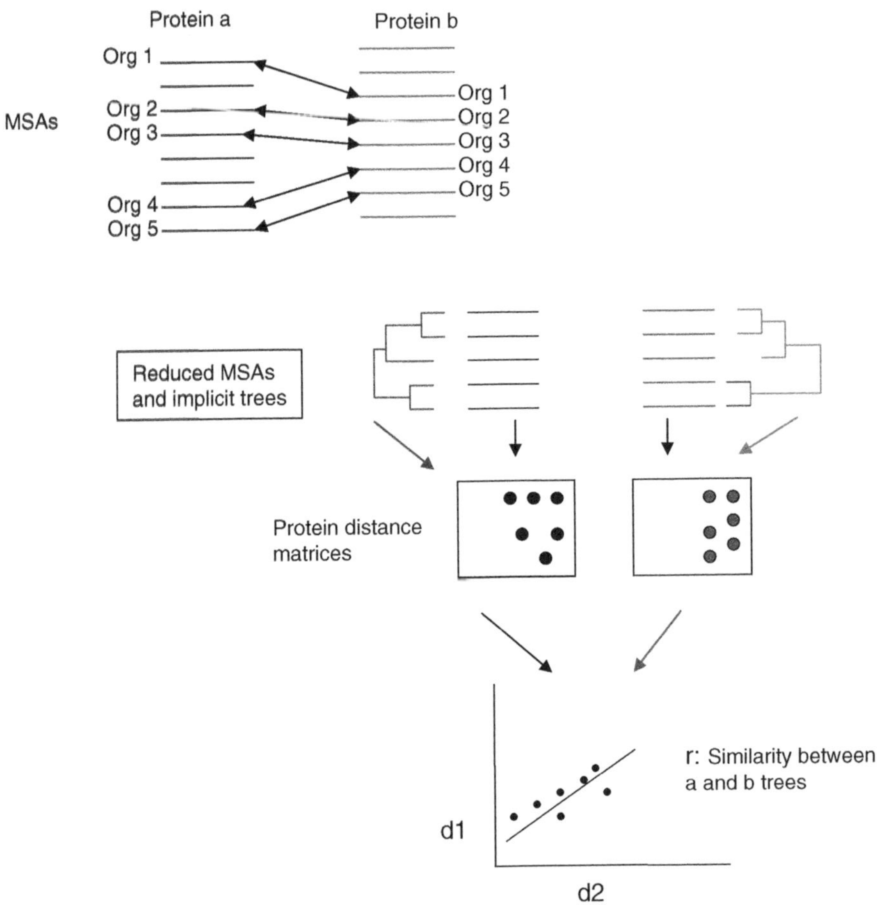

Figure 5.8. Similarity of phylogenetic trees (*mirrortree*). To obtain a quantitative indicator of the interaction between two proteins (*Prot a* and *Prot b*), the MSAs (multiple sequence alignments) of both proteins are reduced to the set of organisms common to the two proteins (*Org 1–Org 5*). Each of the reduced alignments is used to construct the corresponding intersequence distance matrix. These matrices are commonly used to construct the corresponding phylogenetic trees. Finally, the linear correlation between these distance matrices is calculated. High correlation values are interpreted as indicative of the similarity between phylogenetic trees and hence are taken as predicted interactions. Adapted from Valencia and Pazos (2002) with permission.

residues (Göbel et al., 1994; Olmea and Valencia, 1997), to discriminate structural models derived by threading (Olmea et al., 1999; 1997), and to drive *ab initio* folding simulations (Ortiz et al., 1999). For certain proteins, correlated mutations have been demonstrated to be able to select the correct structural arrangement of two proteins based on the accumulation of signals in the proximity of interacting surfaces

(Pazos et al., 1997). This relationship between correlated residues and interacting surfaces has been extended to the prediction of interacting protein pairs based on the differential accumulation of correlated mutations between the interacting partners (interprotein correlated mutations) and within the individual proteins (intraprotein correlated mutations) (Pazos and Valencia, 2002). As in the case of the *mirror tree* method, the main limitation of the in silico two-hybrid approach is the need for complete alignments with a good coverage of species common to the two proteins under study. This limitation arises as a direct consequence of the hypothesis of coevolution, which naturally requires the simultaneous study of the corresponding protein pairs in each genome. On the contrary, compensatory interaction of residues that are expected to lie physically close to each other should provide a better prediction of physical interaction as compared to methods based on general genomic information and indirect functional relationships.

As a result of immense experimental and computational data on a genome- and proteome-wide scale, several research groups have also made tremendous efforts in designing and setting up databases that include computer-controlled information on various interactomes. The following are the most recognized public databases of protein–protein interactions: Biomolecular interaction Network Database (BIND); Database of interacting proteins (DIP); the General Repository for Interaction Data sets (GRID); Molecular Interaction Database (MINT); and a database of predicted functional associations among genes/proteins (STRING). Brief description of these and other databases and relevant references are summarized in Table 5.2.

5.4. HUMAN PROTEIN INTRACTOME

The Y2H system has been applied in a high throughput manner to detect protein–protein interactions for model organisms, including *Helicobacter Pylori* (Rain et al., 2001), *S. cerevisiae* (Ito et al., 2001; Uetez et al., 2000), *Caenorhabditis elegans* (Li et al., 2004), and *Drosophila melanogaster* (Giot et al., 2003). The currently available information on the human interactome network is mainly derived from either literature curated interactions, or from "interologs" (that is, potential interactions based on interactome data for model organisms). A number of research groups have argued that protein interactions are often evolutionary conserved between orthologous proteins from different species (Matthews et al., 2001). However, It is becoming evident that to move toward human protein–protein interaction network, this information needs to be complemented by systematic experimental mapping approaches. Some of these experimental approaches together with efforts based on "interologs," are considered in this section.

5.4.1. Human Intractome Based on Orthologs

Orthologs are genes in different species that originate from a single gene in the last common ancestor of these species. To clarify the terminology used in this text it is

TABLE 5.2. Tools and databases for network analysis. On the basis of the data from Nikolsky et al. (2005) with permission.

Name	Description	URL address
BIND	A curated database of interactions, derived both from the literature and experimental data sets. 8500 interactions are deduced from high-confidence, small-scale experiments from multiple species. (Bader et al., 2003).	http://bind. Ca
DIP	A database of experimentally determined protein–protein interactions, mostly from yeast. Around 10% of DIP reactions are based on high-confidence small scale experiments (Salwinski and Eisenberg, 2003; Salwinski et al., 2004).	http://dip.doe-mbi.ucla.edu/
STRING	A database of known and predicted protein interactions deduced from over 110 genomes, high throughput experiments and gene coexpression (von Mering et al., 2003).	http://string.embl.de
MIPS	This database contains high quality small scale protein–protein interactions both in yeast and in mammals, including several hundred human interactions (Mewes et al., 2002; Pagel et al. (2005)).	http://mips.gsf.de
HPRD	Human proteins reference database provides curated human-specific protein interactions; currently over 22,000 interactions for 10,000 proteins. It also contains seven signaling maps. HPRD is used as a browser for interactions, protein annotations, motifs and domains (Peri et al., 2004).	http://www.hprd.org/
MetaCore database	A manually curated interactions, database for human proteins with known function (Nikolsky et al., 2005).	http://www.genego.com.
MINT and HomoMINT	A searchable interaction database with total of 40,000 interactions, mosly from yeast and fly. Only 3800 interactions include human proteins (Zanzoni et al., 2002).	http://mint.bio.uniroma2.it/mint/
PathArt database	A manually curated database of ~75,000 protein–protein and protein-compound interactions and pathways.	http://jubilantbiosys.com
ResNet database	Automatically extracted and manually validated database of human protein interactions (>30,000), transcriptional regulation, protein modifications, and functional regulations (Daraselia et al., 2004).	http://www.ariadnegenomics.com

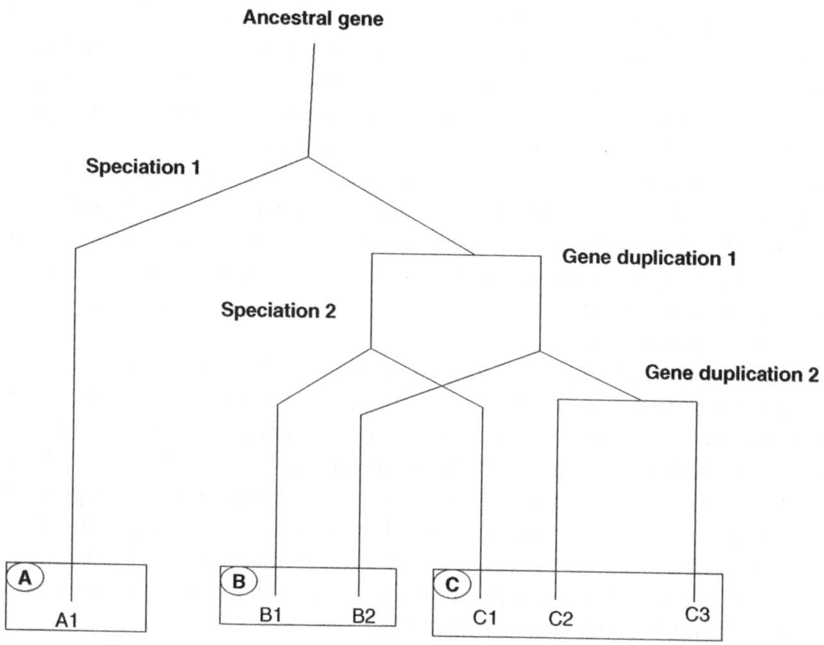

Figure 5.9. Simplified diagram of homology subtypes. Speciation events produce the species A, B and C. The genes A1, B1, C1, C2, and C3 have descended from the ancestral gene following evolutionary events of speciation and gene duplication. Adapted from Jensen (2001) with permission.

helpful to refer to Figure 5.9, which was originally proposed by Fitch (2000). This diagram shows four events of evolutionary divergence, two being events of speciation (i.e. after the two species diverged from each other) and two being events of gene duplication, yielding six contemporary genes in the three organisms, A, B, and C. Determination of orthology or paralogy in a vertical line of descent is simply realized by tracking any pair of genes back to where they join, either at an inverted "Y" (in which case they are orthologs) or at a "horizontal line" (in which case they are paralogs). On the basis of these criteria, A1 has three orthologs in species C, but only C1 is an ortholog of B1. On the contrary, B2 has two orthologs in species C (C2 and C3), whereas B2 and C1 are paralogs. As in vivo experiments with human genes are not feasible, various research efforts have attempted to identify genes in human genome, which may share the same biological function(s) with genes in model organisms. A prerequisite for inferring a function of a human gene from the function of a corresponding gene in a model organism is to establish how the genes have evolved. This is because genes in two species that have directly evolved from a single gene in the last common ancestor are most likely to share the same function. Such genes are called orthologs (Fitch, 1970). Owing to the uncertainty of functional equivalence between the orthologs derived from a single ancestor at the time of speciation, it is important to detect all of them. As these are homologs found in the same genome, they are called paralogs. However, there may also be paralogs that arose from a duplication event before the speciation. These are therefore not orthologs according to the definition. At present, there is no accepted terminology to separate paralogs that were duplicated before a speciation event from paralogs that were duplicated after it. Remm et al. (2001) proposed two new terms, in analogy with the phylogenetic concepts of out-group and in-group. Paralogs predating the speciation event that are not orthologs are denoted out-paralogs. Paralogs that were duplicated after the speciation event, and thus are orthologs, are denoted in-paralogs. Construction of phylogenetic trees is the obvious way to detect orthologs; however, some steps within such approach are difficult to automate and it demands large resources of computing power.

An alternative to phylogenetic methods is to use all-versus-all sequence comparison between two genomes to detect orthologs. This approach is based on the hypothesis that if the sequences are orthologs, they should score higher with each other than with any other sequence in the other genome (Rubin et al., 2000; Wheelan et al., 1999). An extended version of all-versus–all technique has been reported by Remm et al. (2001). The authors introduced an algorithm in-paralog and ortholog identification (INPARANOID) and assessed its performance on a set of previously validated orthologs. It was also applied to the complete set of protein sequences from the *C. elegans and D. melanogaster* genomes.

Lehner and Fraser (2004) used the orthology relationship identified by INPARANOID algorithm to construct a putative human protein-interaction map. On the basis of high throughput interaction data sets from model organisms, the authors constructed what they called a first draft of human protein interaction map. The data was obtained from seven experimental and four computationally predicted protein-interaction maps, and applied an algorithm to identify potential human orthologs. A human protein interaction was included in the prediction list when both interaction partners from a model organism had one or more human orthologs. This strategy generated a human interaction network comprising over 71,000 interactions in more than 6,000 proteins. The sources of these interactions are depicted in Figure 5.10 (a,b). To assess the accuracy of the generated interactions, the authors used the Gene Ontology (GO) annotations (Ashburner et al., 2000). These annotations provide a hierarchical description of gene functions with general functions described by GO annotations at the top levels of the hierarchy, while highly precise functions are described by terms deeper in the hierarchy. Generally speaking, physiologically interacting proteins are expected to have related, but not necessarily identical functions. Therefore, they are expected to share some, but not all GO annotations. One approach to evaluate an interaction data set is to count the proportion of interactions that connect proteins that share common GO terms. The work by Lehner and Fraser (2004) resulted in an impressive network of over 70,000 predicted physical interactions and highlighted a number of aspects, which are considered in here. First, these interactions are inferred from data obtained from model organisms and therefore, the accuracy and coverage of the predicted interactions hinges on following two main elements: the quality and accuracy of the original model organism interaction data sets and the ability to identify the human orthologs

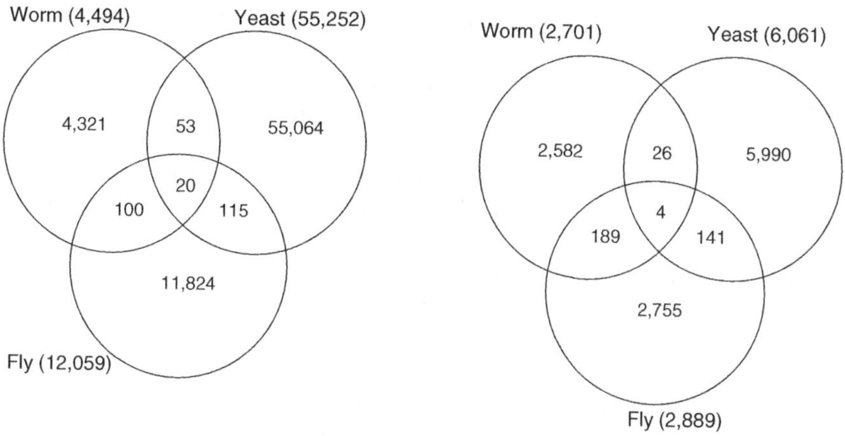

(a) Complete network (71,496 interactions) (b) Core network (11,487 interactions)

Figure 5.10. Sources of predicted human protein interactions: (a) The number of human protein interactions predicted by the interaction maps from each model organism. (b) The number of human protein interactions predicted by the core higher confidence interactions from each organism. Core interactions are those that reconfirmed when retested (worm), or had an interaction score of greater than 0.5 (fly) or were identified more than once in a single assay (yeast, worm). Adapted from Lehrner and Fraser (2004) with permission.

of a model organism protein. It has to be recorded here that the identification of gene orthologs is by no means a trivial task. As it has been mentiond in the previous section, gene duplications can result in a single gene having multiple potential orthologs in a second species. Second, applying a single algorithm across data sets derived from different model organisms has to take into account different degrees in accuracy between these sets. Indeed, Lehner and Fraser (2004) noted that the raw yeast and worm protein interaction data sets were slightly more accurate than the raw data for the fly interaction data set. Such difference necessitated some form of filtering to obtain high-confidence interactions.

5.4.2. Human Interactome Based on Experimental Data

Over the last 2 years, a number of experimental works have provided the first experimental data on human protein–protein interactions. In an important first step toward constructing experimentally-derived human protein interactions map, an Y2H study has been described by Stelzl et al. (2005) in which human protein interactions in more than 3,000 candidate have been described. Prior to this study, the largest human Y2H study reported 755 interactions associated with the Smad signaling system, which is regulated by members of the transforming growth factor β (TGFβ) superfamily (Colland et al., 2004). A high through-put mapping of the same signaling network was also described a year later by Barrios-Rodiles et al.(2005).

To detect interacting pairs of human proteins systematically, Stelzl et al. (2005) screened a protein matrix of 4456 baits and 5632 preys by automated yeast Y2H interaction mating. This screening identified over 3000 interactions among more than 1,700 proteins. The authors used two types of biochemical pull-down experiments to increase confidence in their findings, confirming just over 60% of the 240 pair-wise interactions tested. They also developed a bioinformatics scoring system based on heuristic reliability factors to sort the 3000 putative interactions into sets designated high, medium, and low confidence. The scoring factors to increase confidence in a given interaction can include the observation of a two hybrid signal with three different reporter genes, the interaction found within 3- or 4-node reciprocal interaction clusters or network cliques, the two interacting proteins that share similar functional annotations in the Gene Ontology Database, and the existence of previously identified orthologous interactions in other model organisms (e.g. yeast, fly, worm). We have to bear in mind that the use of this scoring procedure can result in some form of bias toward previously published interactions.

Members of the TGFβ superfamily are secreted signaling molecules that regulate many biological processes such as cell growth, differentiation, and morphogenesis. The disruption of components of the TGFβ superfamily pathways has been associated with a number of diseases, including cancer. The effect of these molecules is transduced through the legend-induced formation of heteromeric receptor complexes of type I and II transmembrane serine-thrionine kinases, which are activated when the type II receptor kinase transphosphorylates type I receptor. In a recent study, Barrios-Rodiles et al. (2005) described a high-throughput strategy to systematically map protein–protein interactions in mammalian cells. This strategy is designated "LUMIER", which stands for luminescence-based mammalian interactome mapping. This strategy uses *Renilla* luciferase enzyme (RL) fused to proteins of interest, which are then coexpressed with individual Flag-tagged partners in mammalian cells. Protein–protein interactions are then determined by performing an RL enzymatic assay on immunoprecipitates using an antibody against the flagged partners. As a proof of principle, the authors applied this strategy for a systematic study of TGFβ superfamily. The main steps in luminescence-based strategy for the detection of protein–protein interactions are depicted in Figure 5.11.

Another important step toward a systematic and comprehensive human interactome project has been provided by the work of Rual et al. (2005). In this study the authors described the initial version of a proteome-scale map of human binary protein–protein interactions. Using a stringent, high-throughput Y2H system, the authors tested pairwise interactions among the products of ~8100 currently available Gate-cloned open-reading frames (ORFs) and detected ~2800 interactions. This data set was verified by an independent co-affinity purification assay. The verification rate was estimated to be ~78%, the generated data set also revealed more than 300 new connections to over 100 disease-associated proteins. Comparison of the proteins within the generated data set with the list of genes associated with human diseases in the Online Mendelian Inheretance in Man (OMIM) database allowed the identification of 424 interacting pairs for which at least one partner had been previously associated with a human disease. This long list included various forms of cancer,

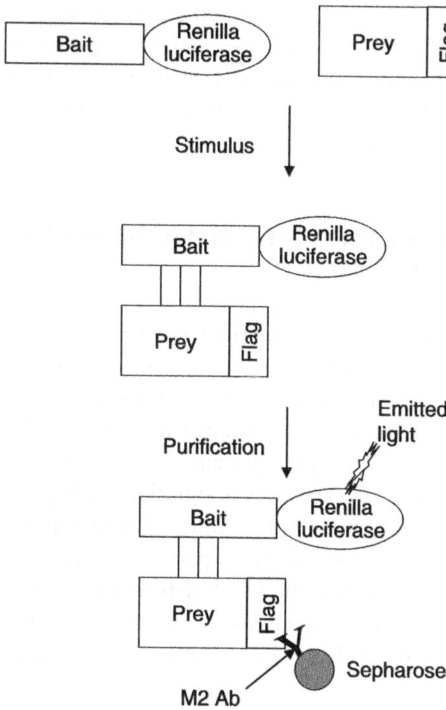

Figure 5.11. The main steps in a Luminescence-based strategy for the detection of mammalian protein–protein interactions. Adapted from Barrios-Rodiles et al. (2005) with permission.

including breast cancer, leukemia, thyroid, and prostate cancer. To gain more insight into the evolution of the resulting interactome, the authors classified proteins into eukaryotic, metazoan, mammalian or human and attempted to address the question of protein evolution. In other words, do proteins specific to different evolutionary classes interact with one another? The authors deduced that the network seems to be enriched for interactions between proteins of the same evolutionary class but not for interactions between proteins from two different evolutionary classes. Although the data set generated by this study is far from comprehensive, it certainly provides clear indications and guidelines for future studies aimed at expanding the human protein-interactions network.

Signal transduction pathways are modular composites of functionally interdependent sets of proteins, which act in a coordinated manner to transform incoming information into phenotypic responses. The pro-inflammatory cytokine tumor necrosis factor (TNF)-α is known to trigger a signaling cascade, converging on the activation of the transcription factor NF-κB, which forms the bases for diverse physiological and pathological events (Ghosh et al., 2002). Bouwmeester et al. (2004) used a combination of TAP, LC/MS-MS, network analysis, and directed functional perturbation studies to investigate the physical and functional map of the human TNF-α/NF-κB signal transduction pathway. These analyses allowed the identification of 221 molecular associations and 80 previously unknown interactors. The same data provided a substantial insight into the logic of this pathway, which future studies may find it applicable to other pathways relevant to human diseases.

5.5. RELATIONSHIP BETWEEN GENE EXPRESSION AND PROTEIN INTERACTION

One of the main challenges of contemporary biology is how to piece together information derived from genome wide gene expression profiling and protein–protein interaction maps. Statistical analysis of large-scale gene expression and protein interaction data have shown that protein pairs encoded by coexpressed genes interact with each other more frequently than with random proteins. Furthermore, the mean similarity of expression profiles is significantly higher for respective interacting protein pairs than for random ones. In other words, protein network analysis will place genes identified in microarray experiments in a broader biological context. Since protein networks reflect the functional grouping of interacting or coordinately induced/suppressed genes, the roles of the subsets of coexpressed genes may be resolved by considering both the gene expression data and protein interaction networks.

To understand complex biological processes, such as cancer initiation and progression, it is necessary to consider differential gene expression in the context of complex molecular networks. The study of such networks requires detailed protein–protein interaction maps. The preliminary versions of such maps have been generated by high throughput methodologies described above and by computational prediction algorithms and, in the case of the human protein reference database (http://www.hprd.org/), by curating known interactions from the literature (Peri et al., 2003). Although human interacome maps are still in their infancy, representing only a fraction of the complete interaction network, a number of groups have begun to evaluate their potential in the understanding of disease progression, treatment response, and biomarkers discovery. In a rare example demonstrating the use of protein interactions network to interpret complex cancer signature, Rhodes and Chinnaiyan (2005) showed how a multiple myeloma signature from Oncomine database can be mapped to the human interactome network.

Protein function annotation is another challenging problem in the postgenomic era. Currently, there are a number of approaches indicated for assigning putative functions to unannotated proteins. An important component in protein function annotation is identifying and characterizing protein–protein interactions. Grigoriev (2001) analysed physical interactions in yeast and observed that proteins encoded by coexpressed genes interact with each other more frequently than with random pairs. In another study concerning yeast, it was shown that interacting protein pairs are more likely to be in the same expression cluster than random pairs (Ge et al., 2001). On a genomic scale, Jansen et al. (2002) attempted to relate the absolute mRNA expression levels and the expression profiles in yeast to protein–protein interactions. The authors noted that permanent complexes, such as the ribosome and proteasome, have a particularly strong relationship with expression, whereas transient complexes do not. The same study subdivided the transient complexes, such as the RNA polymerase II holoenzyme and the replication complex into smaller permanent complexes, which were found to have a strong relationship with gene expression.

In a relatively recent study (Bhardwaj and Lu, 2005), an attempt was made to establish correlation between gene expression profiles and protein–protein interactions within and across different genomes. The authors investigated four following genomes: *E. coli*, *Saccharomyces* (yeast), *Mus musculus* (mouse) and *Homo sapiens* (human). The choice of these species was motivated by the availability of experimental data sets and by the fact that these species cover a wide range of life-forms extending from a prokaryote of few thousand genes to a more complex eukaryote. A number of deductions based on this study are worth considering together with deductions from other studies. First, in the case of *E. coli* the authors found a strong correlation between the expression profiles for interacting pairs compared with their random counterparts, whereas in the other three species the correlation was only slightly more significant than random. To ensure that such favourable results for *E. coli* are not simply because of its small genome size, the same relationship was investigated for another prokaryote, *H. pylori*. According to the authors, such investigation did not give clear results to support or dismiss the results obtained for *E.coli*. Such observation indicates that further studies are needed before reliable conclusions regarding the relationship between protein interactions with gene expression for prokaryates can be drawn. Second, according to the same authors, the inclusion of multiple species has strengthened the correlation between protein–protein interactions and gene expression. In other words, the coexpression of interacting proteins is more conserved than that of random pairs on the genome wide scale of the four investigated species. If these results can be supported by future studies, then they will contribute to the prediction of protein–protein interactions using not only their coexpressiopn properties but also that of their orthologous counterparts. On the basis of this study and others dealing with the same concept, it can be hypothesized that future studies to determine protein–protein interactions will not only rely on their coexpression but also consider their orthologs (if present) in other species.

5.6. GENE SIGNATURES IN CANCER PREDICTION/CLASSIFICATION

Recently, a number of studies have demonstrated the feasibility of using gene expression profiling to predict the outcome of various forms of cancer. Before considering some representative examples, the following general considerations may underline the need for this emerging line of investigation. First, until very recently, the main prognostic factors in breast cancer have been age, tumor size, status of axillary lymph nodes, histological grade, and hormone receptor status. Other factors have also been evaluated for their potential to predict the outcome of disease, but in general, they have only limited predictive power (Isaacs et al., 2001). Second, Chemotherapy or hormonal therapy is known to reduce the risk of distant metastases by approximately 33%; however, almost 75% of patients receiving this treatement would possibly have survived without it (Early Breast Cancer Trialist's collaborative Group, 1998 a, b). These statistics underline an urgent need for strategies, which can accurately select patients who would benefit from adjuvant therapy. To appreciate the emerging role of gene signatures in cancer prediction and/or classification, a number of specific examples will be considered in the following sections.

5.6.1. Breast Cancer

As it has been pointed out in Chapter 3, germ line mutations in the two main breast cancer susceptibility genes, *BRCA1* and *BRCA2*, confer a highly elevated risk of breast and ovarian cancer and account for a significant proportion of inherited breast cancers (Ford et al., 1998). However, it has to be said that many familial breast cancer cases cannot be attributed to mutations in these genes, suggesting a role for additional predisposing genes (Nathanson et al., 2001).

Jönsson et al. (2005) used array-based comparative genomic hybridization (aCGH) to investigate tumor DNA obtained from 14 carriers of *BRCA1* mutations, 12 carriers of *BRCA2* mutations, and 26 cases of sporadic breast cancer. Array-based CGH is known for its resolution, which can only be limited by the number of probes on the array. Furthermore, this approach has the advantage of revealing DNA copy number changes throughout an entire genome. The study by Jönsson et al. (2005) revealed a number of interesting aspects of the genomic profiles in hereditary breast cancer: First, the authors found a striking concordance regarding genomic profiles for two different tumors obtained from the same patient carrying a *BRC2* mutation. Interestingly, these tumors were regarded as independent primary cancers as determined by histopathologic variables. Both tumors were invasive and included extensive ductal carcinoma in situ (DCIS), implying an independent origin rather than a metastatic spread. This latter observation is supported by an earlier study, which used a similar experimental approach to investigate bilateral sporadic breast cancer. This study by Teixeria et al., 2004 reported widely different alterations in the contralateral tumors, which were interpreted as an indication that bilateral cancers can have different origins. These results are of particular interest because the issue of whether multiple, ipsilateral or bilateral breast carcinomas represent multiple primary tumors or dissemination of a single carcinomatous process is still difficult to resolve, especially for individual patients. Second, using hierarchical clustering methods based on ternary data for the 169 discriminating BAC clones, the authors reported that all *BRCA1* tumors were tightly clustered and separated from sparodic cases. On the contrary, *BRCA2* tumors showed higher similarity with their sporadic counterparts. Almost 70% of *BRCA2* tumors were tightly grouped in a subcluster, whereas one tumor clustered together with the sporadic cases and three outlier cases clustered with *BRCA1* tumors. Given the small number of cases investigated, it is difficult to draw definite conclusions regarding these results. These results may indicate that *BRCA2* tumors constitute a more heterogenous group, an observation, which concords with results based on different histopathologic and clinical variables (Lakhani et al., 2000). On the contrary, it could be simply because of a relatively small number of samples investigated.

Various histopathological investigations have indicated that many hereditary breast cancer cases are characterized by mutations in either *BRCA1 or* BRCA2 genes. Hedenfalk et al. (2001) hypothesised that the genes by these two types of tumors may also be different to allow the identification of hereditary breast cancer on the basis of gene-expression profiles. The authors used microarrays of 6,512 cDNA clones of 5,361 genes to investigate samples of primary tumors from seven carriers of the *BRCA1* mutations, seven carriers of the *BRCA2* mutations, and seven patients with sparodic cases of breast cancer. Using supervised learning, they were able to identify a number of differentially expressed genes between *BRCA1*-mutated and *BRCA2*-mutated tumors and use these genes to accurately categorize these samples. Cyclin D1, an important cell cycle regulator known to be overexpressed in certain breast cancers, was one of the genes with increased expression in *BRCA2* mutation–positive tumors, and this finding was confirmed using immunohistochemistry. Interestingly, one sparodic tumor case was classified as having a *BRCA1*- mutated phenotype. Direct sequencing of the *BRCA1* gene in this patient showed no mutation, but the promoter of this gene showed aberrant methylation resulting in silencing of gene expression. Because epigenetic events can be important in oncogenesis, this intriguing finding points to the use of expression profiling for identifying such events in the absence of germline information.

Using complementary DNA (cDNA) microarrays to analyze breast cancer tissues, Perou et al. (2000) identified tumors with distinct patterns of gene expression that they termed "basal type" and "luminal type." These subgroups differ with respect to the outcome of disease in patients with locally advanced breast cancer (Sorlie et al., 2001). In their study, Perou et al. (2000) reported the molecular classification of 65 breast adenocarcinoma specimens from 42 individuals. Hierarchical cluster analysis allowed the definition of three separate subtypes. Based on patterns of gene expression, this heterogeneous tumor class was classified into a known subtype, *ERB-B2*$^+$ and two previously unknown subtypes: estrogen receptor–positive/luminal- like cancers and basal-like cancers. One of the interesting aspects of this study was the presence of 20 primary tumors that were biopsied before and after a 16-week course of chemotherapy and two primary/lymph node metastases pairs. Using gene clustering, the authors showed that paired samples are more highly related to each other than to tumors from other individuals, despite chemotherapy or metastatic evolution. This study also identified gene expression correlates of different cellular features of these tumors. For example, a number of genes known to play roles in cellular proliferation were coordinately expressed by different tumors, and their expression could be correlated with mitotic index. They also identified eight independent gene clusters that seemed to reflect contributions of specific cell types present within tumors such as endothelial cells or B lymphocytes. In a latter study, the same group has extended their findings to a larger set of tumors (Sorlie et al., 2001).

In two successive studies involving different number of patients, van't Veer et al. (2002) and van de Vijver et al. (2002) described the use of gene-expression signatures in breast cancer as predictor of survival. In the first investigation the authors used inkjet-synthesized oligonucleotide microarrays to analyze primary breast tumors of 117 young patients, and applied supervised classification to identify a gene expression signature predictive of a short interval to distant metastases. The analysis was restricted to tumors that were less than 5 cm in diameter from lymph-node-negative patients who were younger than 55 years of age. The authors concluded that a classification system based on 70 genes could outperform all clinical variables in predicting the likelihood of distant metastases within 5 years. A limitation of this study was that the results were derived from and evaluated in two groups of patients selected on the basis of outcome of the disease. To address this limitation and to provide stronger evidence on the utility of gene expression signature as a clinical tool for the assessment of breast cancer metastases, the authors conducted a second study involving 295 young patients with breast cancer. All patients had stage I or II of breast cancer and were younger than 53 years old; 151 had lymph-node-negative disease, and 144 had lymph-node-positive disease. In the second study the predictive power of the previously established 70 genes prognosis profile was evaluated using univariable and multivariable statistical analyses. On the basis of both the studies, a number of deductions were made: First, the prognosis profile performed best as predictor of the appearance of distant metastases during the first 5 years after treatment. According to these authors such finding was not unexpected, because the tumors on which the profile was based had all metastasized within 5 years. Second, the prognosis profile demonstrated to be a strong predictor of the development of distant metastases in patients with lymph-node-positive disease. The importance of such finding lies in the fact that the presence of lymph-node metastases is by itself a strong predictor of poor survival. Within the second study (van De Vijver et al., 2002) there was an observation, which is relevant to consider. The authors suggested that the ability to metastasize to distant sites is an early and inherent genetic property of breast cancer. This observation goes against the prevailing model of metastasis, which holds that most primary tumor cells have low metastatic potential, but rare cells (estimated at less than one in 10^7) within large primary tumors acquire metastatic capacity through somatic mutations (Poste and Fidler, 1980). A partial support for this model comes from the work by Bernards and Weinberg (2002), who hypothesized that metastatic potential is acquired relatively late during multistep tumorigenesis. Currently there is no sufficient evidence in favour or against either of the two ideas. Such evidence if obtained will no doubt impact on the ability to determine the stage at which metastasis starts to manifest. Establishing the stage of metastasis is important because if the metastatic ability of breast cancer is determined early in tumorigenesis, early prognostic testing could be undertaken, which would clearly be beneficial. On the contrary, an early onset of metastatic capability theoretically limits the benefit of early detection and treatment.

The molecular nature of metastasis has been investigated by Ramaswamy et al. (2003). The authors used oligonucleotide microarrays to determine a molecular signature of metastasis in a number of primary solid tumors. This was done by analyzing the gene-expression profiles of 12 metastatic adenocarcinoma nodules of diverse origin, including lung, breast, prostate, colorectal, uterus, and ovary and compared them with the expression profiles of 64 primary adenocarcinomas representing the same spectrum of tumor types obtained from different individuals. On the basis of this study and related interpretation there are a number of elements which merit further consideration. (i) The authors identified an expression pattern of 128 genes that best distinguished primary from metastatic adenocarcinoma. Interestingly, the same gene-expression pattern was also associated with metastasis present in some primary tumors, resulting in a misclassification of these tumors as metastases. This unexpected finding was taken as an indication that a gene-expression program of metastasis may already be present in the bulk of some of primary tumors at the time of diagnosis. To test this hypothesis, the authors analyzed the metastasis-derived gene expression program in several large gene-expression data sets containing molecular profiles of primary solid tumors. Sixty-two stage I/II primary lung adenocarcinomas (Bhattacharjee et al., 2001) were analyzed for expression of the metastases associated genes. Hirarchical clustering in the space of these 128 genes identified two clusters of primary tumors with gene-expression profiles that were highly correlated with the original primary tumor versus metastases distinction. (ii) The identification of gene-expression signature shared by multiple solid-tumor types may raise the hope of finding therapeutic targets common to different types of cancer.

5.6.2. Follicular Lymphoma

Follicular lymphoma is the second most common form of non-Hodgkin's lymphoma, accounting for about 22% of all cases (Armitage and Weisenburger, 1998). The clinical course of follicular lymphoma is variable: In some patients the disease is indolent and slowly progressive over a period of many years, whereas in others the disease progresses rapidly, often transforming into aggressive lymphoma and early death (Horning, 2000; Johnson et al., 1995). Management of this malignancy includes observation, chemotherapy, hematopoietic stem-cell transplantation, and immunologic therapies based on antibodies to B cells (Colombat et al., 2001; Witzig et al., 2002) or idiotype vaccines (Timmerman et al., 2002). At present, there is no conclusive evidence that any of these approaches offer a clinically significant survival advantage. The molecular and cellular mechanisms responsible for the clinical heterogeneity of follicular lymphoma are unknown. The tumor arises from a germinal- centre B cell that in the majority of cases, has acquired a t-translocation that deregulates *BCL2*, a key gene in the regulation of cell death (Montoto et al., 2002). Some tumors subsequently accumulate further oncogenic aberrations that have been associated with transformation to diffuse large-B-cell lymphoma (Lossos and Levy., 2003). However, it is unclear whether these random genetic events account for the clinical heterogeneity of the disease. An understanding of the molecular biology that underlies the survival differences among patients with follicular lymphoma might provide a more accurate and rational method of risk stratification to guide treatment and might suggest new therapeutic approaches as well. Dave et al. (2004) conducted a study to determine whether the duration of survival among patients with follicular lymphoma can be predicted by the gene-expression profiles of the tumors at the time of diagnosis. By whole-genome microarray analysis of gene expression, the authors constructed a multivariate model of survival that revealed aspects of the biology of follicular lymphoma that influenced the duration of survival. Analyses were conducted on fresh-frozen tumor-biopsy specimens and clinical data from 191 untreated patients who had received a diagnosis of follicular lymphoma between 1974 and 2001 were obtained from seven institutions in North America and Europe and studied according to a protocol approved by the National Cancer Institute's institutional review board. The patients had undergone a variety of standard treatments after biopsy, including various chemotherapy regimes (such as those containing anthracyclines and purine analogues) and autologous stem-cell transplantation, or had been under observation. The median age at the time of diagnosis was 51 years (range, 23 to 81), and the median follow-up time was 6.6 years (range, less than 1.0 to 28.2); the median follow-up time among patients alive at last follow-up was 8.1 years. The basic steps to devise a gene expression-based model of survival in follicular lymphoma are depicted in Figure 5.12. The main conclusions of this study can be summarized as follows: First, the gene-expression signatures that predicted survival were unexpectedly derived from nonmalignant cells in tumors. This observation implies interplay between the host immune system and the malignant cells in this form of cancer. Second, the clinical value of the

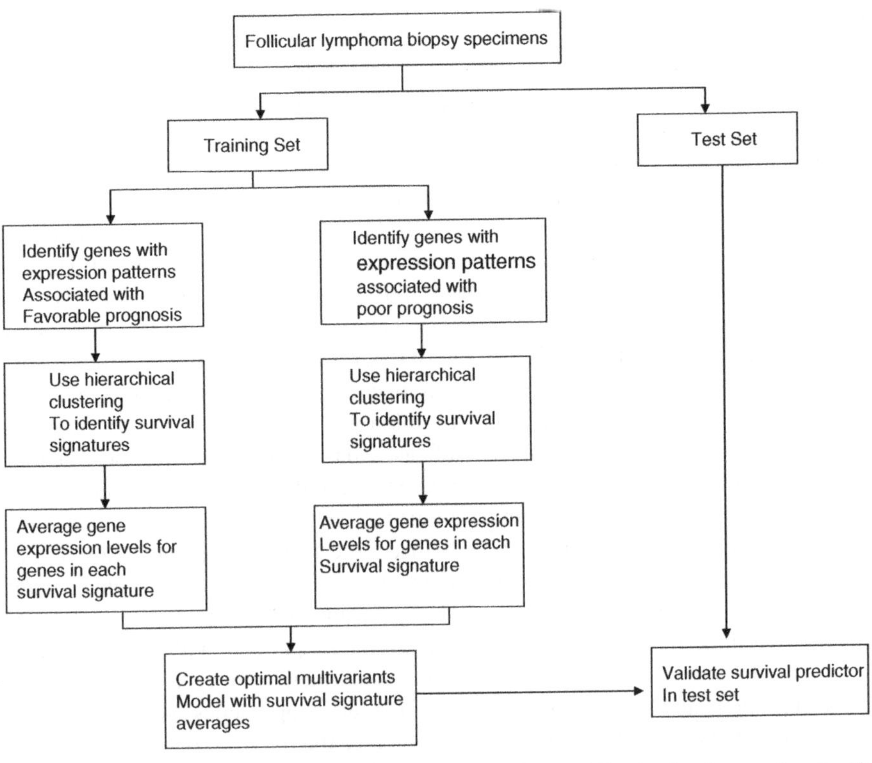

Figure 5.12. An overview of survival signature analysis, used for the development and validation of a survival predictor based on gene expression for follicular lymphoma. Adapted from Dave et al. (2004) with permission.

derived molecular predictor may benefit a subgroup of patients who have indolent form of the disease. This substantial subgroup represents 75% of the overall population of patients with this form of cancer. Given that the median survival following diagnosis is more than 10 years; such molecular predictor can provide a valuable prognostic indicator which can be used for watchful waiting. For the other 25% with less favourable prognosis, patients survive for less than 4 years; this predictor can be used to design clinical trials that have achievable endpoints. Such trials have been difficult to conduct with patients having survival median of over 10 years. Therefore, clinical trials can be designed to enrol patients with the least favourable prognosis, a strategy that would allow assessment of the overall survival. Third, there is considerable clinical evidence that immune responses are important in follicular lymphoma. Indeed in some cases, the lymphoma regresses spontaneously (Horning and Rosenberg, 1984). A similar observation has also been made for melanoma and renal-cell carcinoma, which may imply an effective antitumor immune response (Dave et al., 2004). The work described above provides a molecular indicator to investigate aspects of the immune response to follicular lymphoma that may positively or negatively influence the pace of the disease. The genes in the immune response signatures may be used as markers to identify subpopulation of immune cells that may promote or antagonize the proliferation or survival of the malignant clone.

5.6.3. Lymphocytic Leukemia

One of the challenges in cancer treatment is to target pathologically distinct cancer type. Improvements in cancer classification have thus been central to advances in cancer treatment. For a number of years, cancer classification has relied heavily on morphological appearance of the tumor. However, tumors with similar histopathological appearance can distinctly follow different clinical courses and show different responses to therapy. The emergence of technologies suitable for high throughput genomic and proteomic analyses, in particular DNA microarrays, allowed the generation of global profiles of gene expression in cancer. Known types and subtypes of cancer have been readily distinguished by their gene-expression patterns, and more importantly, new molecular subtypes of cancer have been discovered that are associated with a host of tumor properties, including the propensity to metastasize.

Array-based comparative genomic hybridization (aCGH) is another powerful tool for genome-wide determination of DNA copy number alterations. Initial studies revealed the enormous potential of this approach for identifying candidate disease genes and cancer classification (Pinkel et al., 1998; Fritz et al., 2002). However, the widespread application of aCGH has been hindered by the lack of well characterized, high-resolution clone sets for consistent performance in aCGH assays. An important step in this direction has been achieved by Greshock et al. (2004) who have assembled a set of ~4100 publicly available human bacterial artificial chromosome (BAC) clones evenly spaced at ~1-Mb resolution across the genome, which includes direct coverage of ~400 known cancer genes. This approach together with DNA micraoarrays has been used by different groups to classify cancers and to establish clonal relationship in individual patients. Some representative examples of such studies are considered below.

The use of gene expression profiling for cancer classification has been demonstrated by Golub et al. (1999). Oligonucleotide microarrays were used to study the expression of 6,817 human genes in 72 acute leukemia samples. By using unsupervised learning based solely on the acquired gene expression patterns, these analyses were able to cluster leukemia samples into the known subsets of acute myelogenous leukemia (AML) and acute lymphocytic leukemia (ALL). Furtheremore, using supervised learning, gene sets that were differentially expressed in AML and ALL were used to correctly classify a group of unknown samples into the correct categories. The distinguishing gene patterns included marker genes that were both known, such as myeloperoxidase and terminal transferase, and unknown. Although the distinction between AML and ALL generally is not clinically difficult using modern histopathology and cell surface phenotypes, the study by Golub et al. (1999) provided strong evidence that gene expression profiles in tumor samples can be used for cancer classification. The same study raised a number of relevant questions, which could not be addressed by such analyses. For example, is the clear distinction between the two forms due to the fact that AML and ALL were derived from distinct cellular precursors? In other words, would the same approach deliver such encouraging results when tested for more closely related cancers? Furthermore, class discovery in this case required prior biologic knowledge of AML and ALL to make sense of the observed clusters. The interpretation of new classes discovered with clustering is likely to be more challenging in the absence of known biologic or clinical correlates.

In another study, Armstrong et al. (2002) used both unsupervised and supervised learning to establish the globally distinct nature of mixed-lineage leukemia; a leukemia subset with a decidedly unfavorable prognosis that is defined by a chromosomal translocation involving the mixed-lineage leukemia gene. The identification of differentially expressed molecular markers such as the receptor tyrosine kinase FLT3 by this subtype of leukemia, compared with both ALL and AML, may facilitate novel strategies for molecularly targeted treatment in this treatment-refractory cancer.

5.6.4. Lung Adenocarcinoma

The histopathological sub classification of lung adenocarcinoma still represents a challenging task. This challenge has been emphasized by a study, where independent lung pathologists have concluded that the agreement on such subclassification was only 41% (Sorensen et al., 1993). In the search for molecular markers to classify human lung carcinomas, Bhattacharjee et al. (2001) analyzed mRNA expression levels corresponding to 12,600 transcript sequences in 186 lung tumor samples. This analysis provided strong evidence for biologically distinct subclasses of lung adenocarcinoma. Interestingly, comparison of these results with another study performed with a different set of tumors and expression profiling platform revealed a number of similarities (Garber et al., 2001). For example, the two studies showed strong overlapping in gene expression signatures for previously defined tumor classes such as SCLC and squamous cell lung carcinoma.

There are a number of studies that used microarray analysis to classify NSCLC specimens based on gene or proptein expression profiles (Bhattacharjee et al., 2001; Virtanen et al., 2002; Campa et al., 2003; Yamagata et al., 2003). In these and other studies, a class discovery approach showed that gene expression profiles could both group tumor samples in concordance with their histologic classification as well as the identification of subgroups within histologic subclasses.

Microarray analysis has the potential to predict the survival of patients with NSCLC. Although cluster analysis by Beer et al. (2002) did not perfectly segregate stage I from stage III lung adenocarcinoma, the authors pointed out that the stage I tumors that clustered with stage III samples came from patients that exhibited worse survival. This study showed that the expression of genes that confer a poor prognosis is independent of the stages of the disease at the time of diagnosis. As microarray analysis can identify tumor subsets that share molecular alterations important for cancer progression, incorporating gene expression profiles might provide added therapeutic and prognostic value when combined with traditional staging and histologic analysis. Other studies have also correlated gene- or protein-expression profiles with prognosis (Bhattacharjee et al., 2001; Yamagata et al., 2003; Borczuk et al., 2004). In these studies, subsets of genes or protein peaks differentially expressed in tumors could predict survival differences among patients with AC. In an extension of their study on the use of multiple gene airs to classify lung cancer based on histology, Gordon and associates found that sets of three gene expression ratios could accurately predict prognosis more than 90% of the time (Gordon et al., 2003). The authors then used the most predictive gene pairs to analyze data from the earlier study of Bhattacharjee and coworkers (Bhattacharjee et al., 2001). In this analysis, the set of gene expression ratios could correctly predict survival of patients with stage I AC tumors 64% of the time in low-tumor-content samples, and 74% of the time for high-tumor-content samples (Gordon et al., 2003). The development of prognostic identifiers could guide clinical decision-making, but such application requires validation in prospective clinical trials. The inclusion of genomic and proteomic analyses into future clinical trials, and the analysis of specimens from lung cancer patients receiving standard of care therapy could provide information that might ultimately lead to more accurate, individualized clinical decisions.

5.7. CONCLUDING REMARKS

- Large scale Y2H screens for various species (see Table 5.1) have produced an impressive number of interaction pairs. However, for each species the results covered only a small fraction of the expected number of interactions. For example, the two separate studies with *S.cerevisiae* identified over 3000 potential protein–protein interactions, yet these constitute only 10–15% of the expected total. Likewise, the worm interactome containing over 4000 interactions comprises less than 5% of the expected total of interactions. This observation has to be considered with the fact that various curated data sets compiled from the literature show limited overlap. Such limited overlap among various data sets can be attributed to several factors some of which have been pointed out in this chapter. It has to be clear that highlighting some of the current drawbacks of published protein networks is not intended to undermine the impressive achievements, which these pioneering works have delivered so far. On the contrary, such remarks are intended to underline the fact that far more comprehensive and more reproducible interactions are to be expected in the near future. Having said that, analysis of existing protein interaction maps has suggested that even sparse data can be used to derive initial rudimentary models of biological networks (Uetz and Finley, 2005). However, a complete systems-level understanding of any biological process surely requires more data than current technologies can provide. Many of us tend to think that optimal modeling of a biological system can only be achieved by knowing of all the molecules involved, their various roles, their concentrations, how they fit together, the effect of each molecule on its neighbours, as well as parameters such as how concentrations and interactions change over time. To achieve such a tentalizing objective, we will surely need more time and above all more powerful technologies that outperform existing ones.

- The material discussed in the first part of this chapter has focused on two arguments: protein–protein interactions and gene expression profiling. This choice can be partially justified and extended through the following considerations: (I) Topological features of the protein networks have been shown to reflect the functionality of the interacting genes. A representative example can be found in yeast, where essential genes tend to be well connected and globally centered in the protein network (Jeong et al., 2001; Wuchty and Almaas, 2005). The enormous success of microarrays technology in cancer research has so far provided valuable information on tumor subclasses, marker genes for diagnosis and treatment. However, these analyses do not provide possible association with other co-regulated genes. What I am trying to emphasize is that protein networks will place the genes identified in a broader biological context. Wachi et al. (2005) used interactome-transcriptome analysis to reveal the biological significance of differentially expressed genes in squamous lung cancer that is identified through microarray gene expression profiling. This is one of the rare examples in which predicted human protein interaction map has been used for the analysis of cancer. While microarray analyses have been used extensively for the identification of marker genes for various types of cancer, there is still an evident paucity of investigations employing integrative analysis of the cancer gene signatures. One of the reasons for such paucity lies in the fact that to understand complex biological processes, such as cancer initiation and progression, it is important to consider differential gene expression in the context of complex molecular networks. The study of such networks requires detailed protein–protein interaction maps. A detailed human interactome network that captures the entire cellular network would be invaluable in interpreting cancer signatures. The fact that human interactome maps are still in their infancy, representing only a fraction of the complete interaction network is one of the reasons for the paucity of studies in which gene signatures are interpreted in the context of protein–protein interactions. It can be anticipated that the next few years will witness an expansion of such approach with the direct result of a better understanding of cancer initiation, progression and certainly new classes of cancer biomarkers.

• The pace at which cancer signatures are generated for various types of cancer, coupled with fast advancement in high throughput molecular approaches will impact on future efforts to decipher cancer initiation and progression. Some of these efforts associated with gene signatures in cancer have already given some encouraging results. This statement can be supported by the following considerations: Till very recently most published studies on marker genes have applied genes-expression profiling to single cancer types. We are already witnessing a shift toward developing multiclass classifiers capable of distinguishing between multiple common human malignancies. This approach holds much promise for the uniform, molecular, and database-driven classification of all human tumors. In a demonstration of such type of analysis, Ramaswamy et al. (2003) identified a gene expression signature that was differentially expressed in metastatic tumors of diverse origins relative to primary cancers. This study demonstrated that the metastatic signature was also expressed in a subset of the primary tumors analyzed, leading the authors to hypothesize that such signature may represent a metastatic program that is encoded in primary tumors destined for metastasis. This hypothesis was supported by further analyses to predict time to metastasis in several independent solid tumors data sets for different types of cancer. In all of the data sets analyzed, the metastatic signature was substantially associated with clinical outcome and metastatic conditions, suggesting that metastatic potential is encoded in the primary tumors and is at least in part encoded by a common signature across tumor types. Another example of this type of analyses has been provided by Rhodes et al. (2004). Forty independent data sets from >3700 array experiments have been used to generate 36 cancer signatures representing genes activated in a particular cancer type relative to the normal tissue type from which it arose. Analysis of these signatures identified 67 genes activated in 12 or more of the identified signatures. The authors reported that this 67-gene signature could predict cancer versus normal status in most of the cancer signatures tested, as well as in independent cancer signatures, some of which represented cancer types not included in the original analysis. Second, future efforts will no doubt adopt an integrative molecular analysis of cancer. This emerging approach will be crucial for extracting maximum biological insight from the collective cancer genomic data sets generated by various technologies. According to Rhodes and Chinnaiyan (2005), for such integration to occur, it will be essential to define standards for communicating genomic profiles across diverse experimental systems. The authors envisage an analogous scenario to that currently applied to new DNA sequences, where they can be effortlessly compared to all DNA sequences in Genbank. If such concept is implemented then for example, starting with a new cancer signature, one might find that the signature shares similarity with an in vitro oncogene signature, a transcription factor-binding site profile and a drug-treatment profile. This retrieved information may suggest that the oncogene and transcription factor are active in the new cancer signature and that a specific drug treatment might reverse such signature. Such future effort would require reporting standards similar to those applied to microarray data (Brazma et al. (2001), able repositories and journal requirements for deposition before publication. Currently, there is a vast quantity of genomic profiles as evidenced by an impressive Affymetrix publication list, a fast growing GEO (Gene Expression Omnibus) and ArrayExpress microarray repositories and the Oncomine database of cancer signatures. However, there is still a lack of integrative bioinformatics solutions to allow efficient sharing and analysis of such profiles.

5.8. PROTEIN PHOSPHORYLATION

5.8.1. Introduction

Most of the initial efforts in proteomics have focused on methods capable of high throughput analysis of large number of proteins. Although the identification of proteins in complex mixtures is becoming routine, such identification alone provides only a limited insight into protein functions and signaling pathways. Newly acquired knowledge in genomic and proteomic fields have demonstrated that numerous vital activities of proteins are modulated by post-translational modifications (PTMs) that are not easy to capture by simple assessment of protein-abundances. Furthermore, such modifications cannot be obtained from protein sequences deduced from nucleotide sequences. Over 200 different modifications have been described (Krishna and Wold, 1993). Many of these modifications, such as phosphorylation and glycolysation, have well documented roles in signal transduction, regulation of cellular processes, clinical biomarkers, and therapeutic targets. It is becoming evident that protein phosphorylation is one of the most prevalent intracellular protein modifications that is of central importance in numerous cellular processes including cell differentiation, proliferation, and migration. It is estimated that ~30% of all proteins in a cell are phosphorylated at any given time. However, this number is not reflected in the actual number of phosphorylation sites found so far. For example, the phospho-ELM database (http://phospho.elm.eu.org) currently lists 1703 experimentally verified phosphorylation sites for 556 different proteins derived from eukaryotes, the human protein reference database (http://www.hprd.org) lists 3652 reported phosphorylation sites on 1240 human proteins, and PhosphoSite (http://www. phosphosite.org) lists 6084 nonredundant phosphorylation sites on 2430 human and mouse proteins. However, a recent study by Beausoleil et al. (2004) gave some idea about the expected number of phosphorylation in the cell when the authors performed a large scale phosphoproteomics study on HeLa cell nuclei. Although their method was biased against basic phosphopeptides where His, Lys-Pro, and/or Arg-Pro residues are in close proximity to the phosphoamino acid residue, more than 2000 phosphorylation sites on 967 nuclear proteins were found. Considering the fact that protein phosphorylation analysis is of major interest in numerous laboratories around the world, it is surprising that more information about protein phosphorylation sites has not been gathered since its discovery. In recent years, the optimization of enrichment procedures combined with more powerful detection strategies have made the analysis of the entire complement of phosphorylated proteins in cells a viable option. Although phosphorylation has mostly been described for Ser, Thr, and Tyr, kinases and phosphatases that act on His, Lys, Arg, Asp, and Glu have also been reported. Current literature indicates that Ser is the most abundant phosphorylated amino acid residue, whereas phosphohistidine is currently estimated to be 10- to 100-fold more abundant than phosphotyrosine (Klumpp and Krieglstein, 2002). Having said that, most analytical methods have been developed for the hyroxyl amino acid residues and therefore, the following sections will only refer to such sites of phosphorylation. The sites of phosphorylation are serine, threonine, and tyrosine; however, because of their high content ratio within vertebrate cells the first two sites are known to undergo phosphorylation more often than the tyrosine site. As the direction of biological research starts to shift toward systems rather than component biology, our capability to obtain quantitative data on the phosphorylation of various signaling molecules will represent an important component within the definition of systems biology.

5.8.2. Experimental Approaches for the Detection and Quantification of Protein Phosphorylation

It can be argued that most existing strategies for the detection/quantification of phosphoproteins hinge on two main components: enrichment of phosphoproteins and the method of readout (detection). In designing such strategies a number of intrensic characteristics of phosphoproteins have to be taken into account: First, the stoichiometry of phosphorylation is notoriously low. Second, many of the signaling molecules are present at low abundance within cells and in these cases, enrichment prior to analysis is a prerequisite. Third, the analytical method has to have sufficient dynamic range to allow the capture of both major as well as minor phosphorylation sites. Some of the enrichment protocols for phosphoprotein analysis and methods of detection are considered in the following sections. The increasing use of MS-based methods in the analysis of phosphoproteins renders the present discussion biased toward MS techniques.

5.8.3. Enrichment Strategies

One of the simplest methods for phosphoproteins enrichment is the use of phosphospecific antibodies to immunoprecipitate specific proteins. Currently, there are several commercially available antibodies that bind to phosphorylated tyrosine residues. These antibodies can be used to immunoprecipitate and, therefore, to enrich tyrosine phosphorylated proteins from complex mixtures of proteins such as cell lysates. These antibodies have been demonstrated to be fairly effective at enriching and identifiying low-abundance tyrosine phosphorylated proteins (Pandy et al., 2000). This approach, however, has two drawbacks: First, these antibodies are not effective at enriching phosphopeptides (Marcus et al., 2000). Second, at present there are no suitable antibodies for enriching proteins that are phosphorylated on Ser or Thr residues, and thus alternative methods of enrichment have to be used.

Another enrichment approach uses maniturized immobilized metal affinity chromatography (IMAC). This method is based on the affinity of negatively charged phosphate groups for positively charged metal ions, especially Fe^{3+} and Ga^{3+} immobilized on chromatographic support. IMAC has been successfully used in off-line and online formats for the detection of phosphopeptides using MS (Cao and Stults, 2000; Xhou et al., 2000; Stensaballe et al., 2001). As it is based on the presence of negatively charged phosphate groups, IMAC generally enriches for phosphorylated serine, threonine, and tyrosine residues. However, the specificity of this procedure is hindered by its additional affinity for acidic groups such as aspartic, glutamic acid, and to electron donors (e.g. histidine). Ficarro et al. (2002) have reduced the background effect of such acidic residues by methylester modification of carboxyl groups prior to IMAC, thereby improving subsequent MS-detection of phosphopeptides. These authors reported the identification of 383 phosphorylation sites in 216 peptides starting with 500 μg of yeast protein.

Strong cation exchange (SCX) chromatography has also been used for phosphopeptides enrichment within complex digests (Beausoleil et al., 2004). This approach exploits the difference between the solution charge state of most tryptic phosphopeptides when compared with their unphosphorylated counterparts. As SCX chromatography separates peptides primarily based on charge, phosphopeptides containing a single basic group elute first and are highly enriched.

A more recent and highly attractive approach for phosphoprotein enrichment is based on chemical modification and isotope labeling. Tagging phosphopeptides by specific chemical modification is attractive because it is amenable to large-scale analysis. Methods in which the phosphate moiety is chemically modified, for example, by biotinylation, allow enrichment of phosphopeptides by affinity chromatography and the subsequent identification of the phosphorylated site(s) by mass spectrometry (Schlosser et al., 2001). On the contrary, methods in which peptides are differentially labeled with stable isotopes such as $^{12}C/^{13}C$ or $^{14}N/^{15}N$ allow accurate determination of the abundance of specific phosphopeptides in one sample relative to another by measuring relative MS signal intensity (Blagoev et al., 2003; Weckwerth et al., 2000; Zhang et al., 2002).

The isotope-coded affinity tag (ICAT) method is designed to quantify relative protein amounts in two samples without the need for two-dimensional gel separation. In this approach, all cysteine residues in one sample are modified with a biotinylated "heavy" isotope tag, and that of a second sample with a similar "light" isotope tag; the two samples are then combined and the relative intensity of corresponding heavy and light peptides is determined by MS (Gygi et al., 1999). Phosphoprotein isotope-coded affinity tags are conceptually similar to ICAT, but phosphorylated Ser and Thr residues are tagged instead of cysteine residues. This permits the simultaneous enrichment, quantification, and identification of phosphopeptides via a biotinylated isotope tag (Goshe et al., 2001). While chemical modification methods based on the β-elimination reaction including phosphoprotein affinity tagging cannot modify tyrosine residues, a different approach based on a carbodiimide condensation reaction can be applied to phosphotyrosine (Zhou et al., 2001).

Two methods have described the use of chemical modification of phosphate moiety to enrich phosphopeptides from complex mixtures. In the first study, Oda et al. (2001) have used base-induced elimination of phosphate from Ser and Thr residues to generate a reactive acrylate double bond that is coupled by Michael addition to ethane-1,2-dithiol. This experimental protocol starts with a protein mixture in which cysteine reactivity is blocked through the oxidation with performic acid. Base hydrolysis is used to induce β-elimination of phosphate from phosphoserine and phosphothreionine, followed by the addition of ethanedithiol to the alkene. The resulting free sulfhydryls are coupled to biotin, allowing purification of phosphoproteins by adivin affinity chromatography. Following elution of phosphoproteins and proteolysis, enrichment of phosphopeptides is carried out by a second round of avidin purification. The tryptic peptide mixture is then analyzed by MALDI- TOF-MS or LC-ESI/MS-MS. In these measurements the presence of the labeled peptides can be recognized by monitoring characteristic fragment ions originating from the biotinylated side-chain. This approach can suffer certain interferences which limits its specificity and sensitivity. For example, it requires the blockage of the reactive thiolates of cysteinyl residues via reductive alkylation or performic acid oxidation. Such reactions also result in the oxidation of tryptophan indole rings and methionine residues to mixtures of sulfoxides and sulfones. Furthermore, the base-induced phosphate elimination is not entirely specific and can also affect serine and threonine glycosylation sites.

The second study by Zhou et al. (2001) starts with a proteolytic digest that has been reduced and alkylated to eliminate reactivity of the cysteine residues. Following N-terminal and C-terminal protection, phosphoramidate adducts at phosphorylated residues are formed by carbodiimide condensation with cystamine. The free sulfhydryl groups produced from this step are covalently captured onto glass beads coupled to iodoacetic acid. Elution with trifluoroacetic acid then regenerates phosphopeptides, which are then analyzed by mass spectrometry.

5.8.4. MS Detection of Phosphorylation

Both Matrix-assisted laser desorption ionization (MALDI) and electrospray ionization (ESI) have been used for the identification of phosphopeptides. The first approach uses peptide mass finger printing (PMF). Data are generally generated by using reflector MALDI-TOF analysis of enzymatically digested proteins. The acquired peptide masses are then compared to the theoretically expected tryptic peptide masses for each entry in the database, which can rank various proteins according to the number of peptide matches. More sophisticated scoring algorithms take the mass accuracy and the percentage of the protein sequence covered into account and attempt to calculate a level of confidence for the match (Berndt et al., 1999; Perkins et al., 1999; Eriksson et al., 2000). Other considerations can also be included in the search, such as the fact that larger peptides are less frequent in the database and therefore should be given more weight when matched. In this kind of analysis the mass accuracy of the detected peptides has a strong influence on the specificity of the search. For example, a mass accuracy in the range 10–20 ppm requires at least five peptide masses to be matched to the protein and 15% of the protein sequence needs to be covered to allow unambiguous identification. After a match has been found, a second search is performed to correlate the remaining peptides with the database sequence of the match, taking into account possible modifications. It is worth noting that the MALDI-TOF analysis of phosphopeptides is less straightforward than the identification of intact proteins, which is normally performed in the linear mode. There are a number of reasons behind a more difficult analysis of phosphopeptides. (a) MALDI signals of phosphopeptides are usually suppressed by the abundant signals of their nonphosphorylated counterparts. Further attenuation in the signals is caused by the use of positive ionization mode. (b) PMF is not strictly an MS-MS technique and, therefore, does not deliver direct information on the amino acids sequence, rendering the identification of the phosphorylation site less likely.(c) PMF requires relatively pure samples which necessitates some form of purification prior to MALDI analysis.

Different approaches have been adapted to overcome some of the above drawbacks. For example, a number of research groups have described the use of MALDI-TOF-MS in combination with phosphatase treatment to specifically identify phosphopeptides (Stensaballe et al., 2001; Yip and Hutchens, 1992). This approach attempts to identify phosphopeptides based on a characteristic mass shift owing to the loss of phosphate (80 Da or multiples) after treatment with phosphatase. However, the usual problems associated with analysis of peptide mixtures in MALDI preclude complete sequence coverage of the protein. In another study it was demonstrated how to differentiate between serine or threonine and tyrosine phosphorylation using MALDI-TOF. In the positive ion mode, the tendency for serine or threonine phosphopeptides to show a predominant neutral loss of 98 Da (owing to H3PO4 loss) as compared with a loss of 80 Da (owing to HPO3 loss) can be used to differentiate them from tyrosine phosphopeptides, which generally show only a loss of 80 Da (Annan and Carr, 1996).

Over the last 3 years, analysis of phosphopeptides by MALDI ionization has experienced a notable improvement. Such improvement can be partially attributed to a new generation of mass/charge (m/z) analyzers, which can be coupled to MALDI ionization, including quadrupole-time-of-flight (Q-TOF) (Chernushevich et al., 2001) and TOF–TOF (Medzihradszky et al., 2000). Both configurations allow the acquisition of MS–MS data, the first at relatively low collision energy (>100 eV), whereas the TOF–TOF configuration can deliver sequencing data at a collision energy above 1000 eV.

5.8.4.1. Analyses Using Electrospray Ionization (ESI). Phosphopeptides analysis using ESI can be performed with or without prior chromatographic separation. There are various forms of direct MS analyses, which are described briefly in the following sections.

Electron Capture Dissociation (ECD). The introduction of ECD as a fragmentation method in Fourier transform ion cyclotron resonance mass spectrometers (FTICR-MS) has opened up further experimental possibilities in the analysis of phosphoproteins (Zubarev et al., 1998). The mechanism of ECD is, as yet, poorly understood. In general terms, a peptide captures an electron, which is followed by charge neutralization, leading to fast bond cleavages of the resulting excited radical species. So far, ECD has only been used with Fourier transform ion cyclotron resonance mass spectrometry (FTICR-MS). The reason for this is twofold: First, electron capture by the precursor ion requires at least several milliseconds, which is much longer than the residence time of ions in many types of mass spectrometers (Tsybin et al., 2001). Second, the highest efficiency of this mechanism is for electron energies less than 1 eV, which are difficult to provide in, for example, quadrupole ion trap (Zubarev et al., 2000). Over the last 5 years, ECD combined with FTICR-MS has emerged as a powerful method for the sequencing of proteins and peptides as well as for the study of post-translational modifications (McLafferty, 2001). This approach has also been successfully applied for the exact localization of phosphorylated residues in peptides (Shi et al., 2001; Stensballe et al., 2000). One of the main advantages of ECD is its capability to induce more extensive fragmentation of the peptide backbone compared to conventional collision induced dissociation (CID), thus providing greater sequence coverage (see Fig. 5.13). A further advantage of ECD is its applicability for the identification of phosphorylated residues. In contrast to conventional CID and PSD, no loss of phosphoric acid, phosphate, or water from the parent peptide or the fragments is seen when ECD-based sequencing of phosphopeptides is performed. This allows direct assignment of phosphorylation sites. Owing to the extremely high resolution of FTMS, large peptides and proteins that are not amenable to conventional MS can also be studied (Jensen et al., 1999). This means that unlike other MS methods in which only tryptic peptides are analyzed, studying the whole protein by FTICR will provide a more comprehensive picture of the phosphorylation status of the protein. The most significant limitation of this approach is the requirement for relatively pure samples and the availability of expensive instrumentation and trained personnel.

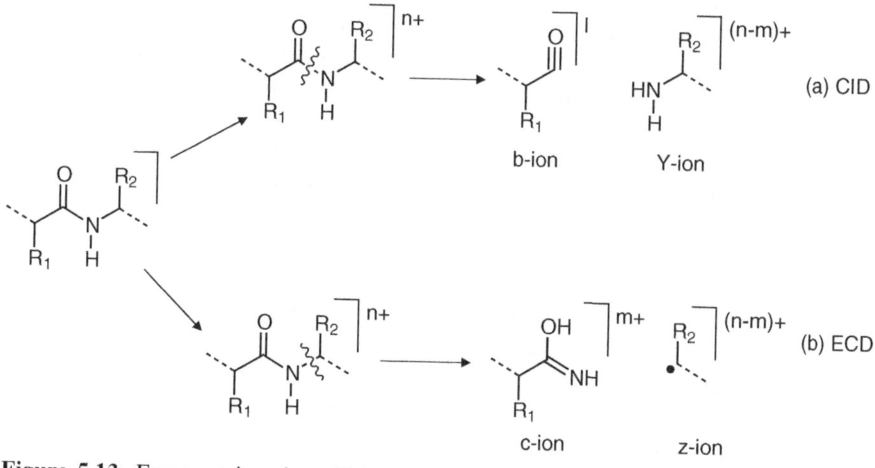

Figure 5.13. Fragment ions in collision induced dissociation (CID) and electron capture dissociation (ECD). Adapted from Zeller and König (2004) with permission.

Precursor Ion/Neutral Loss Scanning. Collision induced dissociation (CID) of phosphopeptides within a tandem mass spectrometer can provide both sequence-specific fragments as well as fragments that are specific for a phosphate group(s). These phosphate-specific fragment ions serve to characterize ions for phosphorylated peptides in precursor-ion scanning experiments. A tandem mass spectrometer operating in negative ion mode is generally used for this purpose. In this method, detection of the specific product ion identifies the corresponding precursor phosphopeptide ion by its mass to charge (m/z) ratio. Subsequent sequencing of the corresponding phosphopeptide requires a switch in polarity and new buffer of the sample, selective MS detection of Ser, Thr, and Tyr phosphorylation is carried out by the use of precursor ion scan of m/z-79 (PO_3^-). These measurements can be carried out on various MS configurations, including triple quadrupole (Annan et al., 2001), hybrid instruments such as Q-TOF (Steen et al., 2001), and more recently on linear ion trap (Williamson et al., 2006). Although all these configurations are capable of performing precursor ion scan, they differ among them in sensitivity and resolution. The latest to be put in the market can be considered a hybrid instrunment, which combines the well established capabilities of a triple quadrupole with an ion trap (Hager, 2002; Le Blanc et al., 2003). When ions losing m/z 79 are detected, the instrument switches from negative to positive ion mode (~0.7 min delay), and enhanced resolution scans are performed to obtain sequencing information. Peptides carrying a phosphate group can therefore, be easily identified by precursor-ion scanning because of the loss of phosphate ($PO3^-$) under alkaline conditions (Carr, 1996; Wilm, 1996). This method is a powerful tool because of its high selectivity and sensitivity and its applicability for serine, threonine, and tyrosine phosphorylated residues. A precursor-ion scanning method that can be performed in the positive mode has also been developed for the specific detection of phosphotyrosine-containing peptides (Steen et al., 2001a, b). This method is based on the ability to selectively detect the immonium ions of phosphotyrosine residues that have a precise m/z ratio. Immonium ions are generated by double cleavage of the peptide backbone. Using newer high-resolution instruments, such as hybrid-TOF mass spectrometers (Chernushevich et al., 2001), these immonium ions can be easily distinguished from other peptide fragment ions. Weak phosphopeptides in the original MS scan can be easily identified using this approach. Once the phosphotyrosine-containing peptides are located, they can be sequenced in the product ion MS/MS mode without any need for switching polarity of the ion source. This scanning method is sensitive and tyrosine phosphorylation sites from subpicomole amounts of gel-separated proteins have been successfully identified. However, the lability of phosphoserine and phosphothreonine residues excludes the use of this method for identification of their phosphorylation.

The second mode of scanning to detect and identify phosphopeptides is commonly termed neutral loss scan. When peptides containing phosphoserine or phosphothreonine residues are subjected to CID, they commonly undergo a gas-phase β-elimination reaction, resulting in a neutral loss of phosphoric acid (loss of 98 Da) or are dephosphorylated (loss of 80 Da). Phosphotyrosines, however, are generally more resistant to this loss. Sequencing of peptides showing this loss can be performed in the same experiment. The resulting MS-MS spectrum would show a spacing of

69 Da due to the presence of dehydroalanine or 83 Da due to dehydroaminobutyric acid. These species can indicate the exact location of phosphorylated serine and threonine residues, respectively (Schlosser et al., 2001; Covey et al., 1991). The drawback of this method is the incidence of false-positive signals as well as the fact that the charge state of the phosphopeptide has to be known in advance.

5.8.4.2. Liquid Chromatography/Mass Spectrometry. Liquid chromatography separation using nanoLC columns containing reversed-phase C18 material alone, or coupled to a second column (e. g. a strong cation exchange resin), is an effective approach for the identification of phosphorylation in a high throughput fashion. The upfront LC step is an efficient way to reduce or even eliminate the ion suppression effect commonly observed in phosphopeptides analysis. In this method, peptides are first loaded onto a nanoLC column (usually 75 μm internal diameter) containing reverse phase C18 material and then eluted by using a gradient directly into a tandem mass spectrometer, which can be an ion-trap, Q-TOF, or a triple quadrupole. The peptides are eluted at a slow flow rate, typically 100–200 nl per minute, and the elution of a peptide generally occurs in a peak lasting ~10–30 s. Hundreds to thousands of peptides can therefore be separated and analyzed using this method. In a variation of this technique, 2D chromatographic separation, first on a strong cation exchange and then on a C18 column, is performed (Washburn et al., 2001). As the peptides elute off the column into the mass spectrometer, a mass spectrum in the survey scan mode (looking at the entire mass spectrum of an eluting peak). Using this scan mode (data-dependent acquisition mode), the instrument can be set to automatically fragment and collect MS/MS data on any number of peaks observed in the MS spectrum based on their intensity, m/z value, or charge state. RP-LC-MS has the advantage of increased sensitivity due to concentration, on-column desalting, and reduced sample handling. However, automatic experiments with nanolitre flows are often disturbed by crystallization at the capillary tip, clogging, and breakdown of signal. It was also observed that phosphopeptides might stick to metal surfaces in HPLC systems (Patterson et al., 2001). Experimental tricks such as "peak parking," where the flow is reduced while a peak of interest elutes, provide more time for manual adjustment and optimisation of parameters (Vissers et al., 2002). With MALDI, experiments can be uncoupled in time; in other words the analyte remains on the target plate and can be further analysed after primary spectra inspection (Keil et al., 2002). Phosphopeptides can be also lost in the desalting step, which is necessary in order to increase sequence coverage and to remove MS-incompatible buffers after IMAC. Small peptides can be too hydrophilic to be retained on RP material, whereas large phosphopeptides could fail to be desorbed from the material due to hydrophobicity (Neubauer and Mann, 1999). Elution of such phosphopeptides is best at basic pH, and binding at acidic pH. For purification of smaller hydrophilic peptides, the use of alternative media such as Poros Oligo R3 RP perfusion chromatography resin, which was originally designed for the purification of oligonucleotides (Köcher et al., 2003) and self-packed graphite powder tips was suggested (Larsen et al., 2002). It is worth noting here that the increased hydrophilicity of the phosphopeptide with a concomitant loss during loading into a RP column is still attracting some debate. In a recent study by Steen et al. (2006), such statement was strongly questioned. The authors used LC/MS analysis to investigate a mixture of peptide/phosphopeptide pairs varying in length from 7 to 17 amino acids, resembling the size of peptides commonly observed in tryptic digests. Cys and Met-containing peptides were excluded to avoid problems with possible oxidation.Water/acetonitrile/0.2% formic acid was used as a mobile phase, while C18 reversed phase was used as the stationary phase. This study concluded that all singly phosphorylated peptides within the analyzed mixture have eluted off the reversed-phase column after the unmodified complement irrespective of the number of basic amino acid residues within these peptides. In other words, although phosphorylation clearly increases the nominal hydrophilicity of peptides with respect to their unphosphorylated counterparts, this effect is compensated under certain LC conditions if the overall net charge of the peptide is decreased by the addition of phosphate groups. In other words, conditions can be used under which phosphopeptides are better retained on the reversed-phase column as long as the number of phosphorylation sites do not exceed the number of basic amino acid residues (Arg, His, and Lys) within the peptides. Figure 5.14 gives an overview of enrichment and MS-based detection methods.

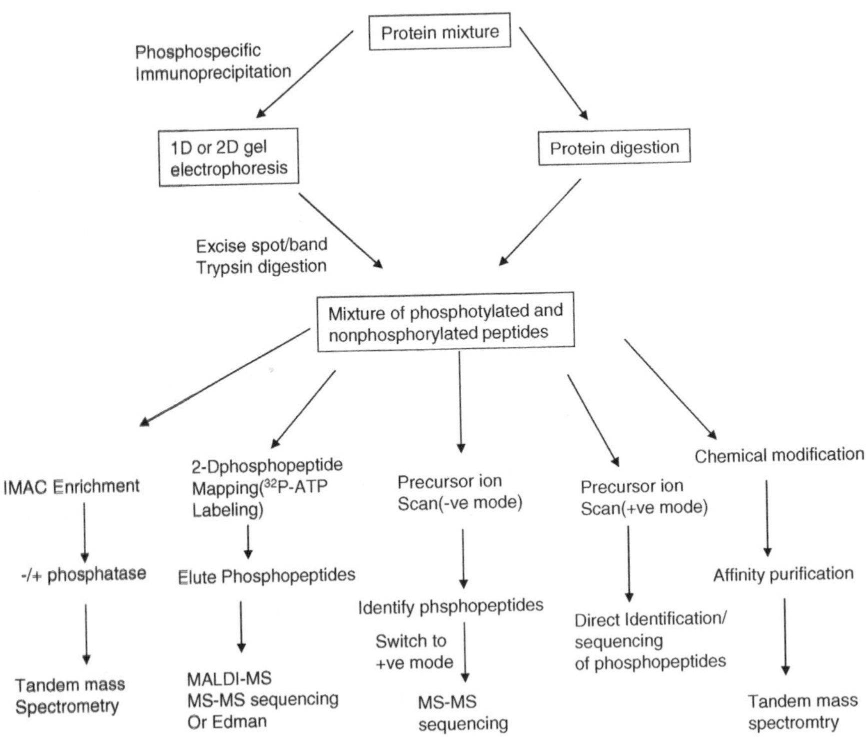

Figure 5.14. An overview of enrichment and MS-based detection methods for the analysis of phosphorylated peptides/proteins. Adapted from Mann et al. (2002) with permission.

5.9. OTHER APPROACHES

Two-dimensional gel electrophoresis (2-DE) is still considered the method of choice for the separation and identification of proteins in complex medium such as cell lysates. However, given the low abundance of phosphoproteins, particularly those containing phosphotyrosine, this is not an efficient approach unless phosphopeptides can be enriched or specifically labeled. In a few cases, however, 2-DE was successfully used in combination with functional information for the analysis of phosphoproteins. In one of these studies, Lewis et al. (2000) identified novel proteins involved in mitogen-activated protein kinase pathways by 2-DE combined with MS based on the kinetics of change in abundance of spots in response to specific kinase activators. Another approach to render 2-DE more efficient in phosphoproteomics is the use of ^{32}P labeling. Such labeling is frequently used as a highly selective and sensitive means of detecting phosphopeptides (Immler et al., 1998; Larsen et al., 2001). With this labeling, direct visualization and quantification of phosphoprotein spots on 2D gels are possible. Using high-resolution narrow-range 2-DE, it is feasible to detect and quantify differentially phosphorylated forms of a protein, which exhibit similar molecular mass but different isoelectric points (Immler et al., 1998). Moreover, ^{32}P labeling coupled to immunoprecipitation allows phosphoproteomic analysis of particular protein complexes or organelles such as the 26S proteasome (Mason et al., 1998). However, the advantages of ^{32}P have diminished with the emergence of MS technologies that enable the direct detection of phosphoproteins as well as identification of the phosphorylation site. In addition, inconvenience and safety issues regarding the handling of radioactive materials are an unavoidable drawback of ^{32}P labeling and preclude its use for human tissue samples.

In a relatively recent article, Wind et al. (2003) have described a combination of one-dimensional gel electrophoresis with laser ablation inductively coupled plasma-mass spectrometry (ICP-MS). In this approach reversed phase at acidic pH is used to separate digested peptides in gel where covalently bound phosphorous mainly represent phosphate esters of serine, threonine, and tyrosine. The method has been tested for a standard mixture of myoglobin, α-casin, and reduced fibrinogen. A special washing step was found to be necessary to remove non-covalently bound phosphate. The authors claimed that the ^{31}P signal was found to contain quantitative information regarding both the relative and the absolute amount of phosphorous present in phosphoproteins. Furthermore, normalizing the ^{31}P signal generated by a single laser ablation trace by the total amount of phosphoprotein applied to the gel, a detection limit of 5 pmol was estimated. The advantage of ICP-MS is that the signal is independent of the chemical form of phosphorous because of the production of singly charged ions of ^{31}P.

Binding assays based on the far Western-blot technique have been used to profile the global tyrosine phosphorylation state (Nollau and Mayer, 2001). In this type of assay, a battery of Src homology 2 domain (SH2) probes is used to detect patterns of specific tyrosine phosphorylated sites. SH2 domains are found in a wide variety of signaling proteins, consist of ~100 amino acids and can be separated from the original proteins without loss of function (Kuriyan and Cowburn, 1997). Their ligand-binding surfaces specifically interact with phosphotyrisine (pTyr) in the context of short linear sequence motifs. The specificity of the interaction is determined by the amino acid composition of the core binding site, with the general motif pYxxψ, where pY stands for phosphotyrosine, ψ for hydrophobic amino acids, and x for selected amino acids important for specific reaction (Songyang et al., 1993; 1994). The basic steps and necessary precautions in the SH2 profiling assay can be briefly summarized as follows: A protein extract is first separated by conventional 1D gel, transferred to a nitrocellulose or polyvinylidene difluoride membrane, and subsequently probed with labeled glutathione S-transferase-SH2 (GST-SH2) fusion proteins. Replicate filters are probed with different SH2 domains having different binding specificities, and the pattern of binding (intensity and apparent molecular mass of specific bands on the gel) for each probe can be compared among different samples. To reduce the background of nonspecific binding and to favor the low-abundance binding partners such as tyrosine phosphorylated proteins, the authors used glutathione-horseradish peroxidase conjugate to label the GST-SH2 fusion protein probe. This labeling procedure was used instead of a more commonly used approach based on anti GST antibody or direct biotinylation (Nollau and Mayer, 2001). In this far-Western filter binding assay, it is essential that different SH2 domains maintain their phosphorylation-dependent and sequence dependent binding specificities.

The rest of this chapter will mainly be concerned with the impact of phosphorylation on the activation of a signaling pathway, which is considered central in the initiation and progression of various types of human cancers.

5.10. THE PHOSPHATIDYLINOSITOL 3-KINASE-AKT PATHWAY (PI3K-AKT)

For the last 15 years, much of the cancer research activities have focused on the central role of *RAS*, the first identified oncogene in neoplastic transformation. Extensive genetic and biochemical investigations of the signaling components of this small GTPase in model organisms led to the model of mitogenic signaling by receptor tyrosine kinases (RTKs) through *RAS* and mitogen-activated protein kinases (MAPKs). In recent years, a second pathway downstream of RTKs that involves PI3K and Akt has gained almost a similar status as an important regulator of mammalian cell proliferation and survival. The dysregulation of several components of this pathway in a wide spectrum of human cancer is one of the reasons why PI3K-Akt has gained such a prominent status in cancer research. Before considering some specific examples on the central role of this signaling pathway in cancer, it would be helpful to give a brief description of the two main components of this pathway.

5.10.1. Phosphatidylinositol 3-Kinase (PI3K)

PI3Ks are heterodimeric lipid kinases that are composed of regulatory and catalytic subunits that are encoded by different genes. Furthermore, the genes that encode the regulatory domains are also subject to differential splicing. One of the main functions of PI3K is to synthesize the second messenger PtdIns (3,4,5) (also know as PIP_3) from PtdIns (4,5) (also known as (PIP_2) (Fruman et al., 1998). With the molecular cloning of PI3Ks, it has emerged that this is a large and complex family that contains three classes with multiple subunits and isoforms. PI3K catalytic subunits are divided into three main classes on the basis of their in vitro lipid substrate specificity and likely mode of regulation (Zvelebil et al., 1997). The class I PI3Ks consists of two subgroups, IA and IB, which transmit signals from tryrosine kinases and G-protein coupled receptors, respectively. Because of its recognized involvement in oncogenesis, the rest of this section will only deal with class IA. The regulatory subunits of class IA PI3Ks are encoded by one of the three genes, α, β, and γ, which are also subject to alternative splicing. The best studied example, p85α, encodes an adapter-like protein that has two *src* homology-2 (SH2) domains and an inter-SH2 domain that binds constitutively to the p110 catalytic subunit. PI3K catalytic activity is tightly regulated in normal cells by various mechanisms. With increasing molecular details, it became clear that PI3Ks were heterodimers with separate regulatory and catalytic subunits, and that the p85 regulatory subunit was a phosphoprotein substrate of many cytoplasmic and receptor tyrosine kinases. It also emerged that p85 is directly associated with many active tyrosine kinases through the physical interaction of its SH2 domain with phosphotyrosine residues. The current prevailing view is that a preformed, inactive p85-p110 complex is present in the cytoplasm of resting cells, poised for activation in response to appropriate cues. For RTKs, this cue comes from ligand-mediated activation of kinase activity and transphosphorylation of the RTK cytoplasmic tail, followed by recruitment of the p85-p110 complex to the receptor by interaction of the SH2 domain of p85 with consensus phosphotyrosine residues on the RTK or with intermediate phosphoproteins, such as insulin receptor substrates (IRS1 and IRS2) (White, 1998). In other words, PI3K becomes active through two events. First, the p110 catalytic subunit is found in close proximity to its lipid substrates in the cell membrane. Second, the RTK-p85 interaction might relieve an inhibitory effect of p85 on p110 kinase activity, possibly due to conformational changes in the p85-p110 complex, involving the SH3 and BCR (breakpoint cluster region homology domain) domains.

5.10.2. Akt (PKB) and Its Activation

Akt is a 57 kDa Ser/Thr kinase and the cellular homologue of the viral oncoprotein v-Akt, which is known to be responsible for a type of leukaemia in mice, and is therefore referred to as c-Akt or Akt (Staal, 1987). The catalytic domain of Akt is most similar to that of cAMP-dependent protein kinase (PKA; 65% similarity) and to protein kinase C (PKC; 75% similarity)–findings that gave rise to two of its additional names, PKB (i.e. between PKA and PKC) and RAC (related to A and C kinase)(Jones et al., 1991). Mammalian genomes contain three *Akt* genes, encoding the isoforms Akt1(PKBα), Akt2(PKBβ), and Akt3(PKBγ)(Datta et al., 1999). Akt2 and Akt3 share >80% homology in amino acid sequence with Akt1, (Vanhaesebroeck and Alessi, 2000). Each gene encodes a protein containing a pleckstrin homology (PH) domain in the N-terminus, a central kinase domain, and

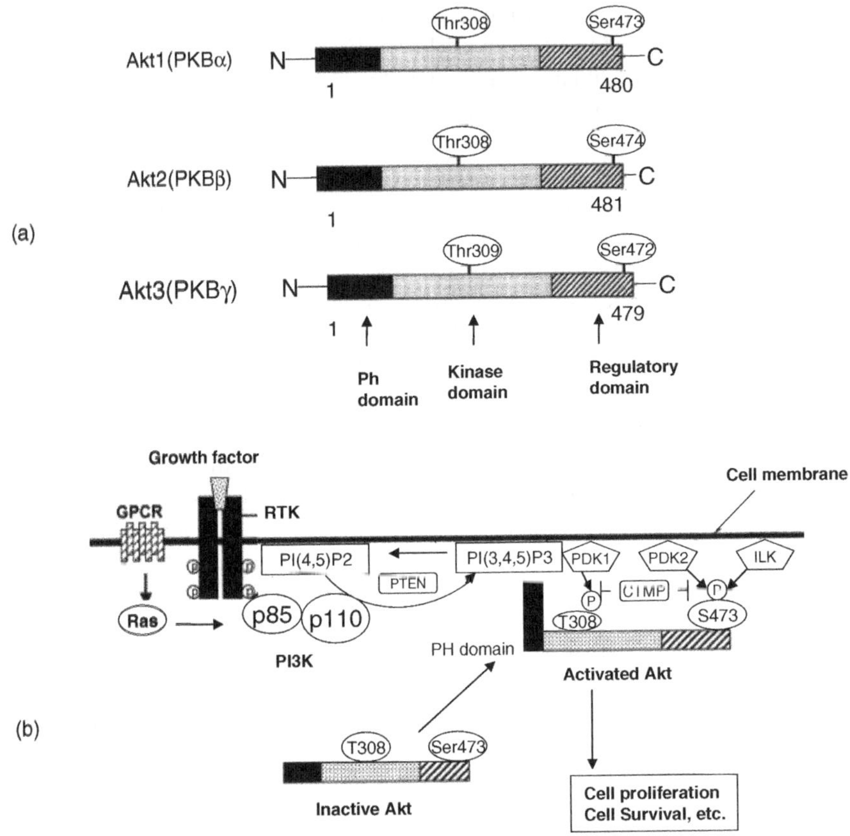

Figure 5.15. (a) Domain structure of the three human Akt isoforms. (b) A model for the regulation of PI3K-Akt signaling pathway. The binding of growth factors to their receptor tyrosine kinase (RTK) or G protein-coupled receptors (GPCR) stimulates the phosphorylation of phosphatidylinositol 3-kinase (PI3K) comprised P85 and P110 subunits. PI3K converts phosphatidylinositole-4,5 bisphosphonate (PI(4,5)P2) to PI(3,4,5)P3, whereas PTEN (phosphatase and tensin homologue) deleted on chromosome 10 reverses this reaction. Akt translocates to the cell membrane and interacts with PI (3,4,5)P3 via its pH domain, being phosphorylated at two residues (Thr308 and Ser473) by phosphoinositide-dependent kinase (PDK) 1, PDK2 and integrin-linked kinase (ILK). Carboxy terminal modulating protein (CTMP) inhibits the phosphorylation. Once active, Akt controls fundamental cellular processes such as the cell cycle and cell survival. Adapted from Osaki et al. (2004) with permission.

a C-terminal regulatory domain (see Fig 5.15). The PH domain of Akt preferentially binds PtdIns (3,4,5)P3 over other PIs. All three mammalian *Akt* genes are widely expressed in various tissues, but Akt1 is most abundant in the brain, heart, and lungs, whereas Akt2 is predominantly expressed in skeletal muscle and embryonic brown fat, and Akt3 is predominantly expressed in the brain, kidney, and embryonic heart (Coffer and Woodgett, 1991; Brodbeck et al., 1999). Fluorescence in situ hybridization has established the chromosomal location of human *Akt* genes to be 14q32 (*Akt1*), 19q13.1–13.2(*Akt2*), and 1q44 (*Akt3*) (Staal et al., 1988; Nakatani et al., 1999; Cheng et al., 1992).

Following the conversion of PtdIns (4,5)P2 to PtdIns (3,4,5)P3 by PI3K on the inner side of the plasma membrane, Akt translocates and binds to the phospholipids (see Figure 5.15). The interaction of the PH domain with PtdIns (3,4,5)P3 is thought to result in conformational changes in Akt, leading to the exposure of its two main phosphorylation sites, Thr^{308} within the kinase domain and Ser^{473} in the regulatory domain (Alessi et al., 1996). Within this model, the PH domain may also mediate the approach of Akt and PDK1 through its heterodimerization. PDK1 assumed to be activated constitutively, phosphorylates Akt at Thr308, thus contributing to the stabilization of the active conformation. Phosphorylation of Akt on Thr308 is necessary and sufficient for its activation (Stokoe, et al., 1997), however, maximal activation requires additional phosphorylation at Ser473 by another kinase called PDK2 (Alessi et al., 1997). This Akt kinase has been characterized biochemically but its molecular identity remains to be determined. There are two relevant observations regarding the phosphorylation of Ser^{473}, which should be underlined. First, sequence scanning of the human genome revealed no PDK1 homologues, which may indicate that PDK2 belongs to a different class of kinases. Second, several recent findings suggest a role for protein kinase integrin-linked kinase (ILK) in the activation, however, there is still no clear evidence that ILK directly phosphorylates Akt at Ser^{473} (Delcommenne et al., 1998; Lynch et al., 1999).

Figure 5.15 shows the domain structure of the three human Akt (PKB) isoforms (a) and a model for the regulation of the PI3K-Akt. (b) As this figure shows, the PI3Ks are heterodimers composed of a catalytic subunit(p110) and an adapter/regulatory subunit(p85), which is activated by receptors with protein tyrosine kinase activity (receptor tyriosine kinase, RTK), and by G protein-coupled receptors (GPCR) (Katso et al., 2001; Vanhaesebroeck and Waterfield, 1999). In this model the activated PI3K takes few seconds to convert the plasma membrane lipid, PtdIns (4,5)P2 to PtdIns (3,4,5)P3. The effect of the latter on cells is mediated through specific binding to at least two distinct protein lipid binding domains, namely the FYVE and plekstrin homology (PH) domains (Pawson and Nash, 2000). The PH domains are globular protein domains of about 100 amino acids found in many proteins, including protein serine/thrionine kinase 3′-phosphoinositide-dependent kinase 1 (PDK1) and Akt. Proteins containing such domains represent crucial mediators for PI3K-induced signaling.

5.10.3. Biological Consequences of Akt Activation

The main biological effects of Akt activation that are relevant to cancer-cell growth can be generally classified into three categories: survival, proliferation (increased cell number) and growth (increased cell size). As it has been pointed out in chapter 3, apoptosis, or programmed cell death, is a normal cellular function that controls excessive proliferation by eliminating unnecessary cells. Cancer cells have devised various mechanisms to inhibit apoptosis and prolong their survival. Evidence has accumulated implicating activated Akt as a major regulator of the apoptotic response in a wide range of cell types (Cantley and Neel, 1999; Vivanco and Sawyers, 2002). The mechanism by which Akt protects cells against death is likely to depend on a number of factors, since Akt directly phosphorylates several components of the cell-death machinery. For example, BAD is proapoptotic member of the Bcl2 family of proteins that promotes cell death by forming a non-functional heterodimer with the survival factor Bcl-X_L. Phosphorylation of BAD by Akt prevents this interaction (Datta et al., 1997). Similarly, Akt inhibits tha catalytic activity of pro-death protease, caspase-9, through phosphorylation (Cardone et al., 1998). In addition, phosphorylation of FKHR, a member of the Forkhead family of transcriptional factors prevents its nuclear translocation (Brunet et al., 1999). Akt is also known to influence cell survival through indirect effects on two central regulators of cell death, nuclear factor of κB (NF-κB) and p53.

Akt also has a recognizable role in proliferation and cell growth. The first role is exercised through signals to the cell-cycle machinery. The latter is regulated by the coordinated action of cyclin–cyclin-dependent kinase (CDK) acomplexes and CDK inhibitors (CKIs). For instance, cyclin D1 levels, which are important in the G1/S phase transition, are regulated at the transcriptional, post-transcriptional and post-translational level by distinct mechanisms. In this scenario, Akt has an important role in preventing the degradation of cyclin D1 by regulating the activity of the cyclin D1 kinase glycogen synthase kinase-3β (GSK-3β). Following phosphorylation by GSK-3β, cyclin D1 is targeted for degradation by the proteasome. Akt directly phosphorylates GSK-3β and blocks its kinase activity, thereby allowing cyclin D1 to accumulate (Diehl et al., 1998). Akt can also negatively influence the expression of CKIs, such as KIP1 (also known as p27) and WAF1 (also known as CIP1 or p21) (Graff et al., 2000). The effects on KIP1 seem to be transcriptional and mediated by FKHR, which represses *CDKN1B* (the gene that encodes KIP1) expression (Dijkers et al., 2000; Madema et al., 2000). Akt can modulate WAF1 activity by affecting its phosphorylation (presumably through intermediate kinases) and binding to proliferating cell nuclear antigen (Rossig et al., 2001).

According to Schmelzle and Hall (2000), the interchangeable use of growth and proliferation is both confusing and incorrect. The authors correctly pointed out that proliferation refers to cell division, whereas growth refers to the synthesis of macromolecules, which results in increased cell size or mass, a process that is enhanced in cancer cells to meet the biosynthetic requirements that are imposed by an enhanced proliferation. Glycogen synthesase kinase-3 (GSK-3), phosphodisterase-3B, mammalian target of rapamycin (mTOR), insulin receptor substrate-1(IRS-1), the Forkhead family member FKHR, cyclin-dependent kinase inhibitors p21 $^{\mathbf{CIP1/WAF1}}$, and p27$^{\mathbf{KIP1}}$ are all targets involved in protein synthesis, glycogen metabolism and cell cycle regulation (Blume-Jensen, 2001). Over the last few years, mTOR has emerged as a central regulator of cell growth. This serine/thrionine kinase (also known as FRAP1) serves as a molecular sensor that regulates protein synthesis on the basis of the availability of nutrients. There is some evidence that Akt enhances protein synthesis by, for example, increasing the phosphorylation of mTOR, which in turn promotes translation of cyclin D mRNA (Muise-Helmericks et al., 1998). mTOR regulates biogenesis by activating p70S6 kinase(RSK), which enhances the translation of mRNAs that have 5' polypyrimidine tracts, and by inhibiting the eukaryotic initiation factor 4E-binding protein-1(4E-BP1)(Nave et al., 1999). At the present time there are still a number of open questions regarding mTOR activation: first, it is not sufficiently clear whether PIK3-Akt is the only stimulus that leads to mTOR activation in cancer cells. It has been argued that mTOR can also function as an ATP sensor (Dennis et al., 2001). Therefore, in tumors that have increased rates of glycolytic metabolism, mTOR might detect the subsequent rise in ATP levels and initiate the signal for increased ribosomal biogenesis that is commonly observed in these cancers (Vivanco and Sawyers, 2002). Second, whether phosphorylation of mTOR by Akt does affect 4E-BP1 and RSK is still to be fully established. In other words, the down stream Akt-mTOR is still controversial. A representation of known connections between Akt and some components of the cell-cycle machinery are given in Figure 5.16 (see also Fig. 5.15).

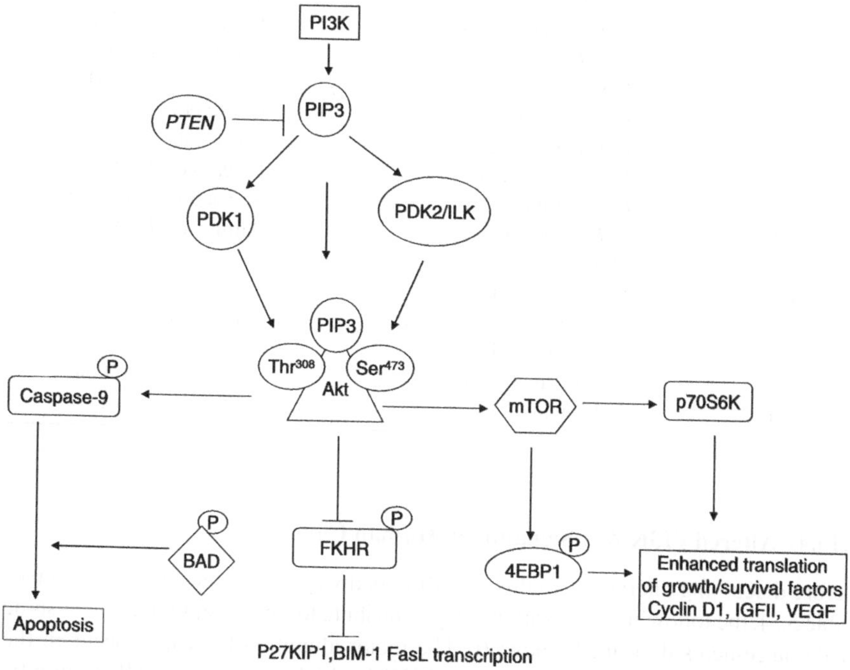

Figure 5.16. PI3K/*PTEN*/Akt signaling pathway. The second messenger PIP3 generated by PI3K recruits Akt and other pH domain containing proteins to the plasma membrane where these proteins may be activated. Once recruited to the membrane, Akt is activated by the phosphorylation at Thr^{308} and Ser^{473} by PDK1 and PKD2, respectively. Activated Akt is assumed to suppress apoptosis by phosphorylating and inactivating the proapotic proteins caspase-9 and BAD. Other events are described within the text. Adapted from Graff (2002) with permission.

TABLE 5.3. PI3K-signaling deregulation in various forms of human cancer.

Cancer Type	Alteration	Reference
Glioblastma	*PTEN* mutation	Wang et al. (1997); Knobbe and Reifenberger(2003)
Ovarian	Allelic imbalance and mutations of *PTEN*	Saito et al. (2000)
	Elevated AKT1 kinase activity	Sun et al. (2001)
	PI3K *p110*α amplification	Shayesteh et al. (1999)
	PI3K *p85*α mutation	Philp et al. (2001)
Breast	Elevated AKT1 kinase activity	Sun et al. (2001)
	AKT2 amplification and overexpression	Bellacosa et al. (1995)
	RSK amplification and overexpression	Barlund et al. (2000 a,b)
	Loss of heterozygosity at *PTEN* locus	Teng et al. (1997)
	PIK3CA mutation	Samuels et al. (2004)
Endometrial	*PTEN* mutation	Yokoyama et al. (2000)
	PTEN silencing	Salvesen et al. (2001)
Colon	*PIK3CA* mutation	Samuels et al. (2004)
Gastric	*PIK3CA* mutation	Samuels et al. (2004)
Melanoma	*PTEN* mutation	Celebi et al. (2000)
	PTEN silencing	Zhou et al. (2000)
Lung	*PTEN* activation	Forgacs et al. (1998)
	PIK3CA mutation	Samuels et al. (2004)
Renal-Cell carcinoma	*PTEN* mutations *MMACI*	Alimov et al. (1999)
Thyroid	*PTEN* mutations	Hsieh et al. (2000)
	AKT overexpression and overactivation	Ringel et al. (2001)
Brain	*PIK3CA* mutation	Samuels et al. (2004)
Lymphoid	*PTEN* mutations	Nakahara et al. (1998)

5.10.4. Altered PI3K-Akt Signaling in Human Cancer

Before discussing alterations in this signaling pathway in specific cancer forms (see Table 5.3), the following observations may contribute to a better appreciation of some of the arguments described below: (i) The major alteration known to occur in the *PI3K* gene is its amplification. The gene *PI3KC*, which encodes the p110α catalytic subunit of PI3K, is located in the chromosome 3q26, a region that is frequently amplified in several human cancers. Findings from a number of studies have confirmed *PIK3C* amplification in ovarian (Shayesteh et al., 1999) and cervix cancer (Ma et al., 2000).However, a more recent study has also reported high frequency mutations of the same gene in different human cancers, including colon, lung, brain, gastric, and breast (Samuels et al., 2004). (ii) So far no modifications or mutations in the *Akt* gene have been reported in mammals. However, a number of studies have reported *Akt* amplifications in various types of human cancers (Bellacosa et al., 1995; Cheng et al., 1996; Knobbe and Reifenberger, 2003). (iii) There is substantial evidence that cell surface receptors are commonly over-expressed or constitutively activated in a wide range of human cancers, which contribute to the activation of a number of downstream signaling pathways (Blume-Jensen and Hunter, 2001). One of the extensively studied examples is ErbB2 tyrosine kinase receptor, which is over-expressed as a result of gene amplification in breast and other forms of cancer (Blume-Jensen and Hunter, 2001). ErbB proteins belong to subclass I of the superfamily of receptor tyrosine kinases (RTKs). The ErbB family contains four members: Epidermal growth factor (EGF) receptor (also designated ErbB1/HER1), ErbB2/Neu/HER2, ErbB3/HER3, and ErbB4/HER4. All four members have in common an extracellular ligand binding domain, a single membrane-spanning region and a cytoplasmic protein tyrosine kinase domain. A family of ligands, the EGF-related peptide growth factors, bind the extracellular domain of ErbB receptors leading to the formation of both homo- and heterodimers. This dimerization is known to stimulate the intrinsic tyrosine kinase activity of the receptors and triggers autophosphorylation of specific tyrosine residues within the cytoplasmic domain. These phosphorylated residues serve as docking sites for signaling molecules involved in the regulation of intracellular signaling cascades (Olayioye et al., 2000; Shoelson, 1997; Weiss and Schlessinger, 1998). (iv) Mutational alterations of *PTEN* and *PIK3CA*, which negatively and positively regulate PI3K activity, respectively have been detected in many forms of human cancer, including glioblastoma, melanoma, prostate, and cervical cancer. *PTEN* located on chromosome 10q23.3 was originally isolated as tumor-suppressor gene in breast cancer and glioblastmas, using traditional positional cloning strategies (Steck et al., 1997; Li et al., 1997). Following its discovery, this gene has been implicated in various forms of human cancers (see Table 5.3). The discovery that one of the primary functions of *PTEN* is a lipid phosphotase of PtdIns (3,4,5)P3, has underlined the importance of the regulation of the latter in cancer. Although *PTEN* might also have activity against protein substrates (Gu et al., 1999; Tamura et al., 1998), mutational studies together with the analysis of heredity cancers have provided a strong indication that PtdIns (3,4,5)P3 phosphotase activity is responsible for the tumor-suppressor function of *PTEN* (Vivanco and Sawyers, 2002). A number of studies have provided convincing evidence that the *PIK3CA* gene, which is located on chromosome 3q26 and encodes for the catalytic subunit, p110α of PI3K, is a major candidate target of chromosomal gain of 3q in human cancers (Volinia et al., 1994). Abnormal genomic amplification of *PIK3CA* has been reported in ovarian, cervical and other forms of carcinoma (Shayesteh et al., 1999; Ma et al., 2000; Singh et al., 2002). Increased copy number of *PIK3CA* was also found to be associated with elevated expression of p110α and high functional PI3-kinase activity and correlated with aberrant cell proliferation and apoptosis, both of which are directly linked to tumor formation. Singh et al. (2002) reported that increased *PIK3CA* expression due to gene amplification is associated with phosphorylated AKT levels. Collectively, these observations suggest that *PIK3CA* is involved in multiple cancer-associated functions, including cell proliferation, apoptosis and survival, and exerts oncogenic

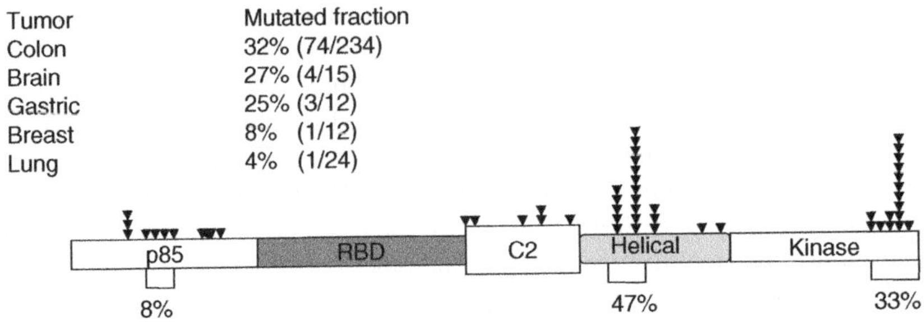

Figure 5.17. Mutations in *PIK3CA* reported in different types of cancer. Arrowheads indicate the location of missense mutations, and boxes represent functional domains (p85 binding domain, Ras binding domain, C2 domain, helical domain, and kinase domain). The percentage of mutations within each region is indicated below boxes, while the fraction of tumors with mutations together with the number of samples is listed above. Adapted from Samuels et al. (2004) with permission.

activities through activation of the PI3-kinase/AKT signaling pathway. In a more recent study, Samuels et al. (2004) reported high frequency of *PIK3CA* mutations in various human cancers (see Fig 5.17). Some specific examples on the alterations of PI3K-Akt in human cancers are considered below.

I. Glioblastomas are the most common and most malignant primary brain tumors, accounting for about 50% of all gliomas (Kleihues et al., 2000). Glioblastomas frequently carry genetic alterations resulting in an aberrant activation of PI3K/Akt signaling pathway. These alterations include phosphatase and tension homology (*PTEN*) mutation, epidermal growth factor receptor (*EGFR*) amplification and rearrangement, as well as carboxyl-terminal modulator protein (*CTMP*) hypermethylation (Knobbe and Reifenberger, 2003; Knobbe et al., 2004). Mutation or homozygous deletion of the *PTEN* tumor suppressor gene is indicated as an important aberration that has been detected in 25–40% of the glioblastomas (Schmidt et al., 1999; Boström et al., 1998). The lipid phosphatase activity seems to be particularly important for the tumor suppressor properties of *PTEN*, with phosphatidylinositol-(3,4,5)-trisphosphate (PIP3) being the major lipid substrate (Stambolic et al., 1998). By dephosphorylation of PIP3, *PTEN* antagonizes the activity of PI3K and thereby inhibits PI3K-dependent activation of the oncoprotein Akt (Vivanco and Sawyers, 2002). PIP3 may also be dephosphorylated by the inositol polyphosphate phosphatase like-1 protein (Inppl1/Ship1), which has been reported to inhibit Akt signaling and to cause cell cycle arrest in glioblastoma cells in vitro (Taylor et al., 2000).

Knobbe and Reifenberger (2003) used real-time PCR and southern blot analysis to investigate the possibility that not only *PTEN*, but also other genes encoding proteins involved in PI3K/Akt signaling could be altered in glioblastomas. The authors performed a molecular profiling of 103 glioblastomas for genetic alterations and aberrant expression of 17 genes related to this pathway. On the basis of these analyses a number of deductions related to PIK3/Akt pathway were made, which are worth considering: (i) *PTEN* alterations were detected in 32% of the glioblastomas investigated, which is in line with previous reports (Knobbe et al., 2002). Interestingly, and in contrast to the frequent *PTEN* alterations, the authors did not detect any homozygous deletions, mutations or loss of expression of the *INPPL1* gene. This gene is considered a candidate tumor suppressor gene for two reasons: first, its gene product, phosphatidylinositol-(3,4,5)-trisphosphate-5-phosphatase (Inppl1/Ship2), dephosphorylates the 5′OH position within the inositol ring of PIP_3 and, like *PTEN*, inhibits PI3K-dependent signaling; second, earlier studies have demonstrated that the overexpression of Inppl1/Ship2 in the glioma cell line U87MG caused a cell cycle arrest in G1 and reduced localization of Akt to the plasma membrane (Krystal et al., 1999; Taylor et al., 2000).

II. Gastric cancer is believed to be one of the most common cancer worldwide, yet the pathogenesis and the molecular genetic events that contribute to this form of cancer remain poorly understood (Rooney et al., 1999). A number of allelotyping studies have shown that ~14–39% of gastric adenocarcinomas exhibit loss of 10q and gain of 3q, in which *PTEN* and *PIK3CA* are located, respectively (Guan et al., 2000; Zur Hausen et al., 2001). To assess the role of mutational inactivation of *PTEN* and/or genomic amplification of *PIK3CA* in gastric carcinoma, Byun et al. (2003) have performed expression and mutation analyses of the genes within 141 human gastric specimens, including 15 cancer cell lines. On the basis of real-time PCR and genomic PCR analyses, the authors demonstrated that mono allelic loss of *PTEN* and increased copy number of *PIK3CA* are frequent and mutually exclusive events in gastric carcinogenesis. Furthermore, down- and upregulation of *PTEN* and *PIK3CA*, respectively were found to be associated with elevated phosphorylation of Akt, an observation, which was attributed to oncogenic activation of the PI3K signaling pathway due to the combined loss of *PTEN* and the amplification of *PIK3CA*. These deductions find partial support in earlier studies, particularly regarding two well documented events: First, the authors reported that 60% of the investigated gastric cell lines and 30% of the primary carcinoma harboured *PIK3CA* amplification, whereas normal and benign tumor tissues lacked such amplification. This observation is in line with earlier studies, which reported that genomic amplification of *PIK3CA*, which encodes the catalytic subunit, p110α of PI3K, has been found in various human cancers, including cervical, ovarian, and squamous cell carcinomas (Shayesteh et al., 1999; Ma et al., 2000; Singh et al., 2002). Furthermore, increased copy number of *PIK3CA* correlates with high kinase activity of PI3K, increased cell growth and decreased apoptosis, which may suggest that this gene is an oncogene that positively regulates the PI3K signaling pathway. Second, a recent study identified mutations of *PIK3CA* in 25% of investigated gastric cancer specimens (Samuels et al. 2004).

III. Prostate: A number of studies have made it clear that *PTEN* (also called *MMAC1*) functions primarily as PIP_3 lipid phosphatase (Myers et al., 1998; Haas-Kogan et al., 1998; Maehama and Dixon, 1998). The strong evidence provided by these and other studies underlined the central role of PIP_3 regulation in cancer. Loss of chromosome 10q, which encompasses *PTEN* (10q23) has been reported to occur in

50–80% of advanced prostatic adenocarcinomas (CaP) (Abate-Shen and Shen, 2000; Pesche et al., 1998). Interestingly, the mutations in this gene have been found in less than 15% of early stage CaP (Dong et al., 1998), while a much higher frequency (~60%) was reported in patients with established metastases (Abate-Shen and Shen, 2000; Vazquez and Sellers, 2000). In human CaP xenografts and cell lines, including the commonly used LNcaP and PC-3 CaP cell lines, the loss of *PTEN* is also frequently reported. One of the results of the loss of PTEN function is the elevation of PIP_2 and PIP_3. The binding of the first phosphoinositide to the PH domain of Akt induces a conformational change, which exposes Thr^{308} residues to phosphorylation by PDK-1. The prognostic value of Akt activation is discussed below.

5.11. PIK3/AKT ALTERATIONS AND PROGNOSTIC BIOMARKERS

There are various lines of evidence suggesting that genetic alterations (e.g. mutation, amplification, overexpression etc.) in PIK3 can be exploited as potential prognostic markers for various forms of human cancers. Within the present discussion I shall also refer to the phosphorylation of Akt as a potential biomarker for certain forms of human cancer. The inclusion of both elements within the same argument is motivated by the fact that Akt is activated by a PI3K-dependent mechanism.

Mutations of PIK3CA in various forms of cancer, including colon, brain, gastric, breast, and lung have been investigated by Samuels et al. (2004). The outcome of this investigation is summarized in Figure 5.17. This figure shows the location of missense mutations in PIK3CA, which are strongly clustered in certain functional domains. If further studies can confirm that mutational activation of this gene is essential for tumor growth, then such strong clustering can be an excellent marker for the early detection of one or more of the investigated tumors. To date, Akt overexpression or activation has been shown to be correlated with poor prognosis in several tumor types, including leukemia, breast cancer, pancreatic cancer, gastric, and hepatocellular carcinoma. To underline this statement a number of specific examples are considered below.

5.11.1. Melanoma

Malignant melanoma is associated with one of the highest mortality rates, particularly in advanced stages. Although early melanomas are curable with surgical excision (Balch et al., 2001), up to 20% of patients will develop metastatic tumors owing to the high capability of the disease for invasion and rapid metastasis of other organs (Houghton and Polsky, 2002). Patients with metastatic melanoma have a poor prognosis, with a median survival of 6–10 months (Jemal et al., 2002). Although the molecular mechanism for drug resistance in melanoma is still poorly understood, it is hypothesized that the therapeutic inefficacy is related to relative inability to induce apoptosis (Soengas and Lowe, 2003). However, unlike other tumor types, melanoma is associated with very low mutation rate of the p53 gene (Ragnarsson-Olding et al., 2002). This infers that the dysregulation of other pathways rather than p53 apoptosis pathway may contribute to distorted apoptosis of melanoma cells, and thus, disease progression. Being a key player of PI3K signaling pathway, Akt has been shown to play a central role in controlling the balance between cell survival and apoptosis (Franke et al., 1997). Upon activation, Akt delivers anti-apoptotic signals by phosphorylating BAD and procaspase (Datta et al., 1997), as well as members of the forkhead family of transcription factors, which in turn induce the expression of proapoptotic factors (Wolfrum et al., 2003; Kops et al., 1999). Furthermore, Akt can promote cell survival by indirectly activating the pro-survival transcription factor NF-κB through the phosphorylation of I-κB kinase (Ozes et al., 1999). In recent years various attempts have been made to correlate the activation and/or overexpression of Akt with the prognosis of various forms of cancer. For example, in a relatively recent study, Dai et al. (2005) have used tissue microarrays and immunochemistry to investigate the role of activated Akt in different stages of human melanocytic lesions. The samples investigated included normal nevi; dysplastic nevi, primary melanomas, and melanoma metastases (total of 292 cases). On the basis of these analyses, a number of observations were made: first, strong correlation between the expression of Akt (phosphorylated at Ser-473) and a poorer 5-year patient survival. This observation is in partial agreement with an earlier study, which showed that the expression of integrin-linked kinase, a kinase that plays an important role in mediating Akt phosphorylation at Ser-473, was correlated with tumor invasion in primary melanomas (Dai et al., 2003). Second, a linear increase in Akt phosphorylation was observed following progression of melanocytic lesions, together with substantial differences between normal nevi and primary melanomas, and between the latter and metastatic melanomas. This stage-specific expression patterns was taken as an indication that Akt activity might be a requirement for the transformation from benign neoplasia to malignancy, and from primary tumors to metastatic stage. Third, one of the interesting observations regarding the investigated cases was the higher percentage of phosphorylated Akt expression in male patients compared to their female counterparts. This difference however was not reflected in the 5-year disease specific survival obtained by Kaplan-Meier survival curve. Higher expression of phosphorylated Akt in male patients was reported in earlier studies and was attributed to thicker male tumors at diagnosis (Hersey et al., 1991; Osborne and Hutchinson, 2001). However, such explanation does not justify the difference reported by Dai et al. (2005), where the average thickness in males and females was not substantially different (2 mm compared to 1.8 mm, respectively). The difference in Akt phosphorylation between the two sexes can be caused by a number of factors, two of which have been cited by other research groups. For example, sex-related hormones may contribute to the signaling pathway involved in this disease. It has been suggested that androgen and estrogen have the ability to activate PI3K through interaction between their respective receptors and the p85α subunit of PI3K, thus leading to the activation of Akt (Simoncini et al., 2000; Sun et al., 2003). Another plausible explanation is that the differential expression of phosphorylated Akt between the sexes is tissue-specific. For example, it has been shown that in male rats, Akt activity was higher in the lacrimal gland compared to that in neuronal cells derived from cortical plate, which was not the case for female rats (Zhang et al., 2003; Rocha et al., 2002). Considering the existing information regarding the higher expression of phosphorylated Akt in male patients, it can be said that the mechanism of such phenomenon and its full clinical significance is still to be established. Future studies, involving higher number of cases and more stringent evaluation criterion would certainly contribute to a better understanding of such effects.

5.11.2. Non-small-cell Lung Cancer (NSCLC)

Clinical outcome for patients with non-small-cell lung cancer (NSCLC) is generally poor, because diagnosis often occurs at late stages and because NSCLC cells are intrinsically resistant to therapy. Early detection of this form of cancer followed by multidisciplinary intervention has the potential of improving survival. However, the overall survival of patients with stage I of the disease remains poor (Mountain, 1997). Some genetic events that confer a poor prognosis have been described (Lu et al., 2004); however, biochemical alterations, including signal transduction pathways that may promote cellular survival are also needed to allow a better understanding of the poor prognosis of the disease.

In a recent study, Tsurutani et al. (2006) used tissue microarrays and immunohistochemical analysis to examine the two phosphorylation sites of Akt in NSCLC specimens (total of 300) and surrounding lung tissue specimens (total 100). The evaluation of the two phosphorylation sites was conducted using phosphospecific antibodies against Thr^{308} and Ser^{473}. The novel element of this investigation is the contemporary evaluation of the two phosphorylation sites, which was not the case in earlier investigations (Balsara et al., 2004). The following were main conclusions of this study: First, the authors reported that less than 73% of NSCLC patients in their study exhibited active Akt in their tumors. In addition, different levels of Akt activation were found to be associated with different histological subtypes of the disease. Both observations were interpreted as an indication that increased phosphorylation at both Thr^{308} and Ser^{473} could be used to predict the overall survival of NSCLC patients, where the inclusion of the first site has improved the prognostic significance of Akt activation. Second, staining for both phosphorylation sites was specific for NSCLC tumor tissue versus surrounding tissue. Furthermore, staining for ser^{473} or Thr^{308} (Single site phosphorylation), was associated with a worse prognosis for stage I patients and patients whose tumors were smaller than 5 cm. These findings have to be considered with existing information regarding the potential of Akt activation as a prognostic marker for some forms of cancer.

For example, Phosphorylation of Akt at Ser^{473} alone has been correlated with poor clinical outcomes in many forms of cancer, including breast, prostate, endometrial, gastric, pancreatic, brain, and melanoma (Dai et al., 2005; Kreisberg et al., 2004; Perez-Tenorio and Stal, 2002; Schlieman et al., 2003). Monitoring only the phosphorylation of Ser^{473} in these diseases has provided valuable prognostic information, however, such prognostic value has not been established in the case of NSCLC. Second, the prognostic significance of Ser^{473} phosphorylation in small and non-small-cell lung cancers is still debatable. This debate is the direct result of discrepancy between existing studies, which attempted to address the prognostic value of Ser^{473} phosphorylation (Mukohara et al., 2003; 2004; Balsara et al., 2004;

Blackhall et al., 2003; Massion et al., 2004). Such discrepancy can be attributed to a number of reasons, including small sample sizes, differences in specimen selection and preparation, and differences in staining or scoring procedures. Tsurutani et al. (2006) argued that an additional reason for such discrepancy is the erroneous assumption that Ser473 phosphorylation alone truly reflects Akt activity. At the present time there is not sufficient data to truly evaluate the advantages of monitoring both, rather than one, of the phosphrylation sites. Future studies, involving much larger sample sizes will no doubt provide further insights into the role of activated Akt in NSCLC and its potential for the prognosis of the disease.

5.11.3. Prostate Cancer

A number of recent works have described the utility of Akt phosphorylation as a potential marker and as a predictor of clinical outcome in prostate cancer. Prognostic markers in prostate cancer are of great importance not only because of the mortality rate associated with prostate cancer but also because of the morbidity associated with current forms of therapy. Numerous ongoing trials are attempting to identify markers to discriminate between patients who require immediate therapeutic intervention and those who are candidates for watchful waiting observation. Although prostate cancer is the most common cancer in males in the United States; we are still looking for a suitable prostate cancer biomarker that distinguishes between tumors with a high potential for recurrence and tumors that are not likely to reoccur (see also Chapter 3). Several studies have combined clinical and pathological parameters, as well as serum markers in an attempt to find predictive models. However, results obtained by these studies were heavily dependent on markers of epithelial differentiation of cancer (Gleason score). The difficulty in predicting disease behaviour is a particular problem in patients with a Gleason score of 7, a group that represents a large proportion of patients with prostate cancer. Ayala et al. (2004) used tissue microarrays to compare the levels of phosphorylation of Akt-1 in prostate cancer and non-neoplastic tissues. The scope of the investigation was to assess the validity of such phosphorylation as a predictor of biochemical recurrence of the disease. Tissue microarrays from 640 patients with triplicate cores of non-neoplastic prostate and benign prostatic hyperplasia (BPH) were immunostained with antibody to Akt phosphorylated at Ser^{473}. On the basis of these analyses a number of deductions were made: First, phosphorylated Akt-1(p-Akt-1) expression was found to be significantly stronger in association with high Gleason grades 8 to 10, than with prostatic intraepithelial neoplasia and all of the other grades of prostate cancer. The same data showed that p-Akt-1 was predominantly expressed in prostate cancer, with $>98\%$ of patients having no expression in their non-neoplastic tissues. Furthermore, high levels of p-Akt-1 expression were found almost exclusively in prostate cancer. Second, the acquired data support the notion that P-Akt-1 can be considered a useful and independent prognostic indicator of biochemical recurrence-free survival. High levels of p-Akt-1 were found to be associated with earlier recurrence, which may imply p-Akt-1 is associated with aggressiveness and disease progression in prostate cancer. Furthermore, multivariate Cox models based on the acquired data indicated that p-Akt-1 index could outperform prostate-specific antigen as a postoperative marker.

The correlation between p-Akt (Ser-473) and poor clinical outcome in prostate cancer was also investigated by Kreisberg et al. (2004). This study was designed to establish whether increased p-Akt combined with decreased phosphorylated signal-regulated kinase (p-ERK) could be used as a predictor of poor clinical outcome for prostate cancer. Postate cancers were obtained after radical prostatectomy from men in the age range, 42–81 years with different ethnic origins. The samples were examined by immunohistochemistry and phospho-specific antibodies. PSA failure (detectable and rising) versus PSA nonfailure (undetectable PSA 5 years after prostatectomy) was used by the authors as a surrogate for clinical outcome. These analyses indicated that increased p-Akt, alone or together with decreased p-ERK, could be a useful predictor of the probability of PSA failure. In other words, the phosphorylation level of these two proteins has the potential for differentiation between clinically aggressive and clinically indolent prostate cancer. These results and those reported by Ayala et al. (2004) merit further considerations. Despite a limited number of studies, dealing with the role of activated Akt in various forms of cancer, the convincing data contained within such studies do indicate some role for this and other proteins in prostate cancer progression. As it has been pointed out in the earlier sections, Akt is a downstream effector of PI3K, which was determined to be the dominant growth-factor-activated cell survival pathway in the androgen-dependent LNCap prostate tumor cell line (Lin et al., 1999).

In fact, Akt activation was markedly increased in an androgen-independent LNCaP cell line isolated from LNCaP xenografts, compared with the parental androgen-dependent cell line. Expression of constitutively active Akt in androgen-dependent LNCaP cells resulted in a 6-fold increase in xenograft tumor growth (Graff et al., 2000). In another study, it was shown that androgen deprivation in androgen-dependent LNCaP human prostate cancer cells stimulated the activation of Akt, which eventually resulted in androgen independence of the cells (Murillo et al. 2001). Malik et al. (2002) and Paweletz et al. (2001) have also reported significantly increased activation (phosphorylation) of Akt in high-Gleason grade prostate cancers. The results by Kreisberg et al. (2004) confirm the importance of Akt activation (phosphorylation) in prostate cancer progression to androgen independence and poor clinical outcome.

5.12. GENERAL OBSERVATIONS

- In the last 5 years, the analysis of the entire complement of phosphorylated proteins in complex mixtures has become a viable option. Despite this impressive achievement there are still a number of evident limitations on the way to routine analysis of phosphoproteins. Some of these limitations are intrinsic of phosphoproteins, while others are likely to be resolved through further optimization of enrichment protocols and the increased use of high performance mass spectrometers. This statement can be partially supported by a number of considerations: First, despite an increased availability of high resolution and high performance mass spectrometers, it is still difficult to acquire complete fragmentation using collision induced disassociation (CID). One direct consequence of this limitation is the inability to localize the exact phosphorylation site(s). A promising and rare exception has been demonstrated through the use of electron capture disassociation (ECD) coupled to FT-ICR mass spectrometry (Sze et al., 2002). Unlike conventional CID measurements, this approach has demonstrated to be capable of providing extensive fragmentation along the backbone of peptides without the loss of phosphoric acid or phosphate. Further optimization of this approach and its wider use will certainly contribute to more reliable identification of the phosphorylation site(s). Second, the two methods for enrichment and quantification of phosphopeptides, which have been developed by Oda et al. (2001) and Zhou et al. (2001), respectively are still considered as the gold standard for MS-based strategies.These successful strategies have demonstrated their potential in a number of studies, yet there are still a number of limitations to be resolved. For example, the β-elimination-based tagging methods do not differentiate between O-glycosylated and phosphorylated serine and thrionine residues. Such limitation is particularly undesirable in the analyses of nuclear and cytosolic proteins, where O-glycolysation is fairly frequent (Zachara and Hart, 2002). Another disadvantage of these two strategies is that they require a large amount of sample, which renders them less suitable for the analyses of low-abundance proteins.

- The human genome encodes over 500 protein kinases as well as over 60 protein phosphotases that can dephosphorylate proteins rapidly with high selectivity (Cohen, 2002; Berwick and Tavare, 2004). Such that, protein phosphorylated in for example, a signal transduction cascade in response to an extracellular stimuli are usually dephosphorylated rapidly back to the original state once the stimuli has been removed (Cohen, 2002). Pioneering methods to study reversible phosphorylation, such as phosphopeptide mapping of proteins from ^{32}P phosphate-labeled cells and the generation of phosphorylation site-specific antibodies have been and continue to be invaluable in the field of signal transduction. In recent years, MS-based methods have become highly popular for the initial identification of phosphorylation. Although knowing the state of phosphorylation at certain residues can provide valuable information regarding protein function, it is ultimately the temporal sum of all modifications that is most likely to define a proteins's biochemical state. In summary, it is reasonable to state that temporal localization of phosphorylation sites is still a formidable challenge, despite considerable advances in instrumentation and enrichment protocols.

- Research efforts over the last two decades have defined the key players in the PI3K-Akt signaling pathway and its importance in various human cancers. The same efforts have also indicated that the true frequency of PI3K-Akt-pathway abnormalities in human cancer is still a challenging task, particularly in the absence of all the factors that may affect its activation. In the previous sections, I have focused on the prognostic potential of abnormalities in this signaling pathway. The main reason for this choice is that the potential of these alterations for therapeutic purposes has attracted tremendous research activities, which could not be adequately covered in few pages. Despite the tremendous amount of research that is needed to fully clarify the deregulation of PI3K-Akt in human cancers, it is already indisputable that alterations in this signaling pathway are very frequent in human cancer. One of the questions which has to be addressed in future studies is how such alterations can be translated into screening, prognosis, and therapeutic targets in cancer? A partial answer to this question has already been provided by a number of studies, some of which have been cited in this text. For example, the clustering of mutations in certain chromosomal regions could make the gene a useful marker for the early detection of certain types of cancer. A representative case has been described for *PIK3CA*, where mutations of this gene in different forms of cancer were found to be clustered in specific chromosomal regions (Samuels et al., 2004). This hypothesis of course needs to be supported by further studies, which also have to deal with the question of how such screening can be designed and conducted in trials involving large numbers of subjects.

Another argument, which has been emphasized in this chapter, is the potential of activated (phosphorylated) Akt in the prognosis of various forms of human cancers. This potential is in line with the biological consequences of Akt activation, which include survival, proliferation (increased cell number), and growth (increased cell size). Phosphorylation of Akt on the Thr^{308} is necessary and sufficient for its activation. However, there is increasing evidence that maximal activation requires additional phosphorylation at Ser473. Whether monitoring two rather than one site of phosphorylation can influence the efficacy of prognosis, is still an evolving argument. Though a number of studies have demonstrated the prognostic value of monitoring a single phosphorylation site, there is yet, a paucity of studies in which both sites are considered.

REFERENCES

Abate-Shen, C., Shen, M. M. (2000) Gene Dev. **14**, 2410.

Alessi, D. R., James, S. R., Downes, C. P., et al. (1997) Curr. Biol. **7**, 261.

Alessi, D. R., Andjelkovic, M., Caudwell, B., et al. (1996) EMBO J. **15**, 6541.

Alimov, A., Li, C., Gizatullin,R., et al. (1999) Anticancer Res. **19**, 3841.

Annan, R. S., Carr, S. A. (1996) Anal. Chem. **68**, 3413.

Annan, R. S., Huddleston, M. J., Verma, R., et al. (2001) Anal. Chem. **73**, 393.

Armitage, J. O., Weisenburger, D. D. (1998) J. Clin. Oncol. **16**, 2780.

Armstrong, S. A., Staunton, J. E., Silverman, L. B., et al. (2002) Nat. Genet. **30**, 41.

Ashburner, M., Ball, C. A., Black, J. A., et al. (2000) Nat. Genet. **25**, 25.

Ayala, G., Thompson, T., Yang, G., et al. (2004) Clin. Cancer Res. **10**, 6572.

Bader, G. D., Betel, D., Hogue, W. V., et al. (2003) Nucleic Acids Res. **31**, 248.

Balch, C. M., Buzaid, A. C., Soong, S. J., et al. (2001) J. Clin. Oncol. **19**, 3635.

Balsara, B. R., Pei, J., Mitsuuchi, Y., et al. (2004) Carcinogenesis **25**, 2053.

Bärlund, M., Monni, O., Kononen, J., et al. (2000a) Cancer Res. **60**, 5340.

Bärlund, M., Forozan, F., Kononen, J., et al. (2000b) J. Natl. Cancer Inst. **92**, 1252.

Barrios-Rodiles, M., Brown, K. R., Ozdamar, B., et al. (2005) Science 307, 1621.

Beausoleil, S. A., Jedrychowski, M., Schwarz, D., et al. (2004) PNAS USA. **101**, 12130.

Beer, D. G., Kardia, S. L., Huang, C. C., et al. (2002) Nat. Med. **8**, 816.

Bellacosa, A., de Feo, D., Godwin, A. K., et al. (1995) Int. J. Cancer **64**, 280.

Bender, A., Pringle, J. R. (1991) Mol. Cell. Biol. **11**, 1295

Bernards, R., Weinberg, R. A. (2002) Nature **418**, 823.

Berndt, P., Hobohm, U., Langen, H. (1999) Electrophoresis. **20,** 3521.

Berwick, D. C., Tavaré, J.M. (2004) Trends Biochem. Sci. **29**, 227.

Bhardwaj, N., Lu, H. (2005) Bioinformatics **21**, 2730.

Bhattacharjee, A., Richards, W. G., Stauton, J., et al. (2001) PNAS USA. **98**, 13790.

Blackhall, F. H., Pintilie, M., Michael, M., et al. (2003) Clin. Cancer Res. **9**, 2241.

Blagoev, B., Krachmarova, I., Ong, S. E., et al. (2003) Nat. Biotechnol. **21**, 315.

Blume-Jensen, P., Hunter, T. (2001) Nature **411**, 355.

Borczuk, A. C., Shah, L., Pearson, G. D., et al. (2004) Am. J. Respir. Crit. Care Med. **170**, 167.

Boström, J., Cobbers, J. M., Wolter, M., et al. (1998) Cancer Res. **58**, 29.

Brazma et al. (2001) Nat. Genet. **29**, 365.

Brent, R., Ptashne, M. (1985) Cell **43**, 729.

Brodbeck, D., Cron, P., Hemmings, B. A. (1999) J. Biol. Chem. **274**, 9133.

Brunet, A., Bonni, A., Zigmond, M. J., et al. (1999) Cell, **96**, 857.

Byun, D-S., Cho, K., Ryu, B-K., et al. (2003) Int. J. Cancer **104**, 318.

Campa, M. J., Wang, M. Z., Howard, B., et al. (2003) Cancer Res. **63**, 1652.

Cantley, L. C., Neel, B. G. (1999) PNAS USA. **96**, 4240.

Cao, P., Stults, J. T. (2000) Rapid Commun. Mass Spectrom. **14**, 1600.

Cardone, M. H., Roy, N., Stennicke, H. R., et al. (1998) Science **282**, 1318.

Carr, S. A., Huddleston, M. J., Annan, R. S. (1996) Anal. Biochem. **239**, 180.

Celebi, J. T., Shendrik, I., Silvers, D. N., et al. (2000) J. Med. Genet. **37**, 653.

Chen, C. Y., Graham, T. R. (1998) Genetics **150**, 577.

Cheng, J. Q., Godwin, A. K., Bellacosa, A., et al. (1992) PNAS USA. **89**, 9267.

Cheng, J. Q., Ruggeri, B., Klein, W. M., et al. (1996) PNAS USA. **93**, 3636.

Chernushevich, I. V., Lobada, A. V., Thomson, B. A., et al. (2001) J. Mass Spectrom. **36**, 849.

Cho, R. J., Campbell, M. J. (2000) Trends Genet. **16**, 409.

Cntley, L. C. (2002) Science **296**, 1655.

Coffer, P. J., Woodgett, J. R. (1991) Eur. J. Biochem. **201**, 475.

Cohen, P., (2002) Nat. Rev. Drug Discov. **1**, 309.

Colland, F., Jacq, X., Trouplin, V., et al. (2004) Genome Res. **14**, 1324.

Colombat, P., Salles, G., Brousse, N., et al. (2001) Blood **97**, 101.

Covey, T., Shushan, B., Bonner, R., et al. (1991) In Methods in Protein Sequence Analysis (Jörnvall, H., et al., eds). Birkhäuser Verlag pp. 249–256 .

Cusick, M. E., Klitgord, N., Vidal, M., et al. (2005) Hum. Mol. Genet. **14**, R171.

Dai, D. L., Makretsov, N., Campos, E. I., et al. (2003) Clin. Cancer Res. **9**, 4409.

Dai, D. L., Martinka, M., Li, G. (2005) J. Clin. Oncol. **23**, 1473.

Daraselia, N., et al. (2004) Proceedings of the 2nd European Workshop on data Mining and Text Mining for Bioinformatics, pp11-18, ECML/PKDD 2004 Committee. (avilable on-line at http://www.informatik.hu-berlin.de/Forschung_Lehre/wm/ws04/#proceedings)

Datta, S. R., Bruner, A., Greenberg, M. E., et al. (1999) Genes Dev. **13**, 2905.

Datta, S. R., Dudek, H., Tao, X., et al. (1997) Cell **91**, 231.

Dave, S. S., Wrigt, G., Tan, B., et al. (2004) N. Engl. J. Med. **351**, 2159.

De Las Rivas, J., Lozano, J. J., Ortiz, A. R. (2002) Genome Res. **12**, 567.

Delcommenne, M., Tan, C., Gray, V., et al. (1998) PNAS USA. **95**, 11211.

Dennis, P. B., Jaeschke, A., Saitoh, M., et al. (2001) Science **294**, 1102.

DeRisi, J., Penland, L., Brown, P. O., et al. (1996) Nature Genet. **14**, 457.

Diehl, J. A., Cheng, M., Roussel, M. F., et al. (1998) Genes Dev. **12**, 3499.

Dijkers, P. F., et al. (2000) Mol. Cell Biol. **20**, 9138.

Dong, J. T., Sipe, T. W., Hyytinen, E. R., et al. (1998) Oncogene **17**, 1979.

Early Breast Cancer Trialist's collaborative Group (1998a) Lancet **352**, 930.

Early Breast Cancer Trialist's collaborative Group (1998b) Lancet **351**, 1452.

Endoh, H., Vincent, S., Jacob, Y., et al. (2002) Meth. Enzymol. **350**, 525.

Enright, A. J., Lliopoulos, I., Kyrpides, N. C., et al. (1999) Nature **402**, 86.

Eriksson, J., Chait, B. T., Fano, D. (2000) Anal. Chem. **72,** 999.

Ficarro, S. B., McCleland, M. L., Stukenberg, P. T., et al. (2002) Nat. Biotechnol. **20**, 301.

Fields, S., Song, O. (1989) Nature **340**, 245.

Fields, S., Sternglanz, R. (1994) Trends Genet. **10**, 286.

Fitch, W. M. (1970) Syst. Zool. **19**, 99.

Fitch, W. W. (2000) Trends Genet **16**, 227.

Ford, D., Easton, D. F., Stratton, M., et al. (1998) Am. J. Hum. Genet. **62**, 676.

Forgacs, E., Biesterveld, E. J., Sekido, Y., et al. (1998) Oncogene **17**, 1557.

Franke, T. F., Kaplan, D. R., Cantley, L. C., et al. (1997) Cell **88**, 435.

Fritz, B., Scubert, F., Wrobel, G., et al., (2002) Cancer Res. **62**, 2993.

Fruman, D. A., Meyers, R. E., Cantley, L. C., et al. (1998) Annu. Rev. Biochem. 67, 481.

Fryxell, K. J. (1996) Trends Genet. **12**, 364.

Garber, M. E., Troyanskaya, O. G., Schluens, K., et al. (2001) PNAS USA. 98, 13784.

Gavin, A. C., Bosche, M., Krause, R., et al. (2002) Nature **415**, 141.

Ge, H., Liu, Z., Church, G. M., et al. (2001) Nat. Genet. **29**, 482.

Giot, L., Bader, J. S., Brouwer, C., et al. (2003) Science **302**, 1727.

Göbel, U., Sander, C., Schneider, R., et al. (1994) Proteins **18**, 309.

Goh, C. S., Bogan, A. A., Joaachimiak, et al. (2000) J. Mol. Biol. **299**, 283.

Golub, T. R., Slonim, D. K., Tamayo, P., et al. (1999) Science **286**, 531.

Gordon, G. J., Richards, W. G., Sugarbaker, D. J., et al. (2003) Cancer Epidemiol. Bioomarkers Prev. **12**, 905.

Goshe, M. B., Conrads, T. P., Panisko, E. A., et al. (2001) Anal. Chem. **73**, 2578.

Graff, J. R., Konicek, B. W., McNulty, A. M., et al. (2000) J. Biol. Chem. **275**, 24500.

Greshock, J., Naylor, T. L., Margolin, A., et al. (2004) Genome Res. **14**, 179.

Grigoriev, A. (2001) Nucleic Acids Res. **29**, 3513.

Gu, J., et al. (1999) J. Cell Biol. **146**, 389.

Guan, X. Y., Fu, S. B., Xia, J. C., et al. (2000) Cancer Genet Cytogenet **123**, 27.

Guarente, L. (1993) Trends Genet. **9**, 362.

Gygi, S. P., Rist, B., Gerber, S. A., et al. (1999) Nature Biotech. **17**, 994.

Hager, J. W. (2002) Rapid Commun. Mass Spectrom. **16**, 512.

Haas-Kogan, D., et al. (1998) Curr. Biol. **8**, 1195.

Han, J. D., Bertin, N., Hao, T., et al. (2004) Nature **430**, 88.

Hanahan, D., Weinberger, R. A. (2000) Cell **100**, 57.

Hartman IV, J. L., Garvik, B., Hartwell, L. (2001) Science **291**, 1001.

Hedenfalk, I., Duggan, D., Chen, Y., et al. (2001) N. Engl. J. Med. **344**, 539.

Hersey, P., Sillar, R. W., Howe, C. G. (1991) Med. J. Aust. **154**, 583.

Herskowitz, J. Rine, J. Strathern (1992) in The Molecular and Cellular Biology of the Yeast Saccharomyces cerevisiae, vol. 2, Gene Expression, E. W. Jones, J. R. Pringle, J. R. Broach, Eds. (Cold Spring Harbor Laboratory, Cold Spring.

Ho, Y., Gruhler, A., Heilbut, A., et al. (2002) Nature **415**, 180.

Horning, S. J. (2000) Ann. Oncol. **1**, 23.

Horning, S. J., Rosenberg, S. A. (1984) N. Engl. J. Med. **311**, 1471.

Hsieh, M. C. et al. (2000) J. Med. Sci. **16**, 9.

Hughs, T. R., Mao, M., Jones, A. R., et al. (2001) Nat. Biotechnol. **19**, 342.

Houghton, A. N., Polsky, D. (2002) Cancer Cell **2**, 275.

Ideker, T., Galitski, T., Hood, L. (2001) Ann Rev. of Genomics and Human Genetics. **2**, 343.

Immler, D., Gremm, D., Kirsch, D., et al. (1998) Electrophoresis **19**, 1015.

Isaacs, C., Stearns, V., Hayes, D. F. (2001) Semin. Oncol. **28**, 53.

Ito, T., Chiba, T., Ozawa, R., et al. (2001) PNAS. USA. **98**, 4569.

James, P, Halladay, J., Craig, E. A. (1996) Genetics **144**, 1425.

Jensen, R. A. (2001) Genome Biol. **2**, 1002.

Jansen, R., Greenbaum, D., Gerstein, M. (2002) Genome Res. **12**, 37.

Jemal, A., Thomas, A., Murray, T., et al. (2002) CA. Cancer J. Clin. **52**, 23.

Jensen, P. K., Pasa-Tolic, L., Anderson, G. A., et al. (1999) Anal. Chem. **71**, 2076.

Jeong, H., Mason, S. P., Barabasi, A. L., et al. (2001) Nature **411**, 41.

Johnson, P. W., Rohatiner, A. Z., Whelan, J. S., et al. (1995) J. Clin. Oncol. **13**, 140.

Johnsson, N., Varshavsky, A. (1994) PNAS USA. **91**, 10340.

Jones, P. F., Jakubowicz, T., Pitossi, F. J., et al. (1991) PNAS USA. **88**, 4171.

Jönsson, G., Naylor, T. L., Vallon-Christersson, J., et al. (2005) Cancer Res. **65**, 7612.

Katso, R., Okkenhaug, K., Ahmadi, K., et al. (2001) Ann. Rev. Cell Dev. Biol. **17**, 615.

Keegan, L., Gill, G., Ptashne, M. (1986) Science **231**, 699.

Keil, O., LeRiche, T., Deppe, H., et al. (2002) Rapid Commun. Mass Spectrom. **16**, 814.

Kitano, H. (2002) Science **295**, 1662.

Knobbe, C. B., Merlo, A., Reifenberger, G. (2002) Neuro Oncol. **4**, 196.

Knobbe, C. B., Reifenberger, G. (2003) Brain Path. **13**, 507.

Knobbe, C. B., Reifenberger, J., Blaschke, B., et al. (2004) J. Natl. Cancer Inst. **96**, 483.

Köcher, T., Allmaier, G., Wilm, M. (2003) J. Mass Spectrom. **38**, 131.

Kops, G. J., De Ruiter, N. D., De Vries-Smits, A. M., et al. (1999) Nature **398**, 630.

Kreisberg, J. I., Malik, S. N., Prihoda, T. J. (2004) Cancer Res. **64**, 5232.

Kleihues, P., Burger, P. C., Collins, V. P., et al. (2000) In: Pathology and Genetics of tumors of the Nervous System. WHO Classification of Tumours. Kleihues, P., Cavenee, W. K. (eds) Lyon, France: IARC Press, pp 29–39.

Klumpp, S., Krieglstein, J. (2002) Eur. J. Biochem. **269**, 1067.

Krishna, R. G., Wold, F. (1993) Adv. Enzymol. Relat. Areas Mol. Biol. **67,** 265.

Krystal, G., Damen, J. E., Helgason, C. D. (1999) Int. J. Biochem. Cell Biol. **31**, 1007.

Kuriyan, J., Cowburn, D. (1997) Annu. Rev. Biophys. Biomol. Struct. **26**, 259.

Lakhani, S. R., Gusterson, B. A., Jacquemier, J., et al. (2000) Clin. Cancer Res. **6**, 782.

Le Blanc, J. C. Y., Hager, J. W., Llisiu, A. M. P., et al. (2003) Proteomics **3**, 859.

Larsen, M. R., Sorensen, G. L., Fey, S. J., et al. (2001) Proteomics **1**, 223.

Larsen, M. R., Cordwell, S. J., Roepstorff, P. (2002) Proteomics **2**, 1277.

Lehner, B., Fraser, A. G. (2004) Genome Biol. **5**, R63.

Lewis, T. S., Hunt, J. B., Aveline, L. D. (2000) Mol. Cell **6**, 1343.

Li, S., Armstrong, C. M., Bertin, N., et al. (2004) Science **303**, 540.

Li, J., Yen, C., Liaw, D., et al. (1997) Science **275**, 1943.

Lin, J., Adam, R. M., Santiestevan, E., Freeman, M. R. (1999) Cancer Res. **59**, 2891.

Lockhart, D. I., Dong, H., Byrn, M. C., et al. (1996) Nat. Biotechnol. **14**, 1675.

Lockhart, D. J. and Winzeler, E. A. (2000) Nature **405**, 827.

Lossos, I. S., Levy, R. (2003) Semin. Cancer Biol. **13**, 191.

Lu, C., Soria, J. C., Tang, X., et al. (2004) J. Clin. Oncol. **22**, 4575.

Lynch, D. K., Ellis, C. A., Edwards, P. A., et al. (1999) Oncogene **18**, 8024.

Ma, J., Ptashne, M. (1987) Cell **51**, 113.

Ma, Y. Y., Wei, S. J., Lin, Y. C., et al. (2000) Oncogene **19**, 2739.

MacBeath, G., Schreiber, S. L. (2000) Science **289**, 1760.

Maehama, T., Dixon, J. E. (1998) J. Biol. Chem. **273**, 13378.

Malik, S. N., Brattain, M., Ghosh, P. M., et al. (2002) Clin. Cancer Res. **8**, 1168.

Marcus, K., Immler, D., Sternberger, J., et al. (2000) Electrophoresis **21**, 2622.

Marcotte, E. M., Pellegrini, M., Ho-Leung, N., et al. (1999) Science **285**, 751.

Mason, G. G., Murray, R. Z., Rappin, D., et al. (1998) FEBS Lett. **430**, 269.

Massion, P. P., Taflan, P. M., Shyr, Y., et al. (2004) Am. J. Respir. Crit. Care Med. (2004) **170**, 1088.

Matthews, L. R., Vaglio, R. P., Ge, H., et al. (2001) Genome Res. **11**, 2120.

Madema, R. H., Kops, G. J., Bos, J. L., et al. (2000) Nature **404**, 782.

McLafferty, F. W., Horn, D. M., Breuker, K., et al. (2001) J. Am. Soc. Mass Spectrom. **12**, 245.

Medzihradszky, K. F., Campbell, J. M., Baldwin, M. A., et al. (2000) Anal. Chem. **72**, 552.

Mewes, H. W., Frishman, D., Güldener, U., et al. (2002) Nucleic Acids Res. **30**, 31.

Miller, I. P., Lo, R. S., Ben-Hur, A., et al. (2005) PNAS USA. **102**, 12123.

Montoto, S., Lopez-Guillermo, A., Ferrer, A., et al. (2002) Ann. Oncol. **13**, 523.

Mountain, C. F. (1997) Chest **111**, 1710.

Mukohara, T., Kudoh, S., Matsuura, K., et al. (2004) Anticancer Res. **24**, 11.

Mukohara, T., Kudoh, S., Yamauchi, S., et al. (2003) Lung Cancer **41**, 123.

Mullen, J. R., Kaliraman, V., Ibrahim, S. S., et al. (2001) Genetics **157**, 103.

Muise-Helmericks, R. C., Grimes, H. L., Bellacosa, A., et al. (1998) J. Biol. Chem. **273**, 29864.

Murillo, H., Huang, H., Schmidt, L. J., et al. (2001) Endocrinology **142**, 4795.

Myers, M. P., et al. (1998) PNAS USA. **95**, 13513.

Nakahara, Y., Nagai, H., Kinoshita, T., et al. (1998) Leukemia **12**, 1277.

Nakatani, K., Thompson, D. A., Barthel, A., et al. (1999) EMBO J. **15**, 6541.

Nathanson, K. L., Wooster, R., Weber, B. L. (2001) Nat. Med. **7**, 552.

Nave, B. T., Ouwens, M., Withers, D. J., et al. (1999) Biochem. J. **344**, 427.

Neubauer, G., Gottschalk, A., Fabrizio, P., et al. (1997) PNAS USA. **94**, 385.

Neubauer, G., Mann, M. (1999) Anal. Chem. **71**, 235.

Nikolsky, Y., Nikolskaya, T., Bugrim, A. (2005) Drug Discov. Today 10, 653.

Nollau, P., Mayer, B. J. (2001) PNAS USA. **98**, 13531.

Novick, P., Osmond, B., C., Botstein, D. (1989) Genetics **121**, 659.

Oda, Y., Nagasu, T., Chait, B. T. (2001) Nat. Biotechnol. **19**, 379.

Olayioye, M. A., Neve, R. M., Lane, H., et al. (2000) EMBO **19**, 3159.

Olmea, O., Valencia, A. (1997) Fold. Des. (1997) **2**, S25.

Olmea, O., Rost, B., Valencia, A. (1999) J. Mol. Biol. **293**, 1221.

Ortiz, A., Kolinski, A., Rotkiewicz, P., et al. (1999) Proteins **S3**, 177.

Osaki, M., Oshimura, M., Ito, H. (2004) Apoptosis **9**, 667.

Osborne, J. E., Hutchinson, P. E. (2001) Br. J. Dermatol. **144**, 476.

Ozes, O. N., Mayo, L. D., Gustin, J. A., et al. (1999) Nature **401**, 82.

Pagel, P., Kovac, S., Oesterheld, M., et al. (2005) Bioinformatics **21**, 832.

Pandy, A., Podtelejnikov, V., Blagoev, B., et al. (2000) PNAS USA. **97**, 179.

Paweletz, C. P., Charboneau, L., Bichsel, V. E., et al. (2001) Oncogene **20**, 1981.

Pandey, A., Mann, M. (2000) Nature **405**, 837.

Patterson, S. D., Aebersold, R., Goodlett, D. R. (2001) In: Pennigton, S. R., Dunn, M. J. (eds), Proteomics, From Protein Sequence to Function. Bios Scintific, Oxford, pp. 87–130.

Pawson, T., Nash, P. (2000) Gene Dev. **14**, 1027.

Pazos, F., Valencia, A. (2001) Proteins Eng. **14**, 609.

Pazos, F., Helmer-Citterich, M., Ausiello, G., et al. (1997) J. Mol. Biol. **271**, 511.

Pazos, F., Valencia, A. (2002) Proteins **47**, 219.

Pellegrini, M., Marcotte, E. M. (1999) PNAS USA. **96**, 4285.

Peri, S., Navarro, D., Amanchy, R., et al. (2003) Genome Res. **13**, 2363.

Perkins, D. N., Pappin, D. J., Creasy, D. M., et al. (1999) Electrophoresis. **20**, 3551.

Perez-Tenorio, G., Stal, O. (2002) Br. J. Cancer **86**, 540.

Perou, C. M., Sorlie, T., Eisen, M. B., et al. (2000) Nature **406**, 747.

Pesche, S., Latil, A., Muzeau, F., et al. (1998) Oncogene **16**, 2879.

Philp, A. J., Campbell, I. G., Leet, C., et al. (2001) Cancer Res. **61**, 7426.

Pinkel, D., Segraves, R., Sudar, D. (1998) Nat. Genet. **20**, 207.

Poste, G., Fidler, I. J. (1980) Nature **283**, 139.
Ragnarsson-Olding, B. K., Karsberg, S., Platz, A., et al. (2002) Melanoma Res. **12**, 453.
Rain, J. C., Selig, L., De Reuse, H., et al. (2001) Nature **409**, 211.
Ramani, A. K., Marcotte, E. M. (2003) J. Mol. Biol. **327**, 273.
Ramaswamy, S., Ross, K. N., Lander, E. S., et al. (2003) Nature Genet. **33**, 49.
Ramaswamy, S., Todd, R. G. (2002) J. Clin. Oncol. **20**, 1932.
Remm, M., Storm, C. E. V., Sonnhammer, E. L. L. (2001) J. Mol. Biol. **314**, 1041.
Rhodes, D. R., Chinnaiyan, A. M. (2005) Nat. Genet. **37**, S31.
Rhodes, D. R., Yu, J., Shanker, K., et al. (2004) PNAS USA. **101**, 9309.
Rigaut, G., Shevchenko, A., Rutz, B., et al. (1999) Nat. Biotechnol. **17**, 1030.
Ringel, M. D., Hayre, N., Saito, J., et al. (2001) Cancer Res. **61**, 6105.
Rocha, E. M., Hirata, A. E., Carneiro, E. M., et al. (2002) Endocrine **18**, 191.
Rooney, P. H., Murray, G. I., Stevenson, D. A., et al. (1999) Br. J. Cancer **80**, 862.
Rossig, L., et al. (2001) Mol. Cell Biol. **21**, 5644.
Ross-Macdonald, P., Coelho, P. S. R., Roemer, T., et al. (1999) Nature **402**, 413.
Rual, J-F., Venkatesan, K., Hao, T., et al. (2005) Nature **437**, 1173.
Rubin, G. M., Yandell, M. D., Wortman, J. R., et al. (2000) Science **287**, 2204.
Saito, M., Okamoto, A., Kohno, T., et al. (2000) Int. J. Cancer **85**, 160.
Salvesen, H. B., MacDonald, N., Ryan, A., et al. (2001) Int. J. Cancer **91**, 22.
Salwinski, L., Miller, C. S., Adam, J. S., et al. (2004) Nucleic Acids Res. **32**, D449.
Salwinski, L., Eisenberg, D. (2003) Curr. Opin. Struct. Biol. **13**, 377.
Samuels, Y., Wang, Z., Bardelli, A., et al. (2004) Science **304**, 554.
Schena, M., Shalon, D., Davis, R. W., et al. (1995) Science **270**, 467.
Schlieman, M. G., Fahy, B. N., Ramsamooj, R., et al. (2003) Br. J. Cancer **89**, 2110.
Schlosser, A., Pipkorn, R., Bossemeyer, D., et al. (2001) Anal. Chem. **73**, 170.
Schmelzle, T., Hall, M. N. (2000) Cell **103**, 253.
Schmidt, E. E., Ichimura, K., Goike, H. M., et al. (1999) J. Neuropathol. Exp. Neurol. **58**, 1170.
Schwikowski, B., Uetz, P., Fields, S. (2000) Nat. Biotechnol. **18**, 1257.
Shayesteh, L., Lu, Y., Kuo, W. L., et al. (1999) Nat. Genet. **21**, 99.
Shi, S. D., Hemling, M. E., Carr, S., et al. (2001) Anal. Chem. **73**, 19.
Shoelson, S. E. (1997) Curr. Opin. Chem. Biol. **1**, 227.
Singh, B., Reddy, P. G., Goberdhan, A., et al. (2002) Genes Dev. **16**, 984.
Simoncini, T., Hafezi-Moghadam, A., Brazil, D. P., et al. (2000) Nature **407**, 538.
Soengas, M. S., Lowe, S. W. (2003) Oncogene **22**, 3138.
Songyang, Z., Shoelson, S. E., McGlade, J., et al. (1993) Cell **72**, 767.
Songyang, Z., Shoelson, S. E., McGlade, J., et al. (1994) Mol. Cell Biol. **14**, 2777.
Sorensen, J. B., Hirsch, F. R., Gazder, A., et al. (1993) Cancer **71**, 2971.
Sorlie, T., Perou, C. M., Tibshirani, R., et al. (2001) PNAS USA. **98**, 10869.
Southern, E., Mir, K., Shchepinov, M. (1999) Nat. Genet. **21**, 5.
Sprinzak, E., Margalit, H. (2001) J. Mol. Biol. **311**, 681.
Staal, S. P. (1987) PNAS USA. **84**, 5034.

Staal, S. P., Huebner, K., Croce, C. M., et al. (1988) Genomics **2**, 96.

Stagljar, I., Korostensky, C., Johnsson, N., et al. (1998) PNAS USA. **95**, 5187.

Stambolic, V., Suzuki, A., de la pompa, J. L., et al. (1998) Cell **95**, 29.

Steck, P. A., Pershouse, M. A., Jasser, S. A., et al. (1997) Nat. Genet. **15**, 356.

Steen, H., Küster, B., Mann, M. (2001a) J. Mass Spectrom. **36**, 782.

Steen, H., et al. (2001b) Anal. Chem. **73**, 1440.

Steen, H., Jebanathirajah, J. A., Rush, J., et al. (2006) Mol. Cell. Proteomics **5.1**, 172.

Stelzl, U., Worm, U., Lalowski, M., et al. (2005) Cell **122**, 957.

Stensaballe, A., Andersen, S., Jensen, O. N. (2001) Proteomics **1**, 207.

Stensaballe, A., et al. (2000) Rapid Commun. Mass Spectrom. **14**, 1793.

Stokoe, D., Stephens, L. R., Copeland, T., et al. (1997) Science **277**, 567.

Sun, M., Yang, L., Feldman, R. I., et al. (2003) J. Biol. Chem. **278**, 42992.

Sun, M., Wang, G., Paciga, J. E., et al. (2001) Am. J. Pathol. **159**, 431.

Sze, S. K., Ge, Y., Oh, H., et al. (2002) PNAS USA. **99**, 1774.

Tamura, M., et al. (1998) Science **280**, 1614.

Taylor, V., Wong, M., Brandts, C., et al. (2000) Mol. Cell Boil. **20**, 6860.

Teixeria, M. R., Ribeiro, F. R., Torres, I., et al. (2004) Br. J. Cancer. **91**, 775.

Teng, D. H., Hu, R., Lin, H., et al. (1997) Cancer Res. **57**, 5221.

Timmerman, J. M., Czerwinski, D. K., Davis, T. A., et al. (2002) Blood **99**, 1517.

Tong, A. H. Y., Evangelista, M., Parsons, A. B., et al. (2001) Science **294**, 2364.

Tsurutani, J., Fukuoka, J., Tsurutani, H., et al. (2006) J. Clin. Oncol. **24**, 306.

Tsybin, Y. O., Hakansson, P., Budnik, B. A., et al. (2001) Rapid Commun. Mass Spectrom. **15**, 1849.

Tucker, C. L., Gera, J. F., Uetz, P. (2001) Trends in Cell Biol. **11**, 102

Uetz, P., Giot, L., Cagney, G., et al. (2000) Nature **403**, 623.

Uetz, P., Finley Jr., R. L. (2005) FEBS Lett. **579**, 1821.

Vanhaesebroeck, B., Waterfield, M. D. (1999) Exp. Cell Res. **253**, 239.

Valencia, A., Pazos, F. (2002) Curr. Opin. Struct. Biol. **12**, 368.

van De Vijver, M. J., He, Y. D., van'T Veer, L. J., et al. (2002) N. Engl. J. Med. **347**, 1999.

van't Veer, L. J., Dai, H., van de Vijver, M. J., et al. (2002) Nature **415**, 530.

Vanhaesebroeck, B., Alessi, D. R. (2000) Biochem. J. **346**, 561.

Vazquez, F., Sellers, W. R. (2000) Biochem. Biophys. Acta 1**470**, M21.

Vidal, M. (1997) In Bartels, P. and Fields S. (eds) The Yeast Two-Hybrid Systems. Oxford University Press, N.Y., pp. 109–147.

Vidal, M., Legrain, P. (1999) Nucl. Acids Res. **27**, 919.

Virtanen, C., Ishikawa, Y., Honjoh, D., et al. (2002) PNAS USA. **99**, 12357.

Vissers, J. P.C., Blackburn, R. K., Mosely, M. A. (2002) J. Am. Soc. Mass Spectrom. **13**, 760.

Vivanco, I., Sawyers, C. L. (2002) Nat. Rev. Cancer **2**, 489.

Volinia, S., Hiles, I., Ormondroyd, E., et al. (1994) Genomics **24**, 472.

von Mering, C., Krause, R., Snel, B., et al. (2002) Nature **417**, 399.

von Mering, C., Huynen, M., Jaeggi, D., et al. (2003) Nucleic Acids Res. **31**, 258.

Wachi, S., Yoneda, K., Wu, R. (2005) Bioinformatics **21**, 4205.

Walhout, A. J. M., Sordella, R., Lu, X., et al. (2000) Science **287**, 116.

Wang, T., Bretscher, A. (1997) Genetics **147**, 1595.

Washburn, M. P., Wolters, D., Yates III, J. R. (2001) Nat. Biotechnol. **19**, 242.

Weckwerth, W., Willmitzer, L., Fiehn, O. (2000) Rapid Commun. Mass Spectrom. **14**, 1677.

Weiss, A., Schlessinger, J. (1998) Cell **94**, 227.

Wheelan, S. J., Boguski, M. S., Duret, L., et al. (1999) Gene **238**, 163.

White, M. F. (1998) Mol. Cell Biochem. **182**, 3.

Williamson, B. L., Marchese, J., Morrice, N. A. (2006) Mol. Cell. Proteomics **5.2**, 337.

Wilm, M., Neubauer, G., Mann, M. (1996) Anal. Chem. **68**, 527.

Wind, M., Feldmann, I., Jakubowwski, N., et al. (2003) Electrophoresis. **24**, 1276.

Winzeler, E. A., Shoemaker, D. D., Astromoff, A., et al. (1999) Science **285**, 901.

Witzig, T. E., Gordon, L. I., Cabanillas, F., et al. (2002) J. Clin. Oncol. **20**, 2453.

Wolfrum, C., Besser, D., Luca, E., et al. (2003) PNAS USA. **100**, 11624.

Wuchty, S., Almaas, E. (2005) Proteomics **5**, 444.

Xhou, W., Merrick, B. A., Khaledi, M. G., et al. (2000) J. Am. Soc. Mass Spectrom. **11**, 273.

Yamagata, N., Shyr, Y., Yanagisawa, K., et al. (2003) Clin. Cancer Res. **9**, 4695.

Ye S.Q., Zhang L.Q., Zheng F., et al. (2000) Anal. Biochem. **287**, 144.

Yip, T. T., Hutchens, T. W. (1992) FEBS Lett. **308**, 149.

Yokoyama, Y., Wan, X., Shinohara, A., et al. (2000) Int. J. Mol. Med. **6**, 47.

Zachara, N. E., Hart, G. W. (2002) Chem. Rev. **102**, 431.

Zanzoni, A., Montecchi-Palazzi, L., Quandam, M., et al. (2002) FEBS Lett. **513**, 135.

Zeller, M., König (2004) Anal. Bioanal. Chem. **378**, 898.

Zhang, L., Li, P. P., Feng, X., et al. (2003) Neorosci. Lett. **337**, 65.

Zhang, X., Jin, Q. K., Carr, S. A., et al. (2002) Rapid Commun. Mass Spectrom. **16**, 2325.

Zhang, M. Q. (1999) Comput. Chem. **23**, 233

Zhang, H., Zha, X., Tan, Y., et al. (2002) J. Biol. Chem. **277**, 39379.

Zhou, H., Watts, J. D., Aebersold, R. (2001) Nat. Biotechnol. **19**, 375.

Zhou, X. P., et al. (2000) Am. J. Pathol. **157**, 1123.

Zhu, H., Klemic, J. F., Chang, S., et al. (2000) Nature Genet. **26**, 283.

Zhu, H., Belgin, M., Bangham, R., et al. (2001) Science **293**, 2101.

Zubarev, R. A., Kellehler, N. L., McLafferty, F. W. (1998) J. Am. Chem. Soc. **120**, 3265.

Zubarev, R. A., Horn, D. M., Fridriksson, E. K., et al. (2000) Anal. Chem. **72**, 563.

Zur Hausen, A., van Grieken, N. C., Meijer, C. J., et al. (2001) Gastroenterology **121**, 612.

Zvelebil, M., MacDougall, L., Leevers, S., et al. (1997) Phil. Trans. R. Soc. London **B351**, 217.

6

ETHICAL ISSUES AND INITIATIVES RELEVANT TO CANCER BIOMARKERS

6.1. INTRODUCTION

Ethical concerns are deeply rooted in the history of mankind. Throughout our history human samples have always had a great symbolic importance in various civilizations. The unprecedented advance in molecular-based medical research has increased the need for both normal and pathological samples obtained from humans at different stages of disease or from normal volunteers for comparison purposes. This new scenario has raised many ethical and social issues involving patient rights, including informed consent, respect of privacy, and protection against potential discrimination. These issues stem from the belief that research participants have rights to appropriate information before, during, and after studies so that they can make informed decisions. The same issues are driven by the fear that the emerging genetic testing may result in discrimination against individuals and certain ethnic groups. Ethical issues can arise during protocol development, obtaining participation, and in the interpretation and notification of text and study results. Additionally, there are ethical considerations concerning the use of biological specimens that are collected and stored for one purpose, and subsequently used for other research purposes (commonly known as secondary uses). A major ethical issue is the protection of participant's privacy and the confidentiality of their test and study results. This means that ethics and regulatory committees need to be well informed about the scope, limitations, and expectations of biomarkers and therapy research in order to be able to respond to social and scientific developments in the use of biomarkers and

Cancer Biomarkers: Analytical Techniques for Discovery, by Mahmoud H. Hamdan

drugs. In the near future, genetic and proteomic information will be increasingly used in population screening to determine individual susceptibility to various forms of disease such as heart, diabetes, and cancer. The vast material dealing with ethical and regulatory issues in biomedical research makes it impossible to cover such material in a single text. In this complex scenario, there is always the risk of bombarding the reader with excessive amount of information with the direct result of reading too much and concluding little. To avoid this pitfall and at the same time give the reader reasonable idea about the question of ethical issues in cancer research, the present discussion will mainly focus on a number of limited arguments, including genetic testing for cancer susceptibility. The inclusion of genetic screening as a representative case is motivated by a number of considerations: First, the number of known genetic mutations that are associated with cancer susceptibility is growing at an exponential rate (Am. Soc. of Clin. Oncl., 2003). Second, genetic testing is now available for the main cancer susceptibility genes, in which rare mutations predispose certain categories of the population to inherited cancer syndromes, such as the heredity breast and ovarian cancer (HBOC), heredity nonpolyposis colon cancer (HNPCC), and familial adenomatous polyposis (FAP). More details on these syndromes can be found at Gene Tests Web site www.geneclinics.com. Third, genetic testing for cancer susceptibility poses complex psychological, social, and ethical concerns.

6.2. BACKGROUND

It is reasonable to state that any ethical analysis of genetic testing for cancer susceptibility must always balance benefits of such testing with the possible harm, which the same test may cause to a donor of a sample or a participant in a test. Such potential harms have different sources. Despite the potential medical benefits, there are several sources of potential harms, beyond those of psychological stress at the individual and family levels. These ethical and social considerations include the following: Breach of privacy, which entails negative economic and social sequences, potential discrimination following disclosure of genetic information to third parties such as insurance companies and employers. In fact, many people would decline a genetic test if employers or health insurers could access the results, and many would opt not to seek medical care or file an insurance claim because they do not want to harm their job prospects. These type of privacy concerns have been documented in the USA, Canada, and Europe (Eamscliffe Research, 2000). Approximately 15% of Americans who are at risk of inheriting a condition reported that they had been asked questions about genetic diseases on job applications; 13% reported that they or a member of their family had been fired or denied a job because of a genetic condition in the family. Also, 22% of those with a known genetic condition reported that they had been refused insurance coverage (Hudson et al., 1995). In Australia, a review of the policies of life-insurance underwriters indicated that they all required that the results of genetic testing be revealed, if known by the applicant (Lynch et al., 2003). Although similar privacy concerns have been documented in surveys in Canada and Europe, the potential impact of genetic testing on access to affordable health insurance is particularly troublesome in the United States, where there is no government-sponsored health-care system. As an increasing number of cancer- predisposing mutations are identified, differences in the prevalence of highly penetrant risk-conferring mutations are emerging. For example, Caucasian women have been found to have a higher frequency of disease-related *BRCA1* and *BRCA2* mutations than African–American women (Newman et al., 1998). Compared with non-Jewish individuals, genetic testing in American and European Jewish populations has also revealed a higher prevalence of distinct founder mutations in *BRCA1, BRCA2* (Shattuck-Eidens et al., 1997; Diez et al., 1999), and *APC* (Woodage et al., 1998) which predisposes to FAP. The appropriate use of racial/ethnic categories in genetic research has been intensely debated (Burchard et al., 2003; Cooper et al., 2003), particularly with respect to using socially defined categories, such as "African–American" in the context of the United States, as a proxy for ancestry. Once a particular group is identified as having a higher prevalence of risk-conferring genotypes, there are increased concerns about discrimination and stigmatization against individuals and communities (Lehmann et al., 2002). For example, mandatory screening for sickle haemoglobin initially resulted in racial stigmatization and discrimination in insurance and employment against African–Americans, regardless of whether or not they had sickle-cell anaemia (Bowman and Murray, 1990; King, 1992). Moreover, African–Americans were targeted, despite the high frequency of sickle haemoglobin in many other communities (Bowman and Murray, 1990). The same could occur in the context of genetic testing for cancer, for example, in Jewish individuals who have a higher probability of carrying cancer-susceptibility mutations (Diez et al., 1999).

6.3. ETHICAL COMMITTEES/ORGANIZATIONS

Certain ancient texts, such as the Hippocratic Oath, have defined various principles that have impacted on some modern bioethics. The crimes committed during World War II, including those attributed to biomedical research, gave a strong impetus to efforts aimed at ensuring the patient's free informed consent and other ethical and social related issues. Currently there is no committee or organization that deals directly with the ethical issues world wide. However, there are a number of national ethic committees which are engaged in debating the problem of research with human samples and biobanks. These committees follow a number of declarations by various world bodies on the handling of human samples destined for biomedical research. The Universal Declaration of Human Rights in 1948 (http://www.un.org/overview/rights.html) was followed by Helsinki Declaration by the World Medical Association in 1964 and its subsequent declarations at Tokyo in 1975, Venice in 1983, and Hong Kong in 1983. The United Nations Educational, Scientific and Cultural Organization (UNESCO, http://unesdoc.unesco.com) has contributed to the elaboration of this framework by formulating the principles of the Universal Declaration on the Human Genome and Human Rights, adopted by the United Nations General Assembly in 1998 (UNESCO Declaration, 1998). One of the articles within this declaration deals with the issue of free informed consent and the rights of those who undergo research, treatment, or diagnosis on their own genome. These declarations have paved the way to various national ethical committees and organizations, some of which are briefly considered below.

In the United Kingdom, a private body, The Nuffield Council on Bioethics (1995), issued a number of recommendations and guidelines regarding the use of human tissues. These recommendations can be summarized as follows:

- Tissues removed in the course of medical treatment should be distinguished from tissues removed from volunteers.
- The consent to treatment should include consent for disposal, storage, or any other ethically acceptable use of removed tissues. In case of volunteers, the information must be explicit about the intended uses of the tissues and about any risks to the donors.
- Research protocols using human materials should be reviewed by ethical committees, and payments to donors should not act as an inducement and should cover only reasonable expenses.
- Tissue banks should operate on a nonprofit basis and not as commercial organizations. In addition, the Council recommended that a central register of tissue banks approved for supplying human tissue for medical treatment and research should be established by the Department of Health.
- Besides aspects on biosafety in sampling, sorting, and handling human tissues, the Council also recommended that human body parts should not be displayed in connection with public entertainment or art.

The misuse of material of human origin that occurr in various countries has driven the regulatory authorities within the United Kingdom to apply a number of restrictive rules. In 2001, The Medical Research Council (MRC) published a number of guidelines based on the Nuffield report, which state clearly that surplus samples should no longer be considered as discarded but rather as gifts (MRC, 2001). Furthermore, at the end of the year 2003, the British House of Commons introduced a bill on the use and storage of human tissues (The United Kingdom Parliament, 2003).

In the USA, a National Bioethics Advisory Commission (NBAC) was established in 1995. Its main priority is to protect the rights and welfare of human research subjects and to address issues related to the management and use of genetic information. 4 years later this commission issued a report on research involving human biological materials; it contained guidelines on ethical issues and policies (Buchanan, 1998, http.//www.Georgetown.edu/research/nrcbl/nbac).

In December 2003, the French National Assembly has adopted a bill on bioethics. This bill regulates various aspects for sampling and storage of tissues and body fluids of human origin for research purposes. The law prohibits the sale of derivatives of the human body; however, the services associated may be invoiced. Owing to the absence of decrees regulating the preparation procedures for the establishment of biobanks, it is not yet necessary to obtain a license for the collection of human material. According to published decrees, the establishment, the conservation, or the transformation of human tissues or cells for scientific purposes will be subject to a license issued by the concerned authority (Centers de Ressources Biologiques, 2003).

In Germany there is no federal law concerning human tissues or biobanks. The question of tissues and cadaver samples is regulated by the common law. In other words, nationwide rules can be enforced by only criminal law. According to the patrimony law "Erbrecht," the human body is not considered an object during life, so that it is not the part of a heritage. As a consequence, a dead body has no owner in Germany. However, on the question of informed consent, a working party of the Medical Ethics Committees published a document in 2003 stating that a "checklist" for the supplier/patient information should be included in the consent to the use of blood or tissue samples and person-related data for research purposes. This committee emphasized that the checklist must be clearly understood by the potential supplier and should include the following: the title of the research project; issues and aims of the research project; overview of the current state of research; description of the study design; procedures; and summary of potential benefits and risks. The consent form must also contain details regarding the type of human material to be used, the aim and duration of its use, and the extent of anonymity of the sample.

Regarding the European Community, the only binding instrument on bioethics is the Convention for the Protection of Human Rights and the Dignity of the Human Being with regard to the Application of Biology and Medicine, the so-called Convention on Human Rights and Biomedicine adopted in 1997 by the Council of Europe. One of the articles of the Convention declares that the human body and its parts shall not, as such, give rise to financial gain. Another article states, "when in the course of an intervention any part of a human body is removed; it may be stored and used for a purpose other than that for which it was removed, only if this is done in conformity with appropriate information and informed consent procedures." On June 6th 2002, a proposal for a European directive defining quality and security standards for the donation, acquisition, testing, handling, storage, and distribution of human tissues and cells was published. However, this directive deals only with tissues and cells that are to be used in or on the human body, in particular for therapeutic purposes. In 2004, The European Community Commission adopted a directive on setting standards of quality and safety for the donation, procurement, testing, processing, storage, and distribution of human tissues and cells. This directive covers all tissues and cells of human origin for application in the human body, except organs, blood, and blood products. Tissues and cells used for research purposes would be covered when administered to the human body but not when used for in vitro research or in animal models (European Commission, 2004). In addition, member states should encourage the procurement of tissues and cells on nonprofit basis.

Despite the limited number of examples cited above, it is evident that ethics and regulatory realities for the use of human material in biomedical research differ widely among different countries. However, by formulating the principles of the Universal Declaration on the Human Genome and Human Rights that was adopted by the United Nations General assembly in 1998, the UNESCO took a first step toward elaborating an ethical framework on an international level.

In the postgenomic era, there are areas in which ethical issues have attracted the attention of the public, including the scientists working in the field of human biobanks, population screening, and the various phases of oncology studies. These three arguments are briefly considered in the following sections.

6.4. HUMAN BIOBANKS

The term biobank refers to an organized collection of biological samples and associated data. Samples and data within these biobanks can be related to a given individual, a family, or to a group within a given population. The data might be collected at the time of sampling or can be added to the database at a later date. This organization of samples can range in the scope from small collections in academic or hospital settings to large-scale national repositories. We have to bear in mind that biobanking is not a new concept, but the attitude of the society toward these structures can be considered as new and still evolving. This newly acquired status of biobanking can be partially attributed to a number of reasons, including an increasing range of applications requiring samples stored within these biobanks, particularly, in genomics and proteomics research, and an increased number of people who think they might benefit by the services provided by these structures. As a consequence, and in line with the emerging trends in science and society relation, the recognized status of biobanks is dramatically changing. From being clinical or academic research tools that were largely ignored by the general public; they have become a subject of intense social debate. Although there are differences in attitude between various countries regarding biobanks, nevertheless, biobanking throughout the world has raised similar ethical and regulatory issues. Some of these issues can be considered as the direct result of ongoing debate regarding various aspects of biobanking. For example, first, there is a mounting demand for providing adequate information to individuals before giving consent to deposit their samples, as well as a rising awareness of the unforeseen research studies that could be carried out using the samples or their associated data. Second, there is a concrete difficulty of reconciling the noncommercial use of human body parts with the growing role of commercial biobanks. Third, there is still an intense debate on how to protect the rights of individual/group without obstructing the progress of research.

As the number of large-scale human biobanks increases (see Table 6.1), there is an obvious need for clear guidelines and regulations regarding ethical issues and the rights of those people whose samples or data are kept within these structures. This situation is further complicated by the increasing number of private biotech companies involved in biobanks and genetics research.

Currently, within Europe there is no directive or convention that specifically covers the collection, use and storage of DNA samples in biobanks. The absence of common rules at the European level has resulted in considerable variation in the domestic law that applies to the use of DNA samples, personal information, and medical records in the countries across Europe (Gibbons et al., 2005). This means that in the secondary research, the use of biological samples without consent may be allowed under one

TABLE 6.1. Some human biobanks within and outside Europe.

Biobank/location	Funding/description
Icelandic Biobank (Reykjavic, Iceland)	A public–private collaboration with deCode (Reykjavik). Blood samples to be collected from 270,000 Icelandic citizens and linked to Iceland health sector Database and genealogical records.
Estonian Genome Project (Tartu, Estonia)	National, population based, supported initially by private money, since the beginning of 2004, the project has received additional funding from the Estonian goverment. DNA to be extracted from blood samples of 1 million Estonian adults and children together with health and genealogical records.
UK Biobank (Manchester, UK)	Publicly funded by the Wellcome Trust, The Medical Research Council, the UK Department of Health. National, population based. DNA, Medical records and lifestyle questionnaires to be acquired from 500,000 UK adult volunteers between 45 and 69 years of age. To begin in 2006, and subjects to be followed for 30 years.
Latvian Genome Project (Riga, Latvia)	Supported by the Latvian Genome Foundation. Pilot Project initiated in 2002 with 60,000 pilot samples
CARTaGENE (Quebec, Canada)	Supported by public money from Genome Canada. DNA to be extracted and stored from over 50,000 adults in Quebec in the age between 25 and 74 years.
Biobank Japan (Kanagawa, Japan)	Initiated in 2003 with public funding from the Japanese ministry of Education, Culture, Sports, Science and Technology. DNA samples to be acquired from 300,000 Japanese individuals of 20+ years of age suffering from 30 common illnesses.
Medical Biobank (Umeä, Västarbotten, Sweden)	Publicly funded by The Swedish National Healthcare System. Contains over 85,000 DNA samples from individuals of 40, 50, 60 years of age in Västarbotten county together medical records.

jurisdiction, but may be prohibited or restricted to certain types of use in another jurisdiction. This potentially could result in a situation where researchers collaborating across Europe may be operating unlawfully if they share research data and samples across borders where different laws are in operation. Scientists have expressed concern that the current regulatory framework for human biobanks within Europe is inadequate. The most recent initiative has been taken by the Council of Europe, which has established a Working Party to draft an Instrument on Research on Stored Human Biological Materials, which may include the consideration of storage of biological materials for research in the future, such as biobanks. The aim of this protocol is to be far reaching so that biobanks will be considered as a type of biological material. However, it is still not clear how such protocol will deal with the issues raised by genetic information. In addition, such protocol would only apply to the countries that were signatories to the Convention on Human Rights and Biomedicine. This would exclude countries like the United Kingdom (Kaye, 2006). At the international level, the international Human Genome Organization (HUGO), The World Health Organization (WHO), and the Organization for Economic Co-operation and Development (OECD) are developing, or have developed guidelines on biobanks, as have a number of European National Ethics Committees. However, these guidelines are not binding for all European countries, and tend to be statements of principles rather than providing models detailing the standards and procedures as to how data and samples should be stored, protected, and used.

6.4.1. Ethical Issues in Biobanking

Over the past few years large-scale biobanks have had to adapt the ethical frameworks that were developed for smaller biobanks, while retaining the ethical principles themselves. This adaptation has been necessary because of a number of developments. First, genetic testing can use samples that were collected initially for general genetic studies. This raises the issue of a secondary and rather unforeseen uses. In such a case it is difficult to establish who should be consulted, and who should decide? Second, population biobanks that were previously used in epidemiology or in anthropology in an academic context are now of utmost interest to various industries (Cambon-Thomsen et al., 2003). Similarly, collection of tissue biopsies that were of no other use than for individual diagnosis or clinical follow-up are now the source of new information for gene expression studies (Reymond et al., 2002). These and other developments associated with biobanking have underlined the need to address a number of issues, both technical and ethical. In the search for clear guidelines to address ethical issues and to protect the rights of the individual without obstructing research, current debate on biobanks and their future role underlines a number of aspects, which ethical and regulatory organizations have to take into account.

- One of the central questions is how to protect the rights of people whose samples and data are in the biobanks (autonomy, confidentiality, and protection of private life), while allowing and encouraging research?
- Samples that are collected for scientific research can be used in the development of commercial products. For example, some samples may be directly involved in drug development, with the extension of various biotherapies (cell therapies, gene therapies, etc.). This means that future regulations have to ensure that such scenario is less likely to occur.
- How to inform donors correctly when one does not know what possible developments there will be over a period of time. In other words, it is difficult to anticipate secondary uses of the sample at the time of its procurement. The

dimensions of informed consent are complex in this matter and are further discussed below.

- Another evolving argument is how to reconcile the logic that considers samples as part of body elements in some contexts, with that which considers samples as sources of data, more than as a body part, in other contexts.
- How to ensure that there is a maximum quality of sample conservation and management, while allowing easy access, without complications?
- How to optimally and openly use the samples for the rapid progress of knowledge, while protecting the rights of priority of the researchers who constituted the collection, balancing the need for recognition of this activity and the interests of companies?
- How to ensure long-term financial sustainability while the use and the interests involved may vary over time? This was one of the points of concern that came out of a survey on biobanks in several European countries (Hirtzlin et al., 2003).
- How to avoid potential unwanted consequences, in terms of stigmatization of specific groups or misuse of results?

Two elements within the above indications, namely informed consent and privacy, merit further considerations.

- *Informed consent.* Although it does not in itself protect an individual, informed consent allows individuals to exercise their legitimate right to decide whether and how their body, its parts, and associated data will be used in research. However, It has to be emphasized that obtaining an informed consent does not necessarily provide ethical justification for putting research subjects or patients at risk. To address some of the limitations of older guidelines on the presumed consent to all possible future uses of samples or data, new ethical frameworks have been created. In this updated frameworks, consent increasingly includes several alternative modalities of involvement for the sample donor. It is commonly agreed that for an informed consent to be meaningful, the participant has to be clearly told about the aim and scope of research and about possible risks and benefits. The World Health Organization (WHO) recommends that the following information be provided in order that the potential donor can make a fully informed decision about whether to agree to his/her samples being used in the proposed research project (WHO, www.who.int/ethics/reproductive-health/hrp/tissue.pdf).
- The nature and amount of the samples to be taken.
- The procedures to be followed to obtain such samples.
- The nature, extent, and duration of any treatment to be provided in the event of complications and injuries related to the procedures used to acquire the samples.
- How the samples will be used in the research?

- The arrangements for disposal of the samples at the end of the research project.
- Who will have access to the samples, and how long will the samples be kept?

Despite the simplicity and clarity of these recommendations, their real life implementation encounters a number of difficulties, which are still to be resolved. For example, the individual might be able to find out how their samples have been or will be used, with the possibility of selectively or definitively opting out of further use of samples or data. Such option, however, is closely linked to how sophisticated is the encoding or encryption system used by a given biobank. For instance, if the samples/data are frequently exchanged then it becomes more difficult to destroy them. In addition, it is sometimes necessary to keep data for follow-up studies, and the use of the original sample as control. In such cases, it is important to inform the subject at the time of consent that withdrawal at a given date will clearly mean that no further data will be generated, and that the remaining sample will be destroyed. The subject has also to be informed that the destruction of the sample does not guarantee the destruction of existing data. The scenario of professional/donor and the question of consent are further complicated by a phenomenon designated "consent fatigue," where some individuals do not wish to receive extensive information (Hoeyer, 2003; Knoppers, 2003).

Informed consent is of course not only limited to individuals, but can also involve groups of people, particularly in population genetics. The intense debate regarding the Human Genome Diversity Project resulted in ethical framework that took into account individual rights as well as cultural differences among communities or groups (Weiss, 1997). This ethical framework for population genetics has been formalized and adapted to large population-genetic studies involving large-scale biobanking (Chadwick and Berg, 2001; Knoppers, 2003). Basically, collective consent is used in small communities but is not adapted to large heterogeneous groups. Instead, a collective debate is held before a project begins and before individual consents are pursued. Within the overall framework, a person can then take a decision that remains at the individual level, but takes into account the dimensions underlined by this collective debate.

It can be argued that the question of secondary uses of stored human samples remains one of the controversial aspect in biobanking (Wendler and Emanuel, 2002, Godard et al., 2003). This is simply because of the impossibility of foreseeing all possible uses at the time of the initial consent by the individual. The main ethical issues are related to the completeness of information given, the need to obtain a new individual consent for each use, and who is going to decide on the issue. Several views have been expressed regarding this issue, which range from denying any use, other than that initially stated, to more flexible attitudes. The latter takes into account a number of considerations: the traceability of the individual identity, the type of further uses that are envisaged in relation to the original objective, the implications of secondary research for the individual, how precisely the use was described at the time of sampling, and the kind of consent that was originally granted.

Privacy. The question of privacy at the individual and/or group level is one of the issues associated with biobanking, which is still to be resolved through binding guidelines and regulations. This issue raises a passionate and justifiable debate because it can lead to unfair discrimination against an individual by a present or potential employer, insurer, or other party on the basis of genetic data derived from a scientific analysis of the individual's biological samples. Therefore, it is necessary to give particular attention to the confidentiality implications of collecting and storing human samples to be used in the research context. In the USA, medical confidentiality rules derive from, and may be enforced under common law and any applicable regulations within a given state. However, the chief American source of pertinent legal authority is the federal Standards for Privacy of Individually Identifiable Health Information (Privacy Rule). This regulation became effective in 2003 after being promulgated by DHHS under the legislative authority of the Health Insurance Portability and Accountability Act (HIPAA) (Annas, 1992). The Privacy Rule requires specific written permission from a patient before anyone may use or disclose "protected health information" (PHI) about that person for nonroutine purposes such as research. These protected information are defined as any "individually identifiable health information" transmitted or maintained by a "covered entity" (e.g., a health care provider, health insurance plan, or data processing firm). In the context of research using stored tissue specimens, the application of the HIPAA Privacy Rule can encounter various difficulties in its application. For example, tissue repositories such as biobanks are not considered covered entitiy subject to the Rule unless they conduct some other kind of activity that brings them within the "covered entity" definition. The Privacy Rule would be implicated only with regard to a covered entity, such as a hospital that discloses PHI for the creation or stocking of the research repository (United States Department of Health and Human Resources, 2004).

An extensive overview of the international situation on protection of genetic data has been published by the British Human Genetics Commission (Crosbie, 2004). This document does not contain clear distinctions between health information and other personal information. Furthermore, the indicated legislations do not seem to distinguish between genetic and other health-related information. In other words, such data protection legislation does not tackle specifically ways to protect personal genetic data; rather, it is general data protection legislation that may nevertheless apply to the collection, storage, and use of personal genetic data.

6.5. LARGE POPULATION SCREENING

A criterion to evaluate the opportunity of conducting population screening had been developed almost 40 years ago (Wilson and Junger, 1968). This criterion was developed when screening primarily focused on detecting early stages or precursors of chronic diseases. With the introduction of testing for genetic susceptibility, particularly for cancer, such criteria had to be refined and expanded. Wald (2001) outlined a number of elements central to population screening, including the identification of persons likely to be at high risk for a specific disorder so that further testing can be done and preventive actions taken; outreach populations that have not sought medical attention for the condition; follow-up and intervene to benefit the screened persons. Existing literature on large population screening seems to underline a simple role, which can be summarized in the following sentence. For a screening test to be applicable to large populations, it has to be inexpensive, reliable, and above all acceptable. Before discussing some of the ethical issues associated with population screening, it would be helpful to consider some established screening protocols relevant to some types of cancer. These considerations are mainly based on the guidelines, which are published annually by the American Cancer Society (ACS).

Each January, the American Cancer Society (ACS) publishes a summary of its recommendations for early cancer detection, including guidelines update, emerging issues that are relevant to screening for cancer, and a summary of the most current data on cancer screening rates for US adults (Smith et al., 2006). Guidelines on the screening for three types of cancer are considered in the following sections.

6.5.1. Screening for Colorectal Cancer

Colorectal cancer is a common condition with a known premalignant lesion (adenoma). There is a relatively long time course for malignant transformation from adenoma to carcinoma and the outcomes are markedly improved by early detection of adenomas and early cancers. Thus, there is great potential to reduce the mortality from this disease by detecting adenomas and early cancers through organized screening. Generally speaking, such screening takes into account the following elements: (a) It is widely accepted that the vast majority of colorectal cancers result from malignant change in polyps (adenomas) occurring in the lining of the bowel 10–15 years before malignant change occurs. The best available evidence suggests that only 10% of 1-cm adenomas undergo malignant change after 10 years. (b) The incidence of adenomatous polyps in the colon increases with age, and although adenomatous polyps can be identified in up to 20% of the population, most of these are small and unlikely to undergo malignant change. The vast majority (90%) of adenomas can be removed at colonoscopy, obviating the need for surgery.

There are a number of screening tests, which have been tried for the early detection of colorectal caner, some of which are briefly considered here. Perhaps the simplest and least expensive is symptom questionnaire, but this has proved predictably insensitive and becomes reliable only when the tumor is relatively advanced. Both digital rectal examination and rigid sigmoidoscopy suffer from the limitation that they only detect rectal or rectosigmoid cancers that are unpleasant and invasive. Colonoscopy is considered the gold standard technique for examination of the colon and rectum. However, besides being expensive, the need for full bowel preparation, sedation, and the small risk of colonic perforation make it unacceptable for population screening. Colonoscopy is, however, the method of choice for screening high-risk patients (hereditary nonpolyposis colorectal cancer and patients with long-standing ulcerative colitis). Since its inception over 10 years ago (Vining et al., 1994), CT colography (virtual colonoscopy) has emerged as one of the most promising techniques for colorectal evaluation. Although a variety of terms and scanning techniques have been used, the basic imaging principles remain the same. In this imaging technique, a thin section, helical computed tomography (CT), is used to generate high-resolution, two-dimensional axial images. Three–dimensional images of the colon, simulating those obtained with conventional colonoscopy, are then reconstructed off-line (Vining et al., 1997). Although it requires full bowel preparation, very expensive CT scanners, and computing facilities, it is minimally invasive, and views of the whole colon can be obtained within 5 min. Data accumulated so far suggest that this technique has a similar sensitivity for large polyps and cancers to colonoscopy or barium enema. Till now, there are no published trials of CT colography in population screening, but the available information suggest that CT colography has the potential to be cost-effective and to reduce the need for colonoscopy in population screening.

ACS guidelines for screening and surveillance for the early detection of adenomatous polyps and colorectal cancer were updated in 2001, and the recommendations for stool blood testing were modified in 2003 by adding fecal immunochemical tests (Smith et al., 2001; Levin et al., 2003). These updated guidelines indicate a number of options for colorectal screening, which may be chosen on the basis of individual risk, personal preference, and accessibility. The ACS recommends that average-risk adults begin colorectal cancer screening at the age of 50, with one of the following options: (a) annual fecal occult blood test (FOBT) or fecal immunochemical test (FIT); (b) flexible sigmoidoscopy every 5 years; (c) annual FOBT or FIT, plus flexible sigmoidoscopy every 5 years; (d) double contrast barium enema (DCBE) every 5 years; or (e) colonoscopy every 10 years. Other tests are being currently evaluated in experimental settings, and also available to a limited degree to the public, are stool DNA testing and computed tomography exams of the colon, also referred to as virtual colonoscopy. Although not recommended at this time, the ACS carefully monitors the accumulation of evidence related to these tests (Levin et al., 2003).

6.5.2. Screening for Early Prostate Cancer

ACS guidelines for testing for early prostate cancer detection were last updated in 2001. Because the current evidence about the value of testing for early prostate cancer detection is insufficient to recommend that average-risk men undergo regular screening, the ACS recommendations emphasize the importance of shared decision making about testing (Smith et al., 2001). The ACS recommends that the prostate-specific antigen (PSA) test and digital rectal examination (DRE) should be offered annually beginning at the age of 50 to men who have a life expectancy of at least 10 years, and that a discussion should take place about the potential benefits, limitations, and harms associated with testing. In men for whom DRE is an obstacle to testing, PSA alone is an acceptable alternative. In the course of deliberating about the guidelines, the ACS Advisory Committee placed strong emphasis on shared decision making between clinicians and patients, stressing that, just as a clinical policy of directly recommending testing is inappropriate, a clinical policy of not offering testing or discouraging testing in men who request early prostate cancer detection tests likewise is inappropriate. In addition, the Advisory Committee also concluded that if men ask the clinician to make the testing decision on their behalf following a discussion about benefits, limitations, and risks associated with prostate cancer testing, they should be tested. Common men at high risk, including men of sub-Saharan African descent and men with a first-degree relative diagnosed before at a younger age (i.e., <65 years), should begin to undergo tests at the age of 45. Men at even higher risk of prostate cancer, due to more than one first-degree relative diagnosed with prostate cancer before the age of 65, could begin to undergo tests at the age of 40, although if PSA is less than 1.0 ng/mL, no additional testing is needed until the age of 45. On the contrary, if PSA is greater than 1.0 ng/mL but less than 2.5 ng/mL, annual testing is recommended. If PSA is 2.5 ng/mL or greater, further evaluation with biopsy should be considered. These guidelines emphasize that informed decision making is no less important for men at high risk. These recommendations for testing do not exclude the need for testing decisions to be preceded by a process of informed decision making. For example, men at high risk should have an opportunity to learn about the potential benefits, limitations, and harms associated with testing for early detection and treatment of early-stage prostate cancer, so that they can make an informed decision with the assistance of a health-care professional. Because PSA is prostate-tissue specific and not prostate-cancer specific, there is no absolute value that is applicable to all men. The range of normal PSA levels has been conventionally considered to be between 0.0 and 4.0 ng/mL, although as Thompson and colleagues have recently shown (Thompson et al., 2005) that there is no cutoff level of PSA at which prostate cancer is not present, but rather a continuum of risk at all levels of PSA values. Levels of PSA less than 4.0 ng/mL, not only increase the sensitivity of PSA, but also significantly diminish specificity, with the accompanying attendant costs and potential harms associated with false positives. Although there is considerable interest in the eventual publication of end results from two contemporaneous randomized trials of prostate cancer screening in the United States and Europe, resolving other challenges related to prostate cancer screening should also be a high priority (Gohagan et al., 1995; de Koning et al., 2002). These include the need for tests that are more effective at measuring the degree of progressivity of prostate cancer (thus avoiding overtreatment where possible), and new approaches to therapy that reduce the risks of serious side effects.

6.5.3. Screening for Cervical Cancer

Cervical cancer provides an excellent example of the benefit of early detection, and subsequent treatment, in reducing the burden of cancer. Experience with this type of cancer has also illustrated the potential of using molecular tests to enhance accuracy and allow the dissemination of early detection. The reading of Pap smears requires expensive cytotechnolgy together with expensive computers that are beyond the capacity of many poor countries.

ACS guidelines for cervical cancer screening were last updated in 2002. The guidelines reflect the current understanding of the underlying epidemiology of cervical intraepithelial neoplasia (CIN), in particular the causal role of human papillomavirus (HPV), and recommends varying surveillance strategies based on a woman's age, her screening history, and the screening and diagnostic technologies she chooses. ACS recommends that cervical cancer screening should begin approximately 3 years after the onset of vaginal intercourse, but no later than the age of 21. Annual screening with conventional cervical cytology smears, or biennial screening using liquid-based cytology, is recommended until the age of 30. At or after the age of 30, a woman, who has had three consecutive, technically satisfactory normal/negative cytology results, may undergo screening every 2–3 years using either conventional or liquid-based cytology. Alternatively, after the age of 30, women who have the same history of normal cytology results may undergo HPV DNA testing with conventional or liquid-based cytology every 3 years. Women who choose to undergo HPV DNA testing should receive counselling and education about HPV and HPV testing. Specifically, the women should be informed about a number of aspects related to such testing: (a) a positive HPV test result does not reflect the presence of a sexually transmitted disease, but rather a sexually acquired infection; (b) almost everyone who has had sexual intercourse has been exposed to HPV and that the infection is very common; and (c) HPV infection usually is not detectable or harmful. Most important, testing positive for HPV neither indicates the presence of cancer, nor will the large majority of women who test positive for an HPV infection develop advanced cervical neoplasia. Average risk women aged 70 and older with an intact cervix may choose to cease cervical cancer screening if they have had no abnormal/positive cytology tests within the 10-year period before the age of 70, and if there is documentation that their three most recent consecutive exams were technically satisfactory and interpreted as normal. However, screening after the age of 70 is recommended for women who have not been previously screened, for whom information about previous screening is unavailable, or for whom past screening is unlikely. The update of the guidelines also addressed screening for cervical cancer in women for whom additional guidance is relevant, including women at higher risk and women who have undergone hysterectomy. Women with a history of cervical cancer or in utero exposure to diethylstilbestrol (DES) should follow the same guidelines that are followed by average risk women before the age of 30, but should continue with that protocol after the age of 30. Women who are immunocompromised by organ transplantation, chemotherapy, chronic corticosteriod treatment, or who are HIV+ should follow US Public Health Service (USPHS) and Infectious Disease Society of America (IDSA) guidelines, which state that they should be tested twice during the first year after diagnosis, and annually thereafter. Furthermore, there is no specific age to stop screening for women with a history of cervical cancer, women with in utero exposure to DES, and women who are immunocompromised (including HIV-positive patients). Women in these risk groups should continue cervical cancer screening for as long as they are in reasonably good health and would benefit from early detection and treatment. Screening with the Pap test is not indicated for women who have had a total hysterectomy, including removal of the cervix, for benign gynaecologic disease.

6.6. GENETIC TESTING FOR CANCER SUSCEPTIBILITY

The full impact of the newly acquired knowledge in the proteomic and genetic fields on the various aspects of cancer is yet to be realized and remains uncertain.

However, it is anticipated that genetic testing for cancer susceptibility could eventually allow physicians to identify individuals who are susceptible to certain types of cancer, and thereby allow them to tailor preventive and therapeutic modalities based on the phenotype information of the individual patient. There are many reasons why potential candidates might decline genetic testing for cancer susceptibility. These include the uncertainty that is associated with positive test results, psychological distress, concerns about family stress, lack of health insurance, and concerns regarding potential discrimination (Lerman et al., 1999; Hadley et al., 2003). It is also interesting to note that relatively few individuals who received genetic testing for breast cancer susceptibility reported that they sought testing as a result of a physician recommendation (Armstrong et al., 2003). This might be in part due to the fact that many physicians are not adequately prepared to recognize familial cancer syndromes or to make appropriate referrals (Sweet et al., 2002). To clarify some of the concerns and anxieties of potential participants in genetic testing, I find it useful to consider the case of genetic testing for HBOC. This is because much of our initial understanding of interest and utilization of predictive genetic testing comes from studies that have offered *BRCA1* and *BRCA2* testing to persons in HBOC families (Biesecker et al., 1993). Other studies have also shown that members of HBOC families who receive education and counselling, more often than not, choose to be tested (Lerman et al., 1996). In addition, testing for germline mutations in *BRCA1* and *BRCA2* genes to assess breast-ovarian cancer susceptibility can be considered a representative case of genetic testing, which can be used to underline ethical issues associated with this kind of testing. Before considering the ethical issues associated with this test, the following general considerations are of interest to the argument.

- Twelve years ago, Mark Skolnick and his colleagues at Myriad Genetics in Salt Lake City (Utah, USA) announced that they had cloned the *BRCA1* gene. This gene *had* been named 3 years earlier by Mary-Claire King when she and her group assigned it to chromosome 17 by linkage analysis, using a large group of families with cases of early-onset breast cancer (Hall et al., 1990). The identification of truncating mutations in the coding sequence of *BRCA1* in families with multiple cases of breast cancer was reported by Miki et al. (1994).
- The discovery that families with a high incidence of male breast cancer were found not to carry *BRCA1* mutations3 has led to the search for other breast cancer genes. *BRCA2* was linked to chromosome 13 in 1994 (Wooster et al., 1994) and was cloned only a year later by the same group. Genetic testing for cancer susceptibility quickly followed.
- In the 12 years since the discovery of *BRCA1* and *BRCA2*, genetic testing for breast and ovarian cancer susceptibility has become integrated into the practice of clinical oncology. The risk of developing cancer is not identical for all carriers of *BRCA1* and *BRCA2* mutations. Risk can be influenced by allelic heterogeneity, modifier genes, and environmental and hormonal cofactors (Narod and Foulkes, 2004).

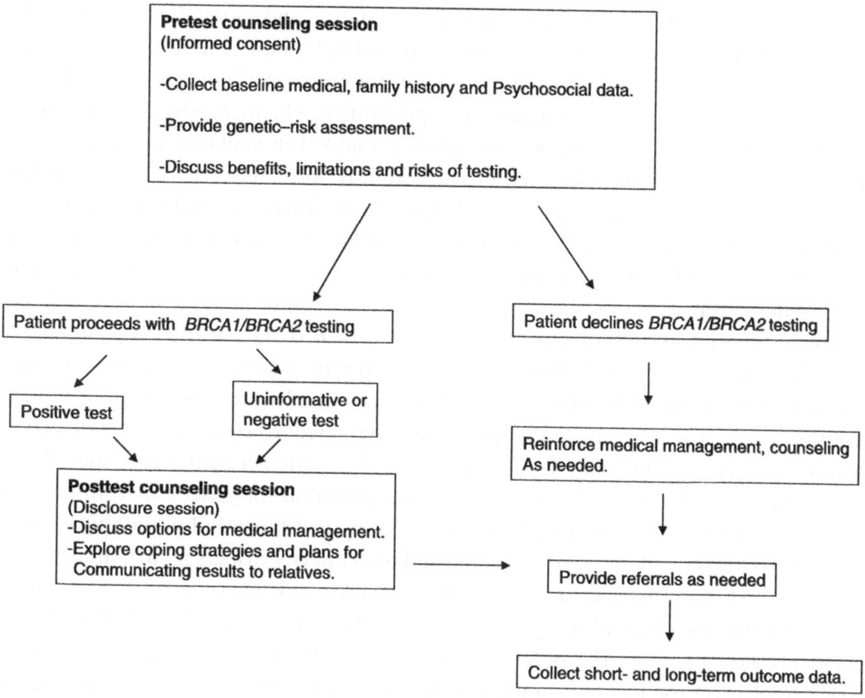

Figure 6.1. Process of genetic testing and counseling for heredity breast and ovarian cancer.

The main steps in genetic testing and counseling for hereditary breast and ovarian cancer (HBOC) have been described by Lerman and Shields (2004), and are schematically represented in Figure 6.1. This scheme shows that before testing an individual's DNA for the *BRCA1* and *BRCA2* mutations that are associated with HBOC, the individual participates in a pretest counseling session to allow the counselor to collect relevant information about the patient. At the same time, the counselor provides information about HBOC and the impact that genetic testing could have on the individual's life. If the individual decides to proceed with the testing, a blood sample will be taken for analysis. The counselor discloses the results of the test at a post-test counseling session, in which medical-management options and coping strategies are discussed. The counselor will provide referrals to oncologists, surgeons, and other specialists if necessary. Despite the potential benefits of this test, there are several negative ethical and social effects which are commonly cited by those who are reluctant to undergo such test. These concerns include breaches of privacy, racial discrimination based on differences in the frequency of risk-conferring alleles, and economic harms associated with job prospects. Some of these ethical and social concerns are considered below.

Discrimination based on race or ethnic ancestry. Race is a contentious topic, as it is an idea that intrudes on the every day life of many people. A debate has recently arisen over the use of racial classification in medicine and biomedical research. In particular with the completion of the first draft of the human genome, some have argued that racial classification may not be useful for biomedical studies, as it reflects a fairly small number of genes that describe appearance. The increasing use of genetic testing, including tests for HBOC, has given an added momentum to the question of race and ethnic origin in biomedical research. Before considering some specific groups, which might be harmed by such test, the following considerations are relevant. First, although clinical evaluation of the *BRCA1* and *BRCA2* genes has been available since 1996, there has been no clear consensus on what specific personal and family history features should prompt consideration of heredity cancer risk assessment. Second, the spectrum and prevalence of *BRCA1* and *BRCA2* mutations in specific populations, including Ashkenasi, African, and other ancestries, is still unresolved. Unfortunately, once a particular group is identified as having a higher prevalence of risk-conferring genotypes, there are increased concerns about discrimination. A history of misuse of population-based genetic screening in some ethnic groups has fuelled increased concern about the potential for genetic discrimination among other ethnic groups. A representative example of misuse of population-based screening is sickle cell anemia, a condition that primarily affects a particular racial group, the African–Americans. Screening for sickle cell anemia initially resulted in significant discrimination against African–Americans (Markel, 1992). More recently, concerns about the potential for group stigmatization and discrimination have been raised for Icelandic individuals (Chadwick, 1999) and Ashkenazi Jews (Davis, 2000). Ashkenazi Jews have an increased frequency of founder mutations in *BRCA1/2* and therefore have an increased rate of inherited breast and ovarian cancers (Abeliovich et al., 1997; Warner et al., 1999). They have also been found to be at high risk for the APC I1307K allele associated with the development of colon cancer (Coughlin and Miller, 1999; Gryfe et al., 1999). The potential benefits of the recognition of these mutations include increased surveillance and the possibility of preventing disease. However, the social and economic consequences of being labeled "at risk" for certain diseases have led some community leaders to discourage genetic testing that targets the Jewish population. Some community leaders have suggested that Jewish women should not pursue genetic testing for *BRCA1/2* because of stigmatization and discrimination against Jews as a group (Post, 1997).

The above examples tell us that the extraordinary power of genetic research on large populations must be balanced with the potential negative effects of that research on the particular populations studied and the potential advantages to individuals who may benefit from genetic information. Knowledge of a predisposition to a disease may give individuals the opportunity to prevent illness through increased surveillance or medical intervention. However, this benefit to individuals within a particular group must be assessed within the context of possible group stigmatization and discrimination. One way to help in achieving this balance is to obtain particular populations input and consent to research within their community (Lehmann et al., 2002). Genetic researchers attempting to do population-based genetic variation research have considered and attempted to minimize genetic discrimination at the level of the individual and the family (Hudson et al., 1995; Geller et al., 1996).

How researchers achieve protection for entire ethnic or social groups, however, remains elusive (Juengst, 1998). Guidelines have been developed to protect aboriginal communities, but their extension into less-cohesive communities, such as Ashkenazi Jews, may not be feasible (Weijer et al., 1999; Davis, 2000). Some have suggested that the social group itself should be involved in the process of research (Foster et al., 1997). This is the process which was undertaken in Iceland, when Decode Genetics proposed to create an electronic database of the country's health records that could be linked to individual's genotype (Ministry of Health, Iceland, 1998; Duncan, 1999).

Privacy in genetic testing. Protection of the privacy of the individual/group in genetic testing is another issue, which needs to be carefully addressed. This means that privacy laws must provide a system that is strong enough to reassure patients that information generated by genetic testing will not be used to discriminate against them. Although reservations about the use of medical information have been in existence for some time, the dramatic increase in genetic testing has prompted fears that patients will suffer discrimination on the basis of genotype, even when it is weakly predictive of phenotype. These fears will intensify as DNA-based testing for hereditary conditions proliferates. In addition to the voluntary testing of adults, there will be an expansion of mandatory screening of newborns. Mandates were originally enacted by states for non–DNA-based screening for inborn errors of metabolism, but these statutes are now being used to expand mandatory screening into DNA-based testing. The number of DNA-based tests that could potentially become mandatory for newborns is extensive. As a consequence of these trends, medical records are likely to contain more and more genetic information with the potential for use in discriminating against individual persons or groups.

In the United States, the Health Insurance Portability and Accountability issued guidelines, which prohibit insurers from excluding an individual from group coverage, or increasing premiums for an individual in a group plan, as a result of past or present medical conditions including those revealed through genetic testing. However, the same guidelines fail to protect individuals from being denied coverage or being charged higher premiums based on genetic information when they purchase coverage in the individual market. In addition, the same guidelines do not include limits on insurers collecting genetic information or requiring genetic testing of those applying for health insurance of any kind. Several nations, including the United Kingdom and Canada, have recently strengthened privacy protection for health information, but few contain sufficient protections to allay patient's concerns. Although the potential for genetic discrimination in health care is considerably less in nations with a government-sponsored health-care system, such as the United Kingdom or Canada, public concern has prompted government action in other arenas. The British government, for example, has issued a moratorium prohibiting insurers from using genetic test results in setting premiums for certain life insurance, long-term care, and income-protection policies until 2006.

There is no doubt that failure to address patient's privacy concerns will continue to undermine various efforts to integrate genetic testing into cancer prevention and treatment.

6.7. ETHICS IN PHASE I ONCOLOGY TRIALS

Investigations of anticancer drugs follow a well-established path, which starts with in vitro experiments followed by studies in animals, and then there is a preliminary clinical introduction in patients suffering from cancer. On the basis of phase I findings, phase II studies are performed to determine the efficacy of the drug, and if they show a reasonable benefit, phase III trials are conducted to compare the new drug with one or more conventional treatments. A demonstration that the new agent is efficacious and safe in a phase III trial often leads to its approval by the Food and Drug Administration (FDA). Phase I trials are closely scrutinized, and their ethics debated, because of the many unknown factors that participants in the trials face and the critical role the trials have in the development of new cancer treatments. The ethical issues of phase I cancer trials have raised deep concerns, partially because these trials involve potentially vulnerable patients who may experience significant risks with limited chance to benefit. Another concern is that those who participate in such trials have unrealistic expectations about their chance to benefit, despite having participated in the informed consent process. Additional concerns can be caused by the uncertainty regarding risks and benefits of cancer drugs in their first stage of human testing. To uphold the principle, "the interests of the subject must prevail over the interests of science and society" as laid down in the Declaration of Helsinki, the effects of these trials on the life of a participant must be rigorously monitored and evaluated. In addition, there has to be a full adherence to ethical and regulatory guidelines/protocols. In other words, a balance is required to ensure that the rights and needs of an individual participating in a clinical trial are not overridden by the trial process. To achieve this, it is essential that the processes involved, such as recruitment, informed consent, decision making, participation, and follow-up, are above all acceptable to the participant. Studies over the last three decades have underlined two fundamental ethical challenges: the risk/benefit ratio and informed consent. An inherent unfavorable risk benefit ratio is one of the common ethical concern regarding phase I oncology studies. Early meta-analyses of phase I trials of anticancer drugs show an overall response rate of about 5% (Von Hoff and Turner, 1991; Itoh et al., 1994; Bachelot et al., 2000). However, these figures have been contested in more recent reviews (Roberts, et al., 2004; Horstmann et al., 2005). To gain some insight into the ethical aspects related to phase I oncology trials, it would be helpful to consider both early and recent works dealing with this aspect.

6.7.1. Risks and Benefits of Phase I Oncology Trials

Risk/benefit ratio is commonly indicated as the major ethical issue in early-phase oncology trials. These trials enrol patients with advanced cancer whose disease is usually refractory to available treatments. Critics of these trials point out that the enrolment of patients with advanced disease in risky research studies, with little chance of direct benefit, is in effect a form of exploitation of vulnerable population (Annas, 1992). Risks and benefits involved in phase I oncology trials have been assessed by Horstmann et al. (2005). These authors reviewed studies that began between 1991 and 2002 and were sponsored by the Cancer Therapy Evaluation Program of the National Cancer Institute, the major sponsor of phase I oncology trials in the United States. Reflecting the full spectrum of phase I oncology trials, this review included trials of chemotherapeutic agents and newer, targeted agents such as anti-angiogenesis factors, vaccines, and gene therapies; trials of combinations of agents, including some already approved by the Food and Drug Administration (FDA). This comprehensive review analyzed 460 trials involving almost 12,000 participants, all of whom were assessed for toxicity and 10,402 were assessed for response to therapy. These meta-analyses came up with a number of interesting conclusions, which are worth considering: First, different types of phase I oncology studies are associated with very different response rates. For instance, the response rate among patients who were treated with immunomodulators was 13.6%, yet the rate was just 3.0% for patients treated with vaccines. Trials that included one or more FDA-approved anticancer agents showed higher response rates than those involving only investigational agents. This means that using a single response rate to describe a given phase I oncology trial can be rather misleading. Second, in an earlier review it was argued that single-agent trials showed that there was a decrease in tumor-response rates over time which was attributed to the use of newer, more specific agents, and changes in trial design (Roberts, 2004). In the review by Hortsmann et al. (2005), response rates per year varied without a clear pattern. In addition, when these rates were grouped in 3-year intervals, there was a decrease in complete or partial responses from 1991 to 2002 but an increase in the rates of stable disease. Little change in the benefit to participants over time was seen when response rates were grouped with stable disease. Third, Hortsmann and colleagues meticulously reviewed the issue of serious toxicity in the protocols they studied. They found a treatment-related death rate of 0.049% at the most. Considering that virtually all participants have a life-threatening disease and have exhausted the conventional options for treatment, toxicity-related death rate of less than 1% suggest that the safety concerns are reasonably addressed within these trials.

The perceptions of cancer patients who are considering participation in phase I clinical trials have been analyzed by various research groups (Daugherty et al., 1995; Kass et al., 1996; Itoh et al., 1997). Results from these studies have shown that patients have high expectations regarding the benefits of experimental therapy and participate largely in the hope of achieving medical benefit and to help other patients and physicians involved in these trials. In a study by Cox (2000), about 50 adult patients with advanced cancer were interviewed to examine their perceptions of participating in early phase anticancer drug trials. The following section describes some of these perceptions. Of the 55 patients included in this study, 22 had received no prior therapy for their disease, and had invariably been told there was no available treatment for their cancer. On the contrary, the 33 patients who had received prior therapy had, in the majority of cases, been told they were now untreatable because of the recurrence of their disease. The authors argued that this situation meant that prior to the offer of trial treatment, many patients experienced an increasing sense of feeling helpless and distressed. The offer of a trial treatment was, therefore, seen as a turning point for these patients. Thirty patients (54%) described the offer of the trial as being "the light at the end of the tunnel," because of the hope it offered. This included the hope that the trial treatment would be a miraculous cure and that it would be better than current treatments, that the patients might achieve relief of symptoms, and that they would live longer. Just over 70% of the group also described how they had information about the clinical trial presented to them in a positive way, both verbally and in the clinician's manner. In addition, all 55 patients recalled being given written information about the trial they had been offered, with just over half of the patients stating that it acted as a reference point for them when they returned home. The patients, reasons for accepting trial participation included the desire to be in expert hands (54%), to help others (52%; this usually encompassed future cancer patients, cancer research and the doctors who were involved in the research), that they felt they had no choice (36%; essentially if they wanted to live they had to do something), and that they had nothing to lose (23%; because they were going to die anyway). However, despite these thoughts, all but one of the patients interviewed in the study said that they would rather persevere than stop the trial and have to deal with the fact that there was nothing else for them. In addition, the responsibility for stopping trial participation was seen by these patients as resting with the doctor. This and other more recent studies on risk/benefit, informed consent, and perceptions of the participants have underlined a number of elements, which also encompass some ethical issues related to these trials.

- Critics and supporters of phase I trials seem to agree that such trials are crucial for the development of much needed new cancer therapies. It is encouraging to note that over the last 10 years there has been a shift in focus toward more targeted drugs, an improvement of supportive care, and enhanced scrutiny of clinical research. Recent publications suggest that the level of risk experienced by cancer patients who participate in phase I treatment appears to have improved.
- Historically speaking oncology phase I trials can be divided into two periods, before and after 1990. Works published in the earlier period report treatment response in 4–6% of the participants and 0.5% death as a result of toxicity. These data are used by critics when doubts are raised about the poor prospect of benefit and the potential for severe harm. On the contrary, supporters of such trial attribute these poor data to the single-agent approach, which does not take into full account the development of new types of anticancer agents, the combination of agents, more optimized trial designs, and the improvement in supportive care. This argument is supported by the results of the trials conducted over the last 10 years, which showed response rates exceeding by 10%.
- Despite a three-decades experience with phase I trials, there is still no articulated procedure (standard) to determine risk/benefit ratio. According to Agrawal and Emanuel (2003), a possible approach would be to elucidate a standard based on socially accepted determinations of risk/benefit ratios already used by the FDA for the approval of anticancer agents. According to these authors, the risk/benefit ratio for phase I oncology trails is not worse than

risk/benefit ratios used by the FDA as a basis for the approval of many chemotherapeutic agents. This argument is supported by a number of examples. For instance, among patients with newly diagnosed stage I breast cancer, for whom 5-year overall survival is greater than 90%, a 2- or 3-drug chemotherapy regime lasting 4–6 months, with its adverse effects, offers an absolute survival benefit of just 1–2% (Fisher et al., 2001; Lippman and Heys, 2001). Yet the vast majority of women choose to receive such treatment. Another example is related to gemcitabine, which is the FDA-approved treatment of choice for metastatic pancreatic cancer, despite 5.4% response rate, because of its demonstrated quality of life benefits (Burris et al., 1997).

- Consent forms, participant's interviews, and other related documents play an integral part in the process of obtaining informed consent. Subjects gain a better understanding of the trial by a consent form and an interview than by an interview alone (Simes et al., 1986). Such forms serve as guides for discussions between the research team and the subject and as a reference that is always available to the subject (Cox, 2000). Because the subject's signature is recorded in these forms, they thus provide legal and symbolic documentation of an agreement to participate. Consent forms are subject to scrutiny by both regulators and third parties, especially when problems arise. The substantial effort expended by investigators, members of institutional review boards, regulators, and others in writing and reviewing consent forms is indicative of their importance as a necessary source of information for subjects. These efforts are mainly designed to ensure that the participant is sufficiently informed before giving his/her consent. However, critics of these trials argue that the patient consent to participate in phase I oncology trials with likely unfavorable risk/benefit ratios is indicative of deficiencies in disclosure, understanding, and voluntariness in the informed consent process (Lipsett, 1982; Kodish et al., 1992; Itoh et al., 1994). The same critics argue that even if patients are given accurate information and understand it, they are vulnerable, and their judgement is likely to be clouded by their illness. As one critic put it, terminally ill patients who consent to phase I oncology studies have "unrealistic expectations and false hopes" (Cheng et al., 2000).

To argue that a decision to participate in such trials is likely to be influenced by the vulnerability of the participant is one of the aspects of informed consent, which raises some passionate debate. Supporters of these trials doubt such vulnerability by putting forward the following considerations: First, the characteristics of patients enrolled in phase I trials are not consistent with what regulations define as vulnerable population; a large majority are white, over 50% are male, and over 50% are college educated (Daugherty et al., 2000). Second, being terminally ill does not necessarily make the patients part of a vulnerable category. Some terminally ill individuals within a group may lack the capacity to make clear decisions because of the illness or associated medication, but that does not mean that the entire group is inherently vulnerable and unable to guard their interests through informed consent. There are various situations where decisions made by terminally ill individuals are accepted without invoking the element of vulnerability. For example, should the consent of patients for life-saving organ transplants be rejected simply because it is made by terminally ill patients who cannot think clearly? In my opinion, the term vulnerable should be strictly used for individual cases and regardless of the type of illness. The use of this term to describe an entire group of terminally ill patients could be demeaning. Most people with advanced cancer are able to make clear and rational decisions. There will always be some individuals who are unable to give adequate and informed consent, just as is true for people without advanced cancer.

- Over the last 10 years, a combination of factors, including several high-profile deaths of research participants and well-publicized problems with the monitoring of research at leading medical centers, has prompted enhanced levels of vigilance by institutional review boards (IRBs) and an enhanced enforcement of regulations protecting human participants (Steinbrook, 2002a,b). The full impact of these changes on the safety of clinical cancer research has yet to be manifested. In the mean time, changes in the types of cancer drugs under study and in the clinical research environment have made participation in single-agent phase I cancer trials safer, particularly with respect to the probability of experiencing a treatment-related death. This enhanced safety will no doubt impact on some future improvements in the risk to benefit ratio. These anticipated improvements have also to be accompanied by more focused efforts toward more targeted drugs and hopefully a much lower dosing. There is indeed a growing realization that optimal dosing for targeted agents may exist at drug levels well below the maximum tolerated dose. Counter to the experience with cytotoxic chemotherapy, toxicity may not be a prerequisite for optimal antitumor activity for the new cohort of agents.

6.8. INITIATIVES RELEVANT TO BIOMARKERS DISCOVERY

One of the challenges of contemporary life science research is the ability to integrate, interpret, and exchange data generated by multiple research disciplines. It is now well recognized that cancer is a multifaceted disease that presents many challenges to clinicians and researchers searching for more effective ways to combat its devastating effects. Among the central tasks of these efforts is the identification of biomarkers for early diagnosis and classification of tumors, and the definition of targets for more effective therapeutic strategies. So far, very few cancer studies have attempted to integrate data sets that were obtained by different profiling techniques. To facilitate the exchange and integration of data generated by different research groups/techniques, a number of large-scale initiatives at the genomic and proteomic levels have been undertaken. Some of these initiatives and their role in the advancing cancer research are considered below.

6.8.1. The Human Proteome Organization (HUPO)

Analyses of the normal human proteome are important in the search for disease biomarkers. The diversity in the technical approaches to tackle this task and the tremendous amount of data generated by a wide range of research activities highlighted the need for infrastructures with adequate resources to facilitate contact, integration, and exchange of knowledge in this area of research. The HUPO organization (www.HUPO.org.) was launched in 2001 (Hanash, 2003) with the aim to accelerate the development in the field of proteomics and to enhance and organize international collaboration in research and education. In its published mission, HUPO underlines three main objectives:

- Consolidate national and regional organizations into a worldwide organization.
- Engage in scientific and educational activities to encourage the spread of proteomic technologies.
- Disseminate knowledge pertaining to the human proteome and that of model organisms.

Some of the above elements have been already put into practice and have started giving their benefits. This statement can be supported by the following examples.

6.8.2. HUPO Initiative Around Biological Fluids

HUPO currently has one biological fluid project, namely the Plasma Proteome Project (PPP), and two organ system projects, namely the Liver Proteome Project (LPP) and the Brain Proteome Project (BPP). Each of these projects has been the focus of numerous activities. The HUPO PPP project is the first to be implemented. This project has a number of objectives: (i) comprehensive analysis of plasma protein constituents in normal humans in large cohorts of subjects; (ii) determination of the extent of variation in plasma proteins within populations in various countries and across various populations from around the world; and (iii) identification of biological sources of variation within individuals over time and assessment of the effect of age, sex, diet, and lifestyle, as well as common medications and common diseases. The pilot phase of this project was designed with the following aims in mind. (a) Compare advantages and limitations of different technology platforms. (b) Compare reference specimens of human plasma by three different methods of anticoagulation, namely potassium-EDTA, sodium-citrate, lithium-heparin, and serum in terms of number of proteins identified and possible interferences caused by the technology platform. (c) Create a global, open source knowledge base/data repository. Given the central role of reference specimens in these large-scale studies, a committee named the Specimen Collection and Handling Committee (SCHC) was created within the PPP in order to evaluate a number of preanalytical variables that can potentially impact on the outcome of results, but are not related to inherent sample (e.g., patient or donor) differences. These include the choice of sample, type and method of collection, and handling issues. This committee in collaboration with tens of laboratories in various countries came up with a number of clear indications and recommendations to be implemented in the pilot phase, which can be summarized in the following points.

- This initial phase serves to establish the advantages and disadvantages of various technologies, including multiple LC–MS/MS, MALDI–MS, FT–ICR–MS, SELDI–MS, depletion of abundant proteins; fractionation of intact proteins on 2-D gels or with LC or IEF methods, protein enrichment, or labeling methods.
- Plasma specimens should be derived using the three most common anticoagulants, namely potassium- EDTA, lithium-heparin, and sodium-citrate.
- Specimens are to be collected from three ethnic groups: Caucasian–American, Asian–American, and African–American. In addition, the National Institute of Biological Standards and Control (NIBSC, UK) provided a lyophilized plasma specimen, which was compared with the frozen specimens. Donors were tested and determined negative for HIV-1 and HIV-2 antibody, HIV-1 antigen (HIV-1), Hepatitis B surface antigen (HBsAg), Hepatitis B core antigen (anti-HBc), Hepatitis C virus (anti-HCV), HTLV-I/II antibody (anti-HTLV-I/II), and syphilis.
- After specimen collection, the impact of processing time and storage temperature on sample integrity should be investigated. The effect of freeze-thaw cycles and the use of protease inhibitors were deemed important and had to be investigated.
- The HUPO PPP elected to use limited pooled specimens in order to reduce the number of variables that could potentially confound the analysis and comparison of methods. While interindividual variation is also important, this finer discrimination between donors was reserved for later phase studies.

The pilot phase of the PPP resulted in a number of publications (Haab et al., 2005; Omenn et al., 2005; Rai et al., 2005) in which a number of aspects relevant to large-scale human plasma/serum proteomic analyses have been underlined. In addition, the data generated by the 35 participating laboratories have created the ideal opportunity to discuss lessons and future directions in the analyses of the plasma proteome. These discussions have identified a number of aspects, which need more focused investigations: (a) to generate guidelines and standardized operating procedures for specimen collection, handling, archiving, and postarchive processing, including the protease inhibitor issue; (b) to increase the use of high-resolution methods to optimize specific immunoaffinity depletion of abundant proteins with minimal non-target losses; (c) to combine separation platforms and MS capabilities with an aim to expand the portion of the plasma proteome that can be profiled with confidence; (d) to achieve quantitative comparisons across specimens, not just compositional analyses; (e) to achieve high concordance in repeat analyses of the same specimen with the same methods; and (f) to overcome the extremely low overlap between protein identification data sets within a large collaboration of this type and, of course, across the literature, especially addressing the discrepancies due to post-MS/MS spectral analysis and peptide and protein database matching.

6.8.3. Early Detection Research Network (EDRN)

The increasing number of potential protein biomarkers has underlined the necessity to validate and ultimately use them in a clinical setting. Such task can be done better as a collaborative effort among the research communities. The National Cancer Institute (NCI) has taken a lead role in this direction by creating the Early Detection Research Network (EDRN). This network brings together national and international experts from academia and industry to promote biomarker discovery and validation and to help in its translation into clinical practice. The EDRN thus serves as an integrated platform designed to accelerate translation of discovery into tools for early detection and risk assessment. A five-phase criterion for the development and evaluation of biomarkers has been established by this network (Sullivan Pepe et al., 2001). The first phase is a preclinical exploratory phase to help identify promising directions. Next is a clinical assay and validation phase necessary to evaluate the ability of the assay to detect established disease. The third is a retrospective/longitudinal phase to determine the putative biomarker's ability to detect preclinical disease and to define a "screen positive" rule. In the fourth phase, prospective screening is developed to identify the extent and characteristics of disease detected by the test and the false-positive rate. In the last phase, a definitive trial is designed (prospective randomized trial) to determine the impact of screening on reducing the burden of disease in the general population. Different experimental techniques and methodologies are used by the EDRN investigators in their pursuit of novel biomarkers. These approaches are used to evaluate different sources of biomarkers. For example, investigations are in progress on the epigenetic mechanisms of hypermethylation, using a panel of genes as a marker for early disease in lung and other cancers. Activities are also underway to detect membrane and secreted proteins in breast cancer through novel signal sequence trapping approaches. The utility of mitochondrial DNA mutations as markers of early detection of some forms of cancer is also evaluated. There is also an effort to build a two-dimensional gel database of lung-specific proteins from human lung samples to help in the detection of lung cancer. EDRN investigators are also involved in the search for prostate-specific markers, through the use of the SELDI platform.

6.8.4. Other Initiatives

Over the last 3 years, the US National Cancer Institute (NCI) in Bethesda, Maryland have been taking inventory of tissue samples from over 60 cancer centers to determine the need for standard operating procedures in biospecimens collection, storage, and data analysis. Developing such standards would be of great help to scientists engaged in cancer research. The biomedical informatics grid associated with this initiative will serve as an infrastructure that connects resources to enable the sharing of data and tools for cancer research.

Information available about proteins continues to increase at a rapid pace. Protein interactions, expression profiles, and structures are being discovered on a large scale which highlight the need for infrastructures capable of handling the organization and access to such information. The Universal Protein Resource (UniProt) can be considered a representative example of such infrastructures (Wu et al., 2006). The foundation of this resource has been laid by the three UniProt consortium members, The European Bioinformatic Institute (EBI), The Swiss Institute of Bioinformatics (SIB), and Protein Information Resource (PIR). UniProt comprises three database components, each of which addresses a different area in protein bioinformatics. The knowledgebase UniProt archive (UniParc), comprising the manually annotated UniProtKB/SwissProt section and the automatically annotated UniProtKB/TrEMBL section, is the main sequence storehouse; its reference clusters, which condenses sequence information and annotation to facilitate sequence similarity searches and analysis of the results (UniRef). The central component of UniProt database is the UniProtKB, which contains richly annotated protein sequence database with extensive cross-references. This component provides an integrated presentation of disparate data, including annotations such as protein name and function, taxonomy, enzyme specific information (catalytic activity, cofactors, metabolic pathway, and regulatory mechanisms), domains and sites, posttranslational modifications, and others. Literature citations provide evidence for experimental data. The UniProtKB contains two sections: UniProtKB/Swiss Prot contain records with full manual annotation or computer-assisted, manually-verified annotation performed by biologists and based on published literature and sequence analysis; UniProtKB/TrEMBL contains records with computationally generated annotation and large-scale functional characterization. UniProt database entries are available for searching, browsing, and retrieval from the Web site (http://www.uniprot.org). More detailed description of this resource has been given by Wu et al. (2006).

The National Center for Biotechnology Information (NCBI) at the National Institutes of Health was created in 1988 with the main goal to develop information systems for molecular biology. In addition to maintaining the GenBank nucleic acid sequence database (Benson et al., 2003), to which data are submitted by the scientific community, NCBI provides data retrieval systems and computational resources for the analysis of GenBank data and a variety of other biological data. The data resources within NCBI have six broad categories, which are available from the NCBI home page http//www.ncbi. nlm.nih.gov and have been described in some details by Wheeler et al. (2006).

6.9. GENOMIC INITIATIVES/RESOURCES

To realize the goal of generating genomics-based cancer-orientated databases, several research groups are participating in various programmes using DNA sequencing as a common platform. As DNA sequencing generates digital data, this technology facilitates the display of data to a wide audience. Currently, there is an array of resources, in the form of databases, tools, and programmes to allow global profiling data and other types of data to be integrated. In addition, there is an expanding number of activities, which contribute to human gene discovery, identify cancer-causing mutations in the genome, and define changes in the transcripts of normal and cancer cells. Some of these large-scale programmes, which are contributing to the unification of genome-wide studies, are briefly discussed.

6.9.1. The Cancer Genome Anatomy Project (CGAP)

Since its inception in 1997, the goal of this project has been to build an interface of genomic and cancer research through the development of information-based resources and physical resources. Overall, the CGAP approach has been to generate data sets without attempting to assign biological function, and to make this information accessible to the community for biological analysis. One of the early and continuing goals of this project has been to build a catalog of genes that are expressed in a wide range of cancer cells as well as in their normal counterparts. A significant emphasis has been placed by the CGAP on producing expressed sequence tags (ESTs) that are derived from cDNA libraries generated from a wide range of cancerous and normal tissues (Strausberg et al., 2000, 2003). As the CGAP is advanced, newer technologies have become available that provide for even higher-throughput sequence-based gene-expression analysis. The CGAP has extensively used the SAGE approach, which allows quantitative analyses of the expressed gene content of cells. Sequence analysis of gene expression (SAGE) is capable of sequencing short tags (usually 14 nucleotides) (Velculescu et al., 1995). Many of these tags are usually concentrated in a single clone to the extent that 30 or more tags can be obtained from a single DNA sequencing reaction. In addition, because the tags are derived from sequences that are adjacent to a specific restriction endonuclease site, the assignment of tags to particular genes and transcripts is greatly facilitated. This is an obvious advantage compared with ESTs, in which tags can come from different regions of a transcript rendering the assignment of two tags to the same or two different genes, particularly in the case in which complete gene sequences are not available. The CGAP researchers have worked with the National Center for Biotechnology Information (NCBI) to build public SAGE database and SAGEmap. The majority of sequencing tags in the SAGE libraries of CGAP are based on experiments that investigated changes in gene expression because of particular mutations or physiological changes. These data sets have been mined to identify genes that are overexpressed in various types of cancer, including breast, ovarian, brain, and pancreatic (Hough et al., 2000; Porter et al., 2001; Riggins, 2001).

6.9.2. The Human Cancer Genome Project (HCGP)

One of the main objectives of HCGP has been to identify protein-coding regions of the human genome. Such objective is based on the rational that transcript sequences are crucial for identifying genes in the human DNA sequence. Therefore, transcript sequencing can be considered as an essential component of the overall human genome sequencing effort. Data within HCGP were generated with a strategy known as open reading-frame (ORF) EST sequencing (Neto et al., 1997, 2000), which provides central protein-coding portions of genes that are typically not spanned by ESTs that are generated from the extremities of transcripts. This methodology involves the

PCR-mediated generation of internal fragments of expressed genes. Clustered EST data allow the detection of transcript variants that are potentially related to malignancy. CGAP and HCGP data have been extensively mined for the global analysis of variants of known human genes that arise as a result of alternative exon splicing (Mironov et al., 1999; Modrek and Lee, 2002), tissue-specific alternative exon splicing (Xie et al., 2002; Xu et al., 2002), and the alternative splicing forms of genes that are differentially expressed in cancer (Correa et al., 2000). Also, examination of the genome-clustered transcript data has shown that it is still unclear whether the extensive variability at the 3′ ends of transcripts that results from the use of alternative polyadenylation sites is significant to the development of malignancy (Iseli et al., 2002).

The CGAP informatics team has built suites of tools that facilitate easy access to the EST data sets and provide for online in silico data analysis to enable the discovery of genes that might be preferentially expressed in particular cancers (Schaefer et al., 2001; Strausberg et al., 2001, 2002). Many studies have shown the value of having the data sets freely accessible online for creative data mining by the entire community. Examples of published analyses of the genes that are expressed in individual tumor types, based on these transcript data, include those for breast (Leerkes et al., 2002; Mitas et al., 2002), prostate (Olsson et al., 2001; Nelson, 2002), colon (De Young et al., 2002), and oral cancers (Patel et al., 2001), as well as genes that are expressed specifically among several cancers of the reproductive organs (Brinkmann et al., 1998). A particularly promising application of this approach is the identification of potential targets of cancer immunotherapy through tissue-restricted gene expression (Vinals et al., 2001; Scanlan et al., 2002).

6.10. ACHIEVEMENTS AND PERSPECTIVES

The past decade has witnessed a remarkable progress in technologies that researchers can use to study thousands of different molecules in a single experiment. These so-called high-throughput analyses allow a multitude of genes or proteins to be analyzed simultaneously. A number of technologies emanating from gene- and protein-based approaches are being applied to discover and evaluate biomarkers. Intensive research efforts in humans as well as in model organisms has, to a large extent, defined the types of change that must occur for the transformation of a normal cell to one that is cancerous (Hanahan and Weinberg, 2000; van Dyke and Jacks, 2002). These include the molecular changes that underlie growth-signal independence, insensitivity to antigrowth signals, evasion of immunosurveillance, apoptosis evasion, unlimited replicative potential, sustained angiogenesis, tissue invasion, and metastasis. Throughout this text, I have attempted to underline the emerging role of high-throughput technologies and their impact on biomarkers discovery. The present microarray analysis is one of the most common ways to detect changes in gene expression in cancer and normal cells, and is quickly replacing differential PCR-based approaches. This approach allows for rapid surveillance of the expression of tens of thousands of genes in one experiment, and can be used to identify changes in gene-expression patterns in normal and cancer cells, or in different types of cancers. SAGE, alternatively, is a comprehensive cloning and sequencing method that can be used to identify and quantify expression of new genes as well as that of known genes. This technique has demonstrated its potential in analyzing expression patterns of low-copy-number genes. Microchip-based approaches are also being developed to identify methylated regions of gene promoters, and gene-specific chips are available that can be used to screen a large number of samples for mutations in oncogenes.

Following a decade of intensive research powered by tremendous investments from genomics companies and government projects racing to clone the human genome, attention has turned to proteomics for the next wave of innovation. Despite the success of DNA microarrays in gene expression profiling and mutation mapping, it is the activity of encoded proteins that directly reflect gene function. For example, posttranslational protein modifications, some of which are cancer-cell specific, cannot be determined from genomic information. Proteomic approaches have begun to provide interesting new cancer markers. Instead of searching for a single molecular marker of cancer, proteomics like genomics allows unbiased quantitative analysis of a large number of proteins in normal and malignant tissues, and in cell populations.

Technical advancements and newly acquired knowledge in cancer research have contributed to what can be described as an emerging approach to biomarkers discovery. This new approach is the result of a number of elements, which can be roughly summarized in the following points:

- A clear shift toward the use of gene/protein patterns rather than the use of a single gene/protein as potential biomarkers. This shift has been facilitated by the availability of parallel rather than serial genomic and proteomic analyses.
- More emphasis on an integrative approach in data analyses. Such emphasis has been underlined by the realization that complex biological processes, such as cancer initiation and progression, require the consideration of differential gene expression in the context of complex molecular networks. The generation of protein–protein networks "interactome" for model organisms and preliminary human protein networks has given a significant impulse to this integrative approach in data analyses.
- An increasing use of array technology in genomic and proteomic research, together with an enhanced knowledge of the molecular mechanisms of cancer, has resulted in a clear shift toward molecular biomarkers. This shift has been accompanied by claims that molecular biomarkers are set to revolutionize the process of prognosis and diagnosis of cancer.
- The increasing use of samples of human origin in proteomic and genomic research underlined the need for clear policies and guidelines on the use of such samples. The emerging trend is how to protect the donor without creating obstacles to efforts aimed at discovering new and safer therapies for a wide range of disease.

Most of the above points have been discussed in various parts of this book; however, at the end of this text I find it relevant to give further considerations to two of the arguments mentioned above.

6.10.1. Molecular Biomarkers

Available techniques to perform high-throughput analysis in genomics and proteomics has led to an ability to conduct discovery-based research, in which large quantities of data can be analyzed without a hypothesis, to search for patterns that usefully discriminate between groups of persons with different diagnosis, prognosis, or response to therapy. These high-throughput technologies have accelerated the discovery of potential biomarkers for various types of cancer, including molecular biomarkers. These molecular biomarkers can bring a change in protein expression, structure, or function; alterations in gene sequences; and expression levels. In addition, DNA-based markers of cancer include mutations, loss of heterozygosity, microsatellite instability, DNA methylation, mitochondrial DNA mutations, and detection of viral DNA.

Although many papers are published each year describing genetic mutations or alterations in expression levels that are associated with various types of cancers, very few of these are developed into reliable molecular markers that can be used routinely in the clinical setting. Our increasing knowledge about cancer initiation and progression indicates that as an epithelial tumor grows, cancer cells are sloughed off the organ epithelium into body fluids such as blood plasma, urine, or saliva. This makes it possible to detect molecular markers such as DNA mutations, methylation patterns, or microsatellite instability in these samples before they are symptomatic. There are several different types of cancer markers that can be detected in serum, urine, or saliva samples. DNA can be analyzed for changes in gene copy number, chromosome translocations, deletions, or loss of heterozygosity. Mitochondrial DNA can also be analyzed for mutations. RNA can be analyzed for expression levels or point mutations, and proteins can be analyzed for structural alterations, changes in enzymatic activity, localization, or expression patterns. One area to which molecular biomarkers are expected to make a substantial contribution is our capability to distinguish between subclasses of the same type of cancer. There is a growing list of investigations along these lines, some of which are considered below.

Squamous cell carcinoma of the cervix (SCCC) is the second most common malignancy in women worldwide (Parkin, 2001). This form of cervical cancer is by far the most common histologic type of cervical cancer. The Pap test is considered to be the most cost-effective cancer-screening test developed to date (Greenberg et al., 1995). This test has dramatically decreased the incidence and mortality rates of cervical cancer by more than 70%, since it was introduced in the United States and many other countries of the world (Eddy, 1990). However, where no Pap screening programs are in place or where a population does not participate in screening programs, the incidence and mortality of the disease remains high. A limitation of the Pap test is that it is morphologically based, and its accuracy can be problematic because of preanalytical processing and interpretive errors. Variation in the reading and classification of the cytologic smears is another drawback of this test. Molecular-based testing for high-risk human papillomavirus (HPV) strains is mostly performed when Pap tests are inconclusive and is generally used in conjunction with liquid-based cytologic methods. These tests are still being investigated in larger studies to further determine their usefulness (Bulkmans et al., 2004). Current guidelines for managing patients with atypical squamous cells call for these cases to be assigned Pap subcategories that distinguish the cases that have a high risk for invasive carcinoma from the cases of undetermined significance (Wright et al., 2002). A molecular test based on multiple diagnostic markers that are associated with the cancer phenotype could potentially identify SCCC with higher specificity than currently available tests. Furthermore, the identification of a subset of the genes expressed in SCCC could be helpful in subcategory assignment.

In a relatively recent study, a panel of genes transcriptionaly upregulated in SCCC were identified by representational difference analysis of cDNA and validated by real-time quantitative reverse transcription-PCR (Sgarlato et al., 2005). This study identified a candidate pool of 65 transcript fragments upregulated in diseased tissue compared with normal tissue. Forty-one transcripts were found to be upregulated in diseased tissues compared with normal tissue in at least one half of the patients by macroarray hybridization. Eleven of those genes were selected for real-time quantitative reverse transcription- PCR analysis, and all were confirmed as transcriptionally upregulated in cancer tissues compared with normal tissues in at least one half of the investigated patients.

In an earlier study, Chen et al. (2003) used cDNA microarrays containing more than 30,000 Unigene clones to examine the gene expression patterns of 34 cervical tissues from different clinically defined stages. The authors reported that global gene expression patterns could separate normal cervical tissues and low-grade squamous intraepithelial lesions from cervical cancers and most of the high-grade squamous intraepithelial lesions (HSILs). Among the top 62 genes (expressed sequence tags) that were overexpressed in tumors and HSIL tissues, 35 were confirmed using in situ hybridization on cervical tissue micorarrays. Many of these genes were overexpressed in high-grade dysplastic and malignant cervical epithelium or in stroma adjacent to the diseased tissues, with cellular proliferation and extracellular matrix-associated genes being the most common. In addition, the extent of gene overexpression was found to increase as the lesions progressed from low-grade squamous intraepithelial lesions to HSILs and finally to cancer.

As it has been pointed out earlier in this text, the prevailing model of metastasis holds that most primary tumor cells have low metastatic potential, but rare cells (estimated at less than one in ten million) within large primary tumors acquire metastatic capacity through somatic mutation (Post and Fidler, 1980). The metastatic phenotype includes the ability to migrate from the primary tumor, survive in blood or lymphatic circulation, invade distant tissues, and establish distant metastatic nodules. This model is primarily supported by animal models in which poorly metastatic cell lines can spawn highly metastatic variants if the process is facilitated by the isolation of rare metastatic nodules, expansion of the cells in vitro, and injection of these selected cells into secondary recipient mice (Fidler and Kripke, 1977; Clark et al., 2000). No direct evidence of this genetic selection model has, however, been documented in human tumors. To study the molecular nature of metastasis, Ramaswamy et al. (2003) analyzed the gene-expression profiles of 12 metastatic adenocarcinoma nodules of diverse origin (lung, breast, prostate, colorectal, uterus, ovary) and compared them with the expression profiles of 64 primary adenocarcinomas representing the same spectrum of tumor types obtained from different individuals. This comparison identified an expression pattern of 128 genes that best distinguished primary and metastatic adenocarcinomas. The interesting element in this study from a biological standpoint is the attempt to look for common gene signatures across data sets and cancer types. This study also demonstrated that the metastatic signature was expressed in a subset of the primary tumors analyzed, leading the authors to hypothesize that such signature may represent a metastatic program that is encoded in primary tumors destined for metastasis. This hypothesis was supported by further analyses to predict time to metastasis in several independent solid tumor data sets for different types of cancer. In all of the data sets analyzed, the metastatic signature was substantially associated with clinical outcome and metastatic conditions, suggesting that metastatic potential is encoded in the primary tumors and is at least encoded in part by a common signature across tumor types. Another example of this type of analyses has been provided by Rhodes et al. (2004). Forty independent data sets from more than 3700 array experiments have been used to generate 36 cancer signatures representing genes activated in a particular cancer type relative to the normal tissue type from which it arose. Analysis of these signatures identified 67 genes activated in twelve or more of the identified signatures. The authors reported that this 67-gene signature could predict cancer versus normal status in most of the cancer signatures tested, as well as in independent cancer signatures, some of which represented cancer types not included in the original analysis.

The above examples and others within this text is an indication that molecular biomarkers hold a great promise for enhancing our ability to establish early diagnosis, prognosis, and subclassification of some types of human cancer. The same class of biomarkers will also contribute to a better prediction of response to therapy and metastasis. This optimistic view does not mean that future challenges facing this class are lesser or easier than those facing other classes of biomarkers. As it has been pointed out, one challenge in assessing research about molecular markers is that the rules of validation for studies of diagnosis and prognosis are not as well developed as for studies of therapy. Therapy is commonly assessed by the experimental method, such as the randomized controlled clinical trial that handles threats to validity arising from problems of heterogeneity, complexity, and bias. Furthermore, the role of the Food and Drug Administration (FDA) and other approval bodies, by overseeing the process of drug development and marketing, provide a way for arbitrating results and an incentive to refine methodology and rules. By contrast, research about diagnosis and prognosis is often conducted using nonexperimental methods of observational epidemiology or clinical epidemiology that are less well developed (Sullivan Pepe, et al. 2001; Ransohoff, 2002) and are not closely overseen by regulatory bodies. Although there is a rough hierarchy of steps by which diagnostic tests can be evaluated (Sullivan Pepe, et al. 2001; Deyo and Jarvik, 2003), the process of validation of these tests is not as well developed or well accepted as for studies of therapy. In other words, it would be useful to draw some lessons from the previous generation of biomarkers, where the initial claims of 100% specificity and sensitivity for some forms of cancer were later found to be rather exaggerated.

6.10.2. Integrative Analysis of Cancer

In the short 10-year history of DNA microarrays, hundreds of large-scale experiments have been done, generating global quantitative profiles of gene expression in cancer. Known types and subtypes of cancer have been readily distinguished by their gene-expression patterns, and more importantly, new molecular subtypes of cancer have been discovered that are associated with a host of tumor properties, including the propensity to metastasize and sensitivity or resistance to various therapies. This tremendous amount of gene analyses is currently accompanied by an equally impressive amount of proteomic data generated by a battery of powerful techniques, including protein microarrays, various separation techniques coupled with a variety of mass spectrometry-based approaches, and various forms of chip technology. The unquestionable complexity of cancer initiation and progression renders an integrative interrogation of different data sets a necessity rather than an option. As it has been pointed out earlier in this text, steps in this direction are already at an encouraging point.

One of the challenges on the road to a global profiling of cancer is the need to capture all the elements of the individual compartments that are profiled, such as the whole transcriptome or the whole proteome. Although this is possible for the transcriptome, other compartments, such as the proteome and metabolome, have numerous features that are difficult to capture and require several different profiling approaches. For example, on the proteomic side it is not possible to assay for protein functional activity, establish protein–protein interactions, and assess protein modifications all with the same platform. The urgent need for an integrative approach in cancer profiling can be underlined by the following considerations: (i) The distinct regulation of RNA and protein levels tells us that the integration of data derived from RNA and protein products that are encoded by the same genes can provide valuable functional information about tumors. This concept has been demonstrated by Nishizuka and Charboneau (2003) who analyzed gene-expression patterns of 60 human cancer cell lines (NCI-60) used by the National Cancer Institute to screen compounds for anticancer activity, and measured levels of 52 cancer-related proteins in these cells. The authors reported that clustered image maps of protein levels have identified two markers that could be used to distinguish colon from ovarian adenocarcinomas. Integration of protein and mRNA data led to the interesting observation that the levels of structural proteins were highly correlated with the levels of their corresponding mRNAs in the NCI-60 cell lines, whereas the levels of nonstructural proteins were poorly correlated with those of their corresponding mRNAs. Gene-expression and proteomic data sets from lung tumors have also been compared and integrated, along with serum samples from the same patients. To determine whether gene-expression profiles could be used in prognosis, mRNA profiles in tumors from

86 newly diagnosed patients, including 67 with early-stage and 19 with advanced-stage lung adenocarcinoma, were measured by oligonucleotide microarray analysis (Beer et al., 2002). A gene-expression index, based on expression of the genes that correlated with survival of the 86 patients, was able to identify low- and high-risk groups among the patients with stage I lung adenocarcinomas. The index included many novel genes that were not previously associated with survival in lung adenocarcinoma. (ii) As already mentioned in the previous section, a study that illustrates the merits of data sharing among different research groups is a meta-analysis of cancer microarray data carried out by Rhodes et al. (2004). In their study, 40 published cancer microarray data sets comprising gene-expression measurements from over 3700 tumor samples were collected and analyzed. A common transcriptional profile that is activated in most cancer types, relative to corresponding normal tissues, was delineated from some of the data sets, providing a metasignature of neoplastic transformation.

I cannot end this text without pointing out some of the achievements and new developments, which give us the hope that wining the battle against cancer is only a matter of time.

- On the technology side, today, the expression of thousands of genes can be simultaneously assessed under different conditions, including disease state and treatment. Powerful technologies including polymerase chain reaction (PCR), serial analysis of gene expression (SAGE), single nucleotide polymorphism (SNP) analyses and microarrays can target almost any DNA, RNA, or protein sequence. Different protein microarray formats have been developed, including tissues, living cells, peptides/small molecules, antibody/antigens, proteins, and carbohydrate arrays. The capability of these formats to provide simultaneous assessment of expression/interaction of 100s and even 1000s of proteins can be considered one of the emerging developments, which is paving the way to new and more powerful strategies in biomarkers discovery. Mass spectrometry based methods for proteomic analysis have been improved on various fronts: new generation of mass spectrometers allows higher mass accuracy, higher detection capability, and shorter cycling times, allowing higher throughput and more reliable data. Two-dimensional chromatography coupled to MS/MS is getting more acceptance as a powerful tool for the analysis of complex protein mixtures. Optimized labeling and fractionation approaches are currently used to reach low-abundance proteins in various biological samples.
- The pace at which cancer signatures are generated for various types of cancer, coupled with fast advancement in high-throughput molecular approaches will impact on future efforts to decipher cancer initiation and progression. Some of these efforts associated with gene signatures in cancer have already given some encouraging results. Till very recently, most published studies on marker genes have applied gene-expression profiling to single cancer types. We are already witnessing a shift toward developing multiclass classifiers capable of distinguishing between multiple common human malignancies. This approach holds much promise for the uniform, molecular, and database-driven classification

of all human tumors. The enormous success of microarrays technology in cancer research provided so far valuable information on tumor subclasses, marker genes for diagnosis, and treatment. However, these analyses do not provide possible association with other co-regulated genes. Protein networks will place the genes identified in a broader biological context. While microarray analyses have been used extensively for the identification of marker genes for various types of cancer, there is still an evident paucity of investigations employing integrative analysis of the cancer gene signatures. One of the reasons for such paucity lies in the fact that to understand complex biological processes, such as cancer initiation and progression, it is important to consider differential gene expression in the context of complex molecular networks. The study of such networks requires detailed protein–protein interaction maps. A detailed human interactome network that captures the entire cellular network would be invaluable in interpreting cancer signatures.

- Research efforts over the last two decades have defined the key players in the PI3K-Akt signaling pathway and its importance in various human cancers. The same efforts have also indicated that the true frequency of PI3K-Akt-pathway abnormalities in human cancer is still a challenging task, particularly in the absence of all the factors that may affect its activation. Despite such anticipated difficulties, there are already rare examples on how such abnormalities can be used to distinguish between various types of cancer. A representative case has been described for *PIK3CA*, where mutations of this gene in different forms of cancer were found to be clustered in specific chromosomal regions.
- The past 10 years have seen a substantial advancement in our understanding of the functional consequences of DNA methylation and its interaction with chromatin structure and the transcriptional machinery. Further insight into what causes DNA methylation patterns to undergo changes in cancer cells has also been acquired. From a clinical perspective, DNA methylation changes in cancer represent an attractive therapeutic target as epigenetic alterations are, in principle, more readily reversible than genetic events. However, the great strength of DNA methylation lies in its potential in the area of molecular diagnostics and early detection of various types of cancer.

REFERENCES

Abeliovich, D., Kaduri, I., Lerer, I., et al. (1997) Am. J. Hum. Gent. **60**, 505.

Agrawal, M., Emanuel, E. (2003) JAMA **290**, 1075.

Annas, G. J. (1992) Health Matrix Clevel. **2**, 119.

Armstrong, K., Weiner, J., Weber, B., et al. (2003) Genet. Med. **5**, 92.

Bachelot, T., Ray-Coquard, I., Catimel, G., et al. (2000) Ann. Oncol. **11**, 151.

Beer, D. G., Kardia, S. L. R., Huang, C-C., et al. (2002) Nature Med. **8**, 816.

Benson, D. A., Karsch-Mizrachi, L., Lipman, D. J., et al. (2003) Nucleic Acids Res. **31**, 23.

Biesecker, B. B., Boehnke, M., Calzone, K., et al. (1993) JAMA **269**, 1970.

Bowman, J., Murray, R., Jr. (1990) *Genetic Variation and Disorders in Peoples of African Origin*, John Hopkins University Press, Baltimore.

Brinkmann, U., Vasmatzis, G., Lee, B., et al. (1998) PNAS USA. **95**, 10757.

Buchanan, A. (1998). http.//www.Georgetown.edu/research/nrcbl/nbac.

Bulkmans, N. W., Rozendaal, L., Snijders, P. J., et al. (2004) Int. J. Cancer **110**, 94.

Burchard, E. G., Ziv, E., Coyle, N., et al. (2003) N. Engl. J. Med. **348**, 1170.

Burris, H. A., Moore, M. J., Andersen, J., et al. (1997) J. Clin. Oncol. **15**, 2403.

Chadwick, R. (1999) BMJ. **319**, 441.

Chadwick, R., Berg, K. (2001) Nat. Rev. Gent. **2**, 318.

Clark, E. A., Golub, T. R., Lander, E. S., et al. (2000) Nature **406**, 532.

Chen, Y., Miller, C., Mosher, R., et al. (2003) Cancer Res. **63**, 1927.

Correa, R. G., de Cavalho, A. F., Pinheiro, N. A., et al. (2000) Genomics **65**, 299.

Centers de Ressources Biologiques (2003) (www.crb-france.org).

Cambon-Thomsen, A., Ducoumau, P., Goumrraud, P. A., et al. (2003) Copm. Funct. Genomics **4**, 628.

Coughlin, S. S., Miller, D.S. (1999) Am. J. Prev. Med. **16**, 99.

Cooper, R. S., Kaufman, J. S., Ward, R. (2003) N. Engl. J. Med. **348**, 1166.

Cox, K. (2000) Psycho-Oncology **9**, 314.

Crosbie, D. (2004) Protection of Genetic Information.www.hgc.gov.uk/business_publications_international_regulations.rtf.

Cheng, J. D., Hitt, J., Koczwara, B., et al. (2000) J. Clin. Oncol. **18**, 421.

Davis, D. (2000) Hastings Genet. Rep. **30**, 38.

Daugherty, C. K., Ratain, M., Grochowski, E., et al. (1995) J. Clin. Oncol. **13**, 1062.

Daugherty, C. K., Banik, D. M., Janish, L., et al. (2000) IRB **22**, 6.

de Koning, H. J., Auvinen, A., Berenguer Sanchez, A., et al. (2002) Int. J. Cancer **97**, 237.

De Young, M. P., Damania, H., Scheurte, D., et al. (2002) In Vivo **16**, 239.

Deyo, R. A., Jarvik, J. J. (2003) Ann. Intern. Med. **139**, 950.

Duncan, N. (1999) BMJ **318**, 1096.

Diez, C., et al. (1999) Br. J. Cancer **79**, 1302.

Eamscliffe Research and Commun, 3rd Wave, Ottawa, Canada (2000).

Eddy, D.M. (1990) Ann. Int. Med. **113**, 214.

European Commission (2004) Brussels, Belgium.

Fidler, I. J., Kripke, M. L. (1977) Science **197**, 893.

Fisher, B., Dignam, J., Tan-Chiu, E., et al. (2001) J. Natl. Cancer Inst. **93**, 112.

Foster, M. W. (1997) Nat. Genet. **17**, 277.

Geller, L., Alper, J., Bellings, P., et al. (1996) Sci. Eng. Ethics **29**, 71.

Gibbons, S. M. C., Helgason, H. H., Kaye, J., et al (2005) Eur. J. Health Law **12**, 103.

Godard, B., Schmidtke, J., Cassiman, J-J., et al., (2003) Eur. J. Hum. Gent. **11**(Suppl.2), 88.

Gohagan, J., Prorok, P., Kramer, B. (1995) Cancer **75**, 1869.

Greenberg, M. D., Sedlacek, T. V., Campion, M. J. (1995) Clin. Obstet. Gynecol. **38**, 600.

Gryfe, R., Di Nicola, N., Lal, G., et al. (1999) AM. J. Hum. Gent. **64**, 378.

Haab, B. B., Geierstanger, B. H., Michailidis, G., et al. (2005) Proteomics **5**, 3278.

Hadley, D. W., Jenkins, J., Diamond, E., et al. (2003) Arch. Intern. Med. **163**, 573.

Hall, J. M., Lee, M. K., Newman, B., et al. (1990) Science **250**, 1684.

Hanahan, D., Weinberg, R. A. (2000) Cell **100**, 57.

Hanash, S. (2003) Nature **422**, 233.

Hirtzlin, I., Dubreuil, C., Preaubert, N., et al. (2003) Eur. J. Hum. Genet. **11**, 475.

Hoeyer, K. (2003) New Genet. Soc. **22**, 229.

Horstmann, E., McCabe, M. S., Grochow, L., et al. (2005) N. Engl. J. Med. **352**, 895.

Hough, C. D., Sherman-Baust, C. A., Pizer, E. S., et al. (2000) Cancer Res. **60**, 6281.

Hudson, K., Rothernberg, K., Andrews, L., et al. (1995) Science **270**, 391.

Iseli, C., Stevenson, B. J., deSauza, S. J., et al. (2002) Genome Res. **12**, 1068.

Itoh, K., Sasaki, Y., Miyata, Y. (1994) Cancer Chemother. Pharmacol. **34**, 451.

Itoh, K., Sasaki, Y., Fujii, H. (1997) Br. J. Cancer **76**, 107.

Juengst, E. T. (1998) Kennedy Inst. Ethics J. **8**, 183.

Kass, N. E., Sugarman, J., Faden, R., et al. (1996) Hastings Center Rep. **26**(5), 25.

Kaye, J. (2006) Eur. J. Hum. Genet. **14**, 245.

King, P. A. (1992) In: Gene Mapping Using Law and Ethics as Guides (Annans, G. J., Ellias, S., Eds) Oxford University Press, New York, pp. 94–111.

Knoppers, B. M. (Ed.) (2003) in Populations and Genetics, Martinus Nijhoff, Leiden.

Kodish, E., Stocking, C., Ratain, M. J., et al. (1992) J. Clin. Oncol. **10**, 1810.

Leerkes, M. R., Cabellaro, O. L., Mackay, A., et al. (2002) Genomics **79**, 257.

Lehmann, L. C., Weeks, J. C., Klar, N., et al. (2002) Genet. Med. **4**, 345.

Lerman, C., Narod, S., Sculman, K., et al. (1996) JAMA **275**, 1885.

Lerman, C., Hughes, C., Trock, B. J., et al. (1999) JAMA **281**, 1618.

Lerman, C., Shields, A. E. (2004) Nat. Rev. Cancer **4**, 235.

Levin, B., Brooks, D., Smith, R. A., et al. (2003) CA. Cancer J. Clin. **53**, 44.

Lippman, M. E., Heys, D. F. (2001) J. Natl. Cancer Inst. **93**, 80.

Lipsett, M. P. (1982) JAMA **248**, 941.

Lynch, E. L., Doherty, R. J., Gaff, C. l., et al. (2003) Med. J. Aust. **179**, 480.

Markel, H. (1992) Am. J. Med. **93**, 209.

Miki, Y. Swensen, J., Shattuck-Eidens, D., et al. (1994) Science **266**, 66.

Ministry of Health, Act on Health Sector Database (1998) Reykjavik, Iceland.

Mironov, A. A., Fickett, J. W., Gelfand, M. S. (1999) Genome Res. **9**, 1288.

Mitas, M., Mikhitarian, K., Hoover, L., et al. (2002) Br. J. Cancer **86**, 899.

Modrek, B., Lee, C. (2002) Nature Genet. **30**, 13.

MRC ethics Series (2001) (www.mrc.ac.uk/pdf-tissue_guide_fin.pdf).

Narod, S. A., Foulkes, W. D. (2004) Nat. Rev. Cancer **4**, 665.

Nelson, P. S. (2002) Ann. NY. Acad. Sci. **975**, 232.

Neto, E. D., Correa, R. G., Verjovski-AL Meida, S., et al. (2000) PNAS USA. **97**, 3491.

Neto, E. D., Harrop, P. R., Correa-Oliveira, R., et al. (1997) Gene **186**, 135.

Newman, B., Mu, H., Butter, L.M., et al. (1998) JAMA **279**, 915.

Nishizuka, S., Charboneau, L. (2003) Proc. Natl. Acad. Sci. USA **100**, 14229.

Olsson, P., Bera, T. K., Essand, M., et al. (2001) Prostate **48**, 231.

Omenn, G. S., States, D. J., Adamski, M., et al. (2005) Proteomics **5**, 3226.

Parkin, D. M. (2001) Lancet Oncol. **2**, 533.

Patel, V., Leethanakul, Gutkind, J. S. (2001) Crit. Rev. Oral Biol. Med. **12**, 55.

Porter, D. A., Krop, I. E., Nasser, S., et al. (2001) Cancer Res. **61**, 5697.

Post, M. (1997) Jewish Bulletin of Northern California, May 23.

Post, G., Fidler, I. J. (1980) *Nature* **283**, 139.

Rai, A. J., Gelfand, C. A., Haywood, B. C. (2005) Proteomics **5**, 3262

Ramaswamy, S., Ross, K. N., Lander, E. S., et al. (2003) Nature Genet. **33**, 49.

Ransohoff, D. F. (2002) J. Clin. Epid. **55**, 1178.

Reymond, M. A., Steinert, R., Escoumou, J., et al. (2002) Dig. Dis. **20**, 257.

Rhodes, D. R., Yu, J., Shanker, K., et al. (2004) PNAS USA **101**, 9309.

Riggins, G. J. (2001) Dis. Markers **17**, 41.

Rights. NY, USA Cox, K. (2002) Patient Educ. Couns. **46**, 31.

Roberts, Jr, T. G., Goulart, B. H., Squitieri, L. (2004) JAMA **292**, 2130.

Scanlan, M. J., Gordon, C. M., Williamson, B., et al. (2002) Int. J. Cancer **98**, 485.

Schaefer, C., Grouse, I., Buetow, K., et al. (2001) Cancer J. **7**, 52.

Sgarlato, G. D., Eastman, C. L., Sussman, H. H. (2005) Clin. Chem. **51**, 27.

Shattuck-Eidens, D., Oliphant, A., McClure, M., et al. (1997) JAMA **278**, 1242.

Simes, R. J., Tattersal, M., Coates, A. S., et al. (1986) Br. Med. J. **293**, 1065.

Smith, R. A., Cokkinides, V., Eyre, H. J. (2006) CA. J. Clin. **56**, 11.

Smith, R. A., von Eschenbach, A. C., Wender, R., et al. (2001) CA. Cancer J. Clin **51**,38.

Strausberg, R. L., Greenhut, S. F., Grause, L. H., et al. (2001) Trends Cell Biol. **11**, 66.

Strausberg, R. L., Buetow, K. H., Greenhut, S. F., et al. (2002) Cancer Invest. **20**, 1038.

Strausberg, R. L., Stimpson, A. J. G., Wooster, R. (2003) Nat. Rev. Genet. **4**, 409.

Strausberg, R. L., Buetow, K. H., Emmert-Buck, M. R., et al. (2000) Trends Genet. **16**,103.

Steinbrook, R. (2002a) N. Engl. J. Med. **346**, 1425.

Steinbrook, R. (2002b) N. Engl. J. Med. **346**, 716.

Sullivan Pepe, M. S., Etzioni. R., Feng, Z., et al. (2001) J. Natl. Cancer Inst. **93**, 1054.

Sweet, K. M., Bradley, T. L., Westman, J. A., (2002) J. Clin. Oncol. **20**, 528.

The Nuffield Council on Bioethics (1995) Human Tissue: ethical and legal issues. London, UK.

The United Kingdom parliament (2003) The Human Tissue Bill (www. Parliament. the-stationary-office.co.uk/pa/cm200304/cmbills/009/2004009.htm.

UNESCO (1998) Universal declaration on the Human Genome and Human.

Thompson, I. M., Ankerst, D. P., Chi, C., et al. (2005) JAMA **294**, 670.

United States Department of Health and Human Resources. Research repositories, databases, and the HIPAA Privacy Rule, Available at, http://privacyruleandresearch.nih.gov/research_repositories.asp. (posted January 12, 2004).

van Dyke, T., Jacks, T. (2002) Cell **108**, 135.

Velculescu, V. E., Zhang, L., Vogelstein, B., et al. (1995) Science **270**, 484.

Vinals, C., Gaulis, S., Coche, T. (2001) Vaccine **19**, 2607.

Vining, D. J., Gelfand, D. W., Bechtold, R. E., et al. (1994) AJR Am. J. Roentgenol. **162**, S104.

Vining, D. J. (1997) Gastrointest. Endosc. Clin. North Am. **7**, 285.

Von Hoff, D. D., Turner, J. (1991) Invest. New Drugs **9**, 115.

Wald, N. J. (2001) J. Med. Screen. **8**, 1.

Warner, F., Foulkes, W., Goodwin, P., et al. (1999) J. Natl. Cancer Inst. **91**, 1241.

Weijer, C., Goldsand, G., Emanuel, E. J. (1999) Nat. Genet. **23**, 275.

Weiss, K. M. (1997) Houst. Law Rev. **33**, 1431.

Wendler, D., Emanuel, E. (2002) Arch. Intern. Med. **162**, 1457.

Wheeler, D. L., Church, D. M., Edgar, R., et al. (2006) Nucleic Acids Res. **32**, D35.

Wilson, J. M. G., Junger, G. (1968) Principles and Practice of Screening for Disease. Public Health paper No. 34, Geneva, WHO.

Woodage, T., King, S. M., Wacholder S., et al. (1998) Nature Genet. **20**, 62.

Wright, T. C. Jr, Cox, J. T., Massad, L. S., et al. (2002) JAMA. **287**, 2120.

Wooster, R., Neuhausen, S. L., Mangion, J., et al. (1994) Science **265**, 2088.

Wu, C. H., Apweiler, R., Bairoch, A., et al. (2006) Nucleic Acids Res. **34**, D187.

Xie, H., Zhu, W-Y., Wasserman, A., et al. (2002) Genomics **80**, 326.

Xu, Q., Modrek, B., Lee, C. (2002) Nucleic Acids Res. **30**, 3745.

ABBREVIATIONS

17-AAG	17-allylaminogeldanamycin.
α_1ACT	α_1-antichymotrypsin.
aCGH	Array-based comparative genomic hybridization.
ACPT	Testicular acid phosphate.
ACS	American Cancer Society.
ADP	Adenosine diphosphate.
AFM	Atomic Force Microscopy.
AKT	Serine/threonine kinase.
ALL	Acute lymphocytic leukaemia.
AML	Acute myelogenous leukaemia.
AQUA	Absolute quantification of proteins.
ARE	Androgen responsive elements.
ASK1	Signal-regulating kinase1.
ATP	Adenosine triphosphate.
BAC	Bacterial artificial chromosome.
BIND	Biomolecular interaction Network Database.
BPH	Benign prostatic hyperplasia.
BPP	Brain Proteome Project.
C2, C3, C5	Cyanine dyes (2, 3, 5)
CA-125, CA-125II, CA 15-3, CA 72-4	Carcinoma-associated glycoprotein antigen.

CARD	Caspase recruitment domain.
Cdc Cdc25B Cdc25C Cdc2 Cdc4	Cyclin-dependent kinases.
cDNA	Complementary DNA.
CEA	Carcinoembryonic antigen.
CFP	Cyan fluorescent protein.
CGAP	Cancer Genome Anatomy Project.
Chk1	Checkpoint kinase 1.
CID	Collision-induced dissociation.
CIN	Cervical intraepithelial neoplasia.
CKII	Casein kinase II.
CT	Spiral computed tomography.
CTMP	Carboxyl-terminal modulator protein.
CZE	Capillary zone electrophoresis.
CTMP	Carboxyl-terminal modulator protein.
2-DE	Two-dimensional electrophoresis.
DAC	5'-aza-2'-deoxycytidine.
DCIS	Ductal carcinoma *in situ*.
DCBE	Double contrast barium enema
DES	Diethylstilbestrol.
DFS	Disease-free.
DHB	2,5-dihydroxy benzoic acid.
DIGE	Differential in gel electrophoresis.
DIP	Database of interacting proteins.
DMH	Differential methylation hybridization.
DNMTs	DNA methyltransferases.
DRE	Digital rectal examination.
4E-BP1	Eukaryotic initiation factor 4E-binding protein-1
ECD	Electron capture dissociation.
EDRN	Early Detection Research Network.
Efp	Estrogen responsive finger protein.
EGFR	Epidermal growth factor receptor.
ELISA	Enzyme-Linked Immunosorbent Assay.
EMEA	European Agency for the Evaluation of Medicinal Products.
ERK	Extracellular signal-regulated kinase.
ESI	Electrospray ionization.
FAP	Familial adenomatous polyposis.
FDA	Food and Drug administration.
FIT	Fecal immunochemical test.
FKHLR1	Forkhead transcription factor 1.
FOBT	Fecal occult blood test.
FRET	Fluorescence resonance energy transfer.

FT-ICR-MS	Fourier-transform ion cyclotron resonance mass spectrometry.
GFP	Green fluorescent protein.
GPCR	G protein-coupled receptors.
GRID	General Repository for Interaction Datasets.
GRP94	Glucose related protein 94.
GST-SH2	Glutathione S-transferase-SH2.
GSK-3β	Glycogen synthase kinase-3β.
HBOC	Heredity breast and ovarian cancer.
HCCA	α-cyano-4-hydroxycinnamic acid.
HCGP	The human cancer genome project.
HIPAA	Health Insurance Portability and Accountability Act.
HNPCC	Heredity non-polyposis colon cancer.
HOSE	Human ovarian surface epithelial.
HPV	Human papillomavirus.
HSPs	Heat shock proteins.
HUGO	The Human Genome Organization.
HUPO	Human Proteome Organization.
IAA	Iodoacetamide.
ICAT	Isotope-coded affinity tag.
ICP-MS	inductively coupled plasma-mass spectrometry.
IEF	Isoelectric focusing.
IGOT	Isotope-coded glycosylation-site-specific-tagging.
ILK	Protein kinase integrin-linked kinase.
IMAC	immobilized metal affinity chromatography.
Inppl1/Ship1	The inositol polyphosphate phosphatase like-1 protein.
IPI	International protein index.
IRS-1	Insulin receptor substrate-1
KSR	Kinase suppressor of Ras.
LCA	α-lactalbumin.
LCM	Laser capture microdissection.
LC/MS-MS	Liquid chromatography/ tandem mass spectrometry.
LIFE	Laser light-induced fluorescence endoscopy.
LPP	Liver Proteome Project.
LUMIER	Luminescence-based mammalian interactome mapping.
MALDI/MS-MS	Matrix-assisted laser desorption association/tandem mass spectrometry
MAPK	Mitogen-activated protein kinase
MAPKAPK-2	MAPK-activated protein kinase-2.
MBP	Myelin basic protein.
MCA–RDA	Methylated CpG island amplification–representational difference Analysis.
MCAT	Mass-coded abundance tagging.
MDR	Multidrug resistance.
M-CSF	Machrophage colony-stumulating factor
MEKK1	Mitogen-activated protein kinase kinase 1.

MINT	Molecular Interaction Database.
MMPs	Matrix metalloproteinases genes.
MPD	Multi-photon detection.
MSAs	Multiple sequence alignments.
MS-AP-PCR	Methylation-sensitive arbitrarily primed PCR.
MSP	Methylation-specific PCR.
MudPIT	Multidimensional protein identification technology.
NBAC	National Bioethics Advisory Commission.
NCBI	National Center for Biotechnology Information.
NCI	National Cancer Institute.
NES1	Normal epithelial cell specific 1 gene.
NF-κB	Nuclear factor-κB.
NIBSC, UK	National Institute of Biological Standards and Control.
NSCLC	Non-small-cell lung cancer.
OECD	Organization for Economic Co-Operation and Development.
OMIM	Online Mendelian Inheretance in Man database.
OPN	Osteopontin.
OS	Overall survival.
OVX1	Lewis X mucin determinant.
PAP	Prostatic acid phosphotase.
PAX5	Embryonic-development gene paired box gene 5.
PCR	Polymerase chain reaction.
PDK1	Phosphoinositides-dependent kinase1.
PI3K	Phosphatidylinositol 3-kinase.
PI3K-Akt	Phosphatidylinositol 3-kinase-Akt pathway.
PIN	Prostate intraepithelial neoplasia.
PIP3	Phosphatidylinositol-3, 4, 5-triphosphate.
PIR	Protein Information resource.
PKB	Protein kinase B.
PMF	Peptide mass finger printing.
PPP	Plasmaproteome project.
PSA	Prostate-specific antigen.
PTMs	Postranslational modifications.
PVDF	Poly vinylidene fluoride.
Q-TOF	Quadrupole-Time-of-flight.
RLGS	Restriction landmark genomic screening.
ROC	A receiver operating curve.
RP-LC	Reverse-phase liquid chromatography.
RTKs	Receptor tyrosine kinases.
RT-PCR	Real-time PCR.
SAGE	Serial analysis of gene expression.
SCC	Squamous cell carcinoma.
SCHC	Specimen Collection and Handling Committee.
SCLC	Small cell lung Carcinoma.
SCX	Strong cation exchange.

SDS-PAGE	Sodium dodecyl sulphate-polyacrylamide gel electrophoresis.
SEC	Size exclusion chromatography.
SEER	Surveillance, Epidemiology and End Results.
SELDI	Surface-enhanced laser desorption ionization.
SGA	Synthetic genetic array.
SH2	Src homology 2 domain.
SIB	Swiss Institute of Bioinformatics.
SILAC	Stable isotope labelling with amino acids in cell culture.
SIMS	Secondary ion mass spectrometry.
SNP	Single nucleotide polymorphism.
SPR	Surface plasmon resonance.
STRING	Database of predicted functional associations among genes/proteins.
TAP-MS	Tandem affinity purification/mass spectrometry.
TGFβ	Transforming growth factor β.
TLSP	Trypsin-like serine prortease.
TNF-α	Tumour nercosis factor-α.
TOF-TOF	Time-of-flight-time-of-flight.
TPR	Tetratricopeptide repeat.
TRAP1	Tumour nercosis factor receptor associated protein1.
TSA	Trichostatin-A.
TVS	Transvaginal sonography.
ULN	Upper limit for normal.
UNESCO	The United Nations Educational, Scientific and Cultural Organization.
USPHS	US Public Health Service.
VICAT	Visible-coded affinity tag.
WHO	World Health Organization.
Y2H	Yeast two hybrid system.
YFP	Yellow fluorescent protein.

INDEX

WILEY-INTERSCIENCE SERIES IN MASS SPECTROMETRY

Series Editors

Dominic M. Desiderio
Departments of Neurology and Biochemistry
University of Tennessee Health Science Center

Nico M. M. Nibbering
Vrije Universiteit Amsterdam, The Netherlands

John R. de Laeter • *Applications of Inorganic Mass Spectrometry*
Michael Kinter and Nicholas E. Sherman • *Protein Sequencing and Identification Using Tandem Mass Spectrometry*
Chhabil Dass, *Principles and Practice of Biological Mass Spectrometry*
Mike S. Lee • *LC/MS Applications in Drug Development*
Jerzy Silberring and Rolf Eckman • *Mass Spectromery and Hyphenated Techniques in Neuropeptide Research*
J. Wayne Rabalais • *Principles and Applications of Ion Scattering Spectrometry: Surface Chemical and Structural Analysis*
Mahmoud Hamdan and Pier Giorgio Righetti • *Proteomics Today: Protein Assessment and Biomarkers Using Mass Spectrometry, 2D Electrophoresis, and Microarray Technology*
Igor A. Kaltashov and Stephen J. Eyles • *Mass Spectrometry in Biophysics: Cofirmation and Dynamics of Biomolecules*
Isabella Dalle-Donne, Andrea Scaloni, and D. Allan Butterfield • *Redox Proteomics: From Protein Modifications to Cellular Dysfunction and Diseases*
Villas-Boas • *Metabolome Analysis: An Introduction*
Mahmoud H. Hamdan • *Cancer Biomarkers: Analytical Techniques for Discovery*

Printed and bound by CPI Group (UK) Ltd, Croydon, CR0 4YY
16/09/2026
14963448-0001